# SOFTWARE RADIO
# ARCHITECTURE

*Library of Congress Cataloging-in-Publication Data:*

Mitola, Joseph
    Software Radios: wireless architectures for the 21st century/Joseph Mitola III.
      p.  cm.
    "A Wiley-Interscience publication."
    ISBN 0-471-38492-5 (Cloth: alk. paper)
    1. Cellular telephones.   2. Radio frequency modulation—Computer programs.   3. Digital
to analog converters.  I. Title.

  TK6570.M6 M55   2000
  621.3845'6—dc21                                 00-027758

Printed in the United States of America

10 9 8 7 6 5 4 3 2 1

# SOFTWARE RADIO ARCHITECTURE

## Object-Oriented Approaches to Wireless Systems Engineering

**Joseph Mitola III**
*Consulting Scientist*

A Wiley-Interscience Publication

**JOHN WILEY & SONS, INC.**

New York · Chichester · Weinheim · Brisbane · Singapore · Toronto

*For Lynné*
*... for Barb & Max*

*and Dad and Mom*
*and my Mentors*
*and the "Alpha's"*

*"In the Public Interest"*

# CONTENTS

# PREFACE

The purpose of this text is to show how to integrate the analog RF and digital aspects of radio with the rapidly emerging large-scale object-oriented software technology needed for open-architecture software-defined radio (SDR).

This is therefore a systems engineering text. It is not a design text. This book will not help you design a better filter for a specific SDR. It will, however, help you make better decisions about how to partition the end-to-end system filtering requirements. It will help you allocate the critical functions of dynamic range and processing capacity in such a way that the filter's design constraints are well founded and that the hardware platform, firmware, and software support the filtering requirements of the software radio. This book will also give you quantitative criteria for deciding whether to host that filter in an analog package, digitally on an applications-specific integrated circuit (ASIC), or in software on the latest digital signal processing (DSP) chip. In addition, if the filter is to be implemented in software, this book will give you the skills to ensure that the software is well structured and performs robustly—even when many tasks are competing for processing resources. The appropriate host for such a filter changes over time. Commercial filter ASICs may become obsolete as DSP processing capacity increases, changing the systems-level tradeoffs. As needs, technology, and team expertise evolve, the effective choice will also change. The effective choice also changes as a function of the top-down design constraints placed on the radio system by the economics of the marketplace and by the larger systems architecture. And the effective choice may be to not implement the filter per se at all, but to procure it as part of an off-the-shelf subsystem. As we migrate to systems on a chip, this means the filter may entail intellectual property that has to be partitioned and protected, and yet has to be leveraged by the rest of the system. A sound systems-level architecture facilitates this process, while an inferior architecture inhibits it. The reduced time to market and cost efficiency of such buy-versus-make choices also require balancing the capabilities and design constraints given competing technical and economic constraints.

Software-radio is therefore an interdisciplinary technology, so this is an interdisciplinary text. The radio-oriented chapters are written for people with strong software background but little background in radio engineering. Software radio is about wideband radio frequency (RF) hardware that is given its "personality" by software. Therefore, the software people have to understand the RF hardware and air interface standards to the degree set forth in this text in order to function effectively on an interdisciplinary team. Similarly, the software-oriented chapters are written for people with strong background in RF, analog radio, or DSP but little background with large-scale software. And software radio is increasingly about complex, large-scale software. One of the revolutionary aspects of software radio is that knowing how to code a radio algorithm in C on a DSP just doesn't give a software engineer the core skills needed to contribute effectively to software radio architecture. In fact, that experience becomes a liability if it causes one to minimize the importance of the new large-scale software engineering methods like CORBA.

In addition, European readers will have to be patient with the tutorial material on SDL, the ITU-standard Specification and Description Language. In teaching the software radios course on which this book is based, I have found that US engineers make little use of formal methods for specifying radio functions. ETSI's emphasis on formal methods and the widespread use of SDL in support of European standards-setting process has not reached across the Atlantic yet. As a result, U.S. practitioners of radio engineering often try to do with pencil and paper what their European counterparts do on a computer— define new air interface standards. This text's treatment of UML extends the SDL material.

This text has several companions. The first is the Special Issue of the IEEE *Journal on Selected Areas in Communication on Software Radios*, published in April 1999 by the IEEE (New York). This *JSAC* is a surrogate graduate-level text. As such, it addresses related graduate-level research topics including mathematical structure of the software radio, virtual radios, advanced digital filter ASICs, smart antennas, and other advanced techniques. The IEEE Press Compendium *Software Radio Technology* by Mitola and Zvonar extends the JSAC with both earlier and more recent technical papers. Prof. Friedrich Jondral's course text in German (U. Karlsruhe) relates fundamental digital radio to SDR.

The dedication of this text to the "public interest" envisions the resulting *affordable*, robust, high-quality radio services as beneficial to the public interest. If coalition partners can cooperate better using software radios in peacekeeping roles, then that serves the public interest. If governments can acquire radio platforms at lower and more predictable cost, then that makes resources available for other public priorities. The focus of this text is the architecture. In this text, architecture is defined as the consistent set of functions, components, and design rules that promote open-architecture evolution of complex radio systems.

The book is organized for ease of access by a variety of readers. Chapters 1–3 provide the high-level background needed for a general understanding of how software radio fits in the larger telecommunications technology. Systems engineers and program mangers should have a solid grasp of chapters 4 and 5 in order to lead architecture evolution. Program managers and software engineers need to pay particular attention to the discussion of complexity drivers. Like any other software-intensive project, software radios are subject to subtle changes of the scope of the software. These chapters attempt to forewarn and thus forearm the team against factors that can drive software complexity and processing requirements out of the bounds of time, personnel, or processing capacity available on the project. The chapters on subsystems (6–12) focus on the requirements that software radio brings to the hardware and software segments. The software-oriented chapters are designed to be useful to hardware-oriented readers and the hardware-oriented chapters are designed to be useful to software-oriented readers. Sufficient basics on signal processing are included to provide a relatively self-contained treatment. The concluding chapters (13–16) provide examples of how to apply software radio architecture to create robust yet affordable multiband multi-mode communications systems.

I really enjoy interacting with those of you who are out there creating SDR systems and propelling the software radio evolution forward. Since you have purchased this text, you probably would like to use the knowledge you gain. A few spreadsheet design tools can help with some important aspects of that task. These are the software radio spreadsheets. You get access to them via the author's software radios web site. The URL is http://ourworld.compuserve.com/homepages/jmitola. The site is for folks who purchased this text. By following the instructions on the site, you can get access to the design aids. I also welcome questions from readers about this text, or anything in the area of software radio technology.

Best regards,

JOE MITOLA

# 1 Introduction and Overview

## I. REVOLUTION AND EVOLUTION

We are now in the midst of another revolution in radio systems engineering. Throughout the 1970s and 1980s, radio systems migrated from analog to digital in almost every respect from system control to source and channel coding to hardware technology. In the early 1990s, the software radio revolution began to extend these horizons by liberating radio-based services from chronic dependency on hard-wired characteristics of the radio, including:

- Radio frequency (RF) band
- RF channel bandwidth and coding
- Propagation media access
- Link layer protocols

Today the evolution toward practical software radios is accelerating through a combination of techniques. These include smart antennas, multiband antennas, and wideband RF devices. Wideband analog-to-digital converters (ADCs) and digital-to-analog converters (DACs) access GHz of spectrum instantaneously. IF, baseband, and bitstream processing is implemented in increasingly general-purpose programmable processors. The resulting software-defined radio (SDR) extends the evolution of programmable hardware, increasing flexibility via increased programmability. The ideal software radio (SWR) represents the point of maximum flexibility in this evolution. In part, the software radio is an ideal that may never be fully implemented. The principles of the software radio nevertheless illuminate tradeoffs among radio architectures. SDR implementations "future-proof" infrastructure against continually evolving standards. Software radio architecture permits one to insert SDR technology gracefully and affordably. For a clear path for product evolution, one must understand the contributions of the ideal software radio to a specific application or market niche. The attempts of researchers to build ideal software radios yield lessons learned from these technology pathfinders. This text assembles these lessons into a coherent process for defining and evolving software radio architecture. It includes insights necessary to invest wisely in SDR-enabling technology. More importantly, it assembles the foundation on which those pursuing this technology can establish a software-radio systems-engineering process through which to navigate the dangerous shoals of this revolutionary evolution of radio engineering.

## II. A SYSTEMATIC EXPOSITION

This text first introduces the fundamental concepts of the software radio. These include the placement of the ADC near the antenna, the criticality of real-time streams, and the mix of implementation alternatives from baseband DSP through a variety of SDR alternatives. It then establishes the commercial and military drivers for an open-architecture for software-defined radios. Before addressing subsystem architectures, it identifies the aspects of the radio systems architecture that drive complexity. This is essential because SDR projects often fail because of unanticipated software complexity. It then covers the architecture principles by subsystem from antenna and RF conversion through DSP and software. It completes the core technical discussion by showing how to balance software computational demand against hardware processing capacity to produce software radios that meet specifications, on time and within budget. The text concludes with an overview of applications, including smart antennas and a mobile disaster-relief case study.

This first chapter provides an overview of the software radio (r)evolution shaping wireless systems engineering today. It introduces the software radio functional architecture. It also explains in more detail how analog, digital, and software radios form an implementation continuum, the software radio phase space. After completing the program of study represented by this text, a top-notch systems engineer will be able to position each project in implementation space as the technical, risk, and economic needs of the application dictate. The goal is to introduce most of the new concepts presented in this text.

It is worth emphasizing that this book does not try to sell the software radio. On the contrary, a software radio approach sometimes yields an ineffective product. One must fully appreciate how analog, digital, and software-intensive approaches complement each other. One may then understand the advantages and pitfalls of each. Ultimately, the reader should be able to decide when, where, and how to apply software radio technology. Thus empowered, each of the many participants in the software radio architecture (r)evolution will be able to contribute with greater impact.

## III. THE IDEAL SOFTWARE RADIO

This section presents a top-down approach to the software radio architecture. The top level components of an ideal software radio handset consist of a power supply, an antenna, a multiband RF converter, and a single chip containing ADC and DAC. The on-chip general-purpose processor and memory that perform the radio functions are illustrated in Figure 1-1.

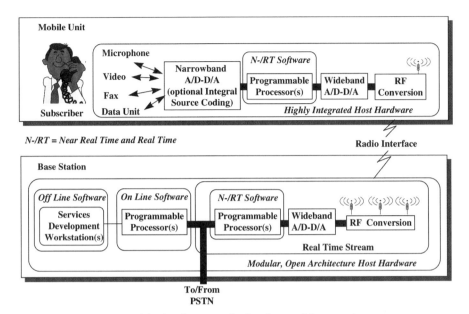

**Figure 1-1** The ideal software radio handset and base station concepts.

The generic mobile software radio terminal interfaces directly to the user (e.g., via voice, data, fax, and/or multimedia) and to the air interface. Driven by convenience and battery life, the mobile unit minimizes dissipated power and manufacturing parts count by maximizing hardware integration. The generic base station interfaces to the air and to the Public Switched Telephone Network (PSTN). With access to the power grid, base stations may employ modular, open-architecture hardware that facilitates technology insertion. Technology insertion opportunities future-proof the wireless infrastructure against the inevitable continuing evolution of air interfaces. Fully instrumented base stations support operations, administration, and maintenance (OA&M), while engineers and researchers may access the SDR network via services development workstation(s). Military base stations (nodes) need to support multiple networks on multiple RF bands with multiple air interfaces (modes). Such base stations may be formed by the co-location of diverse radios on mobile vehicles. These configurations often interfere with each other. The military calls this "cosite interference." The software radio base station attempting to support traffic on multiple channels in the same band can generate self-interference unless transmissions are coordinated or interference is actively cancelled.

The placement of the ADC and DAC as close to the antenna as possible and the definition of radio functions in software are the hallmarks of the software radio. Software radio mobile units and base stations share a common software factory that downloads personalities to the mobile units and updates to the infrastructure. Thus, although software radios use digital techniques, software-controlled digital radios are not necessarily software radios. The key difference is the total programmability of software radios, including programmable RF bands, channel access modes, and channel modulation.

SDR designs use Application-Specific Integrated Circuits (ASICs), Field Programmable Gate Arrays (FPGAs), Digital Signal Processors (DSPs), and general purpose (GP) processor technologies. SDR has become practical as costs per millions of instructions per second (MIPS) of DSPs and general-purpose central processor units (CPUs) have dropped below $10 per MIPS. The economics of software radios become increasingly compelling as demands for flexibility increase while these costs continue to drop by a factor of two every few years. At the same time, absolute processing capacities continue to climb into the hundreds of millions of floating-point operations per second (MFLOPS) to billions of FLOPS (GFLOPS) per chip. At this point, software radio technology can cost-effectively implement commercial first-generation (1G) analog and second-generation (2G) digital mobile cellular radio air interfaces.[1] Over time, wideband third generation (3G) air interfaces will also yield to software techniques on wideband RF platforms. In the interim, SDR implementations will require a mix of hardware-intensive techniques such as ASICs.

In addition, ADCs and DACs available in low-cost chips and single-board open-architecture configurations offer bandwidths of tens of MHz with the dynamic range required for software radio applications. Multimedia requirements for desktop and wireless personal digital assistants (PDAs) continue to exert downward pressure on parts count and on power consumption of such chip sets. This trend will push the ideal software radio technology from the base station to the mobile terminal. Although the tradeoffs among analog devices, low-power ASICs, DSP cores, and embedded microprocessors in handsets remain fluid, cutting-edge base stations are beginning to employ software radio architectures. And new designs for high-end mobile radio nodes such as military vehicular radios are now largely based on some type of software radio approach. The U.S. DoD has spurred on this trend through its Programmable Modular Communications System (PMCS) study and subsequent Joint Tactical Radio System (JTRS) program. Finally, the multiband multimode flexibility of software radios appears central to the goal of seamless integration of personal communications systems (PCS), land mobile and satellite mobile services (including truly nomadic computing), toward which many of us aspire.[2]

---

[1]In this text, the conventional notion of cellular radio is extended to embrace the idea that the propagation of RF from any SDR transmitter defines an implicit RF cell. Its size and shape is determined by the physical placement of antenna(s) and the environment. Antenna height, directivity, path loss, diffraction, and multipath loss shape the cell. A multiband, multimode SDR is uniquely suited to turn such implicit cells into explicitly managed ad-hoc cellular networks.

[2]In fact, the continuing interplay among military and commercial software radios plays an important role in the evolution of SDR technology. For some readers, this may impart a sense that the text skips from military to commercial points of view. The merger of these market segments around common interest in open-architecture SDR platforms is an ongoing process, complete with the common interests and occasional discontinuities highlighted in this text.

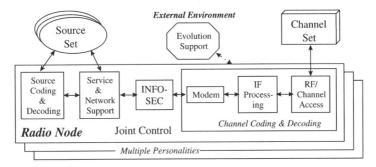

**Figure 1-2** Functional model of a software radio communications system.

## IV. THE SOFTWARE RADIO FUNCTIONAL ARCHITECTURE

Technology advances have ushered in new radio capabilities that require an expansion of the essential communications functions of source coding and channel coding. The new aspects are captured in the software radio functional model.

### A. The Software Radio Functional Model

Multiband technology [1], first of all, accesses more than one RF band of communications channel at once. The RF channel then is generalized to the channel set of Figure 1-2. This set includes RF channels, but radio nodes like PCS base stations and portable military radios also interconnected to fiber and cable; therefore these are also included in the channel set. The channel encoder of a multiband radio includes RF/channel access, IF processing, and modem. The RF/channel access includes wideband antennas, and the multi-element arrays of smart antennas [2]. This segment also provides multiple signal paths and RF conversion that span multiple RF bands. IF processing may include filtering, further frequency translation, space/time diversity processing, beam-forming, and related functions. Multimode radios [3] generate multiple air interface waveforms (modes) defined principally in the modem, the RF chan-nel modulator-demodulator. These waveforms may be in different bands and may span multiple bands. A software-defined personality includes RF band, channel set (e.g., control and traffic channels), air interface waveforms, and related functions.

Although many applications do not require information security (INFOSEC), there are incentives for its use. Authentication reduces fraud. Stream encipher-ment ensures privacy. Both help ensure data integrity. Transmission security (TRANSEC) hides the fact of a communications event (e.g., by spread spec-trum techniques [4]). INFOSEC is therefore included in the functional model although the function may be null for many applications.

In addition, the source coder/decoder pair now includes the data, facsimile, video, and multimedia sources essential for new services. Some sources will be physically remote from the radio node, connected via the synchronous digital hierarchy (SDH) [5], a local area network (LAN) [6], etc., through service and network support (Figure 1-2).

These functions may be implemented in multithreaded multiprocessor software orchestrated by a joint control function. Joint control ensures system stability, error recovery, timely data flow, and isochronous streaming of voice and video. As radios become more advanced, joint control becomes more complex, evolving toward autonomous selection of band, mode, and data format. Any of the functions may be singleton (e.g., single band versus multiple bands) or null, further complicating joint control. Agile beamforming supports additional users and enhances quality of service (QoS) [7]. Beamforming today requires dedicated processors, but in the future, these algorithms may time-share a DSP pool along with the Rake receiver [8] and other modem functions. Joint source and channel coding [9] also yields computationally intensive waveforms. Dynamic selection of band, mode, and diversity as a function of QoS [10] introduces large variations into demand, potentially causing conflicts for processing resources. Channel strapping, adaptive waveform selection, and other forms of data rate agility [11] further complicate the statistical structure of the computational demand. In addition, processing resources are lost through equipment failures [12]. Joint control integrates fault modes, personalities, and support functions on processing resources that include ASICs, FPGAs, DSPs, and general-purpose computers to yield a reliable telecommunications object [13].

In a software radio, the user can upload a variety of new air interface personalities [14]. These may modify any aspect of the air interface, including whether the waveform is hopped, spread, or otherwise constructed. The required resources (e.g., RF access, digitized bandwidth, memory, and processing capacity) must not exceed those available on the radio platform. Some mechanism for evolution support is therefore necessary to define the waveform personalities, to download them (e.g., over the air) and to ensure that each new personality is safe before being activated. The evolution-support function therefore must include a software factory. In addition, however, the evolution of the radio platform—the analog and digital hardware of the radio node—must also be supported. This may be accomplished via the design of advanced hardware modules in an integrated evolution support environment, or by the acquisition of commercial off-the-shelf (COTS) hardware modules, or both.

The block diagram of the radio functional model amounts to a partitioning of the black-box functions of the ideal software radio nodes introduced above into the specific functional components shown in Figure 1-2 and listed in Table 1-1.

Not every implementation needs all subfunctions of this functional model. Thus, one may consider the functional model to be a point of departure for

**TABLE 1-1  Function Allocation of the Software Radio Functional Model**

| Functional Component | Allocated Functions | Remarks |
|---|---|---|
| Source Coding and Decoding | Audio, data, video, and fax interfaces | Ubiquitous algorithms (e.g., ITU [15], ETSI [16]) |
| Service and Network Support | Multiplexing; setup and control; data services; internetworking | Wireline and Internet standards including mobility [17] |
| Information Security* | Transmission security, authentication, nonrepudiation, privacy, data integrity | May be null, but is increasingly essential in wireless applications [18] |
| Channel Coding and Decoding: Modem* | Baseband modem, timing recovery, equalization, channel waveforms, predistortion, black-data processing | INFOSEC, modem, and IF interfaces are not yet well standardized |
| IF Processing* | Beamforming, diversity combining, characterization of all IF channels | Innovative channel decoding for signal and QoS enhancement |
| RF Access | Antenna, diversity, RF conversion | IF interfaces are not standardized |
| Channel Set(s) | Simultaneity, multiband propagation, wireline interoperability | Automatically employ multiple channels or modes for managed QoS |
| Multiple Personalities* | Multiband, multimode, agile services, interoperable with legacy[3] modes | Multiple *simultaneous* personalities may cause considerable RFI[4] |
| Evolution Support* | Define and manage personalities | Local or network support software factory |
| Joint Control* | Joint source/channel coding, dynamic QoS vs. load control, processing resource management | Integrates user and network interfaces; multiuser, multiband, and multimode capabilities |

*Interfaces to these functions have historically been internal to the radio, not plug-and-play.
[3]Legacy refers to modes that are deployed but may be deprecated.
[4]Radio frequency interference.

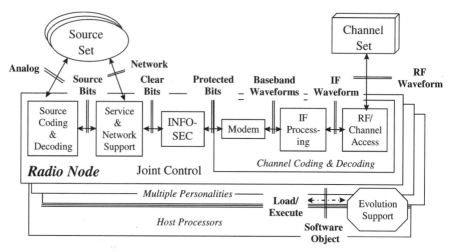

**Figure 1-3**   Standard interface points facilitate development, deployment, and evolution.

the tailoring of SDR implementations. In addition, many of the items in this table may be unfamiliar to some readers. The rest of the text develops the unfamiliar concepts and provides further references to the well-known aspects and standards.

## B. Functional Interfaces

After identifying the functions to be accomplished in a software radio, one must define the interface points among the functional components. Figure 1-3 identifies these interfaces. The notation "RF waveform" is shorthand for air interface. The IF waveform includes most aspects of the air interface, but the signals have been filtered and converted to an IF that facilitates processing.

In addition, IF processing may include A/D and D/A conversion. Baseband waveforms are almost always digital streams (e.g., of data or vocoded voice). They may also be sampled replicas of analog signals, such as digitized FM waveforms. The modem delivers what may be called decoded channel bits ("black" bits in INFOSEC jargon) to the INFOSEC function if one is present. The modem may transform analog IF signals directly to channel bits (e.g., using a despreader ASIC). INFOSEC then transforms these protected bits into clear bits ("red" bits). These bits may be manipulated through a protocol stack in order to yield source bits or network bits. Network bits conform to a network protocol, while source bits are appropriate for a source decoder. The interface to local sources of voice, music, video, etc. includes an analog transducer. Access to remote sources is accomplished via the network interface. In addition to these signal-processing interfaces, there are control interfaces mediated by the user or network (both of which are in the source set in this model). Personalities are downloaded to the radio via the software object interface. The simplest mechanism for maintaining radio software after deployment is the downloading of a complete binary image of the radio. A more flexible approach allows one to download a specific new function such as a specialized voice coder (vocoder).

These interfaces may be thought of as the "horizontal" interfaces of the software radio, since they are concatenated to form signal and control flows among sources and channels. They are further characterized in Table 1-2.

In traditional radio engineering, the definition of such interfaces facilitated the design and development of the radio. Variations in these definitions from one design team to another did not matter, provided the component suppliers and the systems integrator all agreed. The idea of plug-and-play hardware and software modules has become popular in personal computing. The wireless industry seeks to benefit from the adaptation of plug-and-play technology to the software radio. This potentially elevates any functional partitioning to the role of architecture. Plug-and-play requires industrywide agreement on architecture.

## C. Architecture

Since industrywide agreement on anything can be challenging, one should begin with a definition of architecture. The *Random House Unabridged Dictionary* defines *architecture* as "a fundamental underlying design of computer hardware, software, or both [25]". While this is an agreeable definition, it provides no prescription of what "underlying design" entails. The IEEE prescribes that architecture consists of components and interfaces. This leaves one wondering what the components and interfaces are supposed to do. The Defense Information Systems Agency is the U.S. Department of Defense (DoD) agency charged with defining architecture. That agency defines architecture in terms of profiles for communications standards [26]. In its Technical Architecture for Information Management (TAFIM), the DoD characterizes architecture by analogy to "zoning laws and building codes" by which one defines the parameters for the construction of residential and industrial buildings [27].

*1. Functions, Components, and Design Rules*   None of the many possible definitions of architecture suit the purposes of defining architecture for the software radio. One that best relates services, systems, technology, and economics is best suited to the software radio. *Architecture* is therefore defined as a comprehensive, consistent set of *functions*, *components* and *design rules* according to which radio communications systems may be organized, designed, constructed, deployed, operated, and *evolved over time*. This is not inconsistent with the other definitions. But this notion of architecture more clearly addresses partitioning for plug-and-play, and the reuse of functional components. By including functions and design rules, an architecture supports component reuse, even spanning implementations that migrate among hard-

**TABLE 1-2   Top-Level Component Interfaces**

| Interface | Characteristics | Properties |
|---|---|---|
| Analog Stream | Audio, video, facsimile streams | Continuous, infinite dimensional; filtering constraints are imposed here |
| Source Bitstream | Coded bitstreams and packets. ADC, vocoder, text data compression [19] | Includes framing and data structures. Finite arithmetic precision defines a coded, Nyquist [20] or oversampled dynamic range[5] |
| Clear Bitstream | Framed, multiplexed, forward error controlled (FEC) bitstreams and packets | FEC imparts algebraic properties over the Galois fields defined by these bitstreams [21] |
| Protected Bitstream | Random challenge, authentication responses; public key; enciphered bitstreams [22] and packets | Finite dimensional; randomized streams; complex message passing for downloads; if null, this interface reverts to clear bits |
| Baseband Waveform | Discrete time synchronous quantized sample streams (one per carrier) | Digital waveform properties determine fidelity of analytic representation of the signal |
| IF Waveform | Composite, digitally preemphasized waveform ready for up-conversion | Analog IF is continuous with infinite dimensions; digital IF may be oversampled |
| RF Waveform | Power level, shape, adjacent channel interference, etc. are controlled | Analog RF: channel impulse response, spatial distributions via beams and smart antennas [23] |
| Network Interface | Packaged bitstreams may require asynchronous transfer mode (ATM), SS7, or ISO protocol stack processing | Synchronous digital hierarchy (SDH), ATM, and/or Signaling System 7 (SS7) |
| Joint Control | Control interfaces to all hardware and software; initialization; fault-recovery | Loads binary images, instantiates waveforms, manipulates control parameters |
| Software Objects | Download from evolution support systems (e.g., software factory) | Represents binary images, applets; includes self-descriptive languages [e.g., 24] |
| Load/Execute | Software object encapsulation | Downloads require authentication and integrity |

[5]A coded dynamic range is defined by the vocoder. Nyquist–dynamic range results when an analog signal is sampled so as to meet the Nyquist criteria for bandwidth recovery of the sampled signal and has been quantized with sufficient bits of sufficient accuracy to represent the two-tone spurious-free dynamic range of the application. Oversampling above the Nyquist rate can yield additional dynamic range through processing gain—see Chapter 9.

ware and software. A useful architecture partitions functions and components such that (a) functions are assigned to components clearly and (b) physical interfaces among components correspond to logical interfaces among functions. The design rules must ensure that when the hardware and software components are mated, the resulting entity accomplishes the intended functions within the performance bounds established by regulatory bodies, service providers, and users. Accommodating such diverse needs leads to complex radio systems that must be further partitioned in order manage this complexity.

*2. Plug-and-Play*   If an architecture supports plug-and-play, then the design rules have been crafted so that hardware and software modules from different suppliers will work together when plugged into an existing system. Hardware modules will plug-and-play if the physical interfaces and logical structure of the functions supplied by that module are compatible with the physical interfaces, allocation of functions, and other design rules of the host hardware platform. Software modules will plug-and-play if there is a comprehensive but simple interface to the host environment, and if the module offers to the environment the information that it needs in order to employ it as a resource. Software radio architecture, then, defines the partitioning of functions into groups, which may subsequently be allocated to components. It defines the design rules that are appropriate for obtaining the benefits of open architecture. These include the publication of design patterns [28, 29] and interface standards. It also includes the definition of the logical levels of abstraction necessary to simplify comprehensive interfaces by hiding irrelevant details in lower layers.

## D. Levels of Abstraction

Clearly, software radio functions do not all share the same logical level of abstraction. A modem, for example, supports data movement from baseband to IF, data transformation from bits to channel symbols, timing recovery, FEC and the related functions. It is therefore not accurate to think of software radio architecture as merely a collection of functions with associated interfaces. One must also identify the levels of abstraction that naturally partition the hardware and software into radio platforms, middleware,[6] and host communications services, as in Figure 1-4.

In digital radios, the radio hardware platform (radio platform) accomplished most of the radio functions in hard-wired implementations, the parameters of which could be set through a microprocessor from a simple user interface or low-speed data bus. SDR platforms embody GFLOPS of processing capacity that support hundreds of thousands of lines of code (LOC). This software is partitioned into layers as illustrated in Figure 1-4. At the Radio Infrastructure

---

[6]Middleware is software that insulates applications from the details of the operating environment (e.g., the hardware).

| Communications Services | *Applications and related services*  (e.g., over-the-air downloads) |
|---|---|

| Radio Applications | *Air interfaces ("waveforms")*<br>State machines, modulators, interleaving, multiplexing, FEC,<br>control and information flows |
|---|---|

| Radio Infrastructure | Data movement:  drivers, interrupt service routines,<br>memory management, shared resources, semaphores |
|---|---|

| Hardware Platform | Antenna(s), analog RF hardware, ASICS, FPGAs, DSPs, microprocessors,<br>instruction set architecture, operating systems |
|---|---|

**Figure 1-4**   Logical levels of abstraction of the software radio.

level of abstraction, this code moves data among the distributed multiprocessing hardware of the radio platform. At the next level of abstraction, processes thus distributed cooperate to form radio applications. At the highest level of abstraction, applications software deliver communications services to users. Radio applications may incorporate elaborate air interface protocols, and may employ standard wireline data exchange protocols like TCP/IP, so one can envision a much more elaborate vertical protocol slice within this four-level stack.

One must then define interfaces among these levels. One approach is the definition of an applications programming interface (API) from one horizontal layer to the next. The API calls may be thought of as the vertical interfaces among horizontal layers. This approach has been used with reported success on technology pathfinders [30], and will be dealt with in some detail in this text. Not all APIs that have been described conform to the four layers identified above. These four layers, however, are conceptual anchors that help organize the process of evolving the software radio architecture.

One current evolutionary step, for example, is the integration of CORBA [31] into software radio architecture. The Object Management Group (OMG) has defined an Interface Definition Language (IDL) in their Common Object Request Broker Architecture (CORBA). CORBA [34] was developed primarily to define interfaces among software modules that were not originally designed to work together. IDL provides facilities for defining interfaces among software components through the mediation of an Object Request Broker (ORB). Since each new component interfaces to the ORB rather than to the $N$ existing components, the process of integrating a new software component is greatly simplified. CORBA IDL provides a rich technology base from which software radio finds both COTS radio infrastructure and a flexible means for defining interfaces among functional components. Maximum value for software radios requires extending CORBA to define interfaces among functions implemented in hardware. This has considerable benefits in a software fac-

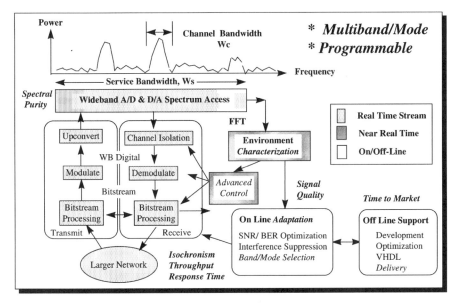

**Figure 1-5**   Signal processing streams of the software radio.

tory that supports both hardware (e.g., ASICs and FPGAs) and software (e.g., DSPs and general CPUs) as function-delivery platforms.

Horizontal interfaces among functional components and the vertical interfaces among layers of abstraction partition the software radio into a matrix of manageable components. These components may then be readily integrated to create a system with desired properties. This text derives this architecture matrix and presents methods that have been proven to ensure the most critical properties of software radios. Among these is the isochronism of the real-time signal-processing streams and the computational stability of the integrated software. Consider each of these in turn.

## V.  BASIC SIGNAL PROCESSING STREAMS

Consider the signal streams of the software radio as illustrated in Figure 1-5. These include a real-time isochronous channel processing stream, a near-real-time environment management stream, an on-line stream to manage radio modes, and off-line data streams that support development tools.

### A.  The Real-Time Channel Processing Stream

The real-time channel processing stream incorporates channel coding (the RF modem functions), INFOSEC if applicable, and radio access protocols (also called internetworking or message processing functions). Channel processing includes discrete-time point operations such as the digital translation of a baseband signal to an IF. Discrete point operations include multiplying a discrete time-domain baseband waveform by a discrete reference carrier to yield sampled in-phase IF samples.

For baseband DSP, the time between samples is on the order of milliseconds to hundreds of microseconds. This allows plenty of time for processing between samples. In the software radio's IF stream, however, the time between samples is on the order of tens of microseconds to hundreds of nanoseconds. Such point operations require hundreds of MIPS and/or MFLOPS to giga-FLOPS with strictly isochronous performance. That is, sampled data values must be computationally produced and consumed within short timing windows in order to maintain the integrity of the signal representation. Input/output (I/O) data rates of this stream approach a gigabit per second per IF ADC or DAC. Although these data rates are decimated through processing, it is challenging to sustain isochronism through DSP I/O interfaces and hard real-time embedded software in this stream.

Isochronous processing therefore should be organized as a hardware pipeline with sequential functions of the stream assigned to serially interconnected processors. Subscriber channels may be organized in parallel, resulting in a multiple-instruction, multiple-data-stream (MIMD) multiprocessing architecture [32]. Processors closer to the RF may be ASICS (e.g., for digital filtering and frequency translation). An important art form in software radio design is the minimization of the hardware footprint subject to the need to accommodate as many subscribers as possible. One of the major contributions of this text is to describe a proven process for accomplishing this balancing act in a way that meets end-to-end specifications in a mathematically predictable way.

## B. The Environment Management Stream

The other shaded boxes of Figure 1-5 comprise the near-real time environment management stream. This stream continuously characterizes radio environment usage in frequency, time, and space. This characterization includes channel identification and the estimation of other parameters such as channel interference levels. The details of this process may be defined by specific signaling and multiple-access schemes. For example, HF Adaptive Link Establishment (ALE) includes probes and responses that characterize several assigned channels. The data is then sent on the channel that is best for the specific subscriber location. The environment management stream typically employs block operations such as fast Fourier transforms (FFTs), wavelet transforms, and matrix multiplication for beam forming. Channel identification results are needed within 540 microseconds to 2 milliseconds for the Global System for Mobile Communications (GSM) [33]. Power levels may be updated in milliseconds. Subscriber locations may be updated relatively infrequently. The block structure of such operations is readily accommodated by a MIMD parallel processor. The interface between this highly parallel environment management stream and the pipelined real-time stream can be challenging. Each stream must include readily identifiable events (such as start of a radio frame) to which the streams may mutually synchronize.

## C. On-line Adaptation

On-line adaptation complements the near-real time dynamics as suggested in Figure 1-5. An air interface mode is a combination of parameters that defines the QoS provided by that mode. Third-generation air interfaces offer a wide range of data rates, for example. Generally, high data rates require high signal to noise ratio (SNR[7]) for a required bit error rate (BER). On-line adaptation bridges across air interface modes, in order to optimize the choice of band and mode subject to the goals and constraints of the user (and/or of the network). As modes become more elaborate, users are confronted with an increasing array of QoS versus price. The burden of choosing RF band and mode in the future will be shared among the user, the network, and the wireless appliance (e.g., PDA). Thus, on-line adaptation is an area in which one can look for increasing research interest as we transition into the complexity of third-generation (3G) wireless.

## D. Off-Line Software Support (The Software Factory)

Off-line tools include systems analysis, enhanced signal processing, and re-hosting of existing software to new hardware or software platforms. These allow one to define incremental service enhancements. For example, an enhanced beamformer, equalizer, and trellis decoder may be needed to increase subscriber density. These enhancements may be prototyped and linked into the channel processing stream in a demonstration facility. Such an arrangement allows one to debug the algorithm(s) and to experiment with parameter settings. One may determine the value of the new feature (in terms of improved subscriber density), as well as its cost in terms of resources impact (e.g., in processing capacity, I/O bandwidth, and time delay).

In an advanced application, a software radio does not just transmit a waveform. It characterizes the available transmission channels, probes the available propagation paths, and constructs an appropriate channel waveform. It may also electronically steer its transmit beam in the right direction, select the appropriate power level, and pick an appropriate data rate before transmitting. Again, in an advanced application, a software radio does not just receive an incoming signal. It characterizes the energy distribution in the channel and in adjacent channels, recognizes the mode of the incoming transmission, and selects the appropriate processing stream. If it has a smart antenna, it also adap-

---

[7]The SNR may be expressed in terms of unmodulated carrier and interference (CIR), signal-to-interference plus noise (SINR), or interference plus distortion.

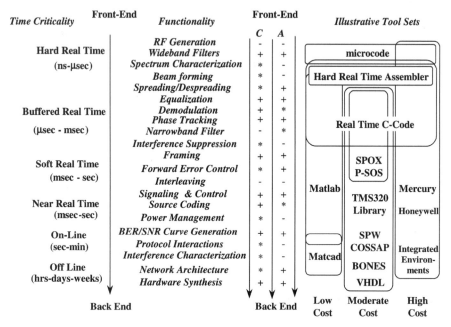

C = Criticality; A = Availability; * = key performance driver; + = important issue

**Figure 1-6**   Complementary views of a software factory.

tively nulls interfering signals, estimates the dynamic properties of desired-signal multipath, coherently combines desired-signal multipath, and adaptively equalizes this ensemble. It may also trellis decode the channel modulation and then corrects residual errors via forward error control (FEC) decoding to receive the signal with the lowest possible BER. Such operations require a family of software components and related tools including those illustrated in Figure 1-6.

The left side of the figure organizes software functions according to time-criticality. Hard real-time software may be delivered as the personality of an ASIC or FPGA. Reduced time criticality means the function is more compatible with true software implementations (e.g., as DSP code). The columns labeled C (criticality) and A (availability) identify challenge areas. Bit interleaving, for example, is not challenging either in terms of criticality or of availability. Interference suppression, on the other hand, is a critical issue as a key performance driver. To the right are three columns of tool sets that represent the sophistication of the software factory. One may develop software-radio products of limited scope (e.g., under 40 k LOC) using the low-cost tools in the first column. As team size grows, or the mix of ASICs, FPGAs, and DSP hardware in the delivery environment becomes more complex, the investment of tens of thousands of dollars per design-seat pays off. The largest, most complex systems benefit from the high-end tool suites costing upward of a million dollars (rightmost column).

The software radio should support incremental service enhancements via software tools in the software factory. These tools should assist in analyzing the radio environment, in defining the required enhancements, in prototyping incremental enhancements via software, and in testing the enhancements in the target radio environment (replete with noise and interference). The tools should make it easy to integrate and test the entire hardware-software system. They should also facilitate the delivery of the service enhancements via software and/or hardware updates, both via conventional OA&M processes and in real time over the air.

Software-based enhancements may be organized around managed objects, collections of data, and associated executable procedures that work together under the overall control of a network management system. These objects may be structured using ORBs to conform to related open-architecture software interface standards (e.g., CORBA). Such enhancements may then be delivered over the air to other software radio nodes. This is the pattern of the software-defined telecommunications network architectures described by NTT [35] and others [36, 37]. A well-integrated set of systems analysis, design, development, and rehosting tools leads to the creation of incremental software radio enhancements relatively quickly, with upgrades provided over the air as software-defined networks proliferate. Technology limitations that require hardware-based delivery (e.g., for vestpocket terminals) are met by mapping critical elements of the service enhancement to hardware (e.g., via VHDL). This leads to a wealth of implementation alternatives.

## VI. IMPLEMENTATION ALTERNATIVES

Implementation alternatives for digital radios, SDR, and software radios may be characterized in the software-radio phase space of Figure 1-7. The phase space compares digital-access bandwidth to the flexibility of the processing platform. These are the two most critical architecture parameters of the software radio. Digital-access bandwidth is approximately half of the sampling rate of the ADC in the isochronous subscriber signal-processing path. Thus, for example, a 5 GHz conversion rate supports nominally a 2.5 GHz analog bandwidth, based on the Nyquist criterion [20]. ADCs with bandwidths of over 6 GHz exist [38], so digitizing RF is not impossible. If all the processing after the ADC were accomplished on a single general-purpose computer, one would have an ideal software radio receiver (the point marked X in the figure). Corresponding digital signal synthesis and up-conversion would yield an ideal software-radio transmitter.

Such extremely wideband ADCs consume substantial power and have a dynamic range of only about 30 dB. These limitations preclude practical implementations of the ideal. In addition, the digital filtering of the 5 giga-sample per second stream to access a given RF band such as 25 MHz of RF spectrum

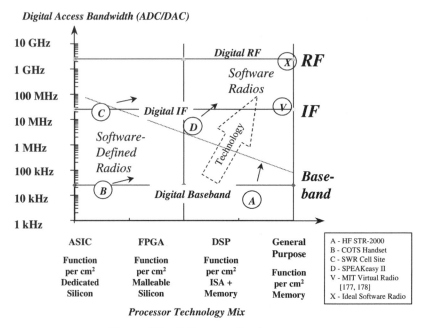

**Figure 1-7**    Software radio phase space.

would require at least 500 gigamultiplications ($5 \times 10^{11}$) per second. This feat is beyond general-purpose computers. Furthermore, there is no single antenna or RF stage that can sustain the analog bandwidth from 2 MHz to 2.5 GHz required as input to the ADC (and conversely for the transmitter). Thus, the ideal software radio is not implementable with today's technology. Why even include it? The ideal software radio represents the point of maximum flexibility for a radio platform. The ideal properties of such a radio represent the best that one could ever achieve, and thus are a useful reference point for measuring progress toward generality and flexibility.

Practical implementations have limited RF coverage due to the narrow-band nature of antennas, RF conversion, and IF processing technology. They also require a mix of digital technologies including ASICs, FPGAs, DSP, and general-purpose processors. The STR-2000 (point A in the figure) was an early baseband HF DSP radio developed by Standard Marine AB. This radio digitized an HF IF signal at a 24 kHz sampling rate. It used twin Texas Instruments (TI) TMS320C30 DSPs to provide a half-dozen standard HF signal formats digitally. This could be accomplished using a general-purpose processor today. COTS handsets (B) minimize size, weight, and power through the use of ASICs. Some handsets demodulate signals in an RF ASIC that creates a digital baseband bitstream directly from analog RF. Combining two such ASICs in a handset enclosure leads to the term "Velcro radio" for this approach [39].

Contemporary software-radio cell-site designs (C) access the allocated up-link[8] RF using a single ADC (e.g., with 25 MHz of analog bandwidth; viz., 70 MHz conversion rate). These designs employ a bank of digital filter ASICs [40] or parallel digital filters [41] to access a hundred or more subscriber channels in parallel. Some implementations incorporate the new high-density FPGAs to provide software-driven configurability in a delivery platform that maximizes throughput for a given technology clock rate [42]. Technologically aggressive designs include SPEAKeasy [3], the military technology pathfinder. SPEAKeasy II (point D in Figure 1-7), which became the baseline for Motorola's WITS 6000 software radio product line [43], incorporated over a GFLOPS of processing capacity for enhanced flexibility. The Virtual Radio (point V in the figure) is the most flexible software radio research implementation reported in the literature [44]. A general-purpose DEC Alpha processor running UNIX accesses a wideband IF digitally. Narrowband AM and FM broadcast receivers and an RF LAN have been implemented purely in software on this platform.

The three fundamental limitations of any SDR implementation, then, are:

1. RF access
2. Digital access bandwidth
3. Digital processing (flexibility and capacity)

The process of plotting an implementation in the software-radio phase space is illuminating. Those that are further to the right *should* be more flexible and easier to extend. But this is sometimes not the case. Systems and software engineering disciplines described in this text are required to capitalize on the flexibility of the hardware. These techniques must be fully employed and systematically practiced throughout the system life cycle. The software design and development process chapters of this book show how to make the touted flexibility a reality. These design techniques also allow one to avoid disaster as a sequence of apparently small incremental requirements added to a simple, stable system yield an unstable "house of cards." Definition of a radio platform is one of the steps that is required to avoid such disasters.

## A. Defining the Radio Platform

One key architecture question is the degree of programmability required for the intended market niche. Contemporary radio designs therefore vary across the dotted line in the phase space. This represents the technology frontier, comprising a mix of ASIC, FPGAs, DSP, and general-purpose processing elements using ADCs and DACs at baseband or IF. Aggressive designs move above and to the right of this line, while conservative designs remain below and to the left. Advancing microelectronics technology moves all implementations

---

[8]The uplink is the link from mobile to base station. The downlink is the reverse link.

**TABLE 1-3  Software Radio Reference Platform Parameters**

| Critical Parameter | Remarks |
|---|---|
| Number of Channels | Number of parallel RF, IF, and /or baseband channels |
| RF Access | Continuous coverage from a minimum to a maximum RF |
| Digital Bandwidth | Bandwidth of the maximum ADC for each RF/IF channel |
| Dynamic Range | End to end, including RF, IF, ADC, AGC, and processing |
| Interconnect Bandwidth | Bandwidth of critical buses, serial ports, backplanes, etc. |
| Timing Accuracy | The precision and stability of system clock(s) |
| Frequency Performance | RF, IF, and local oscillator (LO) accuracy and stability |
| Processing Capacity | MIPS, MFLOPS using standard benchmarks, arithmetic precision (per processor class if appropriate) |
| Memory Capacity | RAM, ROM per processor; mass storage capacity |
| Hardware Acceleration | Parameterize capabilities encapsulated in hardware such as despreader ASICS, FPGAs, and related hybrids |
| Operating Environment | Operating system and related facilities (including CORBA middleware), interfaces (e.g., APIs), and measured determinism |

inexorably upward and to the right over time. At present, handsets—even dual-mode handsets—favor the Velcro approach using RF ASICs with chip-level integration. Some implementations use the VME or PCI bus to facilitate board-level upgrades. Some applications such as law enforcement and general aviation radios are very cost sensitive and therefore typically lag the state of the art by one or two generations (2–8 years).

With such a variety of RF, ADC, and processing hardware implementations, it is extremely difficult to determine whether third-party software intended for one platform will be of any use on another. To address this question quantitatively requires the following steps:

1. Definition of a radio reference platform
2. Characterization of the software processing demand in standard metrics
3. Control of critical hardware and software parameters during development and operations

A radio reference platform is a high-level characterization of the capabilities of the hardware environment of the software radio. Table 1-3 identifies the most critical radio platform parameters that determine the performance of a software radio.

**TABLE 1-4  Notional Mobile SDR Reference Platforms**

| Notional Platform | RF Access (MHz) | Channels | Digital Bandwidth |
|---|---|---|---|
| Low-band PDA | 450–1200 | 3 (traffic, control, rental) | 5 MHz |
| Mid-band PDA | 850–2500 | 3 (traffic, control, rental) | 20 MHz |
| Low-band Military | 30–500 | 4 (voice, 2 data, 1 scan) | 10 MHz |
| Mid-band Military | 88–1200 | 4 (voice, 2 data, 1 scan) | 20 MHz |
| Wideband Military | 800–4000 | 6 (4 JTIDS, 1 voice, 1 scan) | 250 MHz |

If the parameters of Table 1-3 are specified with precision and if the platforms in the family are tested for conformance to the reference platform, then software developed for one member of the family should port readily to another member of the family. The software will not port well (and may not port at all) if special features of the platform beyond the reference set are used.

The specification of a minimum level of capability for each parameter defines a reference platform for a family of software radio implementations. In the late 1990s, most mobile radio hardware could be characterized as a platform with only one or two isolated RF bands (e.g., 900 MHz cellular and 1800 MHz DCS), one or two channels (e.g., traffic and control), and baseband digital processing capability. By 2010, most deployed radio platforms could be defined in terms of broadband RF access, several simultaneous channels supported, and moderate IF digital processing bandwidth. Examples are suggested in Table 1-4. The PDAs will have replaced conventional cell phones in this vision of the future. Given the reference platforms, they will have *b*roadband RF, *m*ultiple parallel data channels, and *w*ide digital processing bandwidth (BMW). In the sequel, the designation BMW-SDR refers to any of the reference platforms in the table.

Reference platforms closer to the ideal software radio make economic and technical sense in infrastructure applications. An industry standard cellular base station, for example, could specify block up and down conversion with 25 MHz bandwidth IF channels (ADCs/DACs with 70 M samples per second and 14 bits of dynamic range). It could specify 100 digital subscriber channels. In such a reference platform, ASIC digital filters (digital receivers) could be allowed (not required) to isolate the RF channels by defining the interfaces to these chips in the hardware acceleration part of the reference platform. DSP- based RF modem software to process the channel waveforms could then be shared among members of this family with confidence. In particular, a new channel modulation such as 16 state quadrature AM (16 QAM), developed for the current generation using the facilities of the reference platform, could be readily ported to the members of the family already in the field. A software download could upgrade the deployed members from a simple air interface based, say, on QPSK, to an improved air interface with a

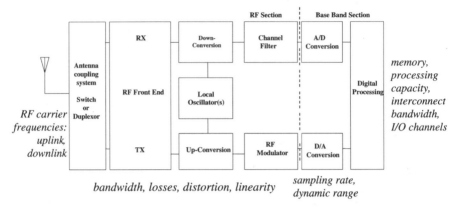

**Figure 1-8**   Reference design for an SDR implementation.

new high-data-rate 16 QAM mode. This would be accomplished with a minimum of touch labor through software downloads, and with no hardware upgrades.

A reference platform need not have an associated block diagram, but it
is often convenient to use such a diagram in the analysis of the feasibility
of a reference model. A reference block diagram is a notional design that
helps one determine the physical parameters of the devices. There are many
dangers in using such a reference design. The uninitiated may wrongful-
ly infer that all implementations must conform to the block diagram. That
would limit competition and creativity. A reference block diagram, then, is
a useful analysis tool, but it need not be part of the reference platform
model.

The reference design of Figure 1-8, for example, has the following drawbacks. First, it implies that ADCs and DACs are the interface between the
digital processing and analog RF sections of the radio. That is often the case.
If, however, the radio access technology approaches the state of the art in bandwidth, the more practical approach may be to demodulate the signal-in-space
using analog components. An ultra-wideband (UWB) [45] communications
system, for example, uses subnanosecond pulses to spread the communications over 2 GHz or more of bandwidth. These pulses are both transmitted
and received with analog circuits, not with DACs and ADCs. A code division multiple access (CDMA) despreader may digitally despread a wideband
CDMA signal in the time domain using a zero-crossing detector in an FPGA.
Although a zero-crossing detector is technically an ADC, the implication of
the reference block diagram is that a Nyquist sampling ADC is used. There
are, however, enough redeeming values to use the block diagram within limits. Its primary value is to associate critical parameters with physical devices
in such a way that one may outline an evolutionary path for the hardware
platform.

## B. Evolving the Radio Platform

A well-conceived reference platform is key to software radio architecture. Reference platforms should be based on trial implementations. One should not overspecify a reference platform so that only one design will conform to the platform. On the contrary, a major goal of using a reference platform rather than a hardware specification is the high degree of implementation flexibility of the reference platform approach. Any reference platform not based on an implementation should be validated by a prototype. In addition, the reference platform should remain valid in spite of anticipated technology advances. This section suggests some approaches to evolving the reference platform. The technical foundation of the reference platform is developed throughout the text.

Software-radio engineering is fraught with pitfalls. Technology breakthroughs are needed for wideband, low loss antennas and RF converters. As these technology advances occur, it is tempting to define a new radio platform for each small incremental advance. Such an approach can lead to the same state one has with digital radios: a deployed collection of many incompatible devices. One should instead define a generic radio platform. The platform should be based on achieving fundamental new capabilities for the customer. With such advanced planning, technology nuances should not yield a proliferation of reference platforms that are incompatible with each other. Instead, the defined platform will motivate technology investment toward well-defined market/customer needs.

It is also challenging to allocate radio platform parameters. For example, one must allocate dynamic range across automatic gain control (AGC), analog third-order intercept point, the ADC's two-tone spurious-free dynamic range, and the processing gain of oversampling integration algorithms. Thus, critical parameters like dynamic range will be distributed across fundamentally different hardware components. One therefore may have to specify the radio platform parameters by component class. This should be avoided, but sometimes it is unavoidable. Suppose, for example, that due to known technology limitations, total dynamic range must be allocated among an RF/IF AGC and an ADC. Its minimum and maximum AGC range should be specified, along with critical control parameters including attack and decay time. The ability of the downstream software to monitor the AGC value must also be specified. Over time, the capabilities of the hardware devices (e.g., accuracy of the ADC) will continue to change. Thus, one may specify an algorithmic relationship among the hardware components (e.g., RF AGC and ADC dynamic range). Consequently, the software that is ported across different implementations of the same platform will achieve the required end-to-end performance.

An accurate estimate of the processing demand of software modules must be balanced against the realizable processing capacity of reprogrammable processor configurations. By specifying processing capacity in terms of standard benchmarks (e.g., SPECmarks [46]), the processing capacity of the reference platform will be well-defined. In addition, sustaining the required data rates across interprocessor interfaces can also be problematic. The tradeoff between large block sizes for efficient data transfer competes with radio applications' needs for low latency. Thus, the reference platform must specify any algorithmic relationship among critical parameters such as interconnect bandwidth and data-transfer block size. As hardware implementations evolve, constraints can be relaxed without redefining the reference platform.

DSP function libraries continue to expand, and block diagram-based *integrated environments* exist. But radio engineers do not yet have the ability to mix and match software radio modules from different suppliers in the way that we can mix and match PCI boards today. Most of the pitfalls can be avoided, however, through the systematic control of critical software radio parameters using radio reference platforms, processing performance management, and other techniques described in this text.

## VII. THE ACQUISITION OF SOFTWARE RADIOS

This section introduces economic aspects of software radios. Since any software-based capabilities benefit directly from Moore's Law, one expects software radios to do the same. Since the radio platform is not purely a computing platform, but will always include some analog hardware, the benefits might not accrue as readily to the service providers who acquire these systems. The commercial cellular industry and the military are the major service providers considered in this text. The commonality of their interests has increased substantially, with open-architecture SDR as their common ground. This section takes an initial look at the acquisition of software radio systems from the perspective of the acquisition manager.

### A. Critical Acquisition Parameters

In 1992 when I introduced the term, almost nobody knew what a "software radio" was. By 1996, six months after the publication of the special issue of the IEEE communications magazine on the software radio, almost every radio vendor on the planet claimed to have one. The term had become an industry buzz-word. By 1999, it had become clear that nobody even *wanted* to have a software radio because it would be unaffordable or inefficient or both. Hence the current focus is on the SDR, the implementable aspects of software radio. Nevertheless, the term "software radio" is often used generically to mean programmable digital radio, SDR, or ideal software radio. This confuses customers, program managers, and investors. We can do better than this. In particular one should be able to characterize a software radio in terms of its critical acquisition parameters. This helps management and potential customers (e.g., the service providers in the commercial and military sectors).

**TABLE 1-5   Four Software Radio Acquisition Parameters**

---

**N: Number of Channels**
   **n**: single channel or multiple channel ($n < 6$),
   **NN**: The full number of subscribers in the RF band.
**PDA: Programmable Digital Access**
   None (0), Baseband (1), IF (2), RF (3)
   **Baseband** bandwidth is defined by single subscriber service (e.g., voice, data modem, video)
   **IF** is that bandwidth which simultaneously supports all NN subscribers in the allocated RF service band (e.g., 12.5 MHz analog FDMA)
**HM: Hardware Modularity**
   None (0), Receiver/Exciter/INFOSEC/Network Modules (1), COTS boards (2), Second Level Modules (ADCs, FPGAs, Receiver Chips, etc.) (3)
**SFA: Software Flexibility and Affordability**
   No air-interface-defining software (0), Single-supplier software (1), Multiple-supplier but single-host platform (2), Multiple-supplier multiplatform software

---

Furthermore, the service providers benefit most from these acquisition parameters if they encourage the use of standard metrics for any would-be software radio. The four key acquisition parameters defined here also characterize the business case for the software radio. These are:

1. The number of air interface channels simultaneously supported (**N**),
2. The level of programmable digital access (**PDA**),
3. The degree of hardware modularity (**HM**), and
4. The scope of software flexibility and affordability (**SFA**).

The first two parameters are included in the radio reference platform, but are emphasized here because of their impact on development risk. Table 1-5 defines these criteria.

## B. Channelization

The number of air interface channels, **N**, includes three groups. These are single channel, multiple channel (i.e., **N** is less than 6), and full access (i.e., **N** is the full number of subscribers in an allocated RF band). Multiple channel nodes are typical of military, civil aviation, and law-enforcement hub applications. The full access class is typical of cellular infrastructure. Single channel software radios provide a baseline of minimum development risk and complexity. Multiple channel nodes require typically distributed multiprocessing. With such small numbers of channels, hardware efficiency per channel is not a major challenge. It becomes a market-discriminator for the full access class, however. This class also carries maximum risk of mismatch between the processing demand offered by the software and the processing capacity deliverable by the hardware. Matching demand to capacity is therefore a focus of this text.

## C. Programmable Digital Access

The level of programmable digital access (**PDA**) is the point in the software radio functional model at which the conversion to digital occurs. This defines the scope over which the radio is reprogrammable. The types of **PDA** are: none (totally analog or fixed function digital radio); baseband programmability; IF programmability; and RF programmability. Some products should have **PDA** = none, notably consumer electronics and low-end pagers. Baseband **PDA** defines a the digital radio, sometimes called the programmable digital radio (PDR). This is not an SDR because the filtering required to get to baseband is implemented in hardware, so even minor changes in the RF structure of the physical layer are unlikely to be accommodated in software alone. IF digitization does not force one to use programmable digital filters to access the baseband subscriber streams, but it should. That is, if one digitizes at IF and then puts the translation from IF to baseband in, say, an ASIC, the implementer has just thrown out the major benefit of IF digitization: air interface (waveform) flexibility. Since Harris, Graychip, and many others produce high-quality, highly programmable, and very power-efficient digital filter chips that accomplish this filtering and translation in a reprogrammable platform, it is not necessary to slip a hard-wired filter into the implementation. But there are modules that seem to justify this approach, such as the massively parallel filter-bank chip. If production cost of infrastructure hardware is the overall driver for the market niche, then one may replace dozens of programmable chips with one or a few nonprogrammable filter bank chips. Such implementations have baseband **PDA**, not IF **PDA**, because although the access at IF is digital, it is *not* programmable. If futureproofing (a commercial-sector driver) or interoperability (a military driver) are more important than absolute cost, then IF **PDA** is more appropriate than baseband **PDA**.

Consider, for example, a Joint Tactical Information Dissemination (JTIDS) radio that hops over 250 MHz [20]. If it uses a fast-tuning LO to dehop the channel to, say, a 3 MHz baseband for despreading, it has 3 MHz IF **PDA**. Since it cannot be programmed for an arbitrary waveform in its 250 MHz agility bandwidth, it does not have true (250 MHz) IF **PDA**. Similarly, a cellular base station that despreads a 20 MHz W-CDMA waveform using an ASIC does not have 20 MHz IF PDA. If that ASIC can also digitize at, say, a 500 kHz sampling rate, then it could process GSM entirely digitally and with complete software flexibility. This would have **PDA** of 250 kHz.

Of course, RF **PDA** is not affordable at present and may not be for some years to come, so contemporary implementations center on IF **PDA**. RF **PDA** is, however, implementable in the laboratory today, so one should not ignore the possibility of a technological breakthrough that will render RF **PDA** the implementation technology of choice.


## D. Hardware Modularity

Hardware modularity (**HM**) identifies the economic impact of the differences in hardware upgrade paths. Architecture may be based on capability-oriented coarse-grain (possibly programmable) radio modules such as receivers and exciters that are specific to an air interface. Alternatively, architecture may be based on technology-oriented coarse-grain modules such as COTS ADC and DSP boards. Finer-grain modules such as FPGA, ADC, and DSP chips are also candidate modules. Finally, the system-on-a-chip approach defines module as a chunk of intellectual property (IP). The granularity of hardware modularity is not prejudicial. In some market segments, it may be preferable for line replaceable modules to be aligned with an air interface and electronically shielded. This is a feature of type 1 **HM**. The key is to explicitly decide what type of modularity is called for by the life cycle evolution in the market segment, and to match that type in the implementation. **HM** types 1–3 are therefore equivalent in the sense that one may not be able to say a priori which approach is preferred over another. The match of the modularity needs to the market segment (e.g., maintenance strategy) determines value to the acquisition manager in this dimension.


## E. Software Flexibility and Affordability

Software flexibility and affordability (**SFA**) characterizes the service provider's ability to acquire plug-and-play software modules from a COTS marketplace. Software that runs on just one radio platform and is available from only the original manufacturer tends to box the service provider into single-source (sometimes very expensive) maintenance and upgrade paths. If the functionality of the unit will not change over its life cycle, then this may be a perfectly acceptable path. This would be a rare occurrence in today's fast-moving marketplaces, however. When radio software is available from multiple suppliers, such as with the Texas Instruments DSP Co-operative, costs tend to be lower. In addition, there are generally software alternatives to the original supplier should they become necessary. Software that runs on many platforms (e.g., Java) and is available from multiple vendors generally gives the service provider a better software product with more flexibility and at a lower cost over the life cycle than the alternatives. Developers amortize their costs over many more units than any other business model, so costs are lower and capability is generally higher than with the other **SFA** business models.

The properties N, PDA, HM and SFA provide insight into software economics. The software radio architecture with maximum economic leverage also requires other aggregate properties such as openness.

## F. Architecture Openness

When functions, interfaces, components, and/or design rules are defined and published, the resulting architectures are called *open*. The full economic benefits of open architecture require the existence of a large commercial base, which sometimes fails to emerge in spite of openness (e.g., publication of the architecture). In the mid-1980s, for example, the government attempted to establish a Government Open Systems Interconnect Protocol (GOSIP) [47]. The government wasted huge amounts of effort and expenditures attempting to shape the marketplace; it lacked the market share necessary to accomplish this. GOSIP was, by the way, the first time in memory that the U.S. DoD/government failed to call the shots in the telecommunications and information processing marketplace that they had dominated throughout the 1970s.

We are poised in a similar situation with respect to open architecture wireless. The SDR Forum is attempting to establish open-architecture standards for SDR-based plug-and-play wireless. The mediation of such a public forum often can help generate industry support for a published architecture. At the same time, the U.S. DoD would like standards that support their needs for interoperability and commonality. The challenge is to balance the generic needs of open architecture versus the specific needs of market segments. The military, commercial wireless, international air transport markets, etc. each might optimize an open architecture in a different direction. The military might emphasize information security, while the commercial sector emphasizes service-delivery. International air transport might want to optimize carrier-differentiating services, while general aviation seeks lower-cost products. Groups like the SDR Forum provide a meeting ground for ideas so that the common ground may be identified.

As system complexity increases, architecture becomes more critical because of its power either to simplify and facilitate system development (a powerful architecture) or to complicate development and impede progress (a weak architecture). Software-defined radios are a powerful technology for the rapid deployment of new services. Some SDR architectures favor the rapid insertion of new technology. Others impart mathematical structure, as shall be developed in subsequent chapters. Industry, however, is far from unanimous on the strategy for open architecture. Some would like to see the SDR Forum define a few critical interfaces (e.g., between smart antennas and core base stations). Others would like to see a more comprehensive treatment (e.g., akin to the functional model presented above). Some see flexibility for future-proofing as paramount while others see cost effectiveness of point solutions as fundamental to growth of the market. Different market segments no doubt need different implementations, so a truly useful industry-wide architecture for the software radio will have to accommodate alternative product solutions. This text therefore does not specify a single architecture. Instead, it frames the technical issues, focusing on the architecture parameters that will drive the economics.

Thus, the reader should better appreciate the strengths and limitations of alternative software radio architectures.

## VIII. BROADER IMPLICATIONS OF THE SOFTWARE RADIO

The prospect of a new technology of multiband, multimode software radios—handsets and infrastructure—has social and political implications. Type certification authorities, for example, are charged with administering the equitable use of radio spectrum. Among other things, they certify that radio equipment meets legally imposed constraints. As discussed below, the software radio introduces new levels of complexity into the type-certification process.

In addition, software radios may operate on any RF band that is within the capabilities of the underlying radio platform, and with any mode for which a software load-image is available. This raises the possibility of truly novel approaches to spectrum management. One of the more interesting is the possibility that software radios could use a spectrum rental protocol to autonomously share spectrum. Another is that by incorporating advanced agent technology, they could evolve their own protocols. Radios capable of such behavior are called "cognitive radios" [48]. This chapter provides a minimal introduction to this area with pointers to the literature. Cognitive radio is an advanced research topic founded on the software radio architecture developed in this text.

### A. Type Certification

The prospect of an evolving radio platform raises questions about type certification. In remarks before the SDR Forum, the U.S. Federal Communications Commission (FCC) [49] described type certification of software radios as presenting "regulatory issues." These include the following:

1. For which service(s) is an SDR approved?
2. Is a new approval needed for each change to an approved SDR unit?
3. How does the FCC enforce the equipment authorization rules for SDRs?
4. How can an unauthorized use of an SDR be prevented?

Regulators rely on a mix of tactics to achieve their goals. Industry is required to obtain licenses for some uses of spectrum, while others are available without a license, provided the manufacturer complies with the regulations. The FCC relies on legal remedies to motivate manufacturers to comply with the rules. They generally specify license requirements in terms of RF power output, modulation, occupied bandwidth, spurious emissions, and frequency stability (e.g., over temperature and voltage supply variations). Analog radios embody these parameters in hardware, so the type certification process has historically focused on the certification of devices. Similarly, digital radios embody these parameters in a mix of analog and digital hardware. Current-generation PDRs with baseband **PDA** embody these parameters in relatively fixed core images that are tightly coupled to the hardware. This is compatible with the current type-certification process as well.

But SDRs with an **IF** **PDA** embody these parameters in software that is loosely coupled to the hardware. Each combination of band and mode has to be certified separately, according to today's process. Over-the-air downloads to the SDR complicate the certification process substantially. In April, 2000, regulators in the United States obtained the advice of industry through comments on proposed rule-making. Industry has the challenge of assisting regulators in defining a certification process that is responsive to the broader social and legal issues, but that does not seriously impede the benefits of SDR technology. Open architecture in some ways exacerbates the certification challenges. A proliferation of software packages enabled by open architecture drives the combinatorial complexity of type certification. Must a service provider certify every possible combination of software modules from every possible vendor? A helpful architecture might have properties that simplify and expedite type certification.

### B. Incremental Download Stability and Type Certification

In addition to defining a partitioning, an architecture may define principles that ensure plug-and-play with desired properties of controllability and reliability. For example, to type-certify an open-architecture SDR, one must guarantee that the properties specified by the regulatory bodies will be preserved *in spite of the software radio's high degree of flexibility*. The need for such guarantees motivates the study of the mathematical properties of the software radio [50]. For example, one may model the statistical demand for computational resources versus processing capacity using queuing theory [51–53]. Real-time performance can be ensured in a fixed architecture using this approach.

The plug-and-play SDR, however, has a *variable architecture* as modules are introduced into the environment and removed. This raises the complexity of the statistics, particularly in complex nodes. In a future 3G cell-site, for example, hundreds of users can invoke dozens of variable-bandwidth services via a pool of shared DSP resources. To make this tractable, there should be a predictable relationship of computational demand between plug-and-play software modules and the host processor environment. This calls for a theory of plug-and-play resource bounds for the software radio within which such predictable relationships will exist. The fact that radio software must run to complete in a short, finite time period that can be specified in advance leads to a proof that radio software need not be Turing-computable [50]. The theory translates into a prohibition on unconstrained While- and Until-loops. These have to be replaced by bounded-While- and bounded-Until-loops that are allowed to run at most $n$ times before generating a protection fault. The related theory of bounded recursion shows how a compiler can calculate $n$ for the programmer so there is no additional programming burden to obtain this protection. Without such protection, While-loops may run forever, consuming unacceptable amounts of time and processing power.

This theoretical advance makes it possible for one to provide a software-engineering environment that can place tight upper bounds on the computational resources of an arbitrary radio-software module. One may therefore prove by induction that a bounded-recursive downloaded module will consume resources that are within tightly specified a-priori limits when loaded into a bounded-recursive system. This can reduce the combinatorial complexity of the type certification of incremental software downloads. Given, for example, $M$ vocoders and $N$ air interfaces, a bounded-recursive software system need test only $M + N$ software configurations, proving the other $MN - (M + N)$ configurations by induction. This supports the incremental download of the $M$ vocoders, reducing download bandwidth on the network. Conventional software has to test all $MN$ integrated load images. Furthermore, a change of non-modular vocoder requires the download of a complete load image, with increased network overhead. This text therefore sets forth the technical issues that underlie this tradeoff between network overhead and download certification complexity.

## C. Spectrum Management Implications

Given that SDRs will continue to become more capable, one can ask whether they might have some fundamental impact on our approach to the use of the radio spectrum. A new research area, *cognitive radio*, suggests that this might indeed be the case [48]. Wireless multimedia applications require significant bandwidth, some of which will be provided by 3G services. Even with substantial investment in 3G infrastructure, the radio spectrum allocated to 3G will be limited. Cognitive radio is a particular extension of software radio that employs model-based reasoning about users, multimedia content, and communications context. Cognitive radio offers a mechanism for the flexible pooling of radio spectrum using a new class of protocols called radio etiquettes. This approach could expand the bandwidth available for conventional uses (e.g., police, fire, and rescue) and extend the spatial coverage of 3G in a novel way. This section characterizes the potential contributions of cognitive radio to spectrum pooling and outlines an initial framework for radio-etiquette protocols.

Figure 1-9 illustrates important aspects of spectrum allocation. Bandwidth that could be made available for the sharing of spectrum, based on current allocations to mobile users, is summarized in Table 1-6.

The literature describes a protocol for spectrum rental among cognitive radios and infrastructure [54]. The effective use of this new protocol requires software radios that always know where they are (e.g., in latitude, longitude, and altitude above mean sea level), and that embed propagation models that include terrain and buildings. In addition, they must know what their users are

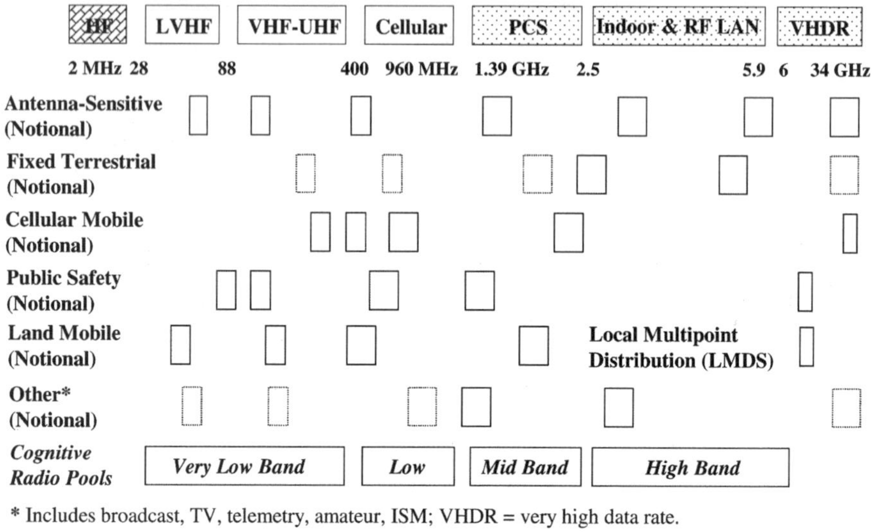

| HF | LVHF | VHF-UHF | Cellular | PCS | Indoor & RF LAN | VHDR |

2 MHz  28          88              400  960 MHz  1.39 GHz  2.5              5.9  6  34 GHz

Antenna-Sensitive (Notional)

Fixed Terrestrial (Notional)

Cellular Mobile (Notional)

Public Safety (Notional)

Land Mobile (Notional)                                    Local Multipoint Distribution (LMDS)

Other* (Notional)

Cognitive Radio Pools    | Very Low Band |   | Low | Mid Band | High Band |

* Includes broadcast, TV, telemetry, amateur, ISM; VHDR = very high data rate.

**Figure 1-9**   Potential spectrum pools.

**TABLE 1-6   Mobile Spectrum Pools**

| Band | $RF_{min}$ (MHz) | $RF_{max}$ (MHz) | $W_c$ | Remarks |
|------|-----------|-----------|-------|---------|
| Very Low | 26.9 | 399.9 | 315.21 | Long-range vehicular traffic |
| Low | 404 | 960 | 533.5 | Cellular |
| Mid | 1390 | 2483 | 930 | PCS |
| High | 2483 | 5900 | 1068.5 | Indoor and RF LANs |

doing (e.g., shopping, which is a low-precedence use, or in need of emergency assistance, which is a high-precedence use). Cognitive radios accomplish this by parsing all incoming and outgoing messages and voice traffic, and analyzing this information to establish the user's priority for use of spectrum. In addition, cognitive infrastructure can offer unused radio spectrum for rent for as little as one second in a microcell. Alternatively, rentals may allow use for minutes to hours in macrocells. The cognitive protocol includes listening for legacy radios to attempt to use the spectrum so that the cognitive radios may politely defer to legacy users. Police, for example, may require the renters to immediately yield the spectrum back to the renting authority. The protocol supports the return of spectrum within 30 milliseconds. Throughput is enhanced if the legacy users can wait for a half-second or more before being guaranteed clear spectrum.

Although cognitive radios may not be practical for years to come, the research points in an interesting direction for spectrum managers. Instead of hard allocations with primary and secondary users, the spectrum managers at some point in the not-too-distant future should be able to delegate the details of spectrum management to the radios themselves. The spectrum managers would then assume the higher-level task of specifying the rules the radios have to follow to insure equitable access that conforms to social, political, and legal norms.

This chapter has provided an overview of software radios. It began top-down by introducing the functional model of the software radio. Next, it introduced the important aspects of software, especially the need for isochronism in multiband, multimode radios that share a pool of processing resources among multiple users. A range of hardware implementations were introduced, differentiated among digital radios, PDRs, SDRs, and ideal software radios. This led to the characterization of acquisition parameters that divide software radios into economics-related classes. Finally, broader implications were presented, including the challenge of type-certifying software radios. The chapter concluded with a brief look into the future of the software radio—the evolution toward the cognitive radio.

## IX. EXERCISES

These exercises are designed to stimulate the serious student to (a) review the foundations, and (b) think further about the key questions of software radio architecture.

1. What are the fundamental limitations of any SDR implementation?

2. Differentiate among the PDR, the SDR, and the ideal software radio.

3. Which SDR applications are most amenable to modular, open-architecture hardware?

4. In which software radio applications is intellectual property (IP) likely to be provided by multiple participants?

5. What major functions does a software radio node perform (e.g., what are the functional components of a generic software radio)?

6. What is the difference between horizontal and vertical partitioning of a software radio? Think of a specific radio system, and describe its potential horizontal and vertical partitions.

7. What services does the software radio obtain from a software factory?

8. What is Turing-computability? Why is it important that radio software need not be Turing-computable?

9. Define architecture. Why is your definition suitable for the software radio?

10. What hardware components comprise a typical radio platform?

11. What are the critical resources that the hardware provides to the software?

12. What is the difference between an open-architecture SDR and one that is closed? Can an architecture be published in an open forum but at the same time effectively closed? How?

13. If you were going to acquire a large, expensive infrastructure based on SDR technology, which acquisition parameters would help you keep costs low over, say, a ten-year life of the equipment?

14. What are the type-certification challenges of the software radio?

15. What impacts could software radio have on spectrum management?

16. In addition to type certification and spectrum management, what other broad implications might arise from the continued proliferation of software radio technology?

# 2 Architecture Evolution

This chapter will convey a deeper understanding of the roots of the software radio. This includes the technical evolution that has resulted in today's emphasis on SDR. And it includes the management motivations toward realizing appropriately tailored implementations. The chapter begins with an introduction to technology-demographics, a method for studying architecture. This includes a historical perspective on radio architecture, which establishes the software radio as a demographic phenomenon. The chapter then characterizes the need for software-radio architecture. The commercial sector and the military sector share many common interests. They also differ in important ways that will be described. Because of the intense interest of these two sectors, and because of the fragmentation of industry into competing groups, there are competing goals for software radio evolution within standards organizations. These goals are considered in some detail since they set the stage for the competition of technical ideas. The chapter concludes with a roadmap of software-radio architecture evolution.

## I. TECHNOLOGY-DEMOGRAPHICS

Demographics is the science of vital and social statistics, as of the births, deaths, diseases, marriages, etc., of populations [55]. Demographics identifies major trends in human history. Demographics, for example, identified the departure of populations from the farms of the United States to the cities after the turn of the century. Demographics later identified the flight of populations from the cities to the suburbs. Technology-demographics,[9] analogously, is concerned with major shifts in technology. The transition of logic and control electronics from vacuum tubes to transistors is one example of such a shift. Identifying these shifts is important because of the fundamental changes in the economics of the related technologies. Those in the expanding part of the industry (e.g., transistors) can attract a disproportionately large amount of capital, for example. They are also faced with intense competition for a scarce pool of those who are skilled in the new technology. Those in the

---

[9]Technology-demographics would be more accurately named *technographics*. This term evokes a misleading meaning, e.g., of computer graphics. The inaccurate term *technology-demographics* evokes the right meaning.

| Era | Generation | Functions | Typical Components | Illustrative Design Rules |
|---|---|---|---|---|
| <1950 | Analog | Transmit and Receive<br>Channel Select, Squelch | Power, Antenna, Packaging<br>[Discrete Analog Baseband] | Channel Allocations, Power Limits<br>Standard Modulations (AM, FM) |
| 1960–70s | Early Digital<br>Microwave | Transmit or Receive<br>Protected Modes<br>Bit Error Rate (BER) Control | Analog + Quadrature Modems<br>Forward Error Control (FEC) | Analog + Operations/Management and<br>Bitstream Multiplex Interfaces<br>Adjacent Channel Power Envelopes |
| 1970–80s | Analog<br>MCR | Analog +<br>Signaling & Control | Analog + Digital Modems<br>+ Embedded Control Processors | Analog + Early Digital +<br>Cell Site and Frequency Plan<br>+ Handoff Protocol |
| 1970–90s | Spread<br>Spectrum<br>(CDMA/FH) | Code Synchronization,<br>Code Management,<br>BER Control | Analog + Early Digital<br>+ De-/Spreading Devices<br>+ Embedded Control Processors | Analog + Code Design<br>+Peer Network Protocols<br>+Digital Voice Channel |
| 1980–90s | TDMA<br>MCR | Analog + Early Digital<br>+ Analog MCR<br>+ Spread Spectrum<br>+ Diversity + Directivity | Analog + Analog MCR<br>+ Spread Spectrum<br>+ Multibeam Antennas | Analog + Early Digital<br>+ Early MCR<br>+ Digital Channel Coding<br>+ Privacy & Authentication |
| >2000 | Future<br>Seamless<br>Multimode<br>Multimedia<br>Networks | + Agile PowerMgt.<br>Data RateMgt.<br>+ BER Agility<br>+ Mode Handover<br>+ Location Reporting | + High Programmability<br>+ Agile Modulators<br>+ Multiband Antennas<br>+ Multiband RF | + Mode Handover Criteria & Protocols<br>+ End-to-End Encryption<br>+ Software Defined Services |

MCR = Mobile Cellular Radio

**Figure 2-1**  Radio architecture evolution: Complexity of functions, components, and design rules increases with each subsequent generation.

waning part of the industry (e.g., vacuum tubes) may survive by carefully focusing on a market niche that is insulated from the larger forces shaping the demographics (e.g., high-power amplifier tubes). They face the challenge of retraining to address the new technology and reducing the workforce in the face of declining market share.

To do justice to the technology-demographics of radio would require a statistical treatment that is beyond the scope of this chapter. However, it is useful to trace the migration of radio architecture in the spirit of demographics. To trace the technical-demographics of architecture requires attention to the elements of architecture: functions, components, and design rules. The trend is clearly toward increasing complexity in each of these areas. This trend leads to the software radio as inexorably as the movement to the suburbs led to shopping malls (for better or worse). This section therefore reviews the evolution of radio leading to the software radio. This should convey a sense of the interplay of technical and economic factors on the decades-long timelines along which software radio technology is evolving.

## A. Functions, Components, and Design Rules

The complexity of functions, components, and design rules of radio architectures has increased with each successive generation of radio as exemplified in Figure 2-1. It reveals a systematic progression of functions, components, and design rules. The functionality of early analog radios was limited to transmit and receive AM or FM, with RF, power, volume, and squelch controls.

An antenna, analog transmitter, receiver, and hardware controls provided necessary and sufficient components. The associated design rules consisted primarily of constraints on the RF emission template including radiated power and out-of-band energy. Citizens band (CB) radio (20 narrowband RF channels at 40 MHz) became very popular with travelers on interstate highways in the United States during the 1970s. Access to CB channels was accomplished using voice protocols such as saying "Break on 19" for permission to join those using channel 19. The channel popular with interstate trucking was constantly jammed with users, while most of the other CB channels were essentially unoccupied. This is an extreme example of what can happen when all radios have identical functionality and attempt to share limited RF channels using manual protocols. Even today, however, for extremely low-cost consumer products, this is the way to go. For example, sportsmen hunting outdoors find walkie-talkie peer networks adequate for their purposes. Many police, fire, and rescue networks are also peer networks in terms of the functional capabilities of the radio equipment. Hierarchical control may be asserted by police headquarters using a voice protocol. If the antenna at headquarters is tall enough, its signal propagates to mobile stations much better than mobile station signals propagate among each other. Thus, headquarters functions as the logical base station of what is functionally a peer network.

First-generation (1G) mobile cellular radio (MCR) systems incorporated the signaling and control functions of wireline telephony into a dedicated, digitally encoded RF signaling channel. This imposed rigorous structure onto the radio network, introducing the mobile wireless network hierarchy. Voice traffic was assigned to a single narrowband (e.g., 25 kHz) analog traffic channel. The networks used the signaling channels to page mobiles, to accept call setup requests, and to direct mobiles to specific traffic channels. This centralized control transformed RF carriers into radio resources to be managed algorithmically.

Historically, frequency-hopped spread spectrum was invented during World War I. The major U.S. investments in direct-sequence spread spectrum (DSSS) occurred early in the Cold War, for example, in the U.S. Defense Advanced Research Projects Agency (DARPA) TEAL WING [56] project, about 30 years ago. This included frequency hopping and direct-sequence spread spectrum. The first large-scale deployment of a hybrid frequency-hopped and direct-sequence spread spectrum radio was the JTIDS radio of the mid-to-late 1980s. JTIDS spread spectrum air interface requires microseconds of timing precision across the network to limit the correlator search in the receiver. The fast frequency-hopping aspect of its interface stresses the state of the art of fast-tuning synthesizers. Hybrid air interfaces like JTIDS therefore imposed the most demanding design rules of that era. In the mid-1980s, the United States dominated the global radio marketplace. U.S. strategic communications technology was unequaled, and U.S. manufacturers, notably Motorola, dominated first-generation mobile cellular radio.

## B. Global Restructuring Through 2G and 3G Mobile Cellular Radio

There is a significant technological breakpoint in Figure 2-1 between spread spectrum and time division multiple access (TDMA). The second-generation (2G) TDMA increased the complexity of commercial mobile cellular radio by an order of magnitude over 1G. The roots of this dramatic transformation were laid in the International Consultative Committee on Telephone and Telegraphy (CCITT, now International Telecommunications Union—Radio, ITU-R). In 1983, the CCITT formed the Gruppe' Speciale Mobile (GSM) [64] special mobile radio study group to define the next generation of mobile cellular radio. The 2G standard employed eight-way time slicing of a single RF carrier in its TDMA air interface. Each RF carrier was modulated to a 270 kHz bandwidth to provide 13 kbps data rate for each of eight users. It also incorporated digital voice coding, digital channel symbols on the traffic channels, equalization, training sequences, precise framing, and other design rules to improve quality of service via the first all-digital commercial air interface. As suggested in Figure 2-1, these techniques significantly increased the complexity of radio networks. The complexity increase was in fact so great that major global suppliers of first-generation cellular equipment like Motorola lagged behind risk-takers in Europe (notably Nokia and Ericsson). By 1996, Ericsson and Nokia were shipping 85% of the 70 million digital handsets manufactured that year. More significantly, dealing with the complexity of GSM forced European radio engineers to develop the radio software-engineering capabilities needed to deal with the increased complexity. The first GSM revenue was generated in Europe in 1993, and by 1998, GSM had become the Global System for Mobile Communications. Over 100 nations are now participants in the GSM Memorandum of Understanding (MoU).

As GSM entered competitive deployment in the early 1990s, the European Community began to focus on precompetitive research for the next generation of mobile cellular radio. In addition to terrestrial mobile cellular, the vision of the European Advanced Communications Technology and Services (ACTS) program embraced satellite mobile radio as well. The related set of radio disciplines became known as wireless. The 3G wireless vision was embraced globally, as illustrated in Figure 2-2. The European strategy for 3G that had been agreed to by 1994 included the following [57]:

"...to develop advanced communication systems and services for economic devel-opment and social cohesion in Europe, taking account of the rapid evolution in technologies, the changing regulatory situation and opportunities for development of advanced trans-European networks and services. The aims are to support Euro-pean policies for early deployment and effective use of advanced communications in consolidation of the internal market, and to enable European industry *to com-pete effectively in global markets*. The work will enable the *re-balancing of public and private investments* in communications, transport, energy use and environment protection, as well as experimentation in advanced service provision. In conjunc-

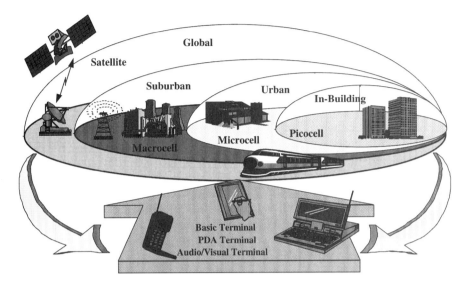

Figure courtesy Stephen Blust and BellSouth Wireless.

**Figure 2-2**   Global vision of third-generation wireless.

tion with the work in the specific programme on information technologies, it will provide a *common technological basis* for applications research and development in the specific programme on telematic systems and will prepare the ground for the development of a European market for information services..." [Emphasis provided by the author]

The "common technological basis" for 3G merged the best of TDMA and CDMA. Qualcomm in the United States had established patent positions on fundamental CDMA technology. Its intellectual property (IP) was the basis for the U.S. CDMA standard, IS-95. Subsequently, Qualcomm led the U.S. thrust in 3G standardization in the ITU, proposing an enhanced version of the U.S. standard [58]. Ericsson, among others, had contravening IP that resulted in part from its participation in ACTS. In September 1999, Ericsson acquired the cellular infrastructure business of Qualcomm. In the exchange, Ericsson acquired the rights to Qualcomm CDMA patents, and Qualcomm acquired the rights to use Ericsson CDMA IP in their chips. Qualcomm's CDMA chip business remained with Qualcomm, as did their U.S. DoD business.

During the early 1990s, Motorola was in the process of forming the Iridium consortium to offer mobile users truly global wireless access via a large constellation of low earth orbit (LEO) satellites. At the same time, the now-defunct Globalstar consortium developed similar services based on conservative geosynchronous (GEO) satellite technology. The INMARSAT consortium had built a successful niche-business in satellite mobile telephony with its GEO satellite constellation. The one-meter apertures and suitcase size of these telephones was acceptable for ships at sea, for example. Iridium proposed to deliver voice and data services in a handheld format. INMARSAT occupied the market niche in which fixed infrastructure is unavailable and awkward terminals are acceptable to the user. Given the monumental success of terrestrial networks, Iridium offered dual-mode handsets capable of interoperating with both the terrestrial network and the Iridium satellite constellation. But its services were limited to voice and low-speed data, at a price of over ten times that of terrestrial networks. In addition, the user had to switch hardware modules in order to change between satellite and terrestrial operation. A software radio implementation would have been less cumbersome but may not have changed the outcome. Although the 3G vision continues to include satellite communications, the September 1999 bankruptcies of Iridium and Globalstar increases the uncertainty of the economic future of mobile satellite services (MSS) in 3G. This uncertainty creates less demand for multimode handsets that include MSS, somewhat reducing the demand for SDR technology.

Service providers need the seamless multimedia functions listed in Figure 2-1 for future networks. These require components that are more sophisticated with significantly increased software complexity, both in the base station infrastructure and in the mobile units. The increasingly complex design rules are necessary to ensure smooth interoperation and service availability in a heterarchical infrastructure with the many time-varying demands of voice, mobile computing, computer-telephony integration, and ultimately multimedia. The industrial process of adopting software-radio architecture ultimately should promote plug-and-play and reduce costs through standardized hardware and software components. While deploying these more complex systems, the service providers must remain cost competitive. Service providers have therefore begun to focus on future-proof infrastructure. Such infrastructure must flex readily across air interface standards as the mix of users dynamically dictates. Software radio architecture therefore must be focused on simultaneously increasing the quality and decreasing the cost of such flexibility. At the same time, software radio architectures must accommodate hardware alternatives and must provide new ways of managing the increasing complexity of continuously evolving standards.

## C. Complexity Equals Software

Radio architectures may be plotted on the dimensions of network organization versus channel data rate as shown in Figure 2-3. These architectures have evolved from early point-to-point and relatively chaotic peer networks (e.g., citizens' band and mobile military push-to-talk radio networks of the 1970s) toward hierarchical structures with improved service quality (e.g., cellular radio). In addition, channel data rates continue to increase for shorter delays, more efficient multiplexing, and/or better robustness in multipath propagation [59]. In a multiple hierarchy (heterarchy), a single radio unit participates in more than one network hierarchy. For seamless multimedia services,

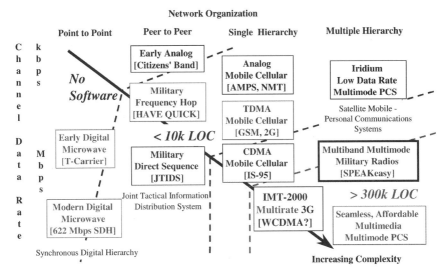

**Figure 2-3**  Network organization and channel data rate vs. code complexity.

a handset might participate in a personal communications systems (PCS) network inside an office building, a second-generation cellular network between the office and home, and a mobile satellite network as the roaming of the user dictates. Military radios and multimedia commercial networks that benefit from multiband multimode technology represent the high end of architecture evolution. But, there also are opportunities to insert software-radio technology more gradually, such as in low-end PDAs.

Considering Figure 2-3 more closely reveals the relationship between network organization and software complexity. Single mode point-to-point and peer network radios require only limited amounts of user-interface and control software. Air interfaces that are more complex, such as JTIDS, require substantial software (or the equivalent embedded in ASIC and/or FPGA digital logic platforms). A few tens of thousands of lines of code (LOC) is typical. Multiband, multimode software radios of the SPEAKeasy II class and beyond, on the other hand, require hundreds of thousands of lines of code to accommodate the complexity of air interfaces in diverse RF bands. This trend toward increased software complexity bears closer examination.

*1. Radios With Minimal Software*  Each of the simpler network structures of Figure 2-3 contributes technology for the software radio. Point-to-point radios included PTT and frequency division multiplexed (FDM) radios of the 1960s and 1970s. They carried 60, 240, and 1920 voice channels or more [60]. T-carrier digital pulse code modulation (PCM) microwave radios superseded them in the high-capacity backbone networks of the 1980s. Current high-capacity PCM systems may employ high-order QAM [20] or multicarrier technology to achieve OC-12 (622 Mbps) and higher data rates. Early

PTT and FDM radios included no software for physical layer and data link functions. During the late 1980s and 1990s, there was a proliferation of embedded processors in such radios. At first 8-bit microcontrollers supported built-in test and parameter configuration. Later 16-bit microprocessors combined with digital ASICs and FPGAs for the "digital radio" revolution of the late 1980s and early 1990s. Today, such radios may embed 10,000 (10 k) LOC in the radio modulator/demodulator (modem), data link protocols, vocoders, and related functions.

For much of the world, the 2 GHz microwave spectrum is now used for cable TV relay or mobile radio. High-capacity backbones once dominated by 2 GHz microwave FDM and then PCM have transitioned to fiber optics in most developed nations. Microwave backhaul (e.g., GSM Bis and Abis links) now connect the base transceiver station (BTS) to base station controllers (BSCs) using frequencies once allocated to fixed telephony. Nevertheless, point-to-point microwave in large part paved the way for today's mobile radios. It contributed adaptive equalizers, forward error control, and trellis coding techniques useful in overcoming fading in mobile channels. QAM and multicarrier techniques from PCM encode many bits per Hz of signal bandwidth if sufficient SNR is available. Each of these techniques for increasing channel data rates may be employed in a mode of a software radio.

In addition, 3G wireless includes data rates up to 2 Mbps. Even more aggressive fourth-generation (4G) visions propose 155 Mbps (OC-3) for wireless local loop, wireless local area networks (LANs), and wireless multipoint distribution [61]. These wireless broadband systems are contemplated in RF bands from 5 to 34 GHz or more. Higher data rates favor early implementation of digital functions in hardware. The function later migrates to software as the DSP technology achieves the required speeds. Although PCM radios employ computationally intensive signal processing in the carrier and timing recovery loops and in the demodulator, the complexity of this code is low. The computational demand stems from the fact that moderate complexity signal processing (e.g., matrix inversion for channel equalization) must be accomplished for each channel symbol at rates of 22.5 to 50 MHz or more. Software complexity for such radios is generally less than 10 k LOC. Most of such code consists of hard-real-time DSP or FPGA code plus microprocessor-based control of hybrid analog/digital functions (e.g., timing recovery using a surface acoustic wave filter that has software-selectable bandwidths).

***2. Moderate Software Complexity***   Software radios may embed technology from military peer networks as also suggested in Figure 2-3. Frequency-hopping radios like the slow-hopped Have Quick radio, for example, were first deployed by the military to reduce jamming vulnerabilities [59]. These peer networks collaboratively select one station as the network control station. JTIDS includes direct-sequence spread spectrum in addition to faster hopping and time division multiplexing to create a robust, high-performance but relatively expensive radio [62]. In addition, however, frequency hopping reduces multipath fading (e.g., in GSM). Spectrum spreading may also be used for multiple access provided the spreading sequences are quasi-orthogonal. This, of course, is the basis for CDMA. In addition, the wider spread spectrum bandwidths make it easier to coherently combine multipath components for enhanced SNR and fade resistance.

Thus, techniques that the military pioneered' for TRANSEC and reduced jamming vulnerability in the 1970s and earlier have now become commercial practice for enhancing QoS in mobile wireless. GSM, for example, uses frequency hopping for improved fade resilience. The IS-95 CDMA system employs spectrum spreading for improved subscriber density and coherent combination of multipath. Smart antennas use beamforming and interference nulling pioneered by the military for jamming suppression. A pattern emerges of technology migration from advanced niche applications such as military systems to broad commercial applications over time. As this migration occurs, the complexity of the resulting commercial radios increases substantially. A single waveform with frequency hop (FH) and other spread-spectrum features typically requires 40 k LOC.

Historically, the migration of new techniques from the military to the commercial sector has taken decades. The competitive mobile wireless marketplace now offers huge financial incentives for rapid cost-effective migration in years or months versus decades. Movement toward open-architecture software radios in contemporary wireless is thus driven in part by the financial incentives of such rapid migration. When software radio implementations are possible, they offer shorter development cycles compared to hardware-intensive approaches. An initial software radio product may be replaced with a more cost-competitive, hardware-intensive product. In addition, an open-architecture approach enables teaming among companies with unique intellectual property, again facilitating the rapid migration from concept to product.

*3. Toward a Million Lines of Code*    First- and second-generation mobile cellular radio and PCS systems are organized into simple hierarchies as also illustrated in Figure 2-3. The RF channel modulation and hence efficiency in use of the spectrum have matured from analog FM frequency division multiple access (FDMA) [63] in first-generation analog systems to time division multiple access (TDMA) [64] in 2G systems. CDMA [65] is characteristic of 3G mobile wireless. The network organization has been that of a single hierarchy. That is, the mobile handset is subordinate to a Base Transceiver Station (BTS), which in turn is subordinate to a Base Station Controller (BSC). The BSC is subordinate to a mobile telephone switching office (MTSO), which in turn is subordinate to the telecommunications management network. Handoff from one transceiver to another operates within one hierarchy.

With the emergence of satellite mobile systems which interoperate with PCS for seamless roaming, handsets may operate in two different hierarchies: the terrestrial and the satellite mobile hierarchy. In the more primitive initial services, the user selects the PCS or satellite mobile mode, allowing the handset to operate in one hierarchy or the other with no handover across hierarchies. But as 3G alternatives become more complex, the handset or network must pick the most appropriate band, mode, QoS and tariff parameters autonomously, leading to some kind of mode-awareness capability distributed among the handset and the hierarchies. Universal Mobile Telephone Service (UMTS) [66] contemplated multimode handsets for the introduction of 3G. One may envision future software radios that can operate on cordless, wireless local loop, macrocellular, satellite mobile, and in-building PCS [67]. So one may have four or five voice radio hierarchies across which seamless handover could be distributed. In addition, there are dozens of possible modes for data including two-way paging, RF LAN, Cellular Digital Packet Data (CDPD), General Packet Radio Service (GPRS), voice modem networks, and 3G alternatives, again leading to the need for seamless heterarchy. CDPD [68] is a data service of the U.S. Analog Mobile Phone System (AMPS) [69], which is evolving to the IS-136 all-digital wireless network. GPRS is a high-speed data service of GSM. Each digital control and data mode adds another 10–40 k LOC, or more, rapidly expanding the radio's software content. In addition, the layers of "infrastructure" code between the operating system and the air interface may comprise from 100 to 400 k LOC. Software radio technology pathfinders require upward of a half-million lines of code today. During the next few years, that complexity will double.

Today's wireless systems engineer must step up to this software challenge. There will be substantial financial rewards for those who can create and evolve architectures that accommodate the insertion of appropriate new technologies from niche disciplines and research centers without having to completely redesign the radio platform. Software radio architecture was conceived to accommodate multiple bands, modes, and hierarchies with only incremental, "mostly software" enhancements. But software radio architecture also enables the insertion of reconfigurable hardware (e.g., FPGAs). Economical incremental upgrades therefore drive software radio architecture. Increased software complexity, however, places software on the critical path of software-radio development. The software-radio architecture should future-proof wireless infrastructure against rapidly evolving air interface standards. In addition, software radio versions of formerly hardware-intensive 1G and 2G waveforms (e.g., GSM) may be integrated into wideband 3G handsets. If the designers can produce flexible despreader ASICs that can both process wideband CDMA in hardware and digitize 2G waveforms effectively, they will assure incremental 3G deployment.[10] To prepare for the next decade of rapid evolution of software-intensive radios, one must understand software radio architecture in depth. The following section therefore reviews the need for this architecture from the commercial perspective.

---

[10]At this point the reader may be somewhat mystified by this discussion. In order to effectively contribute to architecture tradeoffs, one must be exposed to these high-level issues. By the end of the text, the technical approaches involved in these tradeoffs should have been made clear.

## II. COMMERCIAL ARCHITECTURE NEEDS

There is an important difference between the need for a software radio or SDR and the need for a software-radio architecture. Wireless has made substantial progress without an open architecture. Air interfaces and network interfaces constrain radio node performance without unduly constraining implementation alternatives. In addition, there is a difference between an open architecture and a high-quality software-radio architecture. An open architecture is merely a published standard. That is, the publisher of the architecture has decided that it is desirable to forgo intellectual property (IP) rights in exchange for broader industry participation in product development. Such an architecture may admit either a plethora of components or only a very narrowly defined subset of the possible components. It may have powerful design rules that facilitate plug-and-play at low cost. Or it may have such complex and opaque design rules that products intended to work together are constantly in conflict. And it may or may not address the functionality needed for growth in the future. Since the commercial sector's interest in open architecture is based on the success of the PC, it is worth examining that historical precedent.

The computer industry fared well from the 1950s until the 1980s without an open architecture. The Industry Standard Architecture (ISA) of the PC launched the computer industry into a new level of ubiquity and prosperity. This change can be understood in terms of technology demographics. Fledgling Apple had shown that PCs were both possible (using a new microprocessor chip) and economically viable. An open architecture, strictly speaking, was not forced on anyone. Instead, the computer industry leader, IBM, sought to protect its dominant position. Although it was at the time unclear whether the PC would be a success, there was risk to IBM's core businesses if it were a success without IBM leadership. The open-architecture approach marshaled large-scale investments through industry partners. The previously proprietary (closed) architecture of the computer bus and ISA was published by IBM, thus opening the architecture of the PC. Thus, in retrospect, the bottom-up incursion of upstart Apple led to top-down innovation by IBM. The result was a proliferation of third-party graphics boards, peripherals, etc. around the IBM architecture. At about the same time, Microsoft developed DOS for IBM, which was published in the same open-architecture spirit. DOS on the open-architecture PC platform yielded an explosive proliferation of affordable software. The bottom line was that open-architecture hardware and software transformed the computer industry. Before the transformation, the big winners were the systems integrators—IBM, Digital (DEC), etc. After the transformation, the big winners were the ISA-chip makers (notably Intel) and the operating-system suppliers (notably Microsoft).

It is hard to tell whether a similar process might apply to segments of the wireless market. At present, a small handful of integrators dominates the commercial wireless business. For example, Ericsson, Lucent, Motorola, Nokia, Nortel, and Qualcomm each have a significant proportion of the wireless business. These plus only a few other suppliers (e.g., Alcatel) account for the vast majority of the commercial cellular telephone handsets and infrastructure. Are there bottom-up pressures akin to Apple's early success with the PC that would stimulate industry leaders to embrace an open-architecture standard? Is some little company making "personal-handsets" or "personal-cell-phone-infrastructure" based on some new technology? None comes to mind. The industry leaders already make wireless PDAs, for example. So the relationship between the PC industry and the wireless industry is at best not isomorphic. Instead, there is a large and influential set of customers, the commercial service providers, expressing an interest in open architecture. This would be somewhat akin to the "Big 8" accounting firms of the mid-1960s pressuring the computer industry to open up its mainframe architecture. They might not like the costs of supporting different accounting software architectures from IBM, Honeywell, Univac, and GE. Many would argue that the mainframe suppliers could not completely ignore their big customers. But neither would such pressures fundamentally transform the mainframe industry. The technology-demographics just would not be there until some external force appeared on the scene (e.g., the threat of the PC taking away mainframe business).

Nevertheless, leading wireless service providers have been unequivocal about the need to reduce the cost of ownership of wireless infrastructure. BellSouth, notably, has been a leader in the evolution of the SDR.

## A. The BellSouth Software-Defined Radio (SDR)

The need for an SDR migration path was clearly articulated in BellSouth's Software-Defined Radio Request for Information (RFI) [70]. Its release in December 1995 was a watershed event for software radio technology: this was the first public statement of need for software radio technology by a large telecommunications service provider. The SDR RFI includes a comprehensive statement of requirements for future-proof[11] wireless infrastructure. It anticipates SDR infrastructure that can be programmed for new standards and for specific deployments and can be dynamically updated by software uploads after deployment, including over the air via the software-defined network.

As Table 2-1 suggests, wireless service providers have realized that today's infrastructure lacks the flexibility necessary for economical growth of wireless services. With hardware-based infrastructure of the 1990s, many value-added services require new hardware. The first round of offerings of Cellular Digital Packet Data, for example, required dedicated CDPD hardware. When CDPD failed to generate revenue quickly enough, the operators had to bear the costs of two infrastructures. As 3G emerges, additional wireless infrastructures are needed. Add this to dedicated trunk radios and paging systems to yield cumbersome, fragmented commercial infrastructures. These are expensive to

---

[11]Although the concepts were first articulated by BellSouth, the words *future-proof* do not occur in the SDR RFI. The phrase was introduced by the author at the second SDR forum meeting in 1996.

**TABLE 2-1   SDR Functional Highlights**

| | |
|---|---|
| General | Universal interfaces (source coding, channel coding, error control, and protocols) regardless of multitechnology (FDMA, TDMA, CDMA, and hybrids), multiband, and multistandard environments |
| Services* | Seamless internetworking of AM, FM, cellular (analog, TDMA, CDMA), PCS, mobile data and paging |
| Standards* | 30 MHz Special Land Mobile Radio, aviation's APCO 25, HF, VHF/UHF voice/data; voice privacy; GSM, 60 GHz in-building PCS, and FH for the U.S. domestic and global marketplace |
| Flexibility* | Flexible RF, channel, time slot, power, bit rate, equalization, channel coding, and error correction |
| Advances | Adaptive networks, transparent bridging, channel modeling, feedback, adaptive diversity, innovative signaling, and improved quality |
| Growth Path | Velcro → DSP-enabled → multipersonality → variable-personality software radio |

*Lists are illustrative, not exhaustive.

maintain and lack efficient growth paths to accommodate new services and standards. Digital wireless moves providers closer to wireline levels of quality, but the added flexibility and technical advances highlighted in the table are needed to move on to the seamless voice, data, facsimile, and multimedia wireless services to which providers aspire.

The SDR RFI anticipates several stages in evolution toward the SDR. The Velcro-phone allows new services to be added through modular addition of hardware and software components (e.g., RF front ends, modem boards, and related software) in an open-hardware architecture. At this stage, the sound of Velcro ripping apart graphically conveys the continuing high costs of such infrastructure, brought on in part due to the need for touch labor to upgrade the hardware in the field. The second stage, the DSP-enabled SDR, contemplates a mix of DSP and general-purpose computing in which software costs may remain high due to the cost of real-time DSP software. The multipersonality infrastructure envisions relatively seamless, dynamic plug-and-play software on general-purpose computing platforms in a hardware-independent open-architecture software environment. Variable personality radio would employ broadband antennas and RF so that the radio platform can support a wide range of evolving and niche-oriented air interfaces. These stages include multiband, multimode mobile handsets as well as fixed infrastructure.

## B. European Perspectives

Panel C of the International Conference on Universal Communications provided European perspectives on 3G mobile communications [71]. Dr.

Giovanni Colombo [CSELT, Italy] presented an introduction to this panel session, addressing market, R&D initiatives, standardization activities, and migration requirements. Using the frequencies 1885–2025 MHz and 2110–2200 MHz identified in the World Administrative Radio Conference (WARC '92), the European Universal Mobile Telephone Service (UMTS) will be aligned with the worldwide standard for 3G, IMT-2000.[12] This will provide global terminal roaming and other advanced services. Dr. Colombo offered the view that generation "2+" of mobile cellular/PCS systems would augment second-generation GSM digital mobile radio with higher-quality digital speech, data, and LAN capabilities in a personal mobile radio (PMR) package. High "bit-rate" services added to generation 2+ yield generation 3 or UMTS/FPLMTS. In remarks at the First International Workshop on Software Radio, A. Urie (Alcatel) and H. Houmo (Nokia) agreed that affordable, incremental deployment of 3G depends on "SDR handsets" [72].

In the UMTS scenario, CT2 [73] and Digital European Cordless Telephone (DECT) CAI [74] cordless telephones replace first-generation analog cordless for indoor private use up to moderate traffic intensity. DECT bears the higher intensity applications and bridges (via wireless PBXs) to commercial applications. DECT is also envisioned for limited urban applications, such as in shopping malls. Analog cellular, GSM 900, and DCS 1800 (the 1800 MHz version of GSM) all address urban and suburban (outskirts) applications although only analog mobile cellular reaches appreciably offshore, according to Giovanni. Services are envisioned as emerging from POTS to alternating messaging, speech, data file transfer, and high-rate video with data rates from tens of kbps to 1 Mbps. According to this perspective, 1 Mbps rates are limited to indoor wireless LAN applications while hundreds of kbps are available outdoors, nationally and, in a limited way, regionally (e.g., in Europe). Other contemporary European perspectives are more aggressive, contemplating multipoint distribution of video in neighborhoods at data rates up to OC-3 (155 Mbps) and at T1/E1 rates for wireless local loop (WLL) applications [75, 76].

One may infer a European view of how requirements and technology interplay in the marketplace. The driving requirements appear to be:

1. Coverage
2. Interference mitigation
3. Radio resource control
4. Voice, data, and multimedia services
5. Grade of service (GoS)
6. QoS

---

[12]The name Future Public Land Mobile Telecommunications System (FPLMTS) evolved to International Mobile Telecommunications 2000 (IMT-2000).

Again in this synthetic European view, the first-order technological determinants of coverage include the basic radio access technology plus macrodiversity and power control. Interference mitigation is fundamental to high-quality access and smart antennas support these goals, as shall be seen in the sequel. Radio resource control, similarly, centers on variable bit-rate technology such as channel strapping and directional high data rate (including use of the millimeter wave bands). Voice, data, and multimedia services at an acceptable GoS/QoS will require the full range of technologies including transcoding and multibearer wireless. Seamless network architecture must accommodate mode handover, integrated B-ISDN, and other intelligent networking technologies. It remains to be seen whether consumers will be more willing to pay for intelligence in the network or intelligent devices in the home (or office) as the popularity of home answering machines and customer premises voice messaging attests. Some modest level of advanced intelligent network (AIN) is clearly needed, including managed objects in the wireless substructure.

In addition, the European Commission is sponsoring research for the use of 2, 5, 17, 40, and 60 GHz radio for higher data-rate services (1 Mbps through 155 Mbps OC-3). The driving applications are local multipoint distribution services (LMDSs) and wireless local area networks (WLANs). The WLANs offered in the mid-1990s were not a major commercial success. Higher data rates are offered at the higher frequencies. The software-radio derived opportunity to amortize costs of telephone, desktop video teleconferencing, and WLAN in a single adaptable infrastructure may provide the economic benefits needed to propel all three technologies forward.

There is some hope that this vision will in fact materialize. J. Schwarz DaSilva of the European Commission [77] offered a market analysis that estimated 73 million subscribers in 1995 increasing to 170 million by 1998 reaching a growth rate of 170,000 subscribers per month. The European component of that market would reach 12.5 million analog and 15 million digital subscribers by August 1996 with a market penetration of 8% of the population. GSM is the cornerstone of the European approach with deployments in 86 countries, over 191 operators signed to the GSM MoU, and proliferation to DCS 1800 and PCS 1900 in the United States. Juha Rapeli [27] projected 20 M GSM users worldwide in mid-1996 with 1 M per month growth rate. Of these calls, 20 M per month are made by international roamers. These projections turned out to be conservative. To propel the vision forward, Europe spent approximately 9432 MEcus (Millions of European currency units) in a balanced research, technology development, and demonstration program as follows:

- 680 MEcus for telecommunications technologies

- 1932 MEcus for information technologies

- 843 MEcus for telematics applications (e.g., telemedicine)

**TABLE 2-2    European Initiatives (1997–1999) [78, 79]**

| Program | Description |
|---------|-------------|
| AWACS | ATM Wireless Access Communications System |
| COBUCO | Cordless Business Communication System |
| *FIRST* | *Flexible Integrated Radio Systems Technology (an SDR-like project)* |
| *FRAMES* | *Future Radio Wideband Multiple Access Systems (used an SDR approach)* |
| INSURED | Integrated Satellite UMTS Real Environment Demonstrator |
| MEDIAN | Wireless CPN/LAN for Professional/Residential Multimedia Applications |
| MICC | Mobile Integrated Communications in Construction |
| MOMENTS | Mobile Media and Entertainment Services |
| MOSTRAIN | Mobile Services for High Speed Trains |
| *MULTIPORT* | *Multimedia Portable Digital Assistant (needs for an SDR-PDA)* |
| NEWTEST | High-Performance Neural Network Signal Processing |
| *On The Move* | *Multimedia Information Services (needs for an SDR-PDA)* |
| RAINBOW | Radio Access Independent Broadband on Wireless |
| SAMBA | System for Advanced Mobile Broadband Applications |
| SECOMS | Satellite EHF Communications for Mobile Multimedia |
| ABATE | Services/ACTS Broadband Aeronautical Terminal Experiment |
| *SINUS* | *Superconducting Systems for Communications (SDR Implications for RF)* |
| TOMAS | Inter-trial Testbed of Mobile Applications for Satellite Communications |
| *TSUNAMI II* | *Technology in Smart Antennas for Universal Advanced Mobile Infrastructure (SDR technology in several Tsunami projects)* |
| UMPTDUMPT | Using Mobile Personal Telecomms for the Disabled in UMTS Integration |
| WAND | Wireless ATM Network Demonstrator |

In addition, they allocated 540 MEcus for cooperation with third-world countries and international organizations, 330 MEcus for diffusion and "valorization" of research results, and 744 MEcus for training and mobility of researchers. The overall program includes the Flexible Integrated Radio Systems Technology (FIRST) thrust in which Orange UK, SDR Forum members, were participants. Other elements of the program are listed in Table 2-2.

DaSilva also lists smart antennas, superconductivity, and software radio concepts as enabling technologies for UMTS/FPLMTS as pursued through these programs. These technologies respond to the marketplace, which is characterized by DaSilva. "Users want broadband wireless mobile services that ensure full applications portability and multimedia content along with full user mobility across a range of different but fully interoperable network infrastructures such as cable, satellite, fiber, and wireless." Key issues include resource access; worldwide interoperability; terminal and network control; security in control, transmission, and management; interprocess communications across band and mode boundaries; and the allocation of intelligence and resources across the network(s). Innovative use of frequency spectrum, on-demand access to broadband (multimegabit/sec) channels, and adaptability to multimedia content are also goals. In addition, software-controlled RF elements, novel wireless air interfaces, and hardware/software innovation underscore the need for software radio technologies.

Bosco Fernandes, chairman of ACTS Mobile and Personal Communications Domain [77], envisions similar requirements in the marketplace, but he characterizes the "radio interface challenge" as "slotted multiple access (TDMA, CDMA, and OFDMA)." According to Fernandes, multicarrier creates high bandwidth on demand in a single time slot (OFDMA) while TDMA may be either slotted for use by multiple users or dedicated to a single user. Scenarios include bandwidth on demand for more usable bandwidth. For example, GSM could be split eight ways for 13 kbps per user, split four ways for 26 kbps per user, etc., up to 200 kbps per user on a dedicated basis. Wideband CDMA provides the other options for spectrum access and sharing, for a proliferation of modes within an existing spectrum allocation such as GSM.

## C. Asian Perspectives

The first Asian workshop on software radio was held at Keio University, Japan, on April 1, 1998. Technical papers emphasized smart antenna technology [80]. Japanese participation in the First International Software Radio Workshop included a paper from Matshushita/Panasonic on the viewpoint of the terminal manufacturer [81]. At that time, the software radio was regarded as one of the technical possibilities for the multimode terminal. Such a terminal should be capable of personal digital cellular (PDC) [82], Personal Handyphone System (PHS) [82], GSM, IS-95, and W-CDMA. Matshushita envisioned three steps in the evolution of the software radio terminal. The first step, currently in production, is the processing of channel coding and source coding by software. In this step, the baseband modem is implemented in dedicated digital hardware, but bitstream processing (multiplexing, FEC, etc.), control, source codec, and the data terminal interface are implemented in software. In the second step, the structure of the terminal processing channel/source coding and baseband modem are all implemented in software. This architecture is based on analog IF processing with IF ADC and DAC. The benefit of this step is to realize adaptive modulation and adaptive reception schemes. The final step includes digital IF (RF) processing in DSP, CPU, or programmable logic. Only in this final step is the radio reconfigurable by changing software (i.e., an SDR). The second step was on the threshold of being introduced into products, so in response to questions, the author declined to discuss the timetable. Similarly, the timetable for step three was not addressed.

This concept has been under study by the Ministry of Posts and Telecommunications (MPT) of Japan. The primary features of the software radio are flexibility and adaptability. The technical issues in their deliberations consist of wideband RF, wideband high-resolution ADCs and DACs, high-performance digital signal processing (DSP, FPGA, etc.), and software. Panasonic's current DSPs have 600 MHz clock-rates (nominally 600 MIPS). They project 10 GFLOPS by the year 2001–2002, which would support digital IF processing. In their view, the structure of the software greatly depends on how to operate the software radio system. One approach permits the loading of software that is unique to the terminal hardware. The other is to design software to be *not* unique to the terminal hardware. The primary benefit of this latter approach is that new services will be implemented quickly and at the same time for all terminals, at the expense of an increase in processing overhead. Again, in their view, this implies the existence of standard modules, mobile communications tool sets, compiler, and real-time operating system.

In 1999, M. Akaike and M. Muraguchi of Japan organized a session of the International Union of Radio Scientists (URSI) on the software radio [83]. The session addressed the approaches, problems, and potential in the realization of software-radio architecture, specifically the implementation of modulation, demodulation (etc.) in software. Presentations addressed the current research in RF platforms, including increasing the linearity of modulators [84]. There was also an overview of SDR research in Japan [85], and a presentation on cognitive radio, an approach to increasing the computational intelligence of software radios [86].

Also in 1999, the IEICE formed a study group to organize workshops on software radio technology. The first such workshop was held in December 1999. This program was based in part on the success of the First Asian Workshop, and of meetings at Keio and Yokohama Universities. In addition, Asian SDR Forum members as of December 1999 included; NTT DoCoMo, Keio University, Mitsubishi Electric, Toshiba, Kokusai Electric Company, National University of Singapore, Samsung, Kyocera DDI, Sangikyo, Sony Computer Science Laboratory, and Yokohama National University.

These highlights cannot do justice to the breadth and depth of Asian interest in and technology development toward the software radio, but should provide the reader with useful pointers to those who are participating on a global scale.

## D. Regional Differences

Nearly all standards adopted by the ITU, the GSM MoU signatories, etc. accommodate regional differences. This allows manufacturers to provide value-added implementations and enhanced features as summarized in Figure 2-4. The digital microwave air interface, for example, is standardized at the Synchronous Digital Hierarchy (SDH) multiplexer, not at the air interface. Manufacturers therefore employ unique channel codes, interleaving, randomizer and FEC schemes. While this has not inhibited the applications for which the stan-

ITU-R and ITU-T (formerly CCITT)
Regional, National and Local Variations
FCC, TIA, ETSI, ARIB, etc.
Competition among Radio Facilities
   Fixed
   Fixed/Mobile and
   Mobile
      Land Mobile
       Aeronautical Mobile
      Mobile Satellite

- Different RF band allocations
- National use of control bits, randomization, etc. defined by manufacturers
- FDMA, CDMA, TDMA, FPLMTS/UMTS/IMT-2000 regional and local access telephone area (LATA) differentiation
- ICAO & APCO 25 (8 1/3, 12.5, 25 kHz)...

**Figure 2-4**  Software radios address global differentiation.

dards were intended, it provides fertile ground for software radios to provide interoperability across equipment from different manufacturers. Two conventional T- or E-carrier microwave radios from different manufacturers generally cannot form a radio link. Historically, a PTT would acquire point-to-point digital microwave terminals in pairs to establish links. In a peacekeeping situation, however, it might well be advantageous for a military radio to interoperate with the commercial infrastructure. Software radio technology would allow this. As shall be seen in the sequel, software implementation of 1.544 Mbps T-1 and 2.048 Mbps E-1 links are feasible with current DSP technology. Software personalities for these first-level SDH interfaces could include the nuances of randomization, control bits, and specialized channel coding. These details are today proprietary to each manufacturer. The impediments to software radio implementations in this point-to-point microwave radio niche thus include the ownership of IP by many radio manufacturers. In addition, the procurement of point-to-point radios in pairs essentially eliminates the economic motivation for cross-manufacturer air interfaces (including military radios).

On the other hand, software radio technology is an enabler for new ways of thinking about radio communications. What benefits would accrue from software radio in point-to-point microwave? There are several possibilities. For one thing, the enhanced techniques that are now hardware-specific private IP could be used, for example, in the creation of dynamic infrastructure. Dynamic infrastructure is infrastructure that moves (e.g., is portable) and that has dynamic topology. Historically, this has been of military interest. In the past, this meant either acquiring all the radio hardware from one supplier or limiting innovation in the air interface. Using software radio technology, one may support innovation in the air interface on radio platforms from multiple suppliers. Acquisition managers could sustain competition through software value-added in the lower data-rate regimes of SDH (e.g., 1.5–55 Mbps). Operations can be enhanced through software-based IP. Commercial service providers also may find uses for dynamic infrastructure. Rapid build-out of cellular infrastructure requires microwave backhaul from the radio access points to the switching centers. Could the possibility of backfitting new proprietary techniques onto existing infrastructure reduce cost of ownership? Perhaps; if so, then the infrastructure would not be dynamic in the military sense, but the technology required for dynamic infrastructure could reduce cost of ownership of infrastructure in the commercial sector. In part, the amount of terrestrial interference to cellular is increasing in the 2 GHz bands. If microwave interference cancellation techniques are defined in software, then both fixed and mobile radios may rapidly deploy these techniques. Thus, although sought for different reasons, both the military and the commercial sector could well join forces in fostering software-radio-based low-SDH radios. This business evolution is practical only if the hardware and software components conform to a high-quality open architecture.

Figure 2-4 also identifies the International Civil Aviation Organization (ICAO) 8 1/3 kHz standard and the Association of Public-safety Communications Officials (APCO) study group 25 standards (12.5 and 6.25 kHz). The ICAO determined that flight safety was jeopardized in Europe unless aircraft divided each 25 kHz analog voice channel into three 8 1/3 kHz channels. APCO determined that the needs of public safety mandate an initial division of the traditional 25 kHz channel into two 12.5 MHz channels. At some point in the future, and in some highly congested areas, those channels need to be split again to 6.25 kHz. These changes require equipment changes in all aircraft, police cars, and related infrastructure. Such changes have occurred infrequently in the past. In the future, however, technical innovations in waveform design, channel coding, and error control will continue to accelerate. The early adopters of software radio technology should be able to upgrade their systems incrementally when and where needed, and at lower cost than massive reacquisiton of replacement systems and infrastructure. Since the pace of introduction of software radio technology differs according to the need and economics, it is helpful to examine the character of the broad market segments.

## E. Differentiating Market Segments

Figure 2-5 shows how architecture drivers differ across commercial and military market segments. One key applications parameter is the number of simultaneous channel-mode combinations required. As simultaneity increases, more parallel hardware is needed and electromagnetic interference (EMI) problems become more severe. A handset may have two or three modes, but one user is typically using only one at a time. Call waiting and simultaneous voice and data yield an upper limit of probably two simultaneous transmit modes plus possibly an additional GPS receive-only mode. Military manpack and commercial/military avionics, however, require more simultaneous mode-channel

| Market Segment | Simultaneous Channels/Modes | Architecture Drivers | Standards |
|---|---|---|---|
| Handset | 1–2+GPS | Mfg Volume "Velcro" | Chip Level Interfaces |
| Manpack/ Avionics | 4–20 | Size, Weight, Power | PCI PCMCIA |
| Law Enforcement | 20–100+ | Cost | ? |
| PBX, WLL | 20–100+ | Call Quality | ? |
| Base Station/ & Mobile Bases | >100 | Future-Proof DSP Leverage | VME-like +Wideband Bus |

**Figure 2-5**   Market segments and architecture.

combinations within a constrained size, weight, and power envelope. Private branch exchange (PBX) and wireless local loop (WLL) applications require multiple channels but probably only one air interface.

In terms of critical design parameters, commercial infrastructure provides ideal applications of software radio technology, while law enforcement infrastructure falls between the military manpack/avionics and commercial wireless PBX/WLL applications as suggested in the figure. Such infrastructure must support a few dozen simultaneous subscribers and it must operate in bands of mixed transmit and receive frequencies. Its modes include at least standard VHF/UHF voice, modem, facsimile, APCO 25, and transmission security.

Figure 2-5 also suggests how critical design parameters shape the architecture choices and selection of standards. Multiple open-architecture standards recommendations may evolve from standards bodies such as the SDR Forum, one for each significant market niche.

Recent commercial experience underscores the need for increased air interface flexibility. GSM, for example, was adopted by the European high-speed rail system as the standard mobile telephone. Doppler shift in such applications significantly exceeded the capability of existing handsets to track the carrier, resulting in the need for specialized base station and mobile hardware. Software radio handsets and infrastructure would simply invoke a wider bandwidth carrier tracking loop, possibly with other algorithmic enhancements, supporting graceful, low-cost transition into the European high-speed rail application. In the past, such changes were infrequent, so there was no business case for such flexibility. As SDR technology enters production, the tradeoffs will change rapidly, as suggested in Figure 2-6.

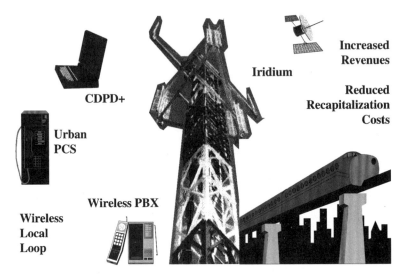

**Figure 2-6**  Commercial drivers center on increased revenues and reduced costs.

The initial introduction of Cellular Digital Packet Data was extremely inefficient. The first CDPD installations had to allocate revenue-bearing channels to CDPD, taking them out of conventional revenue-bearing service. CDPD services were implemented through physically separate hardware since the existing AMPS analog cellular hardware had only limited re-programmability for the CDPD standard. Thus, in addition to lower than expected CDPD revenues, service providers had to support separate hardware infrastructure. A software radio implementation could have included a signal recognition module to detect the presence of the CDPD air interface on any channel followed by invocation of a CDPD channel algorithm to provide the service. The number of CDPD algorithms invoked would be dynamic, reflecting the number of CDPD users instantaneously on the air. This would optimize the mix of conventional and CDPD channels dynamically and transparently. Such anticipated scenarios help establish expectations that software radios can meet as deployment becomes more cost-effective over time.

## III. MILITARY ARCHITECTURE NEEDS

The military shares much with the commercial sector in its need for software radio architecture. There has been a convergence of military and commercial technologies. Radio technologies once used exclusively by the military are now led by the commercial sector. These include frequency hopping, direct-sequence spread spectrum, demand-assigned multiple access, variable data rates, emitter location, and beamforming to null jammers. Only mobile ad-hoc networks with mobile infrastructure are uniquely military. In addition, military

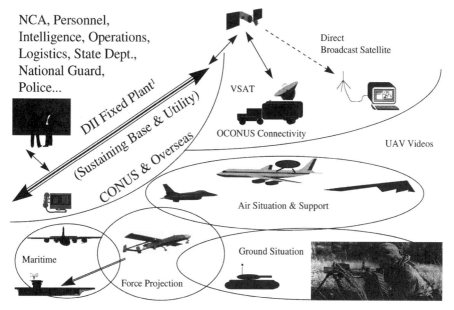

**Figure 2-7**   Illustrative Defense Information Infrastructure (DII) components.

organizations are increasingly embedding commercial technology in mobile applications as well as wireline infrastructure. One could say that in the 1980s there was roughly only 20% common interest between military communications and the commercial sector. Now that level of common interest is more like 80%. This section therefore examines the military needs for software-radio architecture in terms of commonality with commercial interests.

## A.  Defense Information Infrastructures

The Defense Information Infrastructure (DII) of the United States consists of the fixed plant of telecommunications and information processing systems plus the mobile infrastructure that military forces must take with them on deployments around the world. Features of this infrastructure are illustrated in Figure 2-7.

The fixed infrastructure in the United States is acquired and operated by the Defense Information Systems Agency (DISA). The mobile infrastructure is acquired and operated by the uniformed military services (Army, Navy, and Air Force) under their responsibility to organize, train, and equip the military forces. This infrastructure continues to evolve in the direction of greater joint use of the telecommunications and information processing systems by two or more force elements at the same time (e.g., Army and Air Force for air support of ground forces). In addition, this infrastructure increasingly connects deployed elements with elements of the same force supporting from the continental United States (CONUS).

The fixed infrastructure includes conventional telephone service in which the large majority of the telephone capacity is leased from the major commercial service providers including AT&T, Sprint, MCI, and GTE. As illustrated in the figure, this fixed plant now includes new high-data-rate links to deployed forces such as commercial direct broadcast satellite (DBS) service implemented initially using the Hughes 601 geostationary communications satellite. Very small aperture terminal (VSAT) networks also use commercial technology. These augment traditional secure Defense Satellite Communications System (DSCS) and MILSTAR low-data-rate satellite communications (satcom) services. The Global Broadcast Service (GBS) was rapidly created by leasing commercial DBS capacity and configuring commercial receiving equipment for the military application. VSAT services include global sharing of video from unmanned air vehicles (UAVs) such as the Predator medium altitude endurance (MAE) UAV. This fixed infrastructure has supported the National Command Authority (NCA), military operations, and support functions such as personnel, intelligence, logistics, the State Department, the National Guard, police, and others.

### B. Tactical Military Needs

One overarching vision of U.S. military communications domains is conveyed in Figure 2-8. SDR technology may be introduced to reduce the number of different families of radio equipment that must be supported in the field. The primary opportunities appear to be in the terrestrial domain in tactical radios and satellite terminals.

In general, military requirements for radio services include the following (not necessarily in priority order):

- Mobility of both subscribers and infrastructure
- INFOSEC (TRANSEC and COMSEC)
  - TRANSEC for LPI and low probability of detection (LPD)
  - COMSEC, including high-quality encryption
- Ruggedness and reliability in austere operating environments
- Growth from voice and low-speed data to high-speed *tactical internets*
- Improved quality, quantity, timeliness, and suitability of battlespace information
- Interoperability with legacy radios and coalition partners
- Affordability

Mobility to the military means much more than mobility means to the commercial sector. Mobile telephone systems allow the individual subscribers to move supported by massive fixed infrastructure such as the 100 to 300 ft cellular towers that now appear everywhere. This infrastructure is not mobile

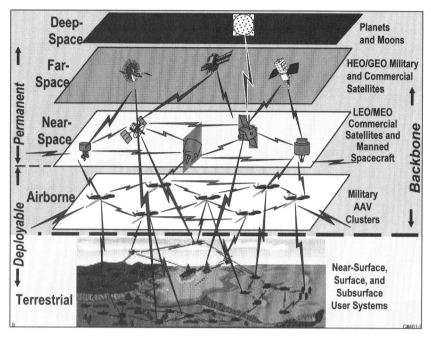

**Figure 2-8**  The scope of military communications [87].

in spite of the mobility of the subscriber. Military applications require mobile infrastructure in addition to mobile subscriber sets—full mobility. The mobile nodes that connect to battlefield wire/fiber, long-haul satellite communications, and/or the fixed commercial infrastructure may be called radio access points (RAPs). The mobile terminals which interface to the RAPs (but not to the long-haul/wireline infrastructure) may be handsets, vehicular-mounted radios, avionics, or autonomous units such as radio relays or remote sensors. Mobility has implications in the home base such as CONUS because of the need to use radio systems in training that are as similar as possible to those to be used in the field. This implies the need for ad-hoc networking, long a military communications research area, and now of importance to commercial wireless [88].

The growing diversity of services includes voice, facsimile, imagery, and video services such as teleconferencing, much of which must be accomplished in a secure environment protected by appropriate authentication and encryption. In 1995, for example, The MITRE Corporation conducted an experiment in video- teleconferencing (VTC) in the Pacific. The ability to conduct a high-level commander's staff meeting among locations separated by several thousand miles every day and for the cost of a few T1 (1.544 Mbps) circuits was a benefit that was well worth the cost. Consequently, over 100 such video-teleconference facilities were in use by 1998 by the U.S. military worldwide. This technology is migrating from point-to-point VTC to collaborative virtual workspace (CVW) environments. With CVW, staff-officers can conduct meetings, exchange ideas, and work together from a workstation using one or two standard voice circuits. The early VTC systems cost tens of thousands of dollars and required T1/E1 data rates. They are therefore appropriate to higher-level organizations that can afford such facilities and bandwidth. CVW, on the other hand, is based on PC technology. It requires only voice-channel bandwidths that are available to most users. Some versions of CVW will operate on a single low-grade voice circuit. The reduction in cost and bandwidth requirements will make such services available "to the foxhole."

In practice, the lowest level of the military hierarchy needing multimedia services may not be the individual soldier. But the U.S. Defense Science Board (DSB) report on the future of the military in 1998 and 1999 affirmed that the individual soldier may in fact benefit greatly from such technology. Transition to such a vision includes measuring the performance of the Force XXI Battle Command, Brigade and Below (FBCB2) [89]. Twenty years ago, few thought that every one in the Army would need or use a Global Positioning System (GPS) satellite receiver, but now we know how valuable they can be to an individual soldier. The diversity of services being made available to individual military subscribers (regardless of echelon in the chain of command) continues to increase, embracing voice, imagery, facsimile, and ultimately selective and appropriately formatted and secure video-teleconferencing.

In military requirements processes, users have begun to articulate their needs for information services in terms of the key parameters: quality, quantity, timeliness (QQT), and suitability. Quality is characterized in terms of BER and link availability. Quantity may be characterized in terms of numbers of different types of communications modes available from a software radio node. Quantity may also be characterized in terms of peak and average data rates available to an individual subscriber. Timeliness is quantified as the time delay between the creation of the information and its availability to the subscriber. Software radio nodes contribute some incremental time delay through processing and queuing delays.

QQT of tactical military information needs to be improved. Quality cannot be measured merely in BER of a disadvantaged radio link. It consists of the net effect of all measures applied to enhance quality such as proper choice of band, channel, and mode; application of FEC with automatic repeat request (ARQ); data quality enhancement, and formatting to enhance the relevance of the information to the user. Opportunities for software radio quality enhancements therefore include agility of any and all of these characteristics of the overall telecommunications system. Timeliness requirements can be established as a function of content. An order for supplies may safely be an hour late in many scenarios; an order to move out in two hours may be an hour late, but its usefulness is degraded by the time delay; an order to shoot down an incoming aircraft cannot be an hour late if the incoming aircraft is five minutes away. Quantity increases may be measured by the number of messages of appropriately high quality received per unit time. It may also be measured by the number of subscribers supported by a RAP of a fixed size, weight, and power. There is a pitfall to measuring quantity as the amount of raw data received per unit time because data and information are not synonymous. If weighted by a figure of merit for information content, raw data can be a meaningful metric of QQT.

Suitability is the degree to which the information being supplied matches the recipient's ability to use it. A SINCGARS user could receive a 10 MB map image at 1.2 kbps. It might be more appropriate for the sender to transmit a line drawing and a few icons that represent the most critical relevant information from that 10 MB map. The icons will arrive in seconds while the full map would take hours and could preclude other critical data from arriving. In order to support new services for some users, the networks will have to become more agile in their ability to tailor content for disadvantaged users (low-data-rate users; those in severe jamming environments; those with few batteries, etc.).

The final area in which there are key military requirements is that of cost effectiveness. No military organization can afford to declare as surplus 1000 radios purchased three years ago for an intended operational life of 20 years. As a result, the new radios must be backwards compatible with older radios, extending and enhancing their contributions over something approaching the intended service life. But radio technology is changing so rapidly that this is a real challenge. So the legacy of discrete radios for HF, VHF/UHF, satellite communications, etc. constitute an existing infrastructure that must be gradually phased out. The software radio insertion goal, then, is to identify insertion points at which the costs saved through technology insertion pay for the replacement of the older inventories. To do this, one must consider radio infrastructure not one band or mode at a time, but as an integrated infrastructure.

But 1990s military radios were dedicated to a single RF band or to a small number of bands and modes focused on the traditional communications mission such as HF, VHF/UHF, or satcom. As a result, supported Army troops equipped with SINCGARS could not communicate in a secure way with supporting Air Force jets equipped with HAVE QUICK. But the trend is toward more joint operations and an increasing need for interoperability across traditionally disparate bands and modes. The Air Force's Rome Laboratories (RL) and the Army's Center for Electronics and Communications (CECOM) therefore joined forces in the SPEAKeasy program to create radio technology to insert in the force structure to fill key gaps in interoperability. At the same time, SPEAKeasy could add increased flexibility through the SDR approach. The SPEAKeasy technology thrusts evolved to the Joint Tactical Radio System (JTRS) program by 1998. European and Asian nations also recognized the need for national information infrastructures that include their own military infrastructure plus interoperability with coalition partners.

## IV. OPEN ARCHITECTURE AND STANDARDS EVOLUTION

Industry organizations such as the SDR Forum are in the process of developing open architecture for SDR [90]. In addition, work of the Object Management Group (OMG), the Telecommunications Industries Association (TIA), the Internet Engineering Task Force (IETF), the Wireless Applications Protocol (WAP) Forum, and the IEEE includes standards relevant to software radio architecture.

### A. The Software-Defined Radio (SDR) Forum

In March 1996, the U.S. government invited industry to participate in what it named the Modular Multifunction Information Transfer Systems (MMITS) forum. It hoped this group would become an industry body to establish open-architecture standards for SPEAKeasy. The initial DoD thinking was that MMITS might be a study group of the VME International Trade Association (VITA) because of the success of VME as an open-architecture standard for the technology pathfinder SPEAKeasy I program. The author was elected to chair the nascent organization, which decided not to align with VITA or any other standards organization per se. Instead, it attempted to function like the ATM Forum, a quick-response consortium that would publish recommendations based on current engineering practice. It planned to delegate the formal standards-setting process to others (OMG, TIA, IEEE, ANSI, ITU, ETSI, etc.). That choice proved to be a wise one. For example, SPEAKeasy II chose the PCI bus, ITT chose PC-104, GEC chose a narrowband control bus, and, in general, there was no consensus on VME or any other backplane as a paradigm for industry cooperation.

In its first year, MMITS established a web site and promoted dialog among government and industry including the FAA, DoD, the U.S. military, and representatives of the UK and France in the dialog. Commercial sector participants included BellSouth, U.S. West, AirNet [91], Orange (UK PCS Supplier), and others in the commercial wireless marketplace. Texas Instruments and Motorola, suppliers of radio silicon, also participated in the MMITS forum. But the business center-of-gravity of the Forum evolved to commercial service providers, military-oriented radio suppliers, and commercial product suppliers, instead of semiconductor manufacturers.

MMITS also began the process of defining open architecture for "plug-and-play digital radios" whether software intensive or more traditional digital radios. The MMITS technical reference model is presented, enhanced, and employed with the author's canonical model in this text as a guide to partitioning radio functions for cost-effective mappings onto modular hardware and software. MMITS is a heterogeneous standards organization. That is, it is concerned with the interfaces among a diverse set of modules, each of which conforms to different hardware and software standards. These include analog at the front end, digital in the middle, user interfaces at the back end, and

**TABLE 2-3  Relevant Product Standards and Organizations**

| Technical Area | Standards Body | Examples |
|---|---|---|
| Analog Hardware | ANSI, TIA, IEEE, ETSI | Antennas, RF connectors, cable, waveguide |
| Interconnect | ANSI, TIA, IEEE | RS-442, LANs, fibre channel, Firewire |
| Bus/Backplane Hardware | VSO, PCI, PCMCIA | VME, Skychannel, Raceway |
| Internetworking | ITU, ISO, ETSI, ARIB, IEEE, TIA, ANSI | TCP/IP, ATM, SDH, SS-7[13] |
| Object-Oriented Software | OMG, Open Group | CORBA, UML[14] |

software interfaces throughout. In June 1999, the MMITS Forum voted to change its name to the Software-Defined Radio (SDR) Forum. It continues its contributions to open-architecture software radios.

Two other classes of standards organization bear on software radios, directly or indirectly: product standards organizations and air interface standards organizations.

## B. Product Standards Organizations

The plethora of hardware standards potentially relevant to SDR can be outlined but not exhaustively enumerated. In part, this is so because product standards have been emerging at breakneck speed. In 1995 and early 1996, for example, the VME standard backplane/bus had a majority of rack-mount open-architecture designs. By late 1996, the PCI bus had taken the lead with a large variety of DSP cards, ADCs, host processors, and other modules necessary for open-architecture software radio. By 1999 compact PCI (cPCI) had increasing popularity, while PC-104 retained a strong market niche. By April, 2000, systems-on-chip using DSP cores had a strong following. The families of product standards that have a continuing relevance to SDR are shown in Table 2-3.

Analog hardware standards may be useful for defining interfaces with antennas. Interconnect and backplanes will probably not be standardized per se, but emerging open-architecture middleware will hide the details of this interconnect technology. Internetworking standards, similarly, have to be accommodated in any viable software radio architecture. Object-oriented standards support the design process (e.g., the Unified Modeling Language, UML [92]). CORBA provides the middleware essential for open-architecture in software radio.

---

[13]Signaling System 7 (SS-7) is the common channel signaling standard for the global PSTN.
[14]UML is the Unified Modeling Language, an open-architecture standard of the OMG.

## C. Air Interface Standards

Finally, air interface standards organizations define channel modulations, frequency allocations, access protocols, and other characteristics of the radio interface needed for interoperability over the air. The ITU has organs that address the radio aspect (ITU-R) and the telecommunications aspect (ITU-T). The European Telecommunications Standards Institute (ETSI) sets radio standards in Europe, while the Telecommunications Industries Association (TIA), the Electronics Industries Association (EIA), and the Institute of Electrical and Electronics Engineers (IEEE) set regional standards in the United States. ARIB sets standards in Asia. And there are numerous other regional, national, and local standards organizations.

The variability of standards creates the need for flexibility of band and mode for a "world phone." A handset that accommodates first-generation, second-generation GSM, and third-generation (3G) CDMA would be such a world phone. Industrial goals for 3G software radio handsets contemplate a mix of ASICs, DSP cores, and general-purpose microcontrollers in 3G handsets. Infrastructure costs are also such a driver that continuing proliferation requires future-proof infrastructure to ensure affordability of future mobile wireless.

In addition, military organizations have evolved dozens of global, regional, and national standards. The US JTRS program, for example, contemplates accommodating over 40 modes in the RF bands from 2 MHz to 2 GHz. Cost-effective delivery of commercial and/or military capabilities can be effected using the software radio architecture. Success of this approach requires the balancing of hardware parameters against software computational demand in the way explained subsequently in this text.

The international standards organizations generally accommodate standards proposed by large constituencies. Thus, it is not likely that air interfaces will become simpler or less technology intensive. In addition, it is unlikely that there will be fewer standards over time, since each standard establishes a market niche. The continued proliferation of new standards therefore gives sustained impetus to the need for software radio architecture capable of supporting continued evolution of air interfaces. In addition, new air interfaces are often so close to the cutting edge of hardware capabilities that the architecture must facilitate direct access to the hardware technology. This seems to contradict the need for hardware independence. As with Group IV facsimile, these goals are not completely incompatible.

## D. The Global Deliberative Process

The United States, Europe, and Asia each have central perspectives on software-radio architecture. These perspectives are reflected in the actions of those who most aggressively push the technology. Participation in the SDR Forum provides a useful gage of interests. U.S. interests currently center on the needs of the cellular service providers and of the military. BellSouth chairs the SDR

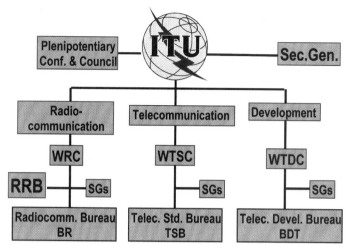

**Figure 2-9** ITU structure. Courtesy, BellSouth.

Forum, for example. The U.S. DoD's needs for the Joint Tactical Radio System (JTRS) are sponsored in the SDR Forum by consortia organized by Motorola (SR21C consortium) and Raytheon (MSRC consortium). The thrust of U.S. interests seems to be affordable infrastructure, with substantial emphasis on open-architecture middleware. The European interests are led by the EC and supported by the GSM MoU committee and ETSI. The thrust of European interests can be summed up as leveraging their dominant position in GSM into success in 3G (and later in 4G). Asian interest is balanced across smart antenna infrastructure and high-performance RF devices for handheld terminals. In general, the U.S. participants seem to be more preoccupied with software while European and Asian participants seem to be more preoccupied with hardware aspects of SDR.

Regulatory bodies such as the ITU and the European Telecommunications Standards Institute (ETSI) enable product development through the establishment of global and regional standards for telecommunications. The ITU, especially, provides the global forum for telecommunications standards setting. Since the international process of standards setting will shape the evolution of SDR architectures, it is worth examining the structure of this body.

The ITU is organized as illustrated in Figure 2-9. The Radiocommunications Sector (ITU-R) administers the World Radio Conference (WRC),[15] the Radio Recommendations Board (RRB), the Radiocommunications Bureau (BR), and related study groups (SGs[16]). ITU-R was originally constituted as the International Consultative Committee on Radio (CCIR) [93]. The Telecom-

---

[15]Formerly the World Administrative Radio Conference (WARC).
[16]Integrated Product Team (IPT) is the U.S. DoD term for study group.

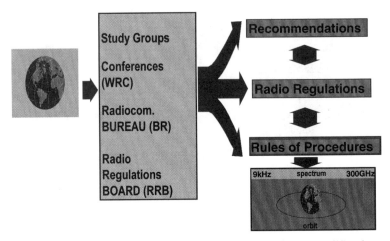

**Figure 2-10**   ITU radiocommunications sector. Courtesy, BellSouth.

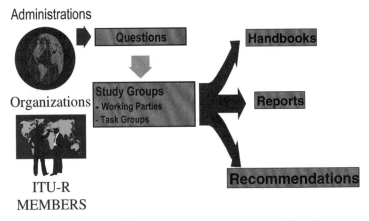

**Figure 2-11**   ITU study groups. Courtesy, BellSouth.

munication Sector (ITU-T) has a similar structure. ITU-T is concerned with
wireline telecommunications.

The ITU operates by considering submissions from world telecommunications administrations of the countries that are members of the ITU. As illustrated in Figure 2-10, the ITU-R initially assigns submissions to study groups. The study groups operate as illustrated in Figure 2-11. The designated SG assigns working parties and task groups to investigate an issue. They produce handbooks, reports, and recommendations. Generally, there is such a diversity of views on an issue that a normalization process is invoked. Recommendations become radio regulations that govern the radio spectrum from 9 kHz to 300 GHz.

In 1996–1998, there were informal discussions of software-radio technology in ITU study groups deliberating the approach to setting 3G standards. Software-radio advocates argued that the basic nature of the regulatory process should change. Instead of defining each air interface, the argument went, one could define a standard framework for software-defined air interfaces. This envisioned framework would include a language by which the radios could share the details of a given air interface. This approach, the argument went, would spur innovation and would speed up the standards approval process. The approach is similar to that taken with the ITU's Series T Recommendations for Group IV facsimile [94]. While Group I-III fax recommendations specified a waveform interface, Group IV specified backoff modes so that a device compliant with Group IV could simulate any of the other waveform interfaces. In addition, the Group IV protocol allowed units to determine whether they were using proprietary extensions such as more efficient compression. If so, fax modems using compatible nonstandard features could discover this fact using the Group IV protocol and could employ those extensions. Although the de facto fax language thus created was very simple, it spurred the creation of a large family of advanced fax products. All of these products are interoperable through the capability exchange and backoff modes defined in the Group IV protocol. The majority of the participants in these early discussions of standards-setting for 3G favored the development of a specific air interface to which hardware could be optimized, rather than the more generic software-defined air interface family. The SDR technology of the early 1990s was not sufficiently mature to support software-radio architecture standardization at that time.

To better appreciate the diversity of perspectives that bear on the standardization of software radio architecture, consider the 3G process illustrated in Figure 2-12. Although ITU-R specifically chartered task group (TG) 8 with the development of IMT-2000 recommendations, its mobile land working party (WP8A) also participated under the overall direction of the Mobile Radio study group. In addition, IMT-2000 envisioned such a range of data rates, tariffs, and QoS that there were numerous ITU-T stakeholders from services and operations to data security, networking, and multimedia services. The IMT-2000 working party (WP 11/3) operated within the Signaling and Protocols study group. It is easy to see how software radio's nearly infinite variety of bands, modes, and data rates would provide the basis for a huge variety of air interface parameter sets with associated data rates, tariffs, and QoS alternatives. The standardization of a software-radio-based family of air interfaces would therefore include at least the participants who engaged in IMT-2000.

Koichi Asatani, standards editor of the *IEEE Communications* magazine, and others, provided an overview of international and regional standards bodies and their interrelationships as of late 1998 [95]. In addition to an overview of ITU activities, Sasaki and Yabusaki [96] present the IMT-2000 standardization process in Japan in-depth. Besides the Ministry of Posts and Telecommunica-

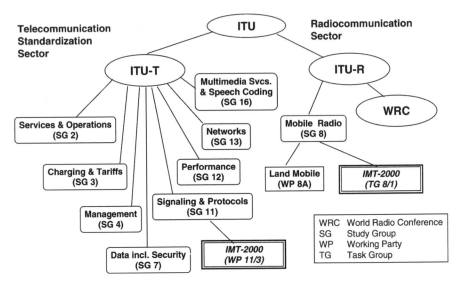

**Figure 2-12**  ITU study groups related to IMT-2000. Courtesy, BellSouth.

tions (MPT), the Association of Radio Industries and Businesses (ARIB), focal
point for the radio aspects, worked with the Telecommunication Technology
Committee (TTC) on the network aspects of IMT-2000. Areas of technical
deliberation included the system, the air interface, satellite systems, wireless
local loop (WLL), the codec, and experimentation. The IMT-2000 set a goal
of 144–384 kbps subscriber data rates over wide areas in order to support
multimedia services like video-teleconferencing. In local areas and indoors,
the goal of 2 Mbps supports wireline-like access. These goals require a wideband waveform, ultimately resulting in 5, 10, and 20 MHz direct-sequence
spread spectrum waveforms with code division multiple access. The 20 MHz
wideband CDMA (W-CDMA) rate requires ASIC implementations for workable 3G handsets. An attempt to support multiple wideband air interfaces so
close to the state of the art could have rendered 3G nonviable. This situation
caused the standards-making process to move toward a single air interface.
Thus, the issues associated with software radio were not central to setting the
3G standards.

On the other hand, the software radio concept offered the GSM proponents
a low-cost means of migrating toward 3G. Given a 20 MHz W-CDMA despreader ASIC, one could easily digitize the 200 kHz GSM subscriber signals
using minimum chip area on the despreader ASIC. The DSP power necessary
for 3G could then be employed to filter the subscriber signals in software,
yielding a GSM SDR mode. In addition, W-CDMA generation would employ
high-speed circuits that could be adapted to generating the GSM waveform.
The GSM MoU committee therefore recommended that software radio technology be employed to provide graceful migration to 3G.

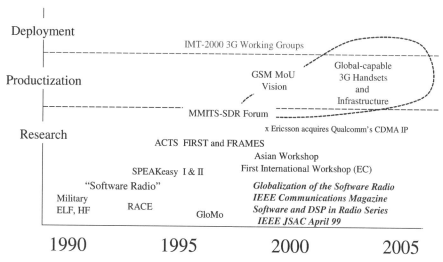

**Figure 2-13**   Architecture evolution roadmap.

## V. ARCHITECTURE EVOLUTION ROADMAP

Software-radio architecture has been evolving from its roots in military communications. It has become a significant force in the GSM MoU, in the EC, in Japan, and in the United States. Those working in electronic warfare in the 1970s and 1980s pioneered wideband techniques [97]. In addition to the traditional emphasis on radar and radar jamming, some developed wideband digital techniques for radio. Those working in HF and ELF, for example, found that ADCs and DSP technology supported what now would be called SDR implementations. The very narrow bandwidths of HF and the low carrier frequencies of ELF made "wideband" ADC technology feasible at 24 kHz ADC sampling rates. The low data rates of both media permitted early-generation DSPs to keep up with the processing demands [98].

As illustrated in Figure 2-13, the concept of the software radio was first published in 1992 in my paper on radio architecture [99]. This led to the May 1995 special issue of the *IEEE Communications* magazine, which was a landmark publication. The issue addressed architecture, ADC, DSP, systems, smart antenna technology, and economics. It did not address the wideband antennas and RF needed to effectively implement practical devices, however. At the time, the U.S. DoD SPEAKeasy I program was the only publicized military software radio. In parallel, the ACTS Flexible Interoperable Radio Systems and Technologies (FIRST) program developed a software-based approach to air interfaces for 3G. Both of these programs used Pentek ADCs, digital receivers, and DSP boards. At the same time, DARPA continued to invest in SPEAkeasy technology through the SPEAkeasy II program, which was awarded in late 1995. Global interest in the technology was further spurred by the 1996 formation of the MMITS Forum. The first European workshop was sponsored by the EC in Brussels in 1997. The first Asian workshop was held a year later in 1998, several months before the First International Workshop. The February 1999 issue of the *IEEE Communications* magazine included a feature topic on the globalization of the software radio, which was well underway at this time. This led to the creation of a feature series on software radio, entitled "Software and DSP in Radio." The GSM MoU vision of the software radio as an enabler for evolution toward 3G was published in 1998, and was discussed at the First International Workshop. At that workshop, the notion of employing SDR in 3G handsets for incremental deployment of 3G with interim measures like GSM EDGE became clear. The industrial roadmap for the period 1999–2003 centered on SDR-enabled 3G.

Military organizations developed further interest in SDR technology. The success of the U.S. SPEAKeasy programs led to the formation of the PMCS study group, which recommended the creation of the JTRS program to acquire future radios with at least SPEAKeasy-class SDR capability. The U.S. DoD has been outspoken in its advocacy of software-radio technology. Other military departments around the world have been less direct in their support of software radio in the open literature. They seem likely to pursue courses of action similar to that of the United States. Some, no doubt, have similar programs and pursuits.

From an academic perspective, research centers at the Massachusetts Institute of Technology [100], Virginia Tech [101], Karlsruhe [102], KTH [54], Keio University, and Yokohama National University [103] developed software radio thrusts in the late 1990s. The IEEE published its *Journal on Selected Areas in Communications (JSAC)* on the software radio in April 1999. This "surrogate graduate text" addressed foundations, systems, smart antennas, and applications. Foundations included a mathematical perspective on the software-radio architecture [50], a survey of ADC technology and challenges [104], designs for massively parallel digital filters [105] [106], and low-power DSP cores [107]. Systems included virtual radios [100], the software realization of a GSM base station with a characterization of processing requirements using standard benchmarks [108]. Other systems included a DSP-based CDMA receiver [109], a novel RF LAN [110], and SPEAKeasy II [111]. Smart antennas included a tutorial on algorithms and complexity [112], smart antennas for high-data-rate QAM signals [113], and active beamforming on transmit [114]. Applications included a novel time-frequency-based communications system [115], trellis coding [116], a CDMA sensor [117], and airborne applications [118].

The way forward seems to include a mix of commercial, academic, and military endeavors. To enable productive paths, one must have a firm grip on the impact of architecture on economics. At a minimum, one must be able to identify the positive contributions of specific architecture proposals that will regularly emerge in the evolution process. In this text, therefore, the term *software-radio architecture* means an open architecture that embraces the variety of implementations up to and including the ideal software radio. This all-encompassing framework admits a variety of commercially driven SDR architectures over time. It also provides a rigorous independent academic perspective on SDR architecture, kind of a hilltop from which to observe the SDR architecture battles that are already underway.

The next step in defining the ideal software-radio architecture is the analysis of specific requirements of a truly ideal software radio. From this perspective, one may prioritize investment strategies to position one self appropriately in the marketplace This requires one to first analyze the communications capabilities that exist at present in each radio band and mode from HF through EHF. Subsequently, one must analyze those aspects of radio that constrain or otherwise shape software-radio architecture.

## VI. EXERCISES

1. Define the functions of a very simple analog radio. What are its components? What software components might be associated with the radio throughout its life cycle? (Remember to include logistics software and other postdeployment support.) To what external interfaces does this radio conform? What design rules are associated with those interfaces? To what internal interfaces do its components conform? (Include the unique families of hardware interfaces.) What other design rules might apply? Now that you have defined functions, components, and design rules, have you defined its architecture? Is it an open architecture?

2. Define the functions of a first-generation cellular telephone handset at a high level. What functions does it share with the corresponding infrastructure? What functions are unique to infrastructure and to the handset? What design rules are unique to each?

3. What additional broad classes of functions were introduced in 2G cellular telephony that were not in the first generation? (Hint: consult [18] to check your answer.)

4. What radio functions were introduced in military radios and later transitioned to commercial applications?

5. What new functions and design rules are essential to 3G that were not needed for prior generations? What functions have been substantially enhanced for 3G? How do these expanded functions constrain implementations? How does the expansion of such functions constrain architecture?

6. What are the primary reasons that the commercial sector might embrace open-architecture SDR? Why might the commercial sector be reluctant to adopt a software-radio architecture? Why might some commercial suppliers decline to participate in open architecture? What broad classes of enterprise in the commercial sector would be naturally attracted to open-

architecture software radio, and which would be naturally repelled by such a proposition?

7. What are the primary reasons that the military sector might embrace open-architecture SDR? For each question in question 6, answer the equivalent question for the military.

8. What common interests do the commercial and military sectors share in the development of open-architecture SDR? How can the academic community support these common interests?

9. What international organizations are concerned with establishing standards for radio? What aspects of networks are strongly related to advances in wireless? Prioritize the impact of software radio technology on these interests. That is, if you had the resources to participate in and contribute to only one or two of the network interests related to software radio, which would be most attractive? Why?

10. What features of radio have historically been private IP? Pick a broad radio application family with which you are familiar, and create a list of private IP and public IP. Which of these IP features are amenable to implementation with SDR technolgy? What advances are necessary in order for each of the others to be implementable in software, or to be somehow positively affected by emerging software radio technology?

11. What are the hurdles currently facing the evolution of SDR? Where in the software radio phase space are these applications? How far have implementations moved in the past year?

# 3 The Radio Spectrum and RF Environment

Radio is the penultimate medium for mobile communications, but it has also been used for many fixed-site applications such as AM/FM broadcast, satellite trunking, point-to-point microwave telephony, and digital TV. Although there are radio applications in very low frequencies (VLF) and extremely low frequencies (ELF), these bands require extensive fixed-site infrastructure whose size and cost is dominated by the mile-long antennas and megawatt-power handling requirements. SDR insertion opportunities in these bands are limited. Therefore, this text is concerned with the bands in which there are major economic opportunities for software-radio technology insertion: HF through extremely high frequencies (EHF).

## I. RF SIGNAL SPACE

Figure 3-1 shows how terrestrial radio uses cluster in RF signal space. This figure shows the notional clusters of the significant band/mode combinations addressed by software radio technology. This space is the two-dimensional cross product of radio frequency and duty cycle.[17] Since coherent bandwidth is proportional to carrier frequency, the figure is also labeled in terms of nominal instantaneous bandwidth. A QPSK encoded T- or E-carrier signal is on continuously for a duty cycle of 1.0. Low-duty-cycle modes such as burst communications and ultra-wideband (UWB) have high peak power as suggested by the additional label on the axis. This is not an exact correspondence, but it shows a trend related to the thermal properties of power-handling devices.

The PTT modes have the duty cycle of voice, which is about 25% during speech epochs. Given conversational pauses, a voice channel is typically occupied less than 10% of the time. The busiest military voice channels are occupied not more than 40% in a full duplex channel such as the typical LVHF military bands. On the other hand, troposcatter radios have high peak power and unity duty cycle. The tropo cluster was positioned to show the high peak power. HF communications may also have high power, but the duty cycle is typically that of voice or low-speed data. As the label on the right side of the figure suggests, the greater the ratio of peak power to minimum power,

---

[17]Duty cycle is the ratio of signal on-time to the elapsed time of an epoch.

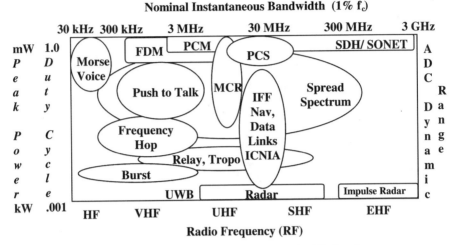

**Figure 3-1**  Communications modes cluster in RF signal space.

the greater the dynamic range requirements on the ADC in the receiver. Since there is no wideband RF or ADC that can encompass all RF with the full dynamic range, designs historically have addressed a single mode. The SDR addresses a few clusters, while the software-radio architecture embraces migration toward the entire signal space. It is therefore essential to consider each of these clusters in detail.

## A. Overview of Radio Bands and Modes

This section provides an overview of radio bands and modes. HF communications consist primarily of voice, narrowband data, and Morse code, some of which is generated by machine and some of which is generated manually. The literature also presents successful research in the use of wideband spread spectrum at HF, including thousands-of-chips-per-bit and millions-of-chips-per-second (MHz) [119]. In addition, HF radar uses direct-sequence spread spectrum in a frequency-hopped pulsed signal structure. Neither of these relatively exotic waveforms are shown in order to focus the figure on the waveforms likely to be encountered in software radios.

LVHF includes spectrum allocated to military users who traditionally have employed half-duplex PTT analog frequency modulated (FM) single-frequency voice modes. Military LVHF also includes many FH spread-spectrum radios. In addition, the literature describes burst signal structures such as meteor burst. These radios transmit data at relatively high data rates for tens to hundreds of milliseconds with high instantaneous data rates, a low duty cycle, and therefore relatively low average data rate.

The LVHF and VHF frequency bands also support frequency division multiplexed (FDM) multichannel radios with typically 4 to 12 radio relay telephony channels for military users. These may also employ pulse code modulation (PCM) for digital telephony as alternate modes of an FDM/PCM dual-mode radio. The FDM mode provides compatibility with older equipment, but the improved quality of PCM makes it the mode of choice for most applications. For long-haul telephony relay, the FDM or PCM signals may use very high-power propagation modes like troposcatter. Thus, the figure shows a high-power cluster for "relay and tropo." These high power modes use constant-duty-cycle FDM and PCM waveforms, an exception to the pattern that higher peak power typically implies lower duty cycle.

Mobile cellular radio (MCR) operates in frequency allocations between 400 MHz and 2.5 GHz, with clusters at 900 and 1800 MHz. There are similar radio services such as special mobile radio (SMR) as low as 40 MHz. The Instrumentation Scientific and Medical (ISM) bands at 2 and 5 GHz support personal communications systems (PCS) and RF LANs. MCR has become popular worldwide for rapid deployment of business and residential telephony in developing economies. MCR avoids the burial of fiber or cable for rapid build-out. Wireless local loop (WLL) has most features of MCR with reduced handset mobility [75].

The military employs specialized radar transponders for the Identification of Friends or Foes (IFF) and for other Integrated Communications, Navigation, and Identification Architecture (ICNIA) functions including tactical data links (e.g., remote radar plan position indicator displays) [120]. Distance measurement equipment (DME) and tactical air navigation (TACAN) also fall into this category of typically moderate duty cycle and moderate to high instantaneous data rate modes. Software radios for the military often must monitor multiple bands and modes for flight safety reasons. They typically require multiple navigation, IFF, and command-and-control communications for redundancy. These modes fall in a cluster of pulsed and lower-duty-cycle/high-peak-power signals.

The Synchronous Optical Network (SONET) [5] carries most backbone telephony in developed nations. Such fibers may be disrupted as much as six times per year per hundred miles of fiber (this rate was an industry rule of thumb in the United States in the early to mid-1990s). Consequently, SONET-compatible high-capacity microwave radios were developed with interoperable data rates of 155 (OC-3) and 622 Mbps (OC-12). Deployments in some infrastructures protect fiber paths, while others cross obstacles where it may be difficult, expensive or impossible to lay fiber, such as extreme terrain and bodies of water. Interoperation with SONET networks connects SDR nodes to the larger PSTN.

Finally, Figure 3-1 shows how radar signals typically emit the highest radiated power and employ the lowest duty cycles of any cluster in RF signal space. Impulse radar can create high-resolution maps of hidden objects (e.g., by penetrating walls). UWB communications use the same subnanosecond pulse technology operating at baseband. Time Domain Corporation's UWB system, for example, encodes data into an impulse train with an average of 40 million pulses per second (PPS). Since UWB communications employ subnanosecond pulses not readily synthesized with current-generation SDR hardware (e.g., FPGAs and DSP chips), UWB is not a focus of SDR standardization. On the other hand, as the underlying digital technology continues to evolve into clock rates over 1 GHz, UWB will ultimately migrate into the domain of the SDR. At today's rate of technology development, UWB will be accessible with SDR technology within 10 years. With the near-term exception of UWB, any of the bands and modes of Figure 3-1 may be implemented using the SDR techniques described in this text.

When used together a mix of modes across multiple radio bands provides a new dimension in QoS, reliability, and efficiency in the employment of the radio spectrum. After considering the top-level characteristics of these bands that are relevant to software-radio architecture, each band is considered in detail.

### B. Dynamic Range-Bandwidth Product

As mentioned earlier, the right side of Figure 3-1 is labeled "ADC Dynamic Range." This highlights the fact that the ratio of lowest to highest power signal in the receiver (total dynamic range) drives the requirements the ADC. As one accesses successively larger chunks of bandwidth, the sampling rate of the ADC must increase to at least 2.0 times the maximum frequency component ($f_{max}$) to satisfy the Nyquist criterion. Sound engineering principles require sampling at 2.5 $f_{max}$. In addition, the larger bandwidths are needed to service multiple subscribers with a single ADC. Narrowband analog receivers employ AGC to accommodate many decades of difference in received signal strength from a high-power nearby subscriber to the weakest, most distant subscriber. Analog receivers also filter high-power interference out of the analog signal-processing band.

The near–far ratio (NFR) is the ratio of the highest-power (presumably nearby) signal to the weakest (presumably most distant) signal. This ratio is 90 dB in GSM. Given a requirement for a 15 dB SNR for BER appropriate to the required QoS, the total dynamic range is at least 105 dB. Any in-band interference can raise this total dynamic range further. As the service bandwidth increases, the probability increases that subscribers and interferers with much higher power will be present in the receiver's RF band. In an HF band from 3 to 30 MHz, for example, the dynamic range of received signals is typically between 120 and 130 dBc (dB relative to full scale). Since ADCs nominally provide 6 dB of dynamic range per bit, one would need an ADC with $130/6 = \sim 22$ bits (at least) to service all potential HF subscribers. Contemporary ADCs with the necessary 70 M samples per second (Msps) sampling rates have only 14 (84 dB) of dynamic range. Thus, it is impossible to access the entire HF band with today's ADC (and DAC) technology. Near-term implementations therefore must tailor the architecture by structuring access to each band so that the communications objectives of SDR applications are met within the numerous constraints of available technology, including the ADC. This tailoring process requires an understanding of the HF and other modes presented below.

To extend this reasoning further, a multiband multimode radio such as SPEAKeasy was intended to service HF, VHF, and UHF military bands (from 2 MHz to 2 GHz). This means both sustaining the high dynamic range of HF and sampling the 2 GHz bandwidth, requiring a 5 GHz sample rate which is 96.9 dB-Hz. A useful *figure of merit*, **F**, for uniform digital sampling using ADCs and DACs is:

$$\mathbf{F} = \text{Dynamic Range (dBc)} + \text{Sampling Rate (dB/Hz)}$$

SPEAKeasy would require **F** = 226.9 dB/Hz (96.9 dB/Hz + 130 dBc), well beyond the state of the art of 140 to 160 dBc/Hz. Although we are making progress in ADC technology, practical engineering implementations of software radios avoid the frontal assault of a single ADC. Instead, the art and science of software radio systems engineering includes the partitioning of the total service bandwidth (e.g., from 2 MHz to 2 GHz) into multiple parallel RF bands. These are partitioned further into multiple parallel service bands (ADC/DAC channels). Each subband would have filtering, AGC, and digital signal processing that match the available ADC technology. The RF signal-space suggests regions within which a single ADC may provide effective sampling. The subbands and modes developed subsequently further refine these regions.

## II. HF BAND COMMUNICATIONS MODES

HF extends from 3 to 30 MHz according to international agreement. The definition of ITU frequency bands is taken from [5]. The length of a full-cycle radio wave in these bands is 100 meters at 3 MHz and 10 meters at 30 MHz, with linear variation between these extremes according to $c = f^*\lambda$, where $c$ is the speed of light, $\lambda$ is the wavelength, and $f$ is the radio frequency. Wavelengths determine the physical sizes of resonant antennas. Antennas resonate well across bandwidths that are less than 10% of the carrier frequency. To cover a full HF band using such a resonant structure would require about ten such antennas. The alternatives are to physically tune the narrowband antennas to operate on a specific subband, or to use a wideband antenna to access more of the band at once. A multiband radio therefore could employ a mix of wideband and tunable narrowband antennas drawn from the conventional antennas described in this and subsequent sections.

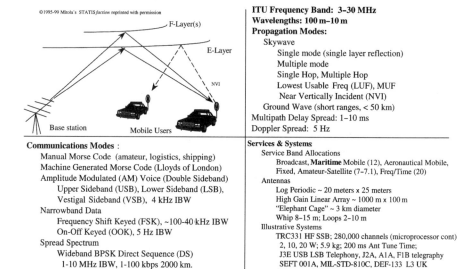

© 1995-99 Mitola's STATIS*faction* reprinted with permission

F-Layer(s)

E-Layer

NVI

Base station

Mobile Users

**ITU Frequency Band: 3–30 MHz**
**Wavelengths: 100 m–10 m**
**Propagation Modes:**
  Skywave
    Single mode (single layer reflection)
    Multiple mode
    Single Hop, Multiple Hop
    Lowest Usable Freq (LUF), MUF
    Near Vertically Incident (NVI)
  Ground Wave (short ranges, < 50 km)
Multipath Delay Spread: 1–10 ms
Doppler Spread: 5 Hz

**Communications Modes :**
  Manual Morse Code  (amateur, logistics, shipping)
  Machine Generated Morse Code (Lloyds of London)
  Amplitude Modulated (AM) Voice (Double Sideband)
    Upper Sideband (USB), Lower Sideband (LSB),
    Vestigal Sideband (VSB), 4 kHz IBW
  Narrowband Data
    Frequency Shift Keyed (FSK), ~100-40 kHz IBW
    On-Off Keyed (OOK), 5 Hz IBW
  Spread Spectrum
    Wideband BPSK Direct Sequence (DS)
    1-10 MHz IBW, 1-100 kbps 2000 km.
  Automatic Link Establishment (ALE) Link Quality Based

**Services & Systems**
  Service Band Allocations
    Broadcast, **Maritime** Mobile (12), Aeronautical Mobile,
    Fixed, Amateur-Satellite (7–7.1), Freq/Time (20)
  Antennas
    Log Periodic ~ 20 meters x 25 meters
    High Gain Linear Array ~ 1000 m x 100 m
    "Elephant Cage" ~ 3 km diameter
    Whip 8-15 m; Loops 2-10 m
  Illustrative Systems
    TRC331 HF SSB; 280,000 channels (microprocessor cont)
    2, 10, 20 W; 5.9 kg; 200 ms Ant Tune Time;
    J3E USB LSB Telephony, J2A, A1A, F1B telegraphy
    SEFT 001A, MIL-STD-810C, DEF-133 L3 UK
    Thomson-CSF, FR.
  Emerging SDR Products

**Figure 3-2**   The HF communications band.


## A. HF Propagation

As Figure 3-2 suggests, HF radio waves are usually reflected from the ionosphere, resulting in communications beyond line of sight (LOS). The ionosphere has several layers from which the waves may reflect. These are identified as the D, E, and F layers in order of increasing altitude. Two or more such
skywaves may be received in what is called *multimode propagation*. These
waves will add (as complex vectors) at the receiver resulting in phase and
amplitude variability. The time differences between two reflected waves (HF
propagation modes) will be about 1 ns per foot of altitude separation. Since
the reflecting layers may be from 1 to 10,000 miles apart, this equates to 1 to
10 ms of delay-spread. In addition, the ionosphere and fixed transmitters on
the earth are typically approaching or receding, imparting Doppler shift onto
the RF carrier. Since the layers of the ionosphere may be moving in different
directions, the Doppler spread at HF is large, typically 5 Hz.

If the RF carrier is too low or too high, it will pass through the ionosphere.
Beyond LOS, reflections from the ionosphere are only possible on radio frequencies between the least usable frequency (LUF) and the maximum usable
frequency (MUF). Specific combinations of RF and antenna configuration can
result in near vertically incident (NVI) propagation in which the waves reflecting from the ionosphere propagate only a few tens of miles. NVI is useful in
mountainous areas for communications between subscribers in adjacent valleys, for example. In addition, HF will reflect from water and from some land-
masses, enabling multihop communications (ionosphere–water–ionosphere–
land).

## B. HF Air Interface Modes

Morse code has been used since the 1800s for ship-to-shore and transoceanic communications. Machine-generated Morse code became popular with the emergence of microprocessors in the mid-1980s. PC-based software readily translates text into Morse. Voice transmission at HF uses amplitude modulation (AM) to accommodate the limited bandwidth of the HF channel. The simple double side band (DSB) AM creates two mirror-image replicas of the voice waveform—one above and one below the carrier, using twice the bandwidth required for the information content. Upper side band (USB) filters the lower of these two voice bands, suppressing any residual carrier. Lower side band (LSB) is the converse of USB. Vestigial side band (VSB) allows a small component of carrier to be transmitted, simplifying carrier recovery in the receiver. Each of these modes is used in HF communications. Voice intelligibility requires only 3 to 4 kHz for the principal formants (sinusoidal information-bearing components of the speech waveform). Consequently, each of these modes may be digitally implemented with an ADC rate of typically 10 to 25 kHz using commodity DSP chips with modest processing power (10 to 25 million instructions per second—MIPS). Thus, the speech-processing niche was one of the first commercial applications of ADCs and DSPs.

Morse code might be thought of as an on-off-keyed (OOK) data mode with the channel code information carried in the duration (pulse width) of the channel waveform—Morse *dits* are three to four times shorter than *daa's*. Because of the relatively low rate at which people can compose and send Morse code, it occupies a bandwidth approximately 5 Hz. This yields a plethora of such narrowband signals packed into the very busy HF bands. Other common HF data modes include frequency shift keying (FSK). The FSK channel code consists of *mark* or *space*, corresponding to a negative or positive frequency shift, respectively. The frequency shift may be as small as a few tens of Hz. Data rates ranging up to 1200 bits per second require FSK shifts of several hundred Hz. An FSK channel symbol is also called a *baud*. It encodes one bit of information. During very short time intervals (from a few milliseconds to a few tenths of a second), the ionospheric transfer function is approximately constant. Higher data rates (e.g., 10 to 40 kbps) may be used for such short intervals to *burst* small amounts of data over long distances using FSK modems. Both standard and burst FSK waveforms can be implemented using commodity DSP chips and low-speed/high-dynamic-range ADCs. HF Automatic Link Establishment (ALE) equipment [121][18] probes the propagation path in a pre-arranged sequence to identify good frequencies on which to communicate. The ALE signals include "chirp" waveforms that linearly sweep the RF channel so that the receiver can estimate the channel transfer function. The two ends of the link negotiate choice of RF based on reception quality.

---

[18]The examples of military communications equipment appearing in this chapter are from [121].

**TABLE 3-1   Software Radio Applications Parameters—Baseband and HF**

| Software Radio Application | Sampling Rate ($f_s$) | Dynamic Range (dB) |
|---|---|---|
| HF Baseband | .5–8 kHz | 24–64 |
| Modems | 8–32 kHz | 48–64 |
| Music | 20–100 kHz | 60–96 |
| HF-IF | .2–10 MHz | 72–120 |
| HF RF | 75 MHz | 130 |

The research literature also describes a long-haul HF telecommunication system using direct-sequence spread spectrum to achieve a data rate of 100 kbps via a 10 MHz spreading sequence [119]. Low grazing angle, nearly optimal choice of transmit and receive frequency, and location and other specialized factors contributed to the success of this experiment, which appears infeasible for general HF communications. The serial modem [122] delivers 1200 to 2400 baud data on HF channels with high reliability. Recently, the SiCom Viper [424] direct-sequence spread-spectrum radio has demonstrated data rates of 19.2 kbps and 56 kbps over skywave HF links on a routine basis by employing cyclostationary techniques in the receiver. This 1 to 2 MHz spread-spectrum signal has an instantaneous SINR of about −50 dB, which it overcomes with processing gain.

The software radio parameters of HF sampling rate and dynamic range depend on the point in the system at which the ADC/DAC operates from baseband through IF to RF, as illustrated in Table 3-1.

**C. HF Services and Products**

Amateur radio (ham), commercial broadcast, aeronautical mobile, amateur satellite, and timing/frequency standards are provided at HF as outlined in Figure 3-2. HF antennas and power amplifiers often dominate the size, weight, and power of HF radio systems. Antennas matched to HF wavelengths are large—some research antennas extend for over a kilometer. Military applications employ circularly disposed array antennas for long-haul communications and location finding using triangulation. Reliable long-haul communications is also possible using small log-periodic antennas (e.g., $20 \times 25$ meters horizontally mounted on a 50 or 100 ft mast). Whip antennas 8 to 15 ft long may also be inductively loaded to match HF wavelengths. And 2 to 10 meter loop antennas measure direction of arrival. Although software radios cannot change the laws of physics that cause HF antennas to be large, they can enhance signals received using smaller, less optimally tuned antennas to achieve quality approaching that of the larger antennas.

Mercury Talk [121] exemplifies the relatively short-range, low-power HF radios. With 2 watts of output power, this radio can close a voice link on a 10 km path. With its 3.5 watt output, it can close a Morse code link over a

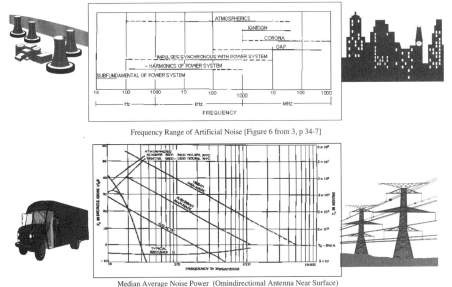

Frequency Range of Artificial Noise [Figure 6 from 3, p 34-7]

Median Average Noise Power  (Omindirectional Antenna Near Surface)
Figures are from *Reference Data For Engineers* © Sams Publishing Co., Inc, used by permission.

**Figure 3-3**   Radio noise and incidental interference.

160 km path. Thomson CSF of France makes the TRC331, another portable HF radio weighing less than 10 kg. Figure 3-2 lists additional narrowband communications standards such radios meet for military interoperability.

## III.  LOW-BAND NOISE AND INTERFERENCE

As illustrated in Figure 3-3 [from 5, p. 34-7], the lower radio bands—HF, VHF, and lower UHF—include significant sources of radio noise and interference. The incidental and unavoidable interference includes automobile ignitions, microwave ovens, power distribution systems, gaps in electric motors, and the like. Cellular bands are dominated by intentional interference introduced by other cellular users occupying the RF channel in distant cells. Unavoidable interference results when tens to hundreds of thousands of military personnel use their LVHF radios at the same time. Thus, high levels of interference characterize these congested low bands.

The noise/interference levels are defined with respect to thermal noise:

$$P_n = kTB$$

where $k$ is Boltzmann's constant, $T$ is the system temperature ($T_0$ is the reference temperature of 273 Kelvin), and $B$ is the bandwidth (e.g., per Hz).

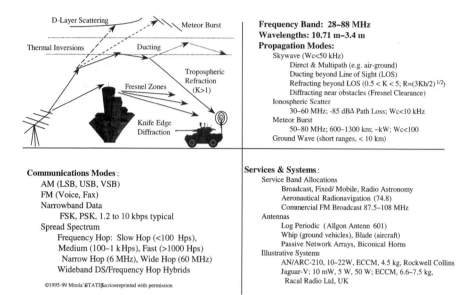

**Figure 3-4**   The LVHF communications band.

In the microwave bands above 1 GHz this thermal noise[19] is a good approximation of the noise background. In urban areas, however, incidental urban
interference dominates thermal noise until about 5 GHz. In the lower bands,
atmospheric noise arises from the reception of lightning-induced electrical
spikes from thunderstorms, etc. halfway around the world. Consequently, this
noise component is much stronger in summer than in winter as illustrated in
the figure. In addition, this noise has a large variance. The short-term (1 ms)
narrowband (1 kHz) noise background varies at a rate of a few dB per second over a range of from 10 to 30 dB, depending on the latitude, time of the
year, and sunspot cycle. High-quality HF receivers track this noise background
independently in each subscriber channel.

## IV.  LOW VHF (LVHF) BAND COMMUNICATIONS MODES

The LVHF band from 28 to 88 MHz has traditionally been the band of ground
armies because of the robust propagation offered among ground-based subscribers in rugged terrain. Amateur radio and the U.S. citizens' band also use
LVHF. The upper edge of this band is defined by the commercial broadcast
band from 88 to 108 MHz. Wavelengths from 10.7 to 3.4 meters admit smaller
antennas than HF, with a $\frac{1}{4}$ wave dipole having a length of 3 to 10 feet, as
summarized in Figure 3-4. Historically, LVHF military users have employed

---

[19]Because of the equation, thermal noise is sometimes called kTB noise.

single-channel half-duplex PTT AM and FM modes. The commercial success of Racal's® Jaguar frequency-hopped radio with its digital vocoding and digital air interface resulted in a proliferation of FH modes for military users during the late 1980s.


## A. LVHF Propagation

Although LVHF frequencies do not reflect from the ionosphere with the reliability of HF, it is possible to scatter these waves from the lower D layer of the ionosphere. D layer scatter at 30–60 MHz with a bandwidth of less than 10 kHz often has only about 8.5 dB greater loss than LOS propagation. In addition, in LVHF, propagation beyond geometric LOS is common due to tropospheric refraction. Since the atmosphere is denser at lower altitudes, the speed of light is less near the ground than at higher altitudes. Since typical LVHF whip antennas provide an omnidirectional radiation pattern with relatively large vertical extent, the waves propagate across significant differences in index of refraction. Therefore, the waves emitted just above the geometric grazing angle propagate beyond the geometric LOS, having been bent down as they traverse the path. This effect can be modeled as an increase in the effective radius of the earth. The approximation of radio horizon is given by:

$$R = \sqrt{4Kh/2}$$

Range is in miles. $K$ is the effective radius of the earth, and $h$ is the altitude of the transmitter in feet. $K$, the effective earth radius, is defined experimentally. $K = 1$ defines geometric LOS propagation. Typically $K = 4/3$ in temperate climates. But $K$ may range from $1/3$ to 3 as a function of climate and weather. At night, particularly in subtropical climates, LVHF waves may propagate by a ducting phenomenon in which the refractive index of the atmosphere exhibits an inversion (air density increases with increasing altitude instead of decreasing). Ducting can extend the range of LVHF two hundred miles or more beyond LOS. Ground-to-air radios also experience skywave multipath scattered from the D layer or refracted through tropospheric ducts.

*1. Diffraction* Knife-edge diffraction is a wave phenomenon in which waves bend around sharp obstructions as if the entire wavefront above the obstacle consisted of point sources. These point sources induce an interference pattern of reinforcement (waves on the average in phase) and cancellation (waves on the average 180 degrees out of phase) called the Fresnel zones. A receiver in the Fresnel zones experiences alternating strong and weak signals as the receiver moves through multiples of a wavelength. VHF radios may maintain reception continuity across Fresnel zones using diversity in space (e.g., multiple antennas) and frequency (e.g., slow frequency hopping) with error control coding.

**2. Reflections from Meteor Trails**   Each minute a dozen meteors penetrate the earth's atmosphere, where they burn up. This creates trails of ionized gas from which radio waves may be reflected. Meteor burst communications use trails that endure for periods of 10 milliseconds to over a second. Meteor burst in the 50 to 80 MHz RF ranges will propagate short bursts of communications over distances of from 600 to 1300 km with radiated power of about 1 kW and with bandwidths of up to 100 kHz. With directional antennas, meteor burst provides a relatively secure way of exchanging low-volume command-and-control data over ranges significantly beyond LOS. LVHF, like HF, may also be propagated via ground wave over short ranges (e.g., 10 km). Ground wave generally suffers large attenuation, with a path exponent of 2.5 to 4. That is, instead of path loss proportional to $1/R^2$, the path loss will be proportional to $1/R^{2.5}$ to $1/R^4$.

## B. Single-Channel-per-Carrier LVHF Air Interface Modes

AM (DSB, USB, LSB, and VSB) and analog-modulated FM voice are common at LVHF. FSK and phase shift keying (PSK) are common data modes. Simple PSK formats such as binary (BPSK) and quaternary PSK (QPSK) offer reliable data service at LVHF from 1.2 kbps to about 10 kbps within the coherence bandwidth of LVHF. The use of digital vocoding and private networks (e.g., TETRA [123]) is increasing in these bands. The analog modes arose in the 1960s. Signal processing was limited to analog frequency translation, filtering, automatic gain control, and simple control circuits. In these modes, each subscriber has a unique RF carrier. Such single-channel-per-carrier (SCPC) modes have historically been preferred by ground-based military forces for squad-level manpack and individual vehicular radios. Contemporary LVHF military radios usually employ FH for TRANSEC. LVHF propagates well in rugged terrain since the waves penetrate vegetation and reflect, refract, and diffract over and around obstacles. This fills in low-lying areas where higher-frequency waves would not penetrate surrounding obstacles.

## C. LVHF Spread-Spectrum Air Interfaces

Spread-spectrum modes include FH, DSSS, and hopped-spread hybrids. Some FH radios hop over subbands of LVHF, employing 1 to 6 MHz hopping bands. Others provide the full 60 MHz hopping agility from 28 to 88 MHz. The narrower hop bandwidths may be implemented digitally via SDR techniques (e.g., using a fixed tuned medium bandwidth RF chain and a 6 MHz ADC/DAC). The 60 MHz hop bandwidths are not accessible using fixed tuned RF, but instead the hops must be heterodyned to a common IF using a fast tuned synthesizer (or two). As ADC and DAC bandwidths and dynamic range continue to improve, SDR radio techniques may extend to wider hop-bandwidths.

The FH radios are typically vocoded. The speech waveform is represented digitally using a vocal tract model such as Linear Predictive Coding (LPC).

LPC-10, for example, was a standard 1200-bit-per-second voice codec used throughout the 1970s and early 1980s. More complex waveforms based on subband coding [124] and adaptive LPC were implemented in DSP chips in the middle to late 1980s. This led to other voice codecs such as Vector Excited Linear Prediction (VELP) and Codebook Excited Linear Prediction (CELP) with better perceptual properties. In addition, many LVHF radios employ slow FH ($<$ 100 hops per second), so that sufficient bits are available per hop dwell to reconstruct a voice epoch. Some coded vocal tract parameters require enhanced error protection because any errors propagate for many bits. Therefore, some LVHF FH radios employ FEC on such speech data. This plus the encryption of the voice bits and hop sequences complicates the transceiver algorithms in SDR implementations of these modes.

### D. LVHF Multichannel Air Interfaces

FM frequency division multiplexing (FM/FDM) for military LVHF applications includes modes with four channels per RF carrier. These meet the connectivity needs of radiotelephony operations of relatively low-echelon military forces. Due to the relatively narrow coherence bandwidths of LVHF, conventional FM/FDM is limited to about 60 channels. These multichannel modes are being supplanted by digitally modulated time division multiplexed (TDM) waveforms such as BPSK or QPSK synchronous PCM. Using 16 kbps delta-modulation or adaptive PCM, one can pack four subscribers into a 64 kbps synchronous BPSK waveform. This mode is more robust in the LVHF propagation environment than four-channel FM/FDM. Other modes of 128 to 256 kbps accommodate other combinations of low and medium data-rate radio relay, depending on the mix of delta modulation, VCELP, CVSD, ADPCM, and other compressive coding waveforms.

### E. LVHF Services and Products

As shown in Figure 3-4, LVHF supports broadcast, fixed, and mobile applications, radio astronomy, aeronautical radio navigation (74.8 MHz), and commercial FM broadcast (87.5–108 MHz). Antenna products include log-periodic arrays for broadband high-gain performance (e.g., the Allgon Antenn 601 [121, p. 597]) and an assortment of whips for ground vehicle applications. In addition, aircraft generally employ blade antennas for aerodynamic compatibility. Passive network arrays and biconical horns [4, p. 613] may also be used for increased gain over relatively narrow access bandwidths. The AN/ARC-210 from Rockwell Collins is an illustrative airborne product that operates in this band. It radiates 10–22 W of power, weighs 4.5 kg, and supports a variety of electronic counter-countermeasures (ECCM) including FH. The Jaguar-V from Racal Radio Ltd., UK [4, p. 69] popularized LVHF FH. This affordable manpack configuration produces power of 10 mW, 5 W, and 50 W with the Jaguar's own advanced FH ECCM in a compact 6.6–7.5 kg package.

TABLE 3-2   SDR Parameters—VHF

| Software Radio Application | Sampling Rate $(f_s)$ | Dynamic Range (dB) |
|---|---|---|
| VHF-UHF BB | 50–150 kHz | 20–60 |
| LVHF-IF (FH) | 12–200 MHz | 66–108 |
| VHF/UHF-IF | 25–500 MHz | 60–96 |
| VHF RF | 650 MHz | 96–120 |

Legend: BB = baseband.

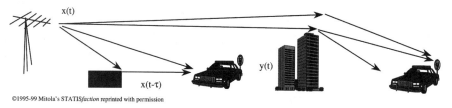

©1995-99 Mitola's STATIS*faction* reprinted with permission

**Figure 3-5**   Basic physics of multipath propagation.

## F. LVHF Software Radio

Software radios operating in LVHF compete with the low battery drain and high output efficiency of customized microprocessor-controlled analog/digital hybrid implementations of these products. The propagation and air interface modes lead to critical SDR parameters, shown in Table 3-2.

Baseband digital processing accommodates single-channel voice and narrowband data communications. LVHF-IF includes multiple-channel radio relays, television, and other radio services. The increased dynamic range reflects the near–far ratio, noise variability, and interference background variations in VHF. RF dynamic range encompasses the entire band. One benefit of operating in LVHF versus HF is the reduction in delay spread by three orders of magnitude from ms to $\mu$sec. In addition to improving the coherent bandwidth of the medium, it reduces the memory requirements and complexity of time-domain equalizer algorithms. A benefit of the reduced noise complexity of LVHF is that simple squelch algorithms (e.g., Constant False Alarm Rate—CFAR) reliably track the LVHF noise floor, while at HF, complex algorithms are required.

## V. MULTIPATH PROPAGATION

LVHF marks the beginning of the LOS bands in which the radio waves can be approximated as traveling in straight lines to the radio horizon. This contrasts with HF, where skywave reflections yield beyond-LOS propagation. Since these waves may reflect from any sufficiently large conductive structure, more than one wave may impinge on the receiver as illustrated in Figure 3-5.

$$y(t) = A(\alpha_1 \cos(\omega_0 t) + \alpha_2 \cos(\omega_0 [t - \tau]))$$
$$= B(\tau) A \cos(\omega_0 t + \theta)$$
$$\text{where} \quad B(\tau) = [\alpha_1^2 + 2 \alpha_1 \alpha_2 \cos(\omega_0 \tau) + \alpha_2^2]^{1/2}$$
$$\theta = -\tan^{-1}[\alpha_2 \sin(\omega_0 \tau)/(\alpha_1 + \alpha_2 \cos(\omega_0 \tau))]$$
$$\text{therefore} \qquad \text{Maximum amplitude} = A[\alpha_1 + \alpha_2]$$
$$\text{Minimum amplitude} = A[\alpha_1 - \alpha_2]$$

©1995-99 Mitola's STATIS*faction* reprinted with permission

**Figure 3-6**   Elementary multipath equations.

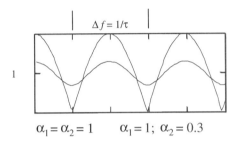

$$\alpha_1 = \alpha_2 = 1 \qquad \alpha_1 = 1; \ \alpha_2 = 0.3$$

**Figure 3-7**   Zones of constructive and destructive interference.

Considering the radiated wave to be a cosine function of time, one can characterize simple multipath in which the direct and reflected paths have amplitudes $\alpha_1$ and $\alpha_2$ as illustrated in Figure 3-6.

Depending on propagation, the amplitudes of the cosine waves may differ. If these amplitudes are nearly identical, then the minimum amplitude $(\alpha_1 - \alpha_2)$ will be nearly zero. This results when the difference in path length is essentially one-half wavelength, yielding cosine waves that are approximately 180 degrees out of phase, a condition known as cancellation or destructive interference. We may also plot the value of $B$ as a function of differential path delay to observe the frequencies at which constructive and destructive interference occur as shown in Figure 3-7.

The literature distinguishes *flat fading* from *selective fading*. This figure can be interpreted to reveal the difference between these two forms of multipath fading. If the bandwidth of the signal is an order of magnitude smaller than $\Delta f$, then as $\tau$ changes, the amplitude of the received multipath signal will follow the shape of the curve in the figure. That is, the entire signal will appear to have the amplitude of the point in the curve corresponding to $\Delta f$. Although multipath induces a small amplitude distortion on the received envelope, essentially the entire signal fades in and out at the same time. So if $\tau$ is a microsecond, $\Delta f$ is 1 MHz. Thus signals with a few kHz of bandwidth fade uniformly in flat fading. If, on the other hand, the signal bandwidth is 2 MHz, then the received signal viewed on a spectrum analyzer

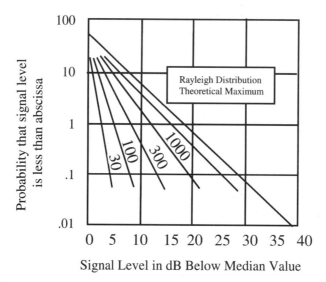

**Figure 3-8** Fading approaches the Rayleigh model above 4 GHz. Fade rate plotted for RF=30, 100, 300, and 1000 MHz.

appears to have a deep null moving as $\tau$ changes over time. The deepest fade is limited to those sinusoidal components that are nearly 180 degrees out of phase while the other components remain unfaded. Such wideband signals are thus subject to so-called selective fading. The multipath contribution to selective and to flat fading are both captured in the equations of Figure 3-6.

As the carrier frequency increases, changes in $\tau$ on the order of one-fifth of a wavelength transition the received signal from deeply faded to moderately faded. Consequently, one may employ more than one antenna spaced appropriately to receive two different signals, selecting the one with highest signal strength to compensate for the faded signal. Diversity reception can be a strong service enabler for SDRs that can employ additional signal processing to combine signals from diversity antennas more effectively than is practicable with analog signal processing.

Instead of the condition described above, there may be more points of reflection and hence more received signals with different received signal strength and time-delay corresponding to different amplitude and phase of the sinusoids at the receiver. In the limit, there may be an infinite number of such sinusoids with uniformly distributed phase and log-normally distributed power, the Rayleigh distribution. Rayleigh's fading model is a very good approximation for the microwave regions above 4 GHz as illustrated in Figure 3-8.

Below 1 GHz, however, the probability that the signal level is less than the abscissa is not as high as the Rayleigh model. Since wavelengths in the microwave region are about a centimeter, water vapor in the atmosphere creates the random delay and amplitude effects characterized in the Rayleigh model. Rice noted that the statistical structure of amplitude varies as a function of the number of strong multipath components, offering a model of amplitude distributions parameterized by the number of such strong paths. As the number of paths with approximately the same phase increases, the amplitude distribution becomes tighter and the variance of the amplitude distribution decreases.

SDR algorithms mitigate fading, for example, by bridging the data clock across deep fades. Coherently combining energy from diversity antennas reduces fade depth. Cyclostationary processing enhances CIR. Because of the statistical structure of fades, the rate of convergence of such algorithms is variable. The processing demands of these algorithms therefore vary as a function of fade depth. Understanding fade mitigation algorithms yields insights into the statistical structure of processing demand imposed by such algorithms. Armed with this understanding, one may design an SDR with sufficient processing capacity and flexibility. By studying collections of such algorithms, one may define an architecture that supports the adaptation of the hardware platform and the insertion of new algorithms as they are developed.

## VI. VHF BAND COMMUNICATIONS MODES

By convention, the very high frequency (VHF) band extends from 30 to 300 MHz. This convention ignores differences in propagation between the LVHF band and VHF above the commercial broadcast band (88–108 MHz). VHF in this section extends from 100 to 300 MHz. This band includes commercial air traffic control (117.975–144 MHz), amateur satellite, and maritime mobile bands as suggested in Figure 3-9. Consequently, SDR accesses to VHF can provide services spanning air, ground, maritime, government, and amateur market segments.

### A. VHF Propagation

VHF includes Fresnel zones, knife-edge diffraction, ducting, and tropospheric refraction like LVHF. VHF has less filling of low-lying and shadowed regions because the shorter wavelengths set up spatially smaller interference patterns. These patterns have smaller angles between successive constructive and destructive interference zones. Wavelengths from one to three meters typical of this band are readily trapped in thermal inversions in the atmosphere in subtropical climates, leading to significant beyond-LOS propagation, particularly at the day–night boundary.

The delay spread of 1 to 10 microseconds allows simple modulation to achieve instantaneous bandwidths of hundreds of kHz. This leads to simple

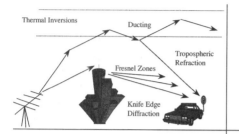

Frequency Band: [30MHz] 100–300 MHz
Wavelengths: [10m] 3 m – 1 m
Propagation Modes:
  Skywave: Radio Line of Sight (LOS)
    Direct & Multipath (e.g.. air-ground)
    Some Ducting beyond LOS
    Refracting, Diffracting beyond LOS
    Ground Wave (short ranges, < 10 km)
  Multipath Delay Spread: 1-10  usec

**Communications Modes :**
  AM (LSB, USB, VSB)
  FM (Voice, Fax)
  Narrowband Data
    FSK, PSK, 75 bps to 9.6 kbps typical
  Multichannel Radio Relay (4–60 Channels)
  Spread Spectrum
    Frequency Hop: Slow Hop (<100 Hps),
    Medium (100–1 kHps), Fast (>1000 Hps)
    Wide Hop (>10 MHz)
    Wideband BPSK/ Hop Hybrids Possible
  ©1995-99 Mitola's STATISfaction reprinted with permission

**Services & Systems:**
  Service Band Allocations [3]
    87.5–108 Broadcasting; 117.975–137 Aeronautical Mobile
    138-144, 148-151 Govt (G); 151–162 Non Govt (NG);
    162-174, 220-222 MHz G, NG mobile radio;
    144–146 Amateur Satellite; 156.7625–156.8735 Maritime
  Antennas
    Log Periodic ( Allgon Antenn  601; [4 p 597] 20–220 MHz)
    Whip, Blade, Discone, Corner Reflectors
    Passive Network Arrays, Biconical Horns ([4, p. 613], -960 MHz)
  Illustrative Systems
    AN/GRC-171(V) 20 W, ECCM(HAVE QUICK), 36 kg,
    225–400 MHz, AM Voice, AM Secure Voice; B has FM.
    Rhode & Schwarz Series 400: 15–300 W; 12/40 channels,
    25, 12.5 or 6.25 kHz Channels, rack mount.

**Figure 3-9**   The VHF band.

receiver architectures (e.g., single-channel push-to-talk with AM conversion or FM discriminator receivers; or FSK mark/space filters for data signals).

## B.  VHF Air Interface Modes

AM, FM, various data modes, and FH spread spectrum such as the U.S./NATO HAVE QUICK I and II slow-frequency-hop air interface are the common modes in VHF as illustrated in Figure 3-9. Wide hops are more practical in these bands because about 120 MHz is available for frequency hopping in the 225–400 MHz VHF and low-UHF bands.

The AM air interface waveform is particularly appropriate for safety-related applications such as emergency communications with aircraft. AM waveforms are audible at negative SNR, extending the range and robustness of unencoded AM voice. FM voice, also a popular military mode, provides greater clarity of voice communications at channel SNRs greater than 7 to 9 dB. Below this SNR, the FM discriminator will not lock to the carrier, yielding only noise. These analog voice modes do not take advantage of today's signal-processing capabilities. Recent research suggests the possibility of extending these modes through wavelet-based digital signal processing [125]. Improvements in components have reduced channel bandwidths from 100 kHz or more in the early days of radio to typically 25 to 30 kHz today, with $8\frac{1}{3}$ and $6\frac{1}{4}$ kHz modes emerging (e.g., APCO 25). Due to congestion of air traffic control radio bands in Europe, for example, these analog AM/FM modes are being constrained to $8\frac{1}{3}$ kHz. This packs three SCPC subscribers into the 25 kHz of spectrum formerly occupied by only a single user.

## C. VHF Services and Products

VHF services include the 87.5–108 MHz commercial FM broadcast bands. Air traffic control uses the 117.975–137 MHz aeronautical mobile band. This band is allocated to civilian air traffic control, while the companion UHF band is allocated to military air traffic control. Consequently, dual-band VHF/UHF avionics radios are common. There are also governmental applications in 138–144 MHz, 162–174 MHz, 220–222 MHz, and 148–151 MHz, and nongovernmental bands from 151 to 162 MHz. The amateur satellite band extends from 144–146 MHz while 156.7625–156.8735 MHz encompasses the maritime mobile band.

VHF antenna products include whip, blade, discone, corner reflectors, passive network arrays, and biconical horns. The high-gain horns, cavity-backed spirals, discones, etc. are relatively large because of the 3 meter wavelength at the low end of VHF. High-gain military antenna products are available for avionics and extensible antenna masts [121]. Some log-periodic antennas such as the Allgon Antenn 601 [121, p. 597] access the subset of VHF from 20–220 MHz. Others span VHF through 960 MHz [121, p. 613]. Such VHF/UHF operation is common for both antennas and discrete analog and programmable digital radios. These radio suites also monitor emergency channels using dedicated transceivers. This includes simultaneous VHF and UHF operation.

Illustrative discrete radio products include general-purpose, single-channel ground-based radios and multichannel radio relays. The AN/GRC-171(V) general-purpose ground-based radio, for example, delivers 20 W of RF power from vehicular power. It includes the HAVE QUICK ECCM/EP (Electronic Protect) mode for interoperability with airborne radios. This radio weighs 36 kg, operates between 225 and 400 MHz, and supports AM voice, AM secure voice, and FM air interfaces. Rhode and Schwarz offer a multichannel radio relay in their Series 400 radio. It produces 15 to 300 watts of power to relay from 12 to 40 channels. Each channel may have 25, 12.5, or 6.25 kHz bandwidth. This rack-mount radio is typical of military radio-relays.

## D. VHF SDR

SDR design for VHF must provide at least the capabilities of the discrete radios, within the price-performance envelope of the associated markets. For the military avionics bands, this means two or more dedicated emergency broadcast receivers. Since one of the features of SDR is the elimination of discrete radios, it may be difficult to obtain type certification for a single SDR to replace two discrete radios. The reliability aspects of two or three discrete radios are well known by the type-certification community. Offering one SDR in place of three discrete radios therefore offers reliability challenges. Ground infrastructure radios have to transmit on both VHF and UHF at the same time in order to interoperate with military and civilian aircraft. This keeps the cost of SDR implementations high. General aviation markets are very price-sensitive. A military avionics SDR priced at $10 k may be affordable, but the price of this product may equal that of the general aviation aircraft. General aviation radios therefore are priced in the range of $1–2 k. Consequently, the introduction of SDR technology into general-aviation markets would tend to lag the introduction into less price-sensitive military markets.

Commercial fleets (e.g., trucking) offer potential SDR insertion opportunities. Many truck fleets, for example, use a GPS-based location system coupled to a satellite-based fleet-tracking system (e.g., OmniTRACKS [420]). In addition, the fleets use CB radio and commercial AM/FM broadcast for local traffic information. Local navigation, wireless on-line maps, and other Intelligent Vehicle Highway Systems (IVHS) are also emerging [421]. Thus, commercial fleets are evolving multiband, multimode capabilities, potentially amenable to SDR insertion.

The algorithm complexity of VHF SDR is similar to LVHF. Most of the modulation formats use SCPC with narrow bandwidths. One potential benefit of SDR technology is the graceful introduction of the new narrowband modulation formats. Digital filtering, both on transmit and receive, makes it relatively easy to manage adjacent channel interference, even in $6\frac{1}{4}$ kHz bands. SDR implementations using baseband DSP also facilitate the introduction of vocoders and packet data in SCPC fleet networks like TETRA.


## VII. UHF BAND COMMUNICATIONS MODES

UHF is clearly the most popular commercial band with the proliferation of MCR and personal communications systems (PCS) between about 400 and 2500 MHz, almost exactly the RF extent of the UHF band (300–3000 MHz).


### A. UHF Propagation

Pure skywave propagates between aircraft and the ground according to square-law path loss. Ground-based MCR/PCS channels scatter and attenuate the hybrid skywave/groundwave with path exponents between 2 (square law) and 4 (fourth law). In addition, losses are nonuniform with range. Loss exponents vary from square law near the antenna, to 2.8 in Rician zones, and fourth law in Rayleigh zones, as distance from the base station increases. In addition, groundwave propagates for short ranges, typically less than 1 km. Multipath delay-spread typically is from 2 to 10 $\mu$secs [126]. *Doppler shift* would be 75 Hz for a 60 mph vehicle using an RF of 840 MHz, typical of MCR applications. The *Doppler spread* is typically about twice the Doppler shift. The Doppler spread establishes the range of frequency offsets that the receiver's carrier-tracking loop must accommodate. Initially, GSM specified a Doppler spread appropriate to ground-based applications. But the adoption of GSM for the Eurorail system quadrupled the Doppler spread requirements, significantly impacting GSM hardware implementations. An SDR handset would merely adjust the carrier-tracking loop bandwidth of the Costas loop algorithm.

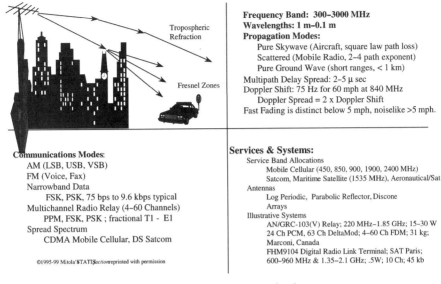

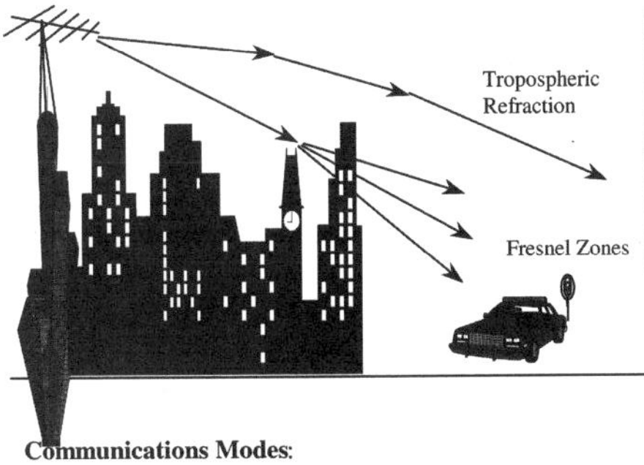
Tropospheric
Refraction
Fresnel Zones

**Frequency Band: 300–3000 MHz**
**Wavelengths: 1 m–0.1 m**
**Propagation Modes:**
Pure Skywave (Aircraft, square law path loss)
Scattered (Mobile Radio, 2–4 path exponent)
Pure Ground Wave (short ranges, < 1 km)
Multipath Delay Spread: 2–5 μ sec
Doppler Shift: 75 Hz for 60 mph at 840 MHz
Doppler Spread = 2 x Doppler Shift
Fast Fading is distinct below 5 mph, noiselike >5 mph.

**Communications Modes:**
AM (LSB, USB, VSB)
FM (Voice, Fax)
Narrowband Data
FSK, PSK, 75 bps to 9.6 kbps typical
Multichannel Radio Relay (4–60 Channels)
PPM, FSK, PSK ; fractional T1 - E1
Spread Spectrum
CDMA Mobile Cellular, DS Satcom

©1995-99 Mitola's STATISfaction reprinted with permission

**Services & Systems:**
Service Band Allocations
Mobile Cellular (450, 850, 900, 1900, 2400 MHz)
Satcom, Maritime Satellite (1535 MHz), Aeronautical/Sat
Antennas
Log Periodic, Parabolic Reflector, Discone
Arrays
Illustrative Systems
AN/GRC-103(V) Relay; 220 MHz–1.85 GHz; 15–30 W
24 Ch PCM, 63 Ch DeltaMod; 4–60 Ch FDM; 31 kg;
Marconi, Canada
FHM9104 Digital Radio Link Terminal; SAT Paris;
600–960 MHz & 1.35–2.1 GHz; .5W; 10 Ch; 45 kb

**Figure 3-10** UHF communications.

The statistical structure of fast fading depends on the speed of the subscriber. The fades are distinct below 5 mph. That is, as the subscriber slowly traverses distances of approximately half a wavelength, the fade has a pronounced time domain null that can be characterized as a distinct event. As the speed of the subscriber increases above 5 mph, however, the subscriber moves rapidly through multipath peaks and nulls. Consequently, the nulls lose their temporal structure. They are better modeled as additive noise.

Sufficient received signal strength for communications in urban areas may require the equalization of a large number of single-hop reflections (see Figure 3-10). The siting of cellular and PCS systems has resulted in the proliferation of a large number of commercial propagation-modeling products. Using 3D building plans accurate to 1 meter, most predict the average received signal strength to within 5 dB. In military applications, reflectors (e.g., trucks, aircraft, and temporary buildings) regularly move, complicating the process of calibrating, updating, and validating the models.

## B. UHF Air Interface Modes

Traditional narrowband air interface modes such as AM (LSB, USB, VSB), FM (voice, fax), and narrowband data (FSK, PSK, 75 bps–9.6 kbps typical) are common throughout UHF. In addition, multichannel radio relays in this band support 60 to 240 channels or more using PPM, FSK, PSK, and QAM channel modulations. Multichannel digital air interface modes include full and fractional T1 and E1. One of the most aggressive and widely known spread spectrum modes, JTIDS, operates over 240 MHz hopping bandwidth in the 1.2 GHz RF band. The U.S. Air Force and NATO publish the HAVE QUICK I and II slow FH air interface used by military aircraft.

In the commercial sector, the most widely deployed spread spectrum air interface today is the CDMA mobile cellular air interface, IS-95. This mode employs a 1.2288 MHz chip rate, supporting 64 subscribers plus signaling and control in 1.25 MHz. The wide bandwidth allows smart antennas to compensate for multipath components [127]. The 3G CDMA recommendations offer alternative spreading rates up to 20 MHz, alternate synchronization schemes, more efficient vocoding, etc. GSM's TDMA format and 1G SCPC FDMA air interfaces are predominant in the cellular bands (450, 900, and 1800 MHz), with the transition to 3G expected between 2001 and 2005.

Relatively simple FH and FH direct-sequence spread spectrum (DSSS) hybrids are also employed throughout the VHF and UHF bands for voice privacy (e.g., by law enforcement organizations).

## C. UHF Services and Products

Service band allocations [5] include mobile cellular (450, 850, 900, 1900, 2400 MHz), fixed-satellite communications, maritime satellites (e.g., 1535 MHz), and aeronautical mobile-satellite communications.

Antenna products for UHF include log-periodic, directional parabolic reflectors, discones, and array antennas. Due to the reduced wavelengths at UHF, the physical size of the antennas is more compatible with avionics and mast-mounted applications. MCR base stations, in particular, employ arrays of relatively high-gain elements to provide the gain necessary to operate with low-gain antennas and low-power mobile handsets. A sectorized cell site, for example, might employ an array of three 8 ft high (by 1 ft wide) antenna elements. These would be arranged in a triangle 30 ft on a side to provide 5 to 8 dB gain over isotropic (dBi), with diversity reception. The handset, on the other hand, might use a helical whip with less than zero dBi gain.

Illustrative military systems in this band include digital multichannel radios such as the GRC-103 [128] and the FHM9104 [129]. The AN/GRC-103 (V) radio relay operates from 220 MHz to 1.85 GHz. It delivers 15–30 watts of power and supports 24-channel PCM, 63-channel delta modulation, and 4 to 60-channel FM/FDM for legacy radio compatibility. This unit weighs 31 kg, not including antenna and cables. The FHM9104 Digital Radio Link Terminal operates in two bands from 600 to 960 MHz and from 1.35 to 2.1 GHz. It dissipates .5 W when multiplexing 10 channels and weighs 45 kg.

## D. UHF SDR

Table 3-3 shows software radio applications parameters derived from propagation and air interface mode considerations.

**TABLE 3-3   Software Radio Applications Parameters—UHF**

| Software Radio Application | Sampling Rate ($f_s$) | Dynamic Range (dB) |
| --- | --- | --- |
| UHF-SHF FDM | .1–25 MHz | 48–96 |
| Cellular Radio | .2–75 MHz | 48–90 |
| UHF Air Nav | 2–25 MHz | 48–90 |
| UHF RF | 5.4–10 GHz | 48–90 |

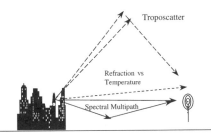

Troposcatter

Refraction vs Temperature

Spectral Multipath

**Frequency Band:  3000–30,000 MHz**
**Wavelengths: .1 m–0.01 m**
**Propagation Modes:**
   "Microwave" Begins at ~ 1 GHz
   Requires Direct or Directionally Scattered Beam
Multipath Delay Spread: .5–1.5 µsec
Multipath Doppler from low-flying aircraft
   may exceed 1 kHz at 15 GHz

**Communications Modes :**
   High Capacity Microwave
      FM/FDM, PPM, PWM
      PCM/ QAM
         1.544, 2.048 ... 45, 90, 155, 622 Mbps
   Troposcatter
      2 and 4.5 GHz bands
      Analog and Digital
   Spread Spectrum
      Satellite CDMA, FH-DS Hybrids

©1995-99 Mitola's STATISfaction reprinted with permission

**Services & Systems:**
   Service Band Allocations
      Radionavigation , Fixed Satellite, Maritime, MetSat,
      Fixed/Mobile Satellite (X-Band), Fixed,  Intersatellite
   Antennas
      Horn, Parabolic Reflector ("Dish")
   Illustrative Systems
      Alcatel TFH950S Digital Troposcatter; 1.7–2.1 & 4.4–5 GHz
      10–1500 W; 2048 kbps; 2 DPSK; 37 kg
      Siemens FM15000; 15.1–15.3 GHz; 100 mW;
      256–1024 kbps; ~10 kg.
      NorTel, AT&T, BT, etc.. LOS Microwave (Developing
      Countries Marketplace)

**Figure 3-11**   SHF communications.

Since cellular SDR is a focus of the subsequent chapters, the discussion of SDR complexity and algorithms is deferred to the software design trade-offs. Because of the need for propagation prediction in mobile wireless, the software-radio architecture encompasses this area.

## VIII.  SHF BAND COMMUNICATIONS MODES

Although SHF, the microwave band, begins at 3 GHz with wavelengths of less than a tenth of a meter, there is a transitional zone between 1 and 3 GHz in which LOS microwave characteristics are present (Figure 3-11). These include the transition of noise from externally generated interference to thermal noise. Since thermal noise is accurately modeled by an additive white Gaussian noise (AWGN) process, thermal or kTB noise is sometimes called AWGN. One should be clear, however, that thermal noise is the physical process, while AWGN is a mathematical abstraction that is not identical to the

TABLE 3-4   Doppler Shift

| Application | F | Reflector | Velocity | Doppler Shift |
|---|---|---|---|---|
| HF | 10 MHz | Ionosphere | 2,000 mph | 3.25 Hz |
| SHF Ground | 1 GHz | Aircraft | 500 mph | 81.4 Hz |
| EHF Ground | 15 GHz | Aircraft | 500 mph | 1.22 kHz |
| EHF LEO | 21 GHz | Satellite | 10,000 mph | 34.2 kHz |

physical process. The ratio of signal to thermal noise defined in the first stage of amplification in the receiver, the low noise amplifier (LNA), usually establishes system sensitivity.

## A. SHF Propagation

Microwave propagation begins at about 1 GHz. Propagation is accurately characterized by LOS ray-tracing algorithms. Consequently, the microwave bands are also called the LOS bands. Propagation effects include Doppler-shifted multipath similar to UHF multipath, but with fewer significant reflections. If one considers the number of reflections that account for 99% of the received signal strength, often there will be > 100 such reflections at the low end of UHF and fewer than 10 at the high end of SHF. Although the specifics are geometry dependent, the trend is notable. For one thing, main beam of SHF radiation is often a narrow pencil beam, reducing opportunities for multipath generation. Sidelobe control is more difficult at UHF, leading to the generation of spurious energy that may be reflected in multipath. In addition, the physical dimensions of buildings and metallic structures are compatible with UHF wavelengths, efficiently reflecting energy as multipath. Finally, SHF energy is absorbed better by the atmosphere, reducing reflections from distant objects. At SHF multipath delay spread ranges from 0.5 to 1.5 microseconds, reduced because there are fewer viable distant reflectors than in the low bands. The radio waves also have less ability to fill in shadowed areas; Fresnel zones have deeper nulls.

Because of the geometric LOS nature of SHF and the ability to build smaller electronic components at higher frequencies, there are many military and civilian radar bands in SHF. Cellular, PCS, and wireless local area networks also are beginning to win frequency allocations in what had historically been the radar bands (e.g., the 5 GHz ISM bands).

## B. Doppler Shift

Doppler shift (e.g., from low-flying aircraft) may exceed 1 kHz at 15 GHz. If a reflecting object moves away with velocity $v$, the distance is increasing, so the wavelength appears to be stretched. Table 3-4 shows the extent of Doppler shift for representative scenarios.

At HF, the plasmas in the ionosphere have an apparent velocity of 2 to 3000 mph or more, inducing Doppler shifts of up to 5 Hz. Ground-based SHF microwave links suffer multipath Doppler with shifts of up to 100 Hz from low-flying aircraft. The Doppler is proportional to the cosine of the angle between the flight path and the microwave LOS. Short-range EHF links experience Doppler of over 1 kHz. In addition, the total Doppler between two moving platforms will be double that shown in Table 3-4. The Doppler shift between a ground-based receiver and a low earth orbit (LEO) satellite operating at EHF decreases from +30 kHz to zero when the satellite is overhead. It then continues to decrease to −30 kHz as the satellite recedes.

## C. SHF Air Interface Modes

Terrestrial SHF air interface modes include high-capacity microwave, troposcatter, and spread-spectrum communications. Also many satellite communications use SHF, as addressed in a subsequent section.

*1. High-Capacity Microwave* Point-to-point microwave radio was initially developed for high-capacity backbone links of the PSTN. In the mid-1980s, digital microwave dominated this market. This technology has generally been superseded by fiber optic links in developed economies. High-capacity microwave retains niche applications in developing economies (where it is cheaper to install surplus microwave equipment than fiber equipment). It is also useful in rough terrain (where it may be impractical to site fiber), and in special terrain (e.g., national parks). Applications requiring mobile infrastructure also remain, of which the military market is probably the largest niche. These radios may also be employed to protect primary fiber links via standby operation. Frequencies formerly used for high-capacity digital microwave are now employed for cable television (CATV) distribution and for microwave backhaul. Backhaul operates between cellular/PCS BTS and the BSC or MTSO. To keep costs low, CATV and backhaul may use analog FM/FDM or commodity digital radios (e.g., T- or E-carrier).

The high-capacity microwave air interface includes legacy analog modes such as FM/FDM, pulse position modulation (PPM), and pulse width modulation (PWM). Newer radios use PCM modulations including BPSK, QPSK, and QAM. QAM and partial response [20] pack more than 1 bit per second in each Hz of allocated bandwidth. QAM uses the many amplitude-phase combinations available through high SNR. Data rates range from 128 kbps for military radios through the OC-12 level of the Synchronous Digital Hierarchy (SDH) illustrated in Figure 3-12.

Packing such high data rates into 20 or 30 MHz frequency allocations requires complex multicarrier-QAM hybrid air interface formats like $4 \times 256$ QAM, four multicarrier signals, each of which carries 256 QAM. This signal requires 40 dB SNR with equalization, FEC, bit interleaving, and randomization. Contemporary receiver products deliver many equivalent GFLOPS of

| U.S., Japanese and European Multiplex Hierarchies | LEVEL | | | |
|---|---|---|---|---|
| | 1 | 2 | 3 | 4 |
| **U.S.** (Bell/T-/DS- Level) | | | | |
| Number of Tributaries | – | 4 | 7 | 6 |
| Number of Voice Channels | 24 | 96 | 762 | 4032 |
| Line Rate (Mbit/sec) | 1.544 | 6.312 | 44.736 | 274.176 |
| Designation | T1/DS-1 | DS-2 | DS-3 | DS-4 |
| **Japan** | | | | |
| Number of Tributaries | – | 4 | 5 | 3 |
| Number of Voice Channels | 24 | 96 | 480 | 1440 |
| Line Rate (Mbit/sec) | 1.544 | 6.312 | 32.064 | 97.728 |
| **European** (CEPT) | | | | |
| Number of Tributaries | – | 4 | 4 | 4 |
| Number of Voice Channels | 30 | 120 | 480 | 1920 |
| Line Rate (Mbit/sec) | 2.048 | 8.448 | 34.368 | 139.264 |
| **Synchronous Digital Hierarchy (SDH)*** | | | | |
| Number of Tributaries | OC-1 | OC-3 | OC-12 | OC-48 |
| Line Rate (Mbit/sec) | 51.84 | 155.52 | 622.080 | 2448.32 |

*ITU-T Recommendations for the Integrated Services Digital Network (ISDN); OC= Optical Communications

**Figure 3-12**   Data rates of the Synchronous Digital Hierarchy.

signal processing, implemented typically in digitally controlled analog–digital hybrid signal processors.

*2. Troposcatter*   As illustrated in Figure 3-11, troposcatter is a long-range transmission mode in which transmitter and receiver are not within LOS of each other. Each radio has LOS access to a point in the troposphere from which radio waves are scattered beyond LOS. Due to the weak coupling between the radio waves and the troposphere, the radios employ very large apertures (e.g., 10-meter diameter dish) with kilowatts of power. Diversity reception is typically mandated, requiring that such large antennas are configured in pairs. Effective isotropic radiated power (EIRP) of 90 dBm may be required to provide sufficient SNR for multichannel relay. The military has been a primary developer and user of troposcatter. A typical use connects headquarters to clusters of geographically dispersed units via trunk circuits. Troposcatter's advantages over satellite communications includes lower acquisition cost and the ability to operate at extreme northern and southern latitudes where satellite communications may be ineffective.

Troposcatter links operate in the 2 and 4.5 GHz bands shown in the table in Figure 3-11. Air interface modes include FM/FDM, PPM, and PWM legacy waveforms as well as PCM. Received SNR generally supports small signal constellations such as BBSK and QPSK but not large constellations such as higher-order QAM.

*3. Microwave Spread Spectrum*   The most widely known military spread-spectrum air interface may be the Joint Tactical Information Distribution System (JTIDS). This mode occupies approximately 250 MHz with a carrier frequency of between 1 and 2 GHz. Reference [20], for example, contains an appendix dedicated to JTIDS. JTIDS employs a 32-chip, direct-sequence spread-spectrum (DSSS) pseudorandom spreading sequence on each data bit. The instantaneous DSSS bandwidth is 3 MHz. These chip bursts are hopped across a 250 MHz agility band at a hop rate exceeding 1000 hops per second. The SPEAKeasy II program considered implementing JTIDS in software radio technology, recommending an architecture in which the waveform is first dehopped to a 3 MHz IF using digitally controlled analog fast-tuning synthesizer and then processed digitally. This is not the ideal software radio, but represents a step toward that objective. JTIDS was not implemented in the SPEAKeasy I or II programs, but is part of the JTRS program.

CDMA is also employed in SHF. Instantaneous CDMA bandwidths of tens of MHz are practical at SHF. Wireless LANs, for example, use 50 MHz CDMA because of its ability to overcome multipath and provide asynchronous multiple access in a single-receiver architecture. Satellite communications also use CDMA with FH-DS hybrids for error performance and transmission privacy.

### D. SHF Services and Products

Service band allocations [5] have been established for radio-navigation, fixed-satellite applications, maritime mobile users, meteorological satellites, fixed- and mobile-satellite communications, fixed terrestrial microwave point-to-point links, and intersatellite communications. Antennas include pyramidal horns and parabolic reflectors. The Rothman lens is a popular 2 GHz terrestrial microwave antenna. The 6 and 11 GHz bands typically employ parabolic reflectors.

Illustrative radio products include the Alcatel TFH950S Digital Troposcatter system [130]. This system operates in the bands from 1.7 to 2.1 GHz and from 4.4 to 5 GHz. It radiates from 10 to 1500 watts of power with data rates to 2048 kbps using 2-state differential PSK. This unit weighs only 37 kg, not including antenna [121, p. 181]. The Siemens FM15000 is a typical military digital-microwave radio. It operates in the 15 GHz band from 15.1 to 15.3 GHz, radiating 100 mW of power into a high-gain antenna (typically a dish). It supports data rates from 256 to 1024 kbps and the unit weighs approximately 10 kg. NorTel, AT&T (Western Electric), BT, NTT, Siemens, Alcatel, and most other major telephone companies have developed point-to-point LOS microwave products or systems.

### E. SHF SDR

Current CDMA receivers are generally implemented in ASICs. Low-power ASICs are essential for handheld terminals. The resulting availability of such components favors their use in dedicated vehicular and avionics nodes as

**TABLE 3-5   SDR Applications Parameters—SHF**

| Software Radio Application | Sampling Rate ($f_s$) | Dynamic Range (dB) |
|---|---|---|
| SHF QAM | 12–100 MHz | 30–72 |
| SHF CDMA | 125–500 MHz | 60–90 |
| SHF Agility | .5–3 GHz | 48–72 |
| SHF-IF | 1–8 GHz | 48–90 |

well. Chip rates of up to a few MHz, however, can be implemented in IF-level software radios. CDMA is a practical mode for an IF-SDR in which the chip rate is less than 2 Mbps, the band allocations is less than 25 MHz, and the near–far ratio is less than 70 dB. This keeps the ADC bandwidth and dynamic range within the state of the art.

Table 3-5 shows typical software radio applications parameters. Thermal noise, limited cochannel interference, and high-SNR air interface modes render these parameters less demanding than those of UHF.

CFAR squelch algorithms designed for AWGN channels are acceptable for RF squelch in the microwave bands. Because of the reduced multipath at SHF compared with the lower bands, receiver complexity may be reduced or data rate may be increased for a given level of receiver complexity. For many deployments in the 11 GHz bands, the geometry admits only the primary path. If there is one reflection from the water or land between transmitter and reflector, the multipath must be equalized, establishing a minimum level of receiver complexity.

In addition, the LOS nature of SHF forces one to point beams accurately in order to close the link. With transmit and receive antenna gains of 30 dB or more and beamwidths of less than 1 degree, one suffers rapid accumulation of pointing loss with small pointing errors. The initial geometric alignment of fixed-installation antennas may be expensive. Alternately, mechanical gimbals or phased arrays that automatically point and track increase system costs. Both of these factors tend to diminish the economic value of SDR technology insertion. If, for example, 70% of the cost of the link is the cost of the six-degrees-of-freedom antenna pedestal, then the introduction of SDR technology benefits at most 30% of the total.

Doppler tracking may require special algorithms and timing, particularly in correcting signals from LEO satellites. A few Hz of Doppler shift in the HF through UHF bands is easily accommodated in the receiver's carrier tracking loop, while at SHF, greater Doppler affects the carrier recovery and equalization algorithms. Consequently, algorithms for carrier recovery, tracking, and equalization must compensate for the range of Doppler shift present. This tends to increase SDR algorithm complexity.

Troposcatter may be supported by SDR technology. The IF bandwidth of less than 20 MHz is readily supported by IF ADCs with dynamic range of 70 dB. Processing demands of troposcatter are modest since the legacy air interfaces such as FM/FDM have low complexity based on analog baseband technology. The economic leverage of SDR technology in a dedicated troposcatter product is minimal, however. The large size of the antenna and the high cost of RF power generation and handling components dwarf the cost of the DSP components. With digital approaches, however, it is possible to equalize the digital troposcatter air interface channel, enhancing received signal quality and/or extending the range of the system.

The point at which one may economically transition from dedicated hardware to SDR technology is a function of application. SHF spans the tradeoffs. Given a 20 MHz frequency allocation in SHF, it is relatively cost effective to implement a 128–256 kbps digital radio using 50 MHz ADCs and contemporary DSP technology. Thus, such a mode may readily be incorporated in a military SDR. On the other hand, a 622 Mbps data rate for the same 20 MHz channel probably requires numerous dedicated ASICs for RF, FEC, interleaving, multiplexing/demultiplexing, and other functions. Data rates between 1 and 10 Mbps are in the area where SDR techniques are emerging.

## IX. ATMOSPHERIC EFFECTS

The atmosphere attenuates electromagnetic radiation through two distinct mechanisms: gaseous attenuation (Figure 3-13a) and absorption due to precipitation (Figure 3-13b). Gaseous attenuation loss is modest up to 12 GHz. Water absorption peaks at 24 GHz, which makes it a candidate for low probability of intercept (LPI) data links. At a range of a few hundred meters and with a one-foot aperture, gain in the direction of the intended receiver will overcome gaseous absorption. At the longer standoff ranges of an unintended receiver, the gaseous absorption reduces sidelobe energy, rendering the link relatively covert. For longer ranges, the "water hole" of minimum absorption at 34 GHz has attracted many military multichannel products and commercial T1/E1 building-to-building data links.

Gaseous absorption increases at 60 and 190 GHz, where the attenuation increases from 0.2 dB/km ($H_2O$) to 15 dB/km for $O_2$, and over 40 dB/km in the 190 GHz water absorption line. Low $O_2$ absorption at 90 GHz has stimulated the development of multichannel communications, radar systems, commercial data links, etc. in this band.

In addition to the presence of gaseous water in the air due to humidity, there are periods of precipitation in most climates. The specific attenuation is a function of frequency, increasing with the intensity of the precipitation. Since precipitation intensities of 0.4 to 4 mm/hr represents a nominal rainy day in temperate regions, attenuation at frequencies below 5 GHz is nominal. In the tropics, however, precipitation ranges from 10 to 50 mm/hr. Typhoons and hurricanes approach precipitation intensities of 100 to 150 mm/hr.

There are few implications of atmospheric absorption for the SDR. The dynamic range requirements in the frequency bands above 5 GHz include the

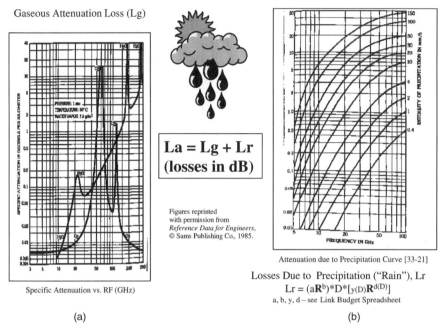

**Figure 3-13**   (a) Gaseous attenuation; (b) precipitation losses.

effects of precipitation. SHF and EHF allow much larger bandwidths than the low bands, increasing the bandwidth requirements of the ADC. Since thermal noise is stronger than interference, the near–far requirements are often less than 70 dB. Consequently, the high-band SDR needs a high-speed ADC with limited dynamic range. Low-band requirements drive the SDR toward the highest possible dynamic range, sacrificing bandwidth if necessary, while high-band requirements drive the SDR toward wider bandwidths. Engineering tradeoffs then result in two or three different types of ADCs for an SDR capable of operating in LVHF through SHF.

## X. EHF BAND COMMUNICATIONS MODES

EHF, the millimeter wave band, spans from 30 to 300 GHz as illustrated in Figure 3-14. EHF is an ideal band for short-haul wideband communications systems. EHF also has potential for satellite-to-satellite data links and for wideband satellite applications.

### A. EHF Propagation

EHF propagation consists primarily of skywave direct-LOS radiation. These signals are easily attenuated by birds, clouds, rain, snow, and ice. Japanese

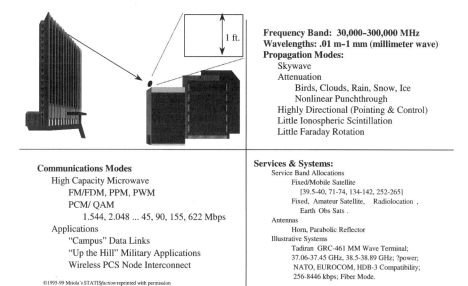

**Figure 3-14**  EHF communications.

researchers have reported a nonlinear punchthrough effect in which the attenuation does not continue to increase with precipitation intensity provided that the microwave power is sufficiently high [422]. EHF beams are directional with beamwidths of fractions of a degree for reasonable antenna apertures (e.g., 1 foot). The difficulty of pointing the beam accurately increases compared to SHF. Pointing accuracy of milliradians may be required. Antenna pointing update rates of tens to hundreds of Hz may be necessary to sustain high-data-rate communications between moving vehicles. Even stationary terminals such as buildings typically must employ closed-loop antenna-pointing control in order to overcome the changes in pointing angles induced by thermal distortion of buildings during diurnal cycles.

## B.  EHF Air Interface Modes

EHF readily supports high-capacity microwave air interface modes including FM/FDM, PPM, and PWM. Contemporary PCM and QAM formats are also supported at data rates from T1/E1 (1.544 and 2.048 Mbps) to T3 (45 and 90 Mbps), and OC-3 (155 Mbps) to OC-12 (622 Mbps). Applications of these air interfaces include "campus" data links that range from 100 m to 1.5 km. The military uses such links for "up the hill" communications from low-lying command centers to radio relay points. They are also used between elevated radio-relay points over a few km ranges. Currently, there is considerable growth in the use of EHF for cellular and PCS BTS-BSC connections.

## C. EHF Services and Products

Service band allocations [5] include several millimeter wave bands [39.5–40, 71–74, 134–142, and 252–265 GHz]. Services include fixed- and mobile-satellite services, fixed terrestrial point-to-point data links, amateur satellites, radio-based location finding, and earth observation satellites. Antennas employed in these bands include horns and parabolic reflectors.

Illustrative systems include the Tadiran GRC-461 [131] millimeter wave terminal operating between 37.06 and 37.45 GHz and from 38.5 to 38.89 GHz. This radio supports NATO, EUROCOM, HDB-3 compatibility with data rates from 256 to 8446 kbps. It also has direct-connect fiber-interoperability modes.

## D. EHF SDR

EHF SDR considerations are similar to those of SHF. The 4G programs may introduce 155 Mbps indoor wireless LANs. These may use phased-array antennas that point beams at desktop computers or mobile PDAs. SDR technology insertion opportunities in these bands may include the incorporation of a 4G LAN into an SDR-based PDA.

## XI. SATELLITE COMMUNICATIONS MODES

Communications satellites [132] operate in the three orbital regimes illustrated in Figure 3-15. Geosynchronous satellites have an orbital period which is nearly identical to the earth's rotational period, resulting in an apparent stationary position above the equator at an altitude of 22,500 miles, approximately. Station keeping and minor orbital imperfections sometimes result in a small bow-tie trace when the satellite's apparent position above earth is plotted in a fixed earthcentric coordinate system. Geosynchronous satellites' tangential visibility ranges between the arctic circles in an oblate elliptical footprint about 8000 miles across. Practical use of such satellites requires the satellite to be positioned a few degrees above the horizon, further limiting the useful footprint below the geometric limits. Such large footprints are realizable with antennas of limited directionality, such as amateur UHF satellites. Many geosynchronous satellites support domestic satellite services, generally through directional beams that limit the radio footprint to an appropriate domestic coverage footprint. One limitation of domestic satellites is the 0.264-second round trip delay between the satellite and the earth. This time delay can be doubled, tripled, or quadrupled for multiple satellite relays. Voice service is problematic with such long time delays since speakers must consciously pause in order to avoid talking over the other participant in the conversation.

Low earth orbit (LEO) satellites orbit at altitudes between 150 and 1500 miles, reducing the round-trip time delay to less than 17 ms. Such satellites

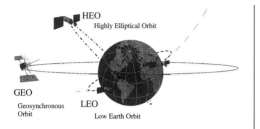

Frequency Bands: **UHF, SHF, EHF**
**Propagation Modes:**
   Geosynchronous (Low Doppler)
   Highly Elliptical (Moderate Doppler)
   Low Earth Orbit (High Doppler)
**Propagation Phenomena:**
   Faraday Rotation
   Ionospheric Scintillation
   Ground Station Sidelobe Control
   Mobile Satellite Rain Fades

**Communications Modes** :
   Heavy Earth Terminal (Symmetrical)
      DOMSAT: Multichannel, Cable Protection
   Mobile Satellite Terminal
      INMARSAT: Voice, LBR Data
   Very Small Aperture Terminal (VSAT)
      Low Data Rate Asymmetrical Terminals
   Direct Broadcast Satellite (Heavy/VSAT)
      Wideband Asymmetrical Terminals
   Spread Spectrum
      Military (SHF, EHF), LEO Proposals

**Services & Systems:**
   Service Band Allocations (GHz)

| | Uplink | Downlink |
|---|---|---|
| UHF | 0.13-0.16 | 0.13-0.16 |
| C Band: | 5.925-6.425 | 3.7-4.2 |
| X Band | 8-12 | 8-12 |
| Ku | 14-14.5 | 11.7-12.2 |
| Ka | 27.5-31 | 17.2-21.2 |

   Illustrative Systems
      Geosynchronous: DOMSATs
      HEO: Molnyia: Trans.-Siberian
      LEO: Iridium

©1995-99 Mitola's STATIS*faction* reprinted with permission

**Figure 3-15**   Satellite communications.

typically have 90-minute orbits with about 10 minutes of visibility above the horizon on each satellite pass. Motorola's Iridium [133, 134] satellite system provided LEO voice telephony and data services with global coverage through a LEO constellation of 66 satellites plus on-orbit spares. Design parameters of such satellites include the number of satellites, the time between service delays waiting for satellite visibility, or the number and duration of satellite outages per day. Such constellations provide good coverage near the equator at the expense of more limited coverage (longer and more frequent outages) at higher latitudes. In addition, continuous coverage at higher latitudes requires a full (and thus relatively expensive) constellation.

Highly elliptical orbits (HEO) provide sustained coverage at higher latitudes with only a few (typically three) satellites: one rising toward apogee, one setting from apogee, and the other out of visibility in perigee. The apogee of such a modest constellation may be placed near the arctic circle to provide high-continuity service to high latitudes such as Siberia as is the case for the Russian Molnyia satellite system.

## A. Satellite Propagation

Satellite frequency bands are traditionally given in terms of lettered band designations [5]. Those used in this text are listed in Figure 3-15. Geosynchronous satellites experience relatively low Doppler shift while highly elliptical satellites near apogee exhibit moderate Doppler of from a nominal few hundred Hz near apogee to upward of 20 kHz descending toward perigee. A C-band HEO satellite, for example, imparts 325 Hz of Doppler shift on a 2 GHz carrier for a nominal 1000 Mph satellite velocity normal to radio LOS from the ground station (radial velocity). The same HEO spacecraft operating at 21 GHz imparts a 17.1 kHz Doppler shift on the carrier for an extreme radial velocity of 5000 Mph. Of course, the radial velocity is zero exactly at apogee for a brief instant while it approaches 17000 Mph at perigee where it is generally not used for communications. Low earth orbit satellites, on the other hand, impart high positive Doppler shifts as the satellite rises above the radio horizon. It rapidly transitions to zero Doppler when directly overhead and then transitions to high negative Doppler shifts just prior to setting on the opposite horizon. Iridium's 785 mi orbit imparts approximately 600 Hz of Doppler on its 3 GHz carrier, resulting in a carrier-tracking requirement of over 1200 Hz, or about 2 Hz/second.

Propagation between satellites and ground terminals must contend with Faraday rotation, ionospheric scintillation, ground station sidelobe control, and mobile satellite rain fades. Faraday rotation is the distortion of electromagnetic wave polarization as the wave transits the plasmas in the ionosphere, resulting in elliptical polarization at the receiver, with the attendant loss of received signal strength with respect to circularly polarized receive antennas. Polarization diversity has limited effectiveness in the downlink. It increases satellite size and weight in the uplink. Ionospheric scintillation is the equivalent of terrestrial multipath. Multiple paths through the ionosphere have path length differences of multiples of a wavelength with nearly equal amplitudes. This leads to alternative constructive and destructive interference that presents an erasure channel in which bursts of erasures occur as the satellite position changes with respect to the propagation paths.

Ground stations of geosynchronous satellite systems must also control spatial sidelobes so as to limit the amount of radiation delivered to typically less than 40 dB below the radiated power. Due to the increased number of positions in geosynchronous orbit, this requirement makes it necessary to employ larger ground station antennas than necessary to achieve the link margin in order to achieve the low sidelobe levels required by international treaty. Active sidelobe cancellation and other advanced techniques may provide additional requirements on the transmission segments of software radios for spectrally and spatially purer waveforms than would otherwise be the case.

During the past five years, the emphasis in satellite communications has been on increasing available bandwidth through the exploitation of Ku and Ka bands in which the higher carrier frequencies (compared with C and X band) support wider instantaneous bandwidths. DARPA and the U.S. National Aeronautics and Space Administration (NASA) co-sponsored the Advanced Communications Technology Satellite (ACTS), which demonstrated 622 Mbps connectivity from a low-cost geosynchronous satellite to a 10 meter parabolic aperture [423]. This geosynchronous satellite positioned above the Americas, provides Ka and Ku transponders. In addition, Japanese and European researchers have employed other satellites to study propagation phenomena.

Japanese researchers, for example, report a high degree of correlation among fade events at Ka and Ku bands, limiting the effectiveness of frequency diversity in overcoming many types of fading [135]. The June 1997 *Proceedings of the IEEE* was dedicated to this subject.

## B. Satellite Air Interface Modes

Domestic satellites (DOMSATs) found application in the 1960s as moderate-to high-capacity trunks for the PSTN. These may be called symmetric applications because transmitting and receiving ground stations typically have antennas that are nearly the same size (often tens of meters). The International Communications Satellite Corporation (INTELSAT) operates transatlantic and other transnational PSTN satellite links that have provided international gateways. This kind of satellite service transitioned from FM/FDM to T/E carrier air interfaces in the 1970s and 1980s. INTELSATs have provided 45 Mbps digital T-carrier service between the United States and Europe. The deployment of extensive OC-3 to OC-48 undersea fiber systems in the late 1980s and early 1990s severely curtailed the international satellite PSTN market. Some developing nations still employ DOMSATs in the PSTN. Compared to fiber, DOMSATs have poorer instantaneous signal quality and suffer from perceptible delays in voice applications. DOMSATs employ heavy earth terminals—parabolic reflectors with diameters of 30 to 100 feet. DOMSATs and international satellites also provide multichannel service to protect fiber and cable service through hot standbys. In addition, news agencies and the entertainment industry employ international satellites extensively for coverage of news and sports, and for the delivery of program content.

Mobile satellite services in wide use include the narrowband International Maritime Satellite (INMARSAT), which provides voice and low-bit-rate (LBR) data via an analog air interface and voice channel modems. Some terminals require a 3-foot Yagi directional antenna to be carefully aligned to the satellite for acceptable signal quality. INMARSAT continues to invest in its satellites and terminal equipment to reduce the size, weight, and power of the terminals.

Satellite services listed in Table 3-6 did not do well in 1999. Iridium and Globalstar, for example, both became bankrupt, and Teledesic was delayed. Iridium is the product of a consortium led by Motorola, for global mobile satellite service. Iridium does not work reliably inside buildings or in automobiles. At a cost of $3 to 7.00 U.S. per minute, this premium service was initially limited to high-end business and governmental users, but failed to attract subscribers.

The Orbcom satellite provides low-data-rate service including worldwide vehicle location services (e.g., for commercial trucking companies). Globalstar and Odyssey have different satellite architectures for providing future satellite telephony in competition with Iridium as outlined in Table 3-6. Odyssey is sometimes called a medium earth orbit (MEO) satellite because its 10,354 km

TABLE 3-6   Emerging Mobile Satellite Services

| Services | Globalstar Mobile Telephony and Data | Iridium Mobile Telephony and Data | Orbcomm Storeand Fwd Messaging | Teledesic High Rate Fixed Svc | Odyssey Mobile Telephony |
|---|---|---|---|---|---|
| Data Rate | 9.6 kbps | 2.4 kbps | 300 bps | 16k – E1 + 1.2G | N/A |
| Modulation | CDMA | TDMA | N/A | N/A [ATM] | CDMA |
| RF | 1/3 GHz | 1/3 GHz | 148/137 MHz | 30/20 GHz | 1.6/2.5 GHz |
| Satellites | 48 | 6 | 36 | 840 | 12 |
| Altitude (km) | 1400 | 785 | 775 | 700 | 10354 |
| Inclination | 52 | 86.4 | 45 | 98 (Sun) | 50 |
| On-board Proc? | No | Yes | No | Yes | No |
| Crosslinks | No | 4@25 Mbps | No | 8@155.52 | No |
| Mass (lbs) | 704 | 1100 | 85 | 747 | 4865 |
| Partners | | Motorola | | McCaw&Gates | TRW |
| On Orbit? | Yes | May 97 | Yes | No | No |

altitude reduces the number of satellites for global (equatorial) coverage to 12. Teledesic proposes to deliver T-carrier Internet access directly to the home. The impressive backers of this initiative include Bill Gates, CEO of Microsoft and McCaw, one of the first and most successful cellular telephone entrepreneurs. It seems doubtful that even an expanding mobile market can support all of these new satellite cellular systems. But the survivors will have a significant need for multimode-multiband mobile radio in the future.

Very small aperture terminal (VSAT) satellite systems now being deployed use large earth terminals at a central hub. The gain of the satellite and its higher power offsets user terminal aperture, reducing it to 0.5 to 1 meter. This asymmetrical arrangement has significantly expanded the market for satellite data services by making the subscriber earth station much easier to deploy and maintain than earlier generations of satellite systems. VSAT air interfaces are digital with burst data services supporting such applications as point-of-sales terminals and centralized bookkeeping for large multinational corporations such as McDonald's restaurants.

Direct-broadcast satellites (DBS) also employ asymmetrical apertures with the heavy aperture in a central hub and VSAT technology terminals on customer premises. The Hughes 601 satellites, for example, deliver over 50 channels of digital television to a 24-inch aperture terminal via a 22.5 MHz satellite transponder. Leased DBS capacity can deliver large amounts of data to thousands of subscribers within the satellite footprint.

In addition to the positive carrier-to-noise ratio (CNR) signals used by the commercial satellite services, spread-spectrum air interface modes are employed in military communications systems. Multiband-multimode military satellite communications (satcom) systems now in the marketplace typically include the U.S. C-band, X-band and K-bands. Such multiband, multimode radios fit the need for long-haul connectivity from military mobile subscriber equipment.

## C. Satellite Services and Products

Service band allocations from reference [3] include the C, Ku, and Ka band uplink and downlink frequencies shown in the figure at the beginning of this chapter. GEO and LEO satellite services and products were also discussed earlier. Military satcom applications include tactical satellite terminals [136].

In the past (with the notable exception of INMARSAT), the large earth terminal apertures, large facility costs, and high maintenance requirements relegated satellite services to governmental users such as DOMSATs and fixed-facility military command and control systems. But the new VSAT and DBS technologies are delivering more bandwidth in a more manageable footprint than ever before, opening up new applications and services that lead to new products. If the mobile satellite successors to Iridium and Orbcom come on-line and expand, each of these modes is a candidate for software radio implementations.

## D. Satcom SDR

Since the cost of DOMSATs is dominated by the large antenna and related facilities cost, SDR technology will not have a first-order impact. It may have the second-order effect of allowing the radio terminal equipment to transition to open architecture (e.g., VME or PCI bus, CORBA software, etc.). But it is hard to imagine the DOMSAT systems in a multiband-multimode application.

Carrier acquisition and demodulation algorithms in the SDR must compensate for satcom propagation extremes. Parameters vary as a function of large, rapidly changing Doppler shifts, reflections from ground clutter, Faraday rotation ionospheric scintillation, and rain fades, in addition to other less significant artifacts of the propagation channel. Bit interleaving, equalization, and trellis coding of the waveform all mitigate propagation effects at the cost of increased receiver complexity. Implications for the software radio include the necessity of maintaining sufficient throughput to accomplish these computationally intensive functions.

## XII. MULTIBAND MULTIMODE SUMMARY

As illustrated in Figure 3-16, each band and mode has advantages and disadvantages. Some bands deliver high data rates appropriate to multimedia needs, but only over short ranges. Other bands bridge long distances but with either lower data rates or higher infrastructure costs. Thus, no single band or mode delivers the capability required.

| Band | Modes | Key Technical Characteristics That Shape Software Radios |
|------|-------|-------------------------------------------------|
| HF | Skywave, NVI | Long range, reliable in mountains/jungles, cheap |
| | FH, Burst | Narrow BW, severe propagation, big antennas |
| LVHF | Beyond LOS | Low cost, general purpose voice, data & relay |
| | FH, Burst | Rayleigh fading, Fresnel zones, interference |
| VHF | Quasi-LOS | Larger coherence bandwidths, cheap |
| UHF | Cellular | Spectrum auctions opening applications |
| | TDMA, CDMA | Spectrum crowding, beamforming, data rate |
| SHF | LOS | Ideal for space communications, large BWs |
| | TDMA | Atmospheric & rain losses significant |
| EHF | Narrow Beams | In-building, campus data links, spatial sharing |
| | Fiber Protect | Very short range, but practical gigabit BWs |

> **No single band or mode delivers reliable, long-haul, high-data-rate, cheap, convenient (unlicensed) service for mobile users.**

**Figure 3-16**   Reliable flexible communications calls for multiband, multimode radios.

The balance of this book, then, describes a set of techniques through which the diverse capabilities of these bands and modes may be integrated through SDR technology and software-radio architecture. The goal is to employ multiple bands and modes seamlessly and efficiently to deliver the services needed for the rapidly growing mobile telecommunications marketplaces.

## XIII. EXERCISES

1. List the primary clusters of band-mode combinations in RF signal space. How can the existence of such clusters inform software-radio architecture? What significant modes are you aware of that the clusters ignore, minimize, or misrepresent? What are the implications for SDR design of those modes that are not addressed by the clusters?

2. Select a band with which you are familiar. What are the propagation phenomena that determine software content in an SDR that encompasses that entire band (e.g., HF, LVHF, etc.)? What modes are used in that band? What specialized modes may account for niche markets? What conventional product would constitute competition for an SDR product for that band? What discriminators should be affordably implementable in the SDR product that an analog radio or PDR would not be able to offer (or offer as affordably)?

3. Select a band adjacent in RF to the band above or below the band in question 2. Answer question 2 for this band. In addition, what benefits

accrue to an SDR product that incorporates capabilities from both bands? Think of a mobile-vehicle applications niche (e.g., interstate trucking, taxi fleets). What additional benefits accrue to the SDR if computer capabilities (e.g., word processing, database, Internet access) that are computationally feasible in the SDR are made available to the end user?

**4.** What are the limits that propagation imposes on a multiband-multimode SDR?

**5.** What design rules should be derived for software-radio architecture from the legacy communications modes in HF? LVHF? the cellular bands?

**6.** What parametric boundaries or constraints tend to partition SDR products? RF band? channel modulation? data rate? Doppler spread? Justify your choices.

# 4 Systems-Level Architecture Analysis

The objective of this chapter is to give the reader practice in addressing software-radio architecture issues at the systems level. The study of systems-level software-radio architecture is first motivated with a realistic case study. The case study includes the critical parameters of most radio architectures. The analysis focuses on those aspects that are significant for software-radio architecture. The balance of the chapter develops the issues raised in the case study.

## I. DISASTER-RELIEF CASE STUDY

This case study considers a mobile communications capability for disaster relief. The capability includes mobile infrastructure, mobile nodes, and handsets. The design emphasis is on defining an open architecture for the infrastructure. Architecture defines components at such a high level of abstraction that one needs a concrete sequence of specific implementations[20] in order to assess the contributions of the architecture. Architecture insight seems to develop with implementation practice. It seems to take a half-dozen design and implementation cycles to develop the intuition necessary to make strong contributions to architecture. This case study therefore should be designed and redesigned by the serious student as the text progresses.

### A. Scenario

The case study addresses the fact that medium-sized urban areas may be decimated by a natural disaster. As illustrated in Figure 4-1, the disaster area may be largely obliterated. The destruction of the Holmstead area in South Florida by hurricane Andrew is a practical example of such a disaster. The populace has enjoyed the use of cellular telephone, but the disaster is assumed to have wiped out the wireless network. At the periphery of the disaster area, connections are available via fiber and/or microwave to the core telecommunications network.

---

[20]To address future implementations, one must often substitute a sequence of designs for the "sequence of implementations" that have not yet been built.

**Figure 4-1** Disaster-relief scenario.

Two software radio problems arise. The first is the design of an SDR product that will meet the need given current technology. The second and more important problem is to define a software-radio architecture within which a family of backwards-compatible SDR products may evolve. This architecture should meet the designer's need for product differentiation and protection of intellectual property. But it also has to entice the rest of industry to participate. The product supplier's first goal in industry participation is to establish product leadership. This includes motivating potential hardware and software suppliers to support the architecture. It must meet customer needs for affordable upgrade paths.

To motivate the design of a radio system, assume that an appropriate national authority has decided that it would like to acquire a capability to rapidly reconstitute communications in such disasters in the future. Sample customers include the U.S. Federal Emergency Management Agency (FEMA), the European Community (EC), and the government of China or Japan. In order to obtain support from these national-scale authorities, a disaster must be of major proportions. Consequently, numerous local, state, and federal institutions converge on the disaster area to look for survivors, set up temporary shelters, prevent crime, and reconstitute the necessities of life. To motivate those who are oriented toward the military sector, mobile infrastructure is the essence of tactical military communications. The exercises explore the possibility of communicating while on the move. Although not strictly a need of the disaster-relief application, communications while infrastructure is moving is a simple extension of the case study. To motivate those who are oriented toward the commercial sector, consider rapid build-out of a developing nation like Thailand of a few years ago.

**TABLE 4-1   Disaster-Relief System Communications Needs**

| Needs Questions | Illustrative Answers |
|---|---|
| Physical Area? | 3–5 local areas of 2–10 km radius each |
| Classes of Subscriber? | Police, fire, rescue, local populace, National Guard |
| Numbers of Subscribers? | 10–20 local and/or national police agencies<br>20–100 fire and rescue squads (10 helicopters)<br>50,000 local populace (including 20 light aircraft pilots)<br>500–3000 National Guard troops with 20–50 aircraft |
| Information Services? | Core: voice, e-mail, tasking/scheduling, databases, fax<br>Growth: video-teleconferencing, telemedicine |
| External Interfaces? | Network: T/E-1 to T/E-3 SDH (microwave, fiber), SS7 |
| Cost? Price? | "A few million dollars" |

To motivate the analysis of architecture, assume that the customer has decided that conventional approaches are too expensive, both in terms of initial acquisition cost and in terms of life-cycle support. The buyers therefore want open-architecture software radio or SDR. They also request concrete evidence that the expected advantages of SDR architecture will be realized in their system.

## B. Needs Analysis

Needs analysis establishes the intuitive relationships among radio system functions, components, design rules, and costs. Systems-level communications needs for a disaster-relief system are summarized in Table 4-1.

The answers to the needs questions define the top-level requirements of the system. Physical area and numbers of subscribers are first-order determinants of the technical needs of wireless infrastructure. There should be design latitude about how many infrastructure nodes are provided. This buyer has specified the physical size and overall communications capability. The fundamental measure of voice traffic is the Erlang [137]. An Erlang is the international unit of traffic intensity that represents an average of one circuit busy out of a group of circuits. Wireless infrastructure provides capacity in Erlangs per square km, at a given Grade of Service (GoS) and Quality of Service (QoS). In this case, there are four major classes of subscriber. Each class brings its own indigenous vehicular and handheld radios and wireless PDAs. These radios establish radio bands and modes that must be supported by the disaster-relief infrastructure. In addition, those people who are providing the communications services will also need local communications. Call these the organization-and-control (OC) users.

Needs analysis examines the general scenario by generating a variety of use-cases. The existence of the OC users as an additional class of users is derived by examining use-cases, detailed vignettes that force one to think about significant details of the application. The analysis of use-cases may be accomplished effectively with few software tools. One might use a database system to record details of entities participating in the scenario. One might use a geospatial information system (GIS) to visualize the distribution of the entities. A spreadsheet tool (e.g., Excel) can perform parametric analysis. A discrete event simulation can characterize queuing delays of message traffic needed to support the e-mail, scheduling, and database services (e.g., OPnet). In addition, UML simplifies some aspects of use-case analysis. UML's use-case view keeps track of external and internal actors and kind of forces one to push through the sometimes-tedious details of a use-case.

The needs analysis for an SDR-based product attempts to limit the needs so that the complexity of the SDR software is minimized. This is because typically over half of the cost of developing an initial SDR product is in the software. To limit the needs is to limit the software complexity. The needs analysis for a software-radio architecture, on the other hand, attempts to define the limits to which the needs could grow in the foreseeable future. This is because architecture is oriented toward providing a growth path, while product design is oriented toward short-term profitability. When customers say they are interested in reaping the benefits of open architecture, they generally have some short-term goal in mind. Some can take a longer-term view, but a course of action that has long-term impact often consists of a sequence of short-term success stories.

The U.S. DoD expresses needs as *requirements*. Through a formalized process, military organizations express, coordinate, and validate their needs. They attempt to prune the needs to the minimum that is operationally acceptable; these are the requirements. In the modernization of the procurement process, the DoD has begun to express requirements in terms of a minimal set (*threshold* requirements), plus a prioritized set of additional needs. There are now laws that encourage the U.S. military departments to acquire products and services more like commercial organizations. Thus, some parts of the DoD acquire commercial communications products, and negotiate warranties in lieu of conformance to military specifications (MIL-SPECs). This evolution drives requirements toward general statements of need as suggested in Table 4-1. In addition, however, military users are continuously striving to balance actual needs (regardless of what the formal requirements specify) against affordability. Thus, as capabilities become affordable, the formal requirements finally embrace what could be recognized as needs all along. Focusing software-radio architecture on needs insulates medium- and long-term architecture evolution from the shorter-term push and pull of the formal requirements process.

The requirements are rarely defined as precisely as a systems designer might like. Consider the cost goal of a few million dollars, for example. The notional buyers of the system are the service providers. They have a top-down sense of the value of the capability. Beyond that, they have to justify budgets based, for example, on cost estimates from industry. The definition of cost, therefore, is an iterative process between the buyer who sets the value and the developers who characterize price as a function of capability. One generally must be satisfied with a rough-order-of-magnitude (ROM) cost goal. Low cost can be a market differentiator. Another competitor might offer a feature-rich product, or one that is more reliable, that costs more. Yet another competitor might offer a product that is compatible with the customer's installed base, or that makes it easier to expand. Any of these approaches can change the cost by 20 to 50% or more. It is therefore essential to adopt a business strategy that can focus on both the short-term SDR design and the participants' goals for long-term architecture evolution.

## C. Exercises

1. What radio bands and modes are implicit in the identification of classes of user? What ambiguities must be resolved before a meaningful design could begin? If discrete radios are packaged with one band/mode per unit, how many units are needed at a base station? If you cannot write an equation for this, what additional assumptions are needed? Make those assumptions and write an equation for the number of units at a base station.

2. Assume SDR units are packaged by RF band. That is, there may be an HF SDR unit covering the band from 2 to 30 MHz, a LVHF SDR unit (30–88 MHz), a VHF aeronautical SDR (100–225 MHz), etc. What is the upper frequency limit of the SDR family for the disaster-relief application? Assume that all modes within a band are defined in baseband software. How many bands must be supported? Which bands could be packaged into a contemporary SDR? Which COTS products might provide the RF coverage needed for such a multiband SDR?

3. Suppose now that you want to define a software-radio architecture that will accommodate an evolution path from the answer to question 2. What are the architecture implications of consolidating multiple RF bands into a single wideband RF? Think of the consolidation of RF bands over time as a design rule for the architecture. What other design rules might one need for architecture that would conflict with this architecture design rule? What technology and marketplace forces will shape the resolution of the conflict(s)? What process might one put in place to assure that an industry-driven SDR architecture evolves to track the realities of these forces?

4. What top-level needs are missing from those provided in this section? For each need you can think of, state an assumed requirement. How might you go about validating your assumption? What computer-based models could you use to explore the requirement? What kinds of short-term implications should be examined for SDR implementation? What longer-term implications should be examined for software-radio architecture?

5. How long should it take to set up or tear down the mobile infrastructure? If this were a military application, would setup and tear-down time be more

critical or less critical? Suppose this were a rapid build-out of wireless infrastructure? What are the implications for software-radio architecture?

**6.** How many people should be in direct support of the communications capability? That is, how many nonrelief personnel will be needed to staff the mobile infrastructure? Is completely unmanned operation feasible once the system has been set up? If not, what operations must be automated for completely unmanned operation?

**7.** Analyze the information services. Could the buyer have specified communications capabilities (e.g., numbers of voice channels, packets per second of data)? Would this be more or less helpful to the systems engineer? What degrees of freedom are provided by specifying communications capabilities in terms of information services versus communications parameters such as number of voice channels? What further analysis is required for systems design?

**8.** Analyze the external interfaces. What further analysis is required for systems design?

**9.** Outline a strawman design of the disaster-relief system using conventional radios, switches, patch panels, etc.

## II. RADIO RESOURCE ANALYSIS

This section develops the process of needs analysis further. It first reviews well-known methods for analyzing radio resources, but from a software-radio perspective. These include spectrum allocation, geographical area coverage, and subscriber distribution over the geographic area. Software-radio resources also include the traffic presented to the radio, the degree of mobility afforded to a subscriber, and the quality of the communications services. To optimize the use of these resources in the pursuit of cost and revenue-generation goals of the service provider, the software radio engineer must quantitatively address several issues. Spectral access, power generation efficiency, and waveform purity complement spatial access. GoS characterizes the availability of the traffic channel to the subscriber. QoS characterizes the expected parameters of that radio channel. All these are necessary in the analysis of software-radio architecture.

### A. Radio Resource Management

Radio resources consist primarily of the RF channels. These channels may bear traffic only, control information (signaling), or a mix of both. In a terrestrial mobile cellular network, the RF channels are reused spatially. Obstacles, Fresnel zones, and locations with excessive interference subtract from the nominal radio resources. These artifacts impart greater than square-law

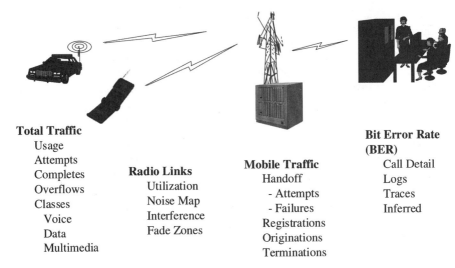

**Total Traffic**
   Usage
   Attempts
   Completes
   Overflows
   Classes
      Voice
      Data
      Multimedia

**Radio Links**
   Utilization
   Noise Map
   Interference
   Fade Zones

**Mobile Traffic**
   Handoff
   - Attempts
   - Failures
   Registrations
   Originations
   Terminations

**Bit Error Rate (BER)**
   Call Detail
   Logs
   Traces
   Inferred

**Figure 4-2**   Radio resource parameters.

losses, with path loss exponents of 2.8 to 4 in some urban areas. In addition, the received signal strength may vary randomly due to environment changes by 10 to 20 dB, and by 30 dB or more due to small changes in multipath reflections and frequency. Thus, there is a time-varying spatial distribution of radio resources as a function of mobile location, obstacles, and infrastructure density and location.

These resources may be characterized further in terms of the parameters illustrated in Figure 4-2. Total traffic offered to the network is a resource in the sense that the number of attempts to use the system represents the maximum available revenue stream. The evolution of software-radio architecture provides opportunities to leverage this resource.

*1. Total Traffic*   Early cellular networks measured offered traffic by monitoring attempts registered in the control channels. Although this is the largest share of lost calls in a well-designed network, it does not measure attempts made from disadvantaged propagation locations where the subscriber cannot access the control channels. Software radio handsets can keep track of such attempts and report them to the network. In addition, they can characterize the offered demand in terms of voice, data, and multimedia traffic that would have been offered. Since the size and frequency of data traffic can be fractally distributed [138], its statistics are more difficult to judge than voice traffic. Thus, specific details on offered video-teleconference opportunities, e-mail traffic, large attachments, etc. gathered at the source by SDR handsets will be of particular help in provisioning 3G networks.

*2. Radio Link Quality*   The mobile traffic supported at a given level of quality (e.g., at a specific BER) is also a resource. In conventional cellular radio, this traffic supplies revenue streams based on voice and data traffic. With a multiband, multimode SDR, this traffic occupies a specific band and mode. If the type of traffic is movable to other available bands or modes, then the SDR network may reassign the traffic to some other band or mode. Third-generation wireless pursues this approach within a specific IMT-2000 band by providing multiple data rates as a function of SNR. With multiband radio, access opportunities are multiplied. A multiband SDR could move the traffic to spectrum rented from the police [425] if the link quality on the cellular networks is not satisfactory. It could also delay the traffic (e.g., a large e-mail attachment) for delivery later to a corporate LAN. In a military setting, this means selecting a different waveform from a library, as a function of traffic, security needs, and dynamic network structure. The useful radio resources, then, include all those bands and modes with sufficient link quality in a specific geographic location that fall within the fundamental limitations of the radio platform: RF coverage, digital access bandwidth, and processing capacity.

Although one would like to measure BER directly, this is often not possible. Service technicians can measure BER under specific conditions, but these conditions may not fully reflect the customer's experience. Future SDRs will have the memory capacity to log BER faults as a function of time and location. Uploading and analyzing logs of fault conditions may then identify causes of low call quality. In applications where revenue generation is of primary importance, this knowledge can be used to selectively enhance the infrastructure. One may manually adjust a beam pattern or introduce a repeater in a Fresnel zone. Smart antennas may adapt to such conditions autonomously, smoothly accommodating minor propagation problems in addition to accommodating increased subscriber density. If network loading is more important than revenue generation (e.g., in military applications), one may redistribute users across bands and modes (e.g., get the right data to the right person at the right time).

*3. Mobile Traffic Profiling*  The mobile traffic that is serviced also must be measured. Standard telephony metrics include arrival rates, call duration (hold time) and class of traffic such as voice, fax, or data. Progress of the channel state-machines may be monitored so that the network operator can identify problem areas. An inordinately large number of handoff failures versus attempts, for example, can signal the need for a gap filler, or improved handoff (to another cell site). A multiband SDR might measure the traffic density in other RF bands when the primary network is lost (e.g., in a deep fade zone). This out-of-band traffic profiling gives the SDR network the information it would need, for example, to plan spectrum rental [425] in lieu of additional build-out of infrastructure. Multichannel SDR nodes have the potential to relay calls on unused channels. Military networks may use this approach to dynamically connect subnetworks that have been cut off in their primary RF band. Amateur radio networks use this polite, inexpensive approach to networking as well. As multichannel SDR nodes proliferate, this mode (sometimes called

Opportunity Driven Multiple Access—ODMA) may be employed either by
the networks or by the nodes to avoid paying for network airtime. The statistics of relay traffic, acquired and shared among SDR nodes, can form the basis
for future planning for relay approaches to spectrum management. In addition,
traffic patterns can reveal attempts to steal airtime. Registration, origination,
and termination patterns therefore provide the planning data necessary for
traffic management, infrastructure provisioning, and identifying potentially
fraudulent use of the radio resources.

*4. The Disaster-Relief Case Study*    A top-down analysis of the disaster-relief
case study identifies the communications resources. Each class of participant
is examined to determine radio equipment and rights to use radio spectrum.
The potential resources identified in this scenario are illustrated in Table 4-2.

This first-level analysis yields a range of numbers of radio units that will
be brought into the disaster area. Each vehicle that carries radio equipment is
referred to as a radio node. Each node has the potential to access its native
allocated or licensed spectrum. Some nodes will have the capability to cover
multiple bands outside of their normal bands of operation. In order to provide
a mesh of connectivity in the disaster area, there must be both some degree
of overlap of radio access, and some baseband switching capability.

Design analysis deals with the question of what radio resources are available to the participants today. For cost-effective product introduction, one must
minimize the hardware and software costs of the system, so one identifies the
minimum radio resources necessary to support the disaster-relief operation.
Architecture analysis, on the other hand, deals with the question of what radio
resources will become available to the participants during a 10- to 20-year
evolution of such designs. The top-down analysis of radio resources for SDR
application in the disaster-relief case study therefore continues with the analysis of the needs and access to the radio spectrum that will become available
over time to the classes of user characterized above.

## B. Modeling Spectrum Use

The spectrum available to the subscribers in a geographical area is a function
of the allocated spectrum, antenna patterns, propagation environment, and the
radio network architecture. Peer networks employ a spatially limited spectrum
because the nodes communicate in a spatial region defined by the radio horizon, including reflections (e.g., from the ionosphere). Hierarchical networks
are not spatially limited because the base station infrastructure permits spectrum reuse within cells that are smaller than the radio horizon. To understand
the way software radio can change one's approach to spectrum reuse, first
review the essential features of spectrum use. Then consider the refinements
introduced by software radio and radio-propagation prediction tools.

*1. A Simple Model of Radio Propagation and Spectrum Reuse*    Ideally, radio
energy propagates in three dimensions so that the carrier-to-noise ratio at the

**TABLE 4-2   Disaster-Relief Communications Resources**

| Parameter | Aspect | Potential Resource |
|---|---|---|
| Physical Area | 3–5 local areas of 2–10 km radius | 3–5 radio cells (or more); 18–150 sq km total area |
| Classes of Subscriber | Police, fire, rescue, local populace, National Guard | APCO radios; cell phones; military radios, wireless trunks, and switches |
| Numbers of Subscribers (by Class) | 10–20 police agencies | 10–20 command nodes (APCO/Tetra) A few special radio types (e.g., U.S. FBI) |
| | 20–100 fire and rescue squads | 20–100 vehicular nodes + 100–1000 handheld |
| | with 10 helicopters | 10 air mobile radio nodes (3 or more radios each) |
| | 50,000 local populace | 500–10,000 cell phones, 500–3000 cordless telephone handsets |
| | including 20 light aircraft pilots | 20 light air mobile nodes (2 or more radios each) |
| | 500–3000 National Guard troops | 50–300 squad radios, 12–80 company radios, 3–10 high-level command network radios, radio relays |
| | with 20–50 aircraft | 20–50 air mobile radio nodes (3 military radios) |
| Classes of Information Services | Voice E-mail • Tasking/scheduling • Databases Fax Video-teleconferencing Telemedicine | Isochronous narrowband traffic Unformatted messages (rescue, local, victims) • Formated (requires client software) • Formated (requires client and server) Hardware or software sources Isochronous MPEG traffic Isochronous wideband traffic |
| External Interfaces | Network: T/E-1 to T/E-3 SDH (microwave, fiber), SS7 | Fiber or microwave interface to the PSTN |

receiver is given by (link budget equation):

$$C/No = 20\log(\lambda/4\pi R) + Pt + Gt + Gr - NF - Lt - kTB$$

where

   $C$ is the power of the carrier
   $No$ is the noise power density in the primary allocation

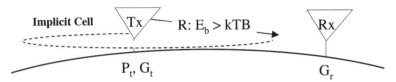

**Figure 4-3**   Implicit cell structure of omnidirectional LOS radio propagation.

$\lambda$ is the wavelength of the RF at the carrier frequency

$R$ is the range, the distance away from the transmitter at which the measurement is taken

$Pt$ is the transmitted power

$Gt$ is the antenna gain of the transmitting antenna

$Gr$ is the antenna gain of the receiving antenna

$NF$ is the noise figure of the receiver, the noise added in amplifying the received signal

$Lt$ is the total of any other losses (e.g., coaxial cable, pointing of antenna beams, etc.)

$k$ is Boltzmann's constant

$T$ is the equivalent temperature of the receiver

$B$ is the bandwidth occupied by the signal

The factor of 20 represents the ideal square-law path loss approximated when transmitter and receiver are in clear LOS of each other (e.g., ground-to-air communications). Depending on the frequency and transmitted power, the range of a transmitter (Tx) may not reach the intended receiver (Rx) as illustrated in Figure 4-3.

When transmitted at sufficiently high power, the radio signal will reach the radio horizon. This is an ideal point, usually beyond the geometric horizon, established by the height of the antennas and the bending of radio waves in the troposphere [139]. Such high-power transmission establishes a pattern of implicit radio cells centered at each transmitter. In this radio use-pattern, all of the users within one another's radio horizon contend for channels within the primary allocation. Normally a spectrum allocation is divided into channels, sometimes with intervening guard-bands to limit adjacent channel interference due to imperfect spectrum-limiting filters (Figure 4-4). Some distant or low-power users will be masked by closer or higher-power users.

Conventional radios are designed to operate in their primary allocation, and may not necessarily access other bands. Nevertheless, advanced channel modulation and coding yields an increasingly large number of alternatives for packing users into spectrum. For example, Figure 4-5 gives an idea of the variety of carrier packing techniques for illustrative spreading rates (in millions of chips per second—Mch/s) available with 3G waveforms. These cdma2000 waveforms were designed to be as compatible as possible with

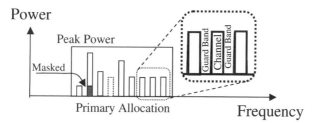

**Figure 4-4**   Contention for channels in a primary spectrum allocation.

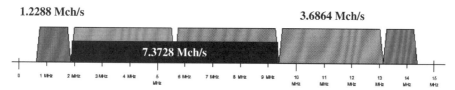

**Figure 4-5**   Illustrative packing of CDMA RF carriers.

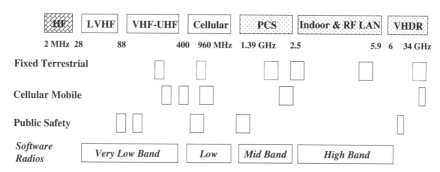

**Figure 4-6**   Software radio bands access multiple spectrum allocations.

cdmaOne. W-CDMA, on the other hand, was designed to be as compatible as possible with GSM. Its spreading rates are compatible with frequency packing in integer multiples of GSM's 200 kHz carrier separation.

Software radios have the technical capability to access any band within a much broader range of radio spectrum. A military radio, for example, might operate in the LVHF band from 28 to 88 MHz exclusively. A police radio, similarly, might operate in the 148–174 MHz VHF band. Thus, a military unit cannot communicate directly with the law enforcement personnel assisting in disaster recovery. A very-low-band software radio, however, would access the spectrum from 28 to 512 MHz, as illustrated in Figure 4-6. Its type certification and authorization to transmit would of course, be limited to specific subbands. But since it can listen across all these bands, it could provide a bridge among otherwise incompatible radios.

**TABLE 4-3   Illustrative Spectrum Efficiency**

| Standard | Wa (MHz) | Wc (MHz) | Rbs (kHz) | Ns | Rbrf (Mbps) | | Rbcell (Mbps) | Efficiency (Mbps/MHz) |
|---|---|---|---|---|---|---|---|---|
| 1G | 12.5 | 0.025 | 9.6 | 1 | 0.024 | 7 | 0.685714 | 0.054857 |
| GSM | 25 | 0.2 | 13.3 | 8 | 0.1064 | 3 | 4.433333 | 0.177333 |
| IS-95 | 1.25 | 1 | 16 | 16 | 0.256 | 1 | 0.32 | 0.256 |
| 3G | 5, 20 | | | | | 1 | | 0.450 (goal) |

Radios have to collaborate to move a masked user to an alternative part of the radio spectrum. The process of discovering the masked user and restructuring spectrum use also requires communications bandwidth, and therefore radio spectrum. In addition, each multichannel SDR may act as a local switching node, forwarding relay traffic around congestion in one band if there is little congestion on another accessible band.

*2. Spectrum Efficiency*   The number of terrestrial radio channels available in a geographic area can be made to vary approximately linearly with the infrastructure density [63]. This requires power reduction so that the carrier-to-interference radio (CIR) is held constant as the number of cell sites increases. Physically, this reuse is possible through limited radio-propagation distances. The reuse factor represents the relationship between the number of channels in the allocated spectrum and the number of channels that can be employed without excessive interference with neighboring cells. A reuse factor of 7 (typical of 1G infrastructure) permits only $\frac{1}{7}$ of the channels of allocated spectrum to be used in a specific cell. GSM's reuse factor is 3, while the CDMA reuse factor approaches 1 (e.g., 65%). The data rate supported per cell, then, is:

$$Rbcell = (Wa/Wc)(Rbrf/\rho)$$

where $Wa$ is the spectrum allocation, $Wc$ is the equivalent spectrum used per RF channel, $\rho$ is the reuse factor, and $Rbrf$ is the data rate per RF channel.

The data rate per RF channel is the product of the data rate per subscriber channel ($Rbs$) and the number of subscribers supported per carrier ($Ns$). $Rbcell/Wa$ is the spectral efficiency. If the units of $Wa$ are MHz, and of $Rbcell$ are Mbps, then units of spectral efficiency are in Mbps/MHz/cell. Illustrative measures of spectrum efficiency are provided in Table 4-3.

Spectrum efficiency has been increasing steadily. The UWC-136 [140], W-CDMA, and CDMA-2000 [141] proposals for 3G all present arguments that those air interfaces will meet the 3G goal shown. The values in the table are rough approximations. The available data rate per channel is reduced by many sources of overhead, which is a function of numerous parameters. These parameters depend on design pragmatics. If, for example, symbol rate,

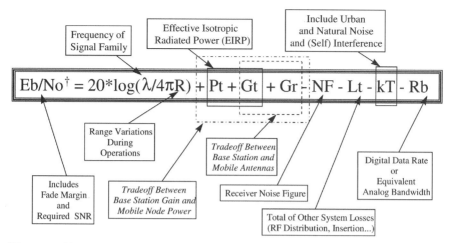

$$Eb/No^{\dagger} = 20*\log(\lambda/4\pi R) + Pt + Gt + Gr - NF - Lt - kT - Rb$$

Frequency of Signal Family

Effective Isotropic Radiated Power (EIRP)

Include Urban and Natural Noise and (Self) Interference

Range Variations During Operations

Tradeoff Between Base Station and Mobile Antennas

Digital Data Rate or Equivalent Analog Bandwidth

Includes Fade Margin and Required SNR

Tradeoff Between Base Station Gain and Mobile Node Power

Receiver Noise Figure

Total of Other System Losses (RF Distribution, Insertion...)

$^{\dagger}$Energy per bit/noise data can be converted to carrier-to-noise ratio; includes required fade margin.

**Figure 4-7**   The link budget.

spreading rate, and Walsh code length are integer multiples, handset ASICs are simplified, possibly with minor loss of spectral efficiency. In addition, an even number of power control groups per frame simplifies the insertion of power control bits [142]. Other factors include loading (fraction of total power that is CDMA power), processing gain (ratio of chip rate to subscriber data rate), Doppler, and duty cycle. The duty cycle can be 25 to 50% for voice, but this is traffic dependent. Internet traffic may be fractally distributed. Differences in these distributions change the number of subscribers that can be accommodated with a given spectrum efficiency.

***3. Link Budget Tradeoffs***   A given air interface mode is characterized by frequency band, bandwidth, and modulation type. These define the efficiency of spectrum use as outlined above. Efficiency of spatial use is determined by the link budget. The transmitter determines radiated power and antenna gain, while the receiver determines receive-antenna gain and receiver sensitivity. These parameters determine the quality of the received signal according to the link budget equation given above and illustrated graphically in Figure 4-7.

This form of the equation is expressed in terms of *Eb/No*, the energy per bit divided by the average noise density. This allows one to express the bit rate explicitly. The link budget determines whether one can close the link, providing the required SNR, with an acceptable rate of signal-loss due to fades. The cellular radio design trades off transmit gain against receive antenna gain and transmit power in the mobile station versus receive gain and radiated power in the base station. Increased gain at the base station means either less antenna gain in the handset or longer battery life due to reduced transmit power.

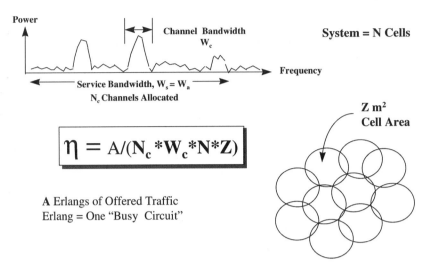

**Figure 4-8**  Efficiency supporting offered traffic in an area.

*4. Spatial Efficiency*   Spatial efficiency may be quantified using the approach illustrated in Figure 4-8 [143]. The spatial efficiency of supporting offered traffic, $\eta$, is the ratio of the offered traffic, $A$ (in Erlangs), to the product of RF spectrum employed and geographic area. RF spectrum employed is the product of the number of subscriber channels supported, $N_c$, times the effective bandwidth required per channel, $W_c$. Geographic area is the product of the effective area per cell, $Z$, times the number of cell sites, $N$. From one perspective, the system designer's goal is to maximize $\eta$ to maximize revenue at minimum cost.

The application of this formula must include inefficiencies and overhead. For example, if eight subscribers share one 200 kHz GSM channel, then each user's effective bandwidth requirement is $200/8 = 25$ kHz. In addition, however, if 100 users share four 200 kHz control channels, then there is an additional $(4*200)/100 = 8$ kHz of overhead-bandwidth required for a total effective bandwidth required of $(25 + 8) = 33$ kHz $= W_c$. Dividing $W_c$ into the allocated bandwidth, $W_a$, yields the number of channels available to bear revenue. The same kind of analysis applies to software-radio architecture. In this case, however, $W_a$ is the accessible bandwidth, and $N_c$ is the potential number of channels accessible in each of the $j$ subbands in $W_a$. Efficiency is given by [spatial efficiency equation]:

$$\eta = A \left/ \left( \sum_{j} (N_{cj} * W_{cj} * N_j * Z_j) \right) \right.$$

for each of $j$ subbands in $W_a$.

With software radio, the emphasis shifts away from the question of effectively using spectrum allocated to one specific purpose. The new optimization question concerns the dynamics of $N_j$. How many broadband SDRs are present in the scene? How many primary users have spare channels for rent? Since BMW-SDRs could forward traffic cooperatively, the shorter-range ISM bands may provide low-cost data paths. Thus, if $N_{cj}$ have overlapping coverage of $A$ in some ISM band, then there is at least one path among any pair of subscribers in area $A$. If that path is in use, what about a path in the $j + 1$ subband? Are any of these channels for rent?

This opportunistic networking approach can be attractive where large numbers of vehicular radios are concentrated in a small physical area, such as at a sports event. Each vehicular radio could become a low-capacity cell site instantaneously. Protocols for such networks have received attention from military researchers [144, 145]. The possibility of BMW-PDAs restructures the spectral efficiency analysis. In addition to efficient packing of users into limited spectrum, the BMW-SDR empowers the user to range across $j$ subbands, dynamically leveling the offered traffic. The shift is from a microview of spectrum packing in one cellular band to a macroview of the spectrum use in a given locale. The military equivalent is a shift away from managing the LVHF band or a VHF LOS band, or the 425 MHz data traffic band in isolation. The new spectrum management question becomes how the mobiles can cooperate with each other to offload busy bands (or vulnerable bands, etc.) and thus to shape traffic across the BMW-SDR's available bands and modes.

In system design trade-studies, one must balance the number of users against the cost of infrastructure and mobile devices. Spectrum may carry an overhead cost from the spectrum auctions process in the United States. Other countries have different approaches to payment for such spectrum. Alternatively, the spectrum may not be encumbered by a tariff, but peak power may be limited to 100 mW or less (e.g., RF LANs in the ISM bands). Thus "free" spectrum can cost more in terms of denser infrastructure than purchased spectrum. Multichannel SDR creates a combinatorially explosive number of possibilities for offsetting these costs using low-power, short-range opportunistic networking (e.g., ODMA). For example, think of a city whose buildings all carry gigabit-per-second fiber LANs. Each street-level window could hold an RF LAN access point with a 10 meter radius in an ISM band. All pedestrian traffic could be "free" in the sense that a BMW-SDR would not have to pay for RF LAN spectrum. Those owning the gigabit-per-second RF LANs and radio access points could set a price for network access.

The spectrum and spatial efficiency analysis provides a useful starting point for analyzing the disaster-recovery system. To extend this analysis, one may model the geometric fine structure of radio cells. Almost no cell site is circular, for example, as discussed in the next section.

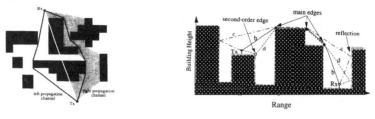

Horizontal and Vertical Artifacts Strongly Influence Path Loss

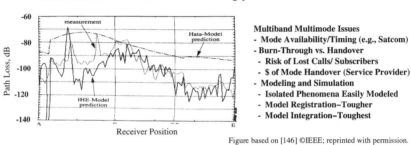

**Multiband Multimode Issues**
- Mode Availability/Timing (e.g., Satcom)
- Burn-Through vs. Handover
- Risk of Lost Calls/ Subscribers
- $ of Mode Handover (Service Provider)
- Modeling and Simulation
  - Isolated Phenomena Easily Modeled
  - Model Registration–Tougher
  - Model Integration–Toughest

Figure based on [146] ©IEEE; reprinted with permission.

**Figure 4-9**   Precise modeling of spatial access.

## C. Modeling Spatial Access

Although air-to-air and ground-to-air propagation has a path loss proportional to $1/R^2$, a path-loss exponent of 2, surface-to-surface applications are characterized by path-loss exponents of 2.5 to 4. Propagation losses are most severe in urban canyons where signals propagate on non-LOS paths by reflection from walls of buildings and refraction over roof edges. These conditions exhibit the higher path-loss exponents. Bertonie et al. [146] model such conditions using the multiple ray-trace approach (the improved Hata model—IHE). The Hata model estimates received signal power in a way that yields an overall shape of the relationship of path loss to receiver position as shown in Figure 4-9. With such limited fidelity, one could predict the approximate coverage of omnidirectional cells in flat terrain, and one could predict the approximate density of infrastructure needed in urban areas. On the other hand, 30 or 40 dB of error between the prediction and the measured received signal strength limited the use of such models. One might estimate how many cell sites would cover a region. The placement of those sites would be based on measurements in the field.

The measured data in Figure 4-9 is representative of urban propagation. It has an irregular fine structure that differs from the smooth Hata model by over 30 dB. The fine structure is not a sample function of a rapidly time-varying stochastic process in which one would expect 20 to 30 dB differences. These measurements are averages reflecting the number and complexity of multipath components. Thus, the mean received signal strength at closely spaced points along the path is irregular. Cell shape also depends on dynamic multipath such

| Propagation Model | Description |
|---|---|
| Deygout | Recursive multiple knife-edge diffraction |
| Walfisch-Ikegami | Empirical, using street width, orientation and average building height |
| Lee Microcell | Empirical, using on building thickness |
| Bertoni | Analytical, with screens of uniform height and separation (Boersma functions) |
| Hata | Smooth parametric estimate |
| Epstein-Peterson | Similar to Hata |
| IHE | Improved Hata, using ray tracing |
| Luebbers | Efficient Geometric Theory of Diffraction (GTD) |
| WiSE (Erceg) | Geometric Optics and Uniform Theory of Diffraction (UTD) |
| 1/r4 | Reference model |

**Figure 4-10**   Illustrative propagation modeling tools.

as from vehicular traffic. The orientation of the mobile station's antenna with respect to the user's body or vehicle and the height and location of the base station antenna also contribute to the irregularities. The original Hata model lacks the fine structure of the observed measurements.

Bertoni's IHE model, on the other hand, begins to capture the fine structure. It explicitly models vertical and horizontal geometric diffraction. As a result, it has substantial agreement with the measurements. IHE has greater maximum deviation from the measurements ($> 35$ dB at a point close to the transmitter) than basic Hata. On the other hand, the total deviation, the product of deviation in dB times distance, is much larger for the Hata model than for the IHE model. Generally, IHE tracks the measurements to within 5 to 10 dB, with crossover points at which model-measurement agreement is exact. IHE fidelity depends on the agreement of the model to the geometry of the site. When buildings, signs, outside wires, and temporary metallic structures are located in the site, the propagation fine structure changes. Major changes can force one to change antennas, install new cells, install repeaters, etc. Additional propagation models are summarized briefly in Figure 4-10. In addition, Erceg recently described an empirical quadratic form of path loss in hilly and flat terrain with light-to-moderate tree density [147].

**Predicted but not realized**

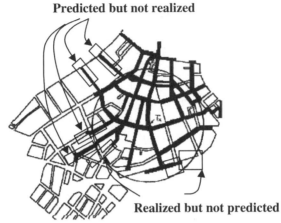

**Realized but not predicted**

Figure based on [148] ©IEEE; reprinted with permission.

**Figure 4-11**    Predictions versus experimental observations [148].

Erceg [148] reports about 5 dB average error with the WiSE tool, which employs the computationally intense techniques shown in Figure 4-10. Figure 4-11 shows how even 5 dB of path-loss error translates into errors in urban coverage. Again, if one were trying to use such a model to place cell sites, one would overlap the sites to compensate for the errors. In this case, the model is fairly consistent in predicting signal that is not present in the experimental data. There were two exceptions, however, as shown in Figure 4-11. The nominally circular shape of the cell site is distorted by terrain and building height. The circle elongates in the uphill direction, for example.

Contemporary commercial siting tools can agree well with measurements as illustrated in Figure 4-12. Some areas exhibit excellent agreement, while in other areas, the difference approaches 20 dB. Such errors can be caused by a failure to account for absorption (e.g., due to trees). On the other hand, a large number of scatterers (e.g., 100), each of which has minimal power (e.g., $-20$ dB compared to the stronger multipath components), can accumulate to an appreciable error.

When static infrastructure is installed, predictions are calibrated to measurements. This, of course, is a labor-intensive process. When the infrastructure is mobile, as in the disaster-recovery scenario, the time and labor required for such calibration are not available. SDR mobile units provide an alternative approach. Calibration and reporting software may be downloaded to SDR nodes over the air. As the initial mobile units are deployed, they may create propagation maps from the transmissions of other mobile units in areas where communication with base stations is not possible. Those maps may then be shared with the mobile base stations so that remedial action may be taken. This can include planning the location of mobile base stations that arrive after the creation of an initial set of maps. It can include

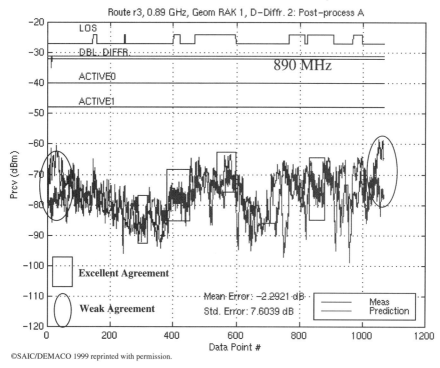

©SAIC/DEMACO 1999 reprinted with permission.

**Figure 4-12** Illustrative performance of the DEMACO commercial propagation tool.

the repositioning of base stations to maximize coverage of critical geography. It can also include the positioning of repeaters, or the tasking of mobile units to act as repeaters. In addition, as the mobiles continue to report measurements in areas of mutual visibility, the propagation models may be recalibrated.

The BMW-SDR allows planning algorithms to change bands and air interface parameters to overcome path impairments. Propagation maps may be set up as a function of the fine-scale propagation conditions. For example, those in valleys or behind obstacles may employ lower carrier frequencies (e.g., LVHF) and higher operating power. Those with excess received signal strength may employ higher carrier frequencies and lower power to clear the lower bands for disadvantaged users. These differences can result in spatial maps in which disadvantaged users employ the best propagation modes while advantaged users relinquish those modes to reduce interference. This results in a series of propagation overlays (Figure 4-13). Assume the typical SDR has three or four channels. Two channels may be used to bridge across two propagation modes. Protocols for linking such layers have been described [149]. In Figure 4-13, two such relays connect nodes A (base) and B (remote) for which there is no direct path.

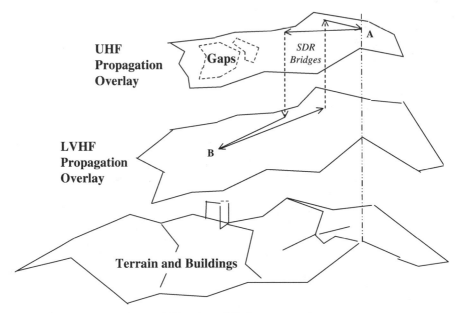

**Figure 4-13**  Use of SDR coverage layers.

Having established the feasibility of a link through the analysis of available spectrum and spatial coverage, one must determine the probability that a link is available when needed. The converse of blocking probability is the well-known *grade of service*.

### D.  Grade of Service (GoS)

The traffic channel is the primary radio resource. Its utilization equals the ratio of the offered load to the available resource for a given time interval. Instantaneously:

$$\rho = d/s$$

where $\rho$ is the utilization, $d$ is the demand for the resource, and $s$ is the supply provided by a server.

Utilization applies to any resource. If $\rho$ is less than 0.5, the demand is met without much waiting in queue due to contention for the resource. As $\rho$ increases above 0.75, the time spent waiting grows exponentially, asymptotically approaching infinity. If $\rho$ exceeds 1.0, then the number of entities waiting for the resource grows linearly with ($d$-$s$). In this situation, the number of calls waiting approaches infinity in the limit. In practice, only a finite number of users offer calls, so the number waiting in line cannot exceed the total number of users minus the number being served. Thus, the infinite queue is an abstraction that models overflow. If the network operator maximizes $\eta$, spatial

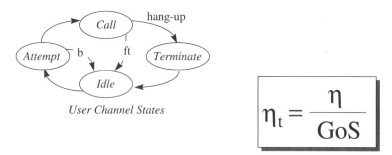

$$\mathrm{GoS} = (1-\alpha)\mathrm{P^b} + \alpha\mathrm{P^{ft}}$$

P$^b$:  probability of blocked call
P$^{ft}$:  probability of forced termination
α:   relative weight of blocking vs terminations

**Figure 4-14**   Blocking and terminations determine grade of service.

efficiency, customers will be unhappy and the network will not be successful because as the offered load increases on a fixed facility (spectrum and cell sites), contention for the facility resources increases. Thus, one must balance offered demand against GoS and available channels.

***1. Channel States***   Since voice calls have well-known statistical structure, a state-model of channel utilization estimates blocking probability as follows. User access to the network is quantified as illustrated in Figure 4-14 and the GoS equation:

$$\mathrm{GoS} = (1 - \alpha)P^b + \alpha P^{\mathrm{ft}}$$

the access parameter is $P^b$, the probability of a blocked call.

   The parameter that represents satisfaction with service is $P^{\mathrm{ft}}$, the probability of a forced termination. GoS, then, is the probability that a call is neither blocked nor terminated while in progress. The state diagram of Figure 4-14 shows how call progress can be interrupted by blocked calls and forced termination. The user is initially in an **Idle** state, not attempting a call and none is in progress. One can think of the state diagram as referring to the "user's channel" although none is assigned prior to a successful call setup. When the user attempts a call, the state transitions from **Idle** to **Attempt**. The network either will admit the call, transitioning to the **Call** state or will not admit the call, transitioning the user back to the **Idle** state. At some time shortly thereafter, the user may again attempt a call. If the network resources are incapable of sustaining the call through to normal termination, then the call will be dropped in progress, a *forced termination* event.

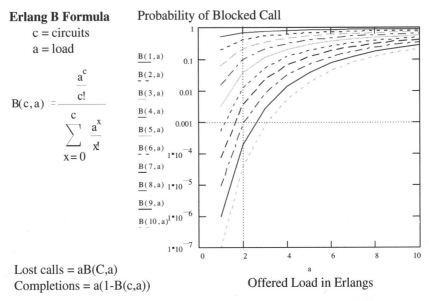

**Figure 4-15**   Erlang B formula predicts call blocking probability.

Blocked calls and forced terminations each penalize the user and thus both must be reflected in GoS. The parameter $\alpha$ of the GoS formula weights the probabilities of blocked calls and forced terminations to reflect the service provider's sense of the market implications. Intuitively, it is annoying to get a network busy signal, but it may be even more annoying to be cut off in mid-sentence. Commercial service providers have characterized these differences in terms of customers lost per 100,000 forced terminations, for example, to support for infrastructure provisioning. If the service provider determines that the rate at which customers change service providers is ten times higher for forced terminations than for blocked calls, the provider might allocate a number of channels to cell handoff. In this case, incoming calls are blocked so that there are channels available for handoff from adjacent cells so calls are not lost to unavailability of channels at a handover event. Similarly, the service provider must provide gap fillers or denser infrastructure if calls are terminated due to low SNR or high CIR. If the necessary radio channels are provided, the subscriber experiences blockages due the statistical structure of call-arrival rates.

*2. Provisioning Against Blockages*   Provisioning is the process of establishing the parameters under which traffic channels are provided to support an expected level of network traffic. The fundamental network resource is the traffic channel, while the critical availability parameter is the probability of a blocked call. The mathematical relationship between these two parameters is the Erlang B formula illustrated in Figure 4-15. This formula applies to uniform probability of call arrival in an arbitrary interval (which generates the Poisson distribution), with exponentially distributed call holding time [150].

Offered load is expressed in Erlangs. One Erlang is the traffic that occupies one network resource (e.g., traffic channel) for the period under consideration. Therefore, an Erlang is an instantaneous concept. If one is considering peak-hour load offered to traffic channels in a network, then one Erlang is 60 channel-minutes of traffic presented in such a way as to block a single traffic channel. This load may be presented as a single 60-minute Internet connection, as 60 "short" one-minute telephone calls, as 15 four-minute conversations, or as any combination of calls which sequentially accumulate to 60 channel minutes. Given $N$ available traffic channels, the probability of a blocked call is just the probability of having to service $N + 1$ or more calls at any given point in time. Under these assumptions, the probability of a blocked call is a function of the load offered (in Erlangs) as shown in Figure 4-15. The crosshairs of the figure show the situation where eight channels are provided in the system ($N = 8$) and two Erlangs are offered, yielding a blocking probability of 0.001. The same load yields a 1% blocking probability when only six channels are provided.

In wireless applications, the channel includes the shared control channels plus the traffic channels for which users are contending. In addition, wireless blockage includes any unavailability of the wireless network resources. Thus, from a subscriber perspective, there is no difference between calls blocked due to contention for a control channel and calls blocked due to contention for a traffic channel. Network operators care about this because they need to know about failed call attempts in order to plan the build-out of infrastructure. Generally, there is a one-to-one relationship between capacity of the control channels and traffic capacity of the network. One may treat the control channel as a fixed overhead per traffic channel. One may then estimate the time a user must spend on a control channel in order to set up a traffic channel. This establishes a demand for the control channels on a per-traffic-channel basis. One then allocates control channels to the necessary fraction of traffic channels. This simple approach to control channel provisioning approximates the behavior of FDMA wireless networks with simple channel-allocation protocols. GSM networks employ virtual control channels of several types with complex authentication procedures. Call reestablishment protocols have been proposed for GSM to enhance customer tolerance of faults.[21] The performance of such measures depends on mobility parameters and intricate details of the call establishment signaling protocol [151]. In general, this leads to the analysis of mobility management. Mobility management [152] includes location management and handoff management, the details of which are beyond the scope of this text. These functions are in the networking aspect of wireless, while this text covers the radio device design, and the physical and link layers of the

---

[21]In this text, the term "fault" refers to any failure to communicate, whether from propagation, handoff failure, unavailability of DSP resources, failure to meet a timing requirement, etc.

networks. These are the primary areas in which there is an evolution from hardware- to software-intensive approaches, and are the areas most critical to the evolution of open architecture.

Some aspects of traffic engineering bear on software radio design, however. In particular, recent research into the fractal nature of LAN traffic [138] suggests that infrequent events occur much more frequently and with much different duration than the uniform/exponential/Poisson model on which the Erlang B formula is based. Exponentially distributed holding times are nice in that the integral over an infinite set of such holding times converges because the longest holding times occur exponentially less frequently, yielding infinitesimal contribution to the integral. Fractal traffic, on the other hand, is distributed logarithmically so that infinite integrals do not converge. One then has to resort to more difficult mathematics in order to model the equivalent of the Erlang B formula. Research in this area is still in progress. One may account for this effect in a simple way. First, use the Erlang B formula for provisioning as above. Then treat the "busy minute" as if it were $N$ times more likely than the exponential distribution predicts. The question of how to set $N$ is addressed in Chapter 13.

The critical step the software-radio designer must take is to slightly over-provision the hardware resources so that processing capacity is available to meet the more-frequent-than-anticipated surges in demand. Although a busy minute may be (formally) predicted to occur only once per century, fractal traffic portends a busy minute every couple of months, and a busy second every couple of weeks. If that busiest second causes the system to crash every couple of weeks, then the product will be rejected by the network operator. A crash once a year might have been tolerated. So these statistics really matter. If the system is designed to robustly and gracefully deal with infrequent overloads, customers and management will be pleased and all will be well. If, on the other hand, one overdesigns for robustness (i.e., hardware overkill), then the system may be unaffordable. The design techniques of this book focus on predictably delivering robust performance without unnecessarily expensive hardware platforms.

Contention for internal processing resources is driven by the statistical demand for the radio system resources of control and traffic channels. Thus the demand patterns for the software-radio resources of DSP chips, software tasks, interconnect, etc. depend on the statistical structure of the use of radio resources. As the number of channels and complexity of the air interface increases, the radio resources demand a complex mix of system resources. Thus, peak demand on a given DSP chip may have a complex relationship to the number of traffic channels in progress. The DSP may set up and tear down channel state machines, log fault conditions, etc. In a well-designed SDR, the time spent waiting for such DSP actions is negligible compared to the time spent accomplishing other tasks. One may wait for 500 ms for the signaling system to authenticate the user. But one cannot afford to wait for the next block of bits from the modem algorithm. In a poorly implemented system, however, resource contention can cause unacceptable delays in processing voice or data traffic. The resource management chapter therefore explains how to effectively manage digital processing resources as a function of the demand for radio resources, in spite of the complexity of some of these relationships. The attention paid to software-radio resource management is thus warranted by the necessity of delivering high GoS in spite of:

1. The statistical structure of offered loads
2. Potentially complex relationships between offered load on the radio resources versus load on the software-radio resources (DSPs, host processors, interconnect, etc.)
3. The statistical structure of software execution times
4. The likelihood of hardware resource failure modes

Provisioning a software radio is similar to provisioning a digital or analog radio. One must provide sufficient channels to meet the GoS given the expected peak demand. Instead of providing physical channels, a software-radio designer provides virtual channels. The degrees of freedom increase substantially. How many virtual channels can one pack into a single DSP, or down a given bus? For a given air interface mode, the answer to that question depends on the complexity of the algorithms that implement the isochronous stream. How complex should those algorithms be? Subsequent sections of this chapter identify the complexity drivers. Subsequent chapters introduce the analysis of complexity, and describe ways of managing that complexity. In general, the better the algorithm, the more processing resources (MIPS, FPGA area, and thus battery power) it takes. One can write a crude modem algorithm in about a hundred lines of code. It will have inferior timing recovery and carrier tracking, though. In addition, if it is implemented on a 16-bit fixed-point processor, its dynamic range will be limited and thus it will have inferior near–far performance. How good, then, does an algorithm have to be? The analysis of quality of service (QoS) provides answers to that question.

## E. Quality of Service (QoS)

While GoS has to do with access to traffic channels, QoS has to do with the technical parameters of those resources. QoS includes data rate and the rate at which data may be corrupted by noise, lost, or delayed by the network. QoS metrics were formalized first in the Integrated Services Digital Network (ISDN) [153]. QoS contracts were formalized in Asynchronous Transfer Mode (ATM) networks [154]. ATM access protocols define the quality of the end-to-end connection to be provided by a network. All such networks add bit errors and delay packets (ATM cells) according to some probability density function. They also will lose cells with a nonzero probability. Usually, the absolute delay through the network is not as critical as the difference between

ATM Contract Negotiated at Entry

| Cell Loss Rate | Cell Delay Spread |
|---|---|
| Errors | Buffering |
| Preemption | Queueing |

| Service | Mode | Wb kbps | Burst Ratio | Burst Length bits | Cell Loss Tolerance log() | Cell Delay Tolerance sec |
|---|---|---|---|---|---|---|
| Voice | PCM/CBR | 64 | 1 | — | -4--6 | .01-.150 |
| | VCELP/VBR... | 16 | 5-15 | 2-3k | -6--8 | .01-.150 |
| Data | PC file transfer | 14.4 | 1 | Kb-Mb | -9 | 10-100 |
| | Trans proc | 64-5k | 40 | 1-3k | -9 | 1-3 |
| Video | Teleconf | 64-14k | 2-5 | 2-10k | -9 | .15-.35 |
| | NTSC | 15k-44k | 2-5 | .5-1.5M | -10 | .04 |

**Figure 4-16**   Quality of service (QoS) negotiated in ATM contracts.

the minimum delay and maximum delay (delay spread) experienced for delivered cells. Cells delivered within a cell delay tolerance may be imparted to the isochronous service stream in a way that preserves sufficient information for the user to be satisfied with the results. Cells delivered outside of this window cannot be so integrated, so users will perceive service degradation.

Most services can tolerate the loss of small amounts of data. Small losses may appear as noise in a voice channel, fax, or picture. Large loss rates impact the system's ability to reconstruct the essential content of the isochronous stream. This results in speech distortion and dropouts, meaningless streaks in a fax, or intermittent loss of video integrity. Different services can tolerate more or less delay spread and data loss as indicated in Figure 4-16. The QoS values in the figure characterize the data bandwidths, burstiness, and loss tolerance of voice, data, and video. These issues are particularly critical to wireless ATM [155]. In addition, recent research has defined common QoS metrics that may be applied uniformly to GSM, wireless ATM, and 3G radio technology [156].

This discussion should sensitize the software radio engineer (particularly those with little background in hard-real-time software) to the way in which time delays in the software will degrade the perceived quality of the service. To ensure a high-quality software radio implementation, the systems architect must establish internal data loss and time delay budgets that are allocated to hardware and software components and then measured and managed throughout the development process.

## F. Review

This concludes the introduction to radio resource analysis. Given a first-order model of offered load and available spectrum, one may use the spectrum efficiency formula to trade off cell site packing density versus blocked and prematurely terminated calls. For fixed-infrastructure applications, this analysis occurs at design time, but for transportable (e.g., military) infrastructure, the measurements, analysis, and corrective actions are part of real-time mobility management. The simple spectral/spatial efficiency model provides a starting point to which one may add important refinements such as the IHE model to quantitatively assess the potential impact of deployments in challenging environments such as urban canyons.

The Erlang B formula, similarly, provides an estimate of the relationship between offered load and availability of system resources. This formula is also useful in estimating the probability that a software radio resource like a bus host is unavailable when needed, as shall be explored in later chapters. GoS defines access to network resources while QoS determines the viability of those resources in support of a given service. The characterization of service classes (voice, data, and video) with respect to ATM cell loss and delay spread provides a starting point for software radio analysis of the degree to which the software radio itself can lose data and increase delay spread internally without impacting perceived QoS. Since data loss and differences in data delivery times will happen on a statistical basis, the software radio designer must be able to characterize the QoS impact of such effects using the data loss and delay spread parameters presented in this section.

The critical design issue for software radio systems, then, is to balance the efficiency of employing spectral, spatial, and infrastructure resources against the GoS and QoS perceived by the user. Greater efficiency requires the use of advanced techniques to overcome locally high user density. A cell site may be larger without impacting on GoS if it employs a beamforming array, for example, but this is more expensive infrastructure. Increased GoS, similarly, requires more cell sites yielding more spectrum reuse, which is also more expensive. On the other hand, greater customer loyalty (due to high GoS) yields lower advertising and other expenses per revenue dollar. Enhanced QoS may be accomplished by dedicating more DSP and general-purpose processing modules, more memory, and/or more expensive interconnect to the task, but such measures also increase the cost of the infrastructure. Military applications are driven by affordability while commercial applications are driven by return on investment and other cost/benefit related parameters. This section has introduced the key parameters that impact the cost/benefit considerations. The larger systems-engineering aspects of this tension between engineering and economics will be taken up again after the design chapters. One must understand the engineering details from a design perspective in order to be able to include the right parameters in the system engineering trade-offs.

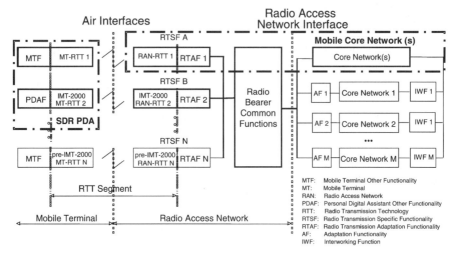

**Figure 4-17**   Third-generation model extended to SDR-RTT.

## G. Exercises

1. List the radio resources of a wireless network. Specifically, what resources are associated with the disaster-recovery application?

2. Consider a four-channel BMW-SDR. List data rates possible for its specific band-mode combinations. Describe the alternatives for packing a 1 Mbit file transfer into the data capacity of that radio.

3. In general, how would you determine how many base stations are needed for a disaster-recovery scenario? Assume that the base stations must be self-contained in a van or pickup truck. What design features directly determine the number of such mobile base stations needed for a given scenario? How could you write a specification that would limit the number of base stations in an unambiguous way without stating a specific number?

4. Describe the packet size for a wireless data mode (or assume 1 kbit per packet if unknown). Is the QoS for voice acceptable with a packet loss rate of $10^{-1}$ or $10^{-8}$?

## III. NETWORK ARCHITECTURE ANALYSIS

The radio resources provided by SDR nodes are employed in a network architecture. The simplest networks impart few performance requirements on the nodes. Moderately complex second-generation networks increase node complexity and impose timing requirements. Third-generation systems increase this complexity by offering a larger variety of QoS alternatives. SDR networks stretch the 3G model to include multiband multimode aspects of the radio transmission technology (RTT), as illustrated in Figure 4-17. This network architecture diagram draws a notional boundary between the radio transmission technology (e.g., GSM, CDMAone, SINCGARS) and the other functions of the mobile terminal or PDA. A modem algorithm is part of RTSF.

In military applications, much of the core network is mobile, employing high-capacity microwave RTT trunking. The plan for logical network topology in such applications reflects the plan for physical movement on terrain of the supported users. As multichannel SDRs enter the commercial market to take advantage of free or low-cost air time offered by corporate LANs, the network dynamics in the commercial sector could meet or exceed that of today's most dynamic military networks. Thus, Figure 4-17 has to be interpreted in terms of both static and dynamic radio-access networks.

This section analyzes those aspects of network architecture that constrain software-radio architecture. As mentioned earlier, network architecture is moving toward a heterarchical organization of multiple interlocked hierarchies. MT-RTT 1 (Figure 4-17), for example, could be a mobile-satellite service (MSS). The future SDR PDA might access the MSS plus single hierarchies and peer networks. The relevant ideas in network architecture are first considered, followed by analysis of typical commercial and military networks.

## A. Network Hierarchies

Radio networks may be organized as peer networks, hierarchies, or heterarchies. In peer networks, every node is identical. The simplest peer network is the point-to-point link in which only two radios constitute the entire network. In more complex peer networks a variable number of radios share the radio resources (typically push-to-talk channels). In military peer networks, one of the nodes is designated as the control station, but there is no physical difference between the radio at the control station and the out-stations. Other networks may have no control station at all, such as an ad-hoc walkie-talkie network used by a party of hunters in the wilderness. In still other peer networks, such as citizens band (CB), there is no designated control station, but manual procedures are intended to control the radio resources. The resources are shared to the degree that the users follow the procedures. CB allocates a specific channel for brief contact while on the move, with a rule that long conversations are taken to some other channel by mutual agreement. In practice, much of the mobile chatter occurs on this single overused control channel while most others go unused. Still other peer networks such as JTIDS employ digital control channels. The nodes, then, implement radio-resource management algorithms, allocating certain logical channels to control and others to different types of services. The functions in such a peer network are self-contained within the radio nodes. That is, any peer radio could perform any network functions.

Hierarchical networks, on the other hand, consist of different types of nodes that perform different functions. A GSM mobile cellular network, for example, includes the BTS where transmitters and receivers are located. BSCs manage

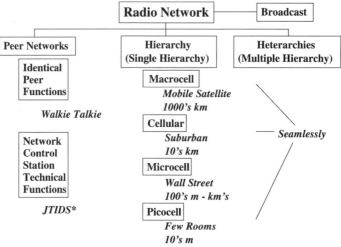

*Joint Tactical Information Dissemination System [1, p.647].

**Figure 4-18**   Radio network architectures.

multiple base stations and transcode the data. The GSM Transcoder and Rate Adaptation Unit (TRAU), for example, decodes GSM's 13 kbps RPE-LTP voice traffic and recodes it to standard 64 kbps PCM for the wireline network. This adapts the unique requirements of the air interface to the standards of the wireline network. Mobile switching centers (MSC) set up calls, authenticate users, manage privacy keys, interface to the larger network's signaling and control system, and perform other related functions.

Peer networks employ spatially limited spectrum because the only communications connectivity is via a specific band of spectrum that can be called the primary allocation. Hierarchical networks concentrate traffic in the primary allocation at a base station, and then transfer that traffic throughout the network via a trunking system. The trunks may also use radio spectrum.

As illustrated in Figure 4-18, first-generation cellular networks employed cells with diameters of tens of kilometers. As the number of subscribers grows, it becomes necessary to reduce cell size to accommodate more users. Sectorized cells are typically divided into three 120-degree sectors in order to accommodate roughly three times the subscriber density. Alternatively, one may deploy microcells with smaller diameters. Cells in urban canyons may have diameters of only hundreds of meters. Personal communications services (PCS) and networks (PCNs) may reuse frequencies within large buildings with picocells that are tens of meters in diameter. Finally, satellite mobile telephone systems such as INMARSAT introduces mobile macrocells or megacells, cells the size of the satellite-beam footprints, hundreds of kilometers across. The term *macrocell* (umbrella cell) also applies to a hierarchy of cells overlapped at the same cell site to ensure continuity of service to rapidly moving vehicles.

**TABLE 4-4   Multiple Independent Hierarchies**

| Domain | Mode | Range | Mobility | Air Time | Data Rate (kbps) |
|---|---|---|---|---|---|
| Home | Cordless | 10–30 m | Pedestrian | Free | 16–144 [2048] |
| Commuting | Cellular | 10–30 km | Vehicular | $0.3–2/min | 9.6, 16 [384] |
| Shopping | DECT | 50–250 m | Pedestrian | Free–$1/min | 8–144 [> 384] |
| Work | W-PBX | 10–30 m | Pedestrian | Free | 9.6, 16, 32 [2000] |
| Work | RF LAN | 1 room | Fixed | Free | 2–10 Mbps [155 4G] |
| Travel, sport | Satellite | 3000 km | Vehicular | $1–5/min | 2.4–8 |

*3G specifications indicated by [numbers in brackets].

A cell site on an urban interstate highway system, for example, might employ a 10 km diameter macrocell for rapidly moving subscribers in vehicles on the highway. The same site simultaneously supports a 3 km diameter cell for pedestrian and slow-moving traffic. All of these cell families overlap somewhat, so the terminology is not precise, but this description is representative of contemporary usage.

With the proliferation of service providers and air-interface standards, the industry is moving toward a heterarchical arrangement, a set of interlocking hierarchies. Consider the set of radio access mechanisms available to the consumer (Table 4-4). The customer of the future at some point could be attracted to a single handset that interoperates across all these diverse modes and service providers. The SDR-enabled PDA could provide the platform for integrating what are today multiple independent hierarchies, each of which have different handsets and different user interfaces. For the consumer, the SDR-PDA should offer access driven by the user's goals. Some consumers may require low cost. Business users may place more of a premium on wideband connectivity to important events (for reasonable cost). The business user, for example, may want the PDA to send an important five-line e-mail from the train while commuting. The 2 MB attachment, however, should be delayed for ten minutes until the PDA is within range of the corporate RF LAN. Third-generation service providers may prefer to enhance their revenue streams by servicing that 2 MB attachment, so the SDR-PDA can become the consumer advocate in load shaping. An SDR handset can offer the consumer a range of alternative QoS/cost profiles, negotiating with multiple networks on behalf of the consumer, while today's simple dual-mode handset does whatever the network tells it. The SDR PDA may result in a shift of emphasis from the services provided by "the network" to the network accesses provided by "the SDR server" in the user's hand.

The trend toward alternative hierarchies brings with it an increase in the complexity of the software radio. The relatively blind handoff from one cell site to another employed in first-generation analog systems has been supplanted by soft handoff in CDMA systems. Handover to other modes, however, requires order-of-magnitude increased complexity if a subscriber's services are to be seamlessly shifted from one air interface, hierarchy, and service provider to another without losing call continuity or disrupting multimedia services. Increased complexity of services means increasing complexity of software interactions both with other software and with hardware. For example, the software radio designer must decide whether to dedicate hardware to each air interface mode or to timeshare programmable antennas, RF conversion, and ASICs, across multiple modes as DSP and host processors are shared. The air interfaces imposed by the network environment constrain these tradeoffs as discussed in the subsequent sections.

## B. Commercial Networks

Commercial hierarchical networks define the functional interfaces to which SDR products must conform. First-generation cellular networks [63] used 25 or 30 kHz analog traffic channels with digital control channels. In the United States, these networks first evolved to Interim Standard IS-54 (Digital AMPS) [69]. Three IS-54 digital voice-channels shared a TDMA channel formerly occupied by one analog traffic channel. Enhanced signaling and control led to the IS-136 [157]. The U.S. delegation has proposed that IS-136 evolve to UWC-136, a third-generation network [158]. U.S. digital networks lagged the European GSM network, however. GSM was defined between 1983 and 1992, with initial deployments in 1993. U.S. IS-54 digital networks were not deployed until 3 to 5 years later. Qualcomm developed the first CDMA digital network in the early 1990's [159], standardized as IS-95 in 1993 [160]. Qualcomm's CDMA technology provides the foundation in intellectual property (IP) for wideband CDMA (W-CDMA) third-generation radio transmission technology [161, 162].

*1. GSM Case Study*   GSM provides a convenient tutorial example of a single-hierarchy wireless network. The value of the worldwide installed base of GSM systems exceeds $80 B, with over 100 million subscribers. Although much more complex than prior networks, the enhanced voice quality, interoperability across national borders in Europe, and expanded services of GSM yielded the success story of 2G wireless. Its network features provide a widely published foundation for analyzing SDR constraints imposed by the network [33]. Figure 4-19, for example, shows the major components of a standard GSM network. The MS supports the air interface. User services beyond dialing include the display of short message service (SMS) packets, storage and retrieval of telephone numbers, and delivery of data streams to attached laptop computers.

   The GSM base station subsystem includes multiple BTS, transcoder(s), and the BSC. Multiple BSSs are managed by each Mobile Services Switching Center (MSC), the primary component of the Network and Switching Subsystem (NSS). The system operator manages the GSM network via the Operation Subsystem (OSS), the major component of which is the Telecommunications

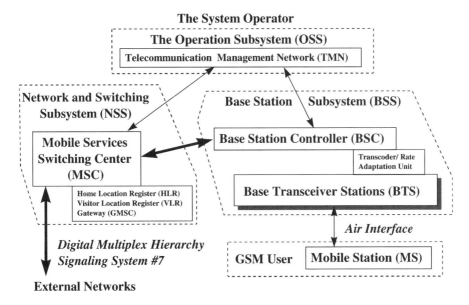

Global System for Mobile Communications [10].

**Figure 4-19**  GSM network architecture.

Management Network (TMN). This partitioning distributes physical, network, transport, and applications aspects of the network into physically distinct subsystems. The BTS contains the antennas and RF subsystems of the GSM fixed network. In some configurations, the BTS translates the radio signals to a convenient microwave LOS relay called the Abis link. This partitioning places the BTS control subsystems, the TRAU, and the BSC units at central location. This approach can reduce the cost of the BTS. Alternatively, the BTS and BSC functions may be integrated into a single "compact base station" subordinate to an MSC. In early GSM systems, the TRAU transformed the 64 kbps DS0 [5, 6, 169] subscriber channel of the PSTN into the 13 kbps RPE-LTP data stream of the GSM network. TRAUs now support half-rate, enhanced full-rate, and other vocoders. The TRAU is one of the more computationally intensive DSP components of the GSM BSS. The BSC handles packet assembly, call control, and handoff of mobile subscribers among BTSs. Each MSC supports up to 12 BSC groups while each BSC supports up to 512 BTS and over 23,000 traffic channels.

***2. SDR Feasibility for GSM***  The primary constraint imposed by a network is the support of its air interface. In addition to the TDMA waveform structure, GSM imposes equalization, near–far performance, and other constraints on the radio [33]. Network constraints including signaling, INFOSEC, and packet services impose further constraints in the timing of operations from the

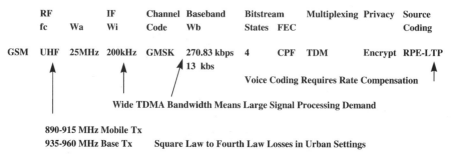

| | RF | | IF | Channel | Baseband | Bitstream | | Multiplexing | Privacy | Source |
|---|---|---|---|---|---|---|---|---|---|---|
| | fc | Wa | Wi | Code | Wb | States | FEC | | | Coding |
| GSM | UHF | 25MHz | 200kHz | GMSK | 270.83 kbps | 4 | CPF | TDM | Encrypt | RPE-LTP |
| | | | | | 13 kbs | | | | | |

Voice Coding Requires Rate Compensation

Wide TDMA Bandwidth Means Large Signal Processing Demand

890-915 MHz Mobile Tx
935-960 MHz Base Tx     Square Law to Fourth Law Losses in Urban Settings

[Implicit Multiplexing]: [PTT] = Push To Talk; [CA] = Circuit Assigned; [CD] = Code Division.

CPF: GSM specifies various modes of Convolutional, Parity, and Fire codes [8].

**Figure 4-20**   GSM base station critical parameters.

modem to the applications layer of the protocol stack. Each of these imposes constraints on an SDR implementation. One must accommodate the memory and processing requirements of the software modules that implement the functions. In addition some constraints, such as near–far performance, impose end-to-end constraints on the RF, ADC, DACs and digital signal processes throughout the system. Managing the SDR technology so that it meets such constraints is the focus of the design-oriented chapters of this text. The first-order feasibility of an SDR approach, however, is determined by examining the critical parameters of the air interface, as follows.

Figure 4-20 shows the critical parameters of the GSM air interface that are relevant to software-radio technology insertion. First, the RF carrier frequencies $f_c = \{890\text{–}915\}$, $\{935\text{–}960\}$ MHz raises the question of SDR RF band structure. This is accommodated conventionally in two narrow bands. One band supports downlink transmission, and the other uplink reception. Since GSM employs frequency domain duplexing (FDD), $f_c^{\text{uplink}}$ and $f_c^{\text{downlink}}$ are separated by 45 MHz. This separation facilitates transmitting and receiving at the same time using separate RF/IF chains. Alternatively, an SDR could employ one wideband antenna and RF section for transmission, say, from 800 to 1600 MHz, and another virtually identical section to receive across the same RF band. A single wider bandwidth antenna and diplexer then must be used. With such an approach, one may simultaneously transmit and receive hundreds of noninterfering channels in an appropriate digital architecture. Such an approach demands the suppression of self-generated interference, which in turn demands extreme linearity and spurious-free performance in the RF and IF components. Active cancellation of the transmitted signals may also be necessary [426]. One continues the top-down analysis by examining the requirements of the digital architecture.

The access bandwidth $(Wa)$ is the second critical parameter. Since the GSM uplink and downlink bandwidths are each 25 MHz, a full-service IF SDR requires a minimum ADC bandwidth of 25 MHz. This corresponds to a well-engineered sampling rate of 70 MHz ($> 2.5 \times W_a$ which is 40% more than the Nyquist sampling rate of $2 \times Wa$). Having acquired the 25 MHz bandwidth via the ADC, IF filtering may be accomplished digitally. The critical parameters of this process are the input sampling rate (e.g., 70 MHz), the output bandwidth, and the dynamic range. Filter shape (e.g., quality of Q-factor) is also important since it determines the number of taps in a digital delay-line implementation, which in turn defines the equivalent computational complexity of the filter. The design chapters address the detailed analysis of this processing stage. Initial feasibility requires one to have a digital filter with 70 MHz input sampling rate, a filter shape compatible with the air interface standard, and sufficient dynamic range to support the near–far requirements of the application. The output bandwidth of this IF filter is determined by the next critical parameter, IF bandwidth, $W_i$. First-generation analog cellular standards used IF bandwidths of 25 or 30 kHz per RF carrier, which is also the baseband bandwidth. Since GSM employs time division multiple access, eight TDMA channels are multiplexed on a single RF carrier, resulting in a 200 kHz carrier spacing with approximately 230 kHz bandwidth for 95% of primary signal power per RF carrier. A digital IF filter for GSM, then, could accept a 70 MHz input sampling rate, translate the signal to digital IF or baseband, and shape the extracted subscriber channel within a $\sim 200$–230 kHz bandwidth. There is an implementation tradeoff between a bank of many parallel digital filters, which are now commodity parts, versus a massively parallel digital filter bank tailored to the air interface. Again, the detailed analysis of these tradeoffs is deferred to subsequent chapters.

Next, a digital demodulator must determine the timing of the channel symbols, accommodate carrier variations (e.g., Doppler), and produce an isochronous baseband bitstream. In GSM, the 270.833 kbps digital TDMA stream is recovered in bursts of from up to eight subscribers per carrier. GSM's embedded training sequence must be equalized for each such burst. This imposes the largest computational demand of the demodulation process. Each individual subscriber, however, has a sustained data rate of only 13 kbps. Thus, if the digital architecture fans out sampled-analog bursts to NDSP chips for demodulation, the aggregate output data rate is only $N*13$ kbps. One DSP could find burst boundaries and synchronously distribute alternate bursts among eight DSP chips, one per subscriber. This design was appropriate for GSM using DSP chips with 20 to 50 MFLOPS of processing capacity. Alternatively, a single processor may timeshare demodulation among eight software tasks. This design is more appropriate to GFLOP-class processors.

The bitstream parameters characterize the internal data rates that must be supported by the downstream software tasks in order to provide isochronous service. Downstream tasks include framing, demultiplexing, processing signaling data, supporting protocol stack(s), and voice coding and decoding (vocoding). GSM's A5 stream cipher, for example, imposes additional processing demand for each bit of the 13 kbps subscriber bitstream. The TRAU function, furthermore, must deliver 64 kbps DSO streams to the SDH interface of the

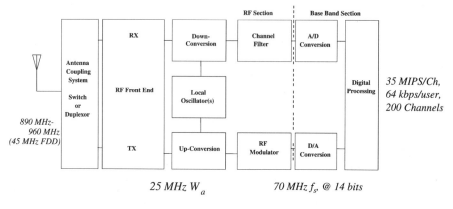

**Figure 4-21**    Reference design for GSM SDR base station.

wireline network. One may summarize the top-level design decisions of such first-cut architecture analysis in an annotated reference-design block diagram (Figure 4-21).

The overall process of allocating these top-level parameters iterates between an analysis of the air interface and an assessment of the available implementation technology. In addition to the device-oriented parameters like ADC sampling rate, a good reference design addresses the critical end-to-end parameters of the telecommunications system. These include allocating the QoS metrics to the hardware and software components. Achieving end-to-end dynamic range, for example, has historically been challenging in digital radio and SDR implementations. The RF AGC, ADC dynamic range, digital filtering, and arithmetic precision of the signal-processing algorithms all contribute to end-to-end dynamic range. The effects of each must be balanced through a dynamic-range allocation process. In addition, the dynamics of such a series of AGC functions must be controlled. If the RF AGC is suppressing a signal at the same time that the IF filter is suppressing the signal, the signal-processing algorithms may overcompensate, distorting the signal and losing BER in spite of large input signal energy. This kind of condition can be avoided through attention to implementation pragmatics discussed in the design chapters.

*3. GSM SDR Architecture Evolution*    The complexity and sophistication of the GSM system continues to advance with the new features highlighted in Figure 4-22 [164]. Supporting this evolution underscores the difference between a mere design and an architecture. A mere GSM digital-radio design requires redesign at each incremental evolution of the implementation technologies. A robust GSM SDR architecture, on the other hand, supports affordable technology insertion. In addition, the physical and network layers of the

| Service Evolution | Technical Issue | Technical Impact |
|---|---|---|
| Speech | Density | Half rate & 1/4 rate codecs |
| | Quality | Enhanced full rate (<16kbps)* |
| Coverage | Density | Antenna diversity, space division multiple access (SDMA)** |
| | Quality | Handover algorithms* |
| | | Enhanced demodulation/decoding** |
| | Cell Selection | Umbrella vs. microcell |
| | | Dynamic processing gain available* |
| UIC Trains | Speed (Doppler) | 250 kmh@900MHz > 400kmh@1900 |
| | | Adapted demodulator* |
| | Group Calls | Single speaker in conference |
| High Data Rate | 100kbps+ | Channel strapping, handset** |
| Packet Services | GPRS | Dedicated channels, managed by the radio uplink (BTS/BSC) |
| Toward PCS | Band Overlap | DCS1800, GSM900, DCS1900, DECT |
| | | Multiband multimode handsets* |
| Mobile Satellite | Interoperation | Multiband multimode handsets* |
| Service Insertion | Generic Standard | Toward the software defined network |

\* = Software Radio DSP Impact; \*\* = Large DSP/ MIPS Impact  GPRS=General Packet Radio Service; Issues Raised in [5].

**Figure 4-22**   GSM evolution.

air interface evolve. A robust software-radio architecture supports incremental software-based feature enhancement of these layers. It retains its integrity and facilitates such evolution.

For example, the enhanced full-rate GSM codec not only improves the speech quality, it also makes the speech data rate more compatible with the 64 kbps DS0 PCM employed in the PSTN. Half and quarter-rate codecs are needed in applications like low-cost handsets with low average power-drain and small handset size. Such codecs also enhance network revenues by packing more subscribers into the GSM TDMA frames during peak utilization. ASIC-based digital vocoder designs minimize power but must be redesigned to accommodate such changes. FPGA-based designs can be reprogrammed, but they may run out of logic devices or chip area given the increased complexity of new vocoder algorithms. DSP and general-purpose software implementations may run out of processing capacity or memory. Thus, there is no panacea. In order for an SDR-based approach to rise to the level of architecture, there must be clear technology insertion points so that increasingly complex algorithms may be inserted on increasingly powerful hardware platforms (Figure 4-23).

To underscore this point, consider the evolution of GSM for the high-speed European trains. Since the Doppler shift of these trains exceeded the RF carrier-tracking loop of conventional base stations, enhanced digital hardware was deployed for the trains. In an SDR architecture, the Doppler tolerance of the carrier-tracking loop would be a parameter. A simple software update would have upgraded existing SDR infrastructure with no changes in the base-station hardware production lines. Deployed base stations could be upgraded

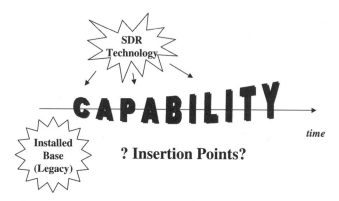

**Figure 4-23**   SDR technology insertion challenges.

by an over-the-air download. This improves cost of ownership for the service provider but reduces hardware sales opportunities for the infrastructure equipment provider.

GSM is also being deployed in support of PCS in the 1800 and 1900 MHz bands. This offers the possibility of designing SDR GSM base-station product lines that can be programmed for worldwide deployment. Such equipment could also be reprogrammed for maintenance support across geographic regions. In order to determine whether to migrate from a successful digital base-station product line to an SDR-enhanced product line, one must consider the marketplace benefits of the SDR approach. There are also pitfalls, including the performance of SDR hardware, the cost of software maintenance, and the potential loss of revenue since network operators no longer need to purchase new hardware for each significant new change in an air interface standard. Feature-based pricing offers alternative revenue streams. The related issues that the insertion of SDR technology brings may be addressed in an analysis of alternative business models.

The GSM handset is an obvious target for software-radio technology insertion. There is a tradeoff between the inherent flexibility of software radio nodes and the low cost of single-mode inflexible radios. Historically, cost tradeoffs (covered in detail in later chapters) favor single- or dual-mode digital handsets based on production ASIC chip sets over the DSP-intensive SDR. The cost of a dual mode handset with physically distinct antenna coupling, RF conversion, and demodulation ASICs is only 125% the cost of a single-mode handset. When the number of different modes exceeds four, a wideband SDR implementation typically becomes more cost-effective than the ASIC-based approach.

Base-station economics, similarly, have favored digital base stations in which the baseband functions are programmable, but the IF processing and demodulation is accomplished in dedicated hardware. As the ADC, digital filter and DSP technology becomes more affordable, the tradeoffs begin to favor

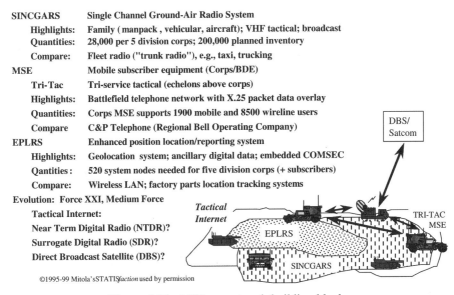

SINCGARS          Single Channel Ground-Air Radio System
   Highlights:    Family ( manpack , vehicular, aircraft); VHF tactical; broadcast
   Quantities:    28,000 per 5 division corps; 200,000 planned inventory
   Compare:       Fleet radio ("trunk radio"), e.g., taxi, trucking
MSE               Mobile subscriber equipment (Corps/BDE)
   Tri-Tac        Tri-service tactical (echelons above corps)
   Highlights:    Battlefield telephone network with X.25 packet data overlay
   Quantities:    Corps MSE supports 1900 mobile and 8500 wireline users
   Compare        C&P Telephone (Regional Bell Operating Company)
EPLRS             Enhanced position location/reporting system
   Highlights:    Geolocation  system; ancillary digital data; embedded COMSEC
   Qantities :    520 system nodes needed for five division corps (+ subscribers)
   Compare:       Wireless LAN; factory parts location tracking systems
Evolution:  Force XXI, Medium Force
   Tactical Internet:
   Near Term Digital Radio (NTDR)?
   Surrogate Digital Radio (SDR)?
   Direct Broadcast Satellite (DBS)?

©1995-99 Mitola's STATISfaction used by permission

**Figure 4-24**   Military network building blocks.

SDR base stations. Initial products introduced during 1996–98 (e.g., AirNet's wireless PBX [165]) were not competitive with the more conventional designs that use analog IF and channel modems. By 1999, Lucent and others were emphasizing the software architecture and programmability of their infrastructure products [427].

### C. Military Networks

Military users, similarly, have to work in a variety of networks including peer networks, military hierarchies, and even the local commercial hierarchies (e.g., to work with local organizations during a crisis or natural disaster).

*1. Ground-based Military Networks*   In the United States, the army is responsible for terrestrial telecommunications infrastructure on the move. That architecture now includes SINCGARS, the Mobile Subscriber Equipment (MSE), and the Enhanced Position Location Reporting System (EPLRS) [166]. This set of systems is on the path to evolve toward a two-level *tactical Internet*. The MSE is the radio component of the TRI-TAC mobile switching system. The MSE provides the hierarchical multichannel links between deployed subscriber groups and higher echelons as outlined in Figure 4-24. This is the cornerstone of the "upper" tactical Internet.

SINCGARS and ELPRS provide wireless services to the mobile subscriber. This is the cornerstone of the "lower" tactical Internet. SINCGARS connects geographically adjacent subscribers via secure push-to-talk voice and packet data services. SINCGARS interoperates with MSE at specified radio access points that allow single-channel radio traffic to be multiplexed into trunks for long-haul connectivity. EPLRS is a data network that also reports the position of each group of subscribers. EPLRS can deliver ancillary digital data with differentiated classes of service similar to JTIDS. MSE also supports packet data. All of these systems employ high-quality communications security (COMSEC). The U.S. Army continues to evolve the next-generation data system referred to as the *tactical Internet*. The Near-Term Digital Radio (NTDR) and Surrogate Digital Radio (SDR) are interim radio nodes that were deployed with the army's experimental Task Force XXI to characterize the operational and technical needs for communications in the future. NTDR provides hundreds of kilobits per second per user, while SINCGARS provides only a few kbps. The U.S. Army has also identified the need for a high-data-rate radio that can link tactical operations centers (TOCs). Ultimately, the United States appears to be headed toward the SPEAKeasy-class radio JTRS.

*2. The Joint Tactical Radio System (JTRS)*   In the summer of 1997, the Programmable Modular Communications System (PMCS) integrated process team (IPT) recommended the consolidation of the more than 200 nomenclatured U.S. radio families into a single program, JTRS, under the joint management of the three U.S. military services. The Joint Tactical Radio (JTR) mission needs statement (MNS) and Operational Requirements Document (ORD) express the vision for the functional capability of the JTRS. This vision generally follows the architecture framework of the SDR Forum. Future JTRS radio nodes such as a manpack or vehicular radio would have to support a wide variety of bands and modes including those listed in Table 4-5.

In addition, It is clear from this list of modes that JTRS radios will have relatively complex software. SPEAKeasy II planned two-dozen modes but succeeded in implementing only six plus GPS in the first two years. The program encountered more than a 2 : 1 growth in development cost, because of the inherent difficulty of systems and software engineering, design, development, and integration. The SPEAKeasy II software consisted of over 300,000 LOC. The design techniques presented in this book make it possible to deal with this growing complexity successfully.

### D. Mode Parameter Analysis

One approach to dealing with a proliferation of modes is to determine their parametric similarities. One may cluster the modes according to the critical software radio parameters given above. Figure 4-25, for example, clusters the JTRS parameters according to RF, bandwidth, and data rate.

From the cluster plot, it is easy to see that an SDR with 10 MHz of bandwidth and 2 Mbps of data rate readily supports the majority of the modes. Since some modes such as JTIDS employ fast frequency hopping and/or other ECCM techniques, the agility bandwidths may be understated. JTIDS, for example, hops over an available 240 MHz in the 1250 MHz band. In order to

**TABLE 4-5   JTRS Bands, Modes, and Data Rates**

| Modes/Capabilities | Frequency Band | Bandwidth | Data Rates |
|---|---|---|---|
| HF ISB w/ALE | 2–30 MHz | 3–12 kHz | 4.8/9.6 kbps |
| HF SSB w/ALE | 2–30 MHz | 3 kHz | 2.4/9.6 kbps |
| Link 11 (TADIL-A) | 2–30 & 225–400 MHz | 3 & 25 kHz | 2.25 kbps |
| STANAG 4285 (HF) | 2–30 MHz | 3 kHz | 2.4 kbps |
| STANAG 4529 | 2–30 MHz | 1.24 kHz | 1.8 kbps |
| ATC HF Data Link | 2–30 MHz | 3 kHz | 0.3, 0.6,1.2,1.8 kbps |
| SINCGARS | 30–88 MHz | 25 kHz | 16 kbps |
| SINCGARS SIP/ASIP | 30–88 MHz | 25 kHz | 16 kbps |
| VHF MSRT | 30–88 MHz | 25 kHz | 16 kbps |
| VHF FM | 30–88 MHz | 25 kHz | 16 kbps |
| VHF for ATC | 118–137 MHz | 8.33 kHz | N/A |
| VHF AM | 120–156 MHz | 25 kHz | 16 kbps |
| VHF FM LMR | 136–174 MHz | 12.5 & 25 kHz | 25 kHz: 16 kbps |
| ATC VHF Data Link | 118–137 MHz | 25 kHz | 31.5 kbps |
| UHF AM/FM PSK | 225–400 MHz | 25 kHz | 16 kbps |
| HAVE QUICKI/II | 225–400 MHz | 25 kHz | 16 kbps |
| UHF DAMA Satcom | 225–400 MHz | 5 and 25 kHz | 0.075, 0.3, 0.6, 1.2, 2.4, 4.8, 9.6, 16 kbps |
| UHF SATCOM MDR | 225–400 MHz | 5 and 25 kHz | 5 kHz: 7.2 kbps; 25 kHz: 32 kbps |
| STANAG 4231 | 224–400 MHz | Not published | Not published |
| Link 4A (TADIL-C) | 225–400 MHz | 25 kHz | 5 kbps |
| Link 11B (TADIL-B) | 225–400 MHz | 25 kHz | 0.6, 1.2, 2.4 kbps |
| JTT/CIBS-M | 225–400 MHz | 5 and 25 kHz | 19.2 kbps |
| SATURN | 225–400 MHz | 25 kHz | N/A |
| High-Capacity LOS (HCLOS) | 225–440 MHz and 1350–2690 | 50 MHz | 0.256, 0.512, 0.768, 1.5, 2.5, 4,8 Mbps |
| UHF LOS HDR | TBD | TBD | TBD |
| UHF FM Public | 403–512 MHz | 5, 12.5, & 25 kHz | 25 kHz: 16kbps |
| EPLRS | 420–450 MHz | 3 MHz | 57 & 114 kbps VECP |
| Cellular Radio | 800–900 MHz | 12.5–30 kHz | 2.4–9.6 kbps |
| GPS—Commercial | L1: 1575.42 MHz | C/A 2.046 MHz | N/A |
| GPS—U.S. Government | L1: & L2: 1227.6 MHz | 20.46 MHz | N/A |
| Link 16 (TADIL-J) | 969–1206 MHz | 3 MHz | 236 & 118 kbps FEC |
| Mode S Level 4 | 1030/1090 MHz | 3 & 3.5 MHz | N/A |
| INMARSAT A, B, C, M | 1525.0–1660.5 MHz | Service specific | Various |
| DWTS | 1350–1850 MHz | 125 kHz | 144,256,288,512,1024; 1.544, 2048 kbps |
| Soldier Radio | 1.75–1.85 GHz | 25Kz | 16kbps |
| WDW | Vendor proposed | TBD | TBD |
| VMF to Link16 | N/A | 31 kHz | 15.2, 28.8, 57.6 kbps |
| COBRA | TBD | TBD | TBD |

Legend: Mobile Subscriber Radio Terminal (MSRT), Joint Tactical Terminal (JTT), Common Integrated Broadcast Service Module (CIBS-M), Land Mobile Radio (LMR), Digital Wideband Transmission System (DWTS), Wideband Digital Waveform (WDW), UHF DAMA Satcom is DAMA, DASA MIL-STD-188-181/182/183 compliant; Medium Data Rate (MDR); STANAG 4231 is a UHF SATCOM system; High Data Rate (HDR), To Be Determined (TBD); Link 16 and TADIL J are the air interface of the JTIDS radio system.

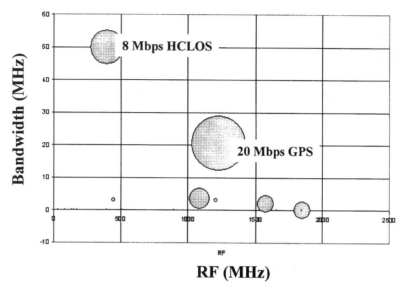

**RF (MHz)**

**Figure 4-25**   Mode clusters.

accommodate this aspect of the waveform digitally, an ADC would have to sample at about 600 MHz. This is more than an order of magnitude greater than the 10 MHz required for most JTRS radios.

Mode analysis may also be applied to commercial infrastructure and wireless products like PDAs. With conventional radio design, one is given a specific band and mode for which to build a product. In SDR design, one is given a small set of modes within one band for which to design a flexible module. Multiple modules may comprise the SDR family. In software-radio architecture evolution, one must define a framework within which these digital radio products, SDR families, and technology insertion may be mutually supportive. In addition to requirements, mode analysis can be applied to the analysis of product families in the architecture definition process.

## IV. ANALYZING THE PROTOCOL STACKS

The previous sections on network architecture analysis were concerned with the physical nodes of the network (mobile units and base stations) and their node-to-node interfaces, the air interfaces. This section is concerned with analyzing the internal logical structure of the nodes. That logical structure consists of the vertical components of the protocol stacks overlaid on the horizontal components (antennas, RF, modem, INFOSEC module, etc.). The vertical nature of the protocol stack arises from the increasing level of abstraction of capabilities available from the physical layer to the applications layer of the ISO/OSI protocol stack [6] as illustrated in Figure 4-26.

| Layer | # | Definition | Software Radio Functions |
|---|---|---|---|
| Application | 7 | User Services | Voice, Fax, Data, Multimedia, Location Finding Services* |
| Presentation | 6 | Translation and Remapping Syntax, Control Code Translations | Encryption, Virtual Terminals Object Request Broker* |
| Session | 5 | Connection Between Applications Overload Control, Checkpointing | Connection Allocation Radio Network Reconfiguration* |
| Transport | 4 | Group & Broadcast, Peer-Peer (5 ISO Classes of Transport) | Reliable Message Transfer Bandwidth Management* |
| Network | 3 | Segmenting, Routing & Integration | Transparent Data Transfer Background Packet Modes* |
| Datalink | 2 | Framing, Addressing & Error Control | Link (Point-to-Point) Interface Adaptive Quality Management* |
| Physical | 1 | Mechanical, Electrical Functional, Procedural and Timing | Transmission Segments ALE, Peer Contact* |

* Unique to software radio and software-defined network architectures.
Stallings, *Handbook of Computer-Communications Standards, Volume 1, The Open Systems Interconnection (OSI) Model* [7].

**Figure 4-26**   ISO protocol stack.

Software-radio architecture analysis addresses all levels of this protocol stack. In peer networks, the radio node supports all layers of the stack, while in hierarchical networks, nodes of different classes (e.g., MS, BTS, BSC, and MSC) each support different subsets of the layers. A software-radio architecture that spans these diverse horizontal components therefore has to accommodate differences in the vertical layers. In a multiband, multimode radio, there will also be significant differences in protocol from one mode to another. Therefore, if the radio is to provide cross-band and cross-mode services, the protocols have to be mapped to each other. This section reviews the seven layers so that subsequent sections can address software-radio architecture analysis related to protocols.

Layer 7, the applications layer, encapsulates the services provided to the user, which include voice, fax, data, multimedia, and teleconferencing. For 3G and beyond, applications will generally also include location finding. For military command, telemedicine, telecommuting, and collaborative computing, the applications layer also will include shared voice/whiteboard, and digital video. The mapping of protocol stacks on this layer should identify local facilities (e.g., GPS location-estimation) and network servers (e.g., databases) needed to support classes of applications.

Layer 6, the presentation layer, provides translation and mapping with control code translations. Encryption also may be provided at this layer. Object Request Brokers (ORBs) may appear to be encapsulated as layer 6/5 services (e.g., supporting remote procedure calls—RPCs) that connect applications from diverse vendors. Layer 5 provides reliable connections between applications, including overload control and checkpointing. Software radios may conduct different sessions on different bands and modes, transparently delivering connectivity to the upper layers in spite of unavailability of a lower-layer communications paths. To do this effectively, applications-specific data exchanges must be introduced between historically isolated layers [428]. Layer 4 reliably transfers messages with group addressing, broadcast, and peer-to-peer connectivity. ISO defines five classes of transport [6], a mix of which may be supported by a software radio. Software radios may also support bandwidth agility in layer 4, trading off data rate for BER or SNR as appropriate to the characteristics of the propagation path. This radio-generated agility may conflict with transport parameters of the application, so wireless-aware applications allow different parameter profiles for different classes of bearers.

The lower three layers of the protocol stack have significant implications for software radios. Layer 3, the network layer, segments the messages, routes them across the network, and reintegrates the packets into messages in the receiver. Software radios not only support standard services, but they also can provide background routing services, which the military calls *transparent bridging*. Layer 2, the data link layer, frames the messages into data bursts appropriate for the air interface. It also adds forward error control and retransmits packets as necessary. Layer 1, the physical layer, provides the RF channel modulation, mechanical and electrical interfaces with appropriate data sequencing and timing. Media Access Control (MAC) handles the layer 1, 2, and 3 aspects of access to a shared medium (e.g., for wireless LANs). Digital radios compensate for the anomalies of the physical layer through propagation channel estimation and equalization. Some modes like HF ALE and TETRA can also select the best available channels in order to construct a reliable virtual channel from a set of physical channels. Software radios have the potential to autonomously select the band and mode, extending the ALE virtual channel architecture to any RF band and air interface within the capability of the radio platform. This level of agility substantially complicates network control, node power management, probing for channels, and negotiating QoS parameters on each mode used. To better appreciate the nature of the protocol mappings, consider the software-radio implications of the protocol layers.

### A. Mapping SDR Applications to Protocol Stacks

In order to accommodate telecommunications applications on wireless devices, the software radio architecture must provide resources in the quantities and availability appropriate to the service. Table 4-6 identifies the typical protocol-related applications characteristics that can be analyzed to characterize the necessary software-radio resources. The objective of the analysis is not to design the protocol stack. Instead, the objective is to analyze those characteristics and parameters that constrain SDR resources including internal data bus capacities, memory (RAM, ROM, and mass storage), and processing capacity.

**TABLE 4-6  Typical Applications-Layer Characteristics**

| Applications Class | Characteristics and Parameters |
| --- | --- |
| All | Number of channels, underlying mode and bit rate, protocol profile (e.g., WAP over GSM & GPRS) |
| Location-aware | Location accuracy, update rate, number of mobiles |
| Voice | Source code, bit rate, frame rate |
| Facsimile | Page-buffer space, number of simultaneous channels, proprietary protocols to be supported (e.g., beyond Group IV) |
| Packet Data | Protocol (e.g., V.xx, X.25, TCP/IP, MPLS [167], ATM...), queue space, total delay, and delay spread |
| E-mail | Directory server, domain name server, gateways, proxies, security features, application constraints (e.g., Eudora, Outlook, Netscape) |
| File Transfer | Protocol (e.g., FTP), delay tolerance, maximum file size |
| Database | Size, query language (e.g., SQL or other), update rate, and latency |
| Voice Mail | Number of users, speech storage capacity, simultaneity |
| Multimedia | Mix of media (e.g., voice and shared whiteboard), delay tolerance, BER/FEC by class of service (e.g., line drawings vs. voice) |
| VTC | Source coding, profile (e.g., capability within H.320) |

Interoperability among applications can be achieved via CORBA [31]. If an application has an interface to the CORBA ORB, it can work with any other such application. The software-radio architecture has to include the facilities necessary to ensure that CORBA works for the radio applications. Thus, it includes design rules about how to apply CORBA and its IDL. In addition, real-time performance of isochronous applications has to be ensured explicitly in the architecture. This means augmenting CORBA with domain management facilities that express the unique constraints among radio applications. These constraints must include aggregate delay tolerance, data loss rates, BER, and data rates in addition to the format compatibility provided by CORBA. Thus augmented, CORBA can link distributed applications to radio transport resources implemented in a mix of lower-layer protocols. The lower-layer protocols might include a mix of dialup V.xx modems (using TCP/IP), GPRS, or X.25. Using CORBA for applications interoperability can also provide a mechanism for linking lower-layer protocols together in support of distributed applications.

Teleconferencing can be workable with disadvantaged terminals like wireless PDAs if one maps content to mode. In other words, a disadvantaged user might get a complete background image update only once each minute, but would get a full whiteboard update whenever it changes. The disadvantaged user might be given a choice between higher speech quality and image update rate. The network architecture has to specify that a mapping exists between ad-

**TABLE 4-7   Illustrative Applications/Protocol Constraints Matrix**

| Application | Voice | VTC | E-mail |
|---|---|---|---|
| Voice | PTT-IP isochronism | Voice priority vs. image | Voice annotations |
| VTC | MPEG recoding | Secure recoding | E-mail .mpg clips |
| Email | Background data transfer mode (AVD) | Background data transfer mode (AVD) | Map to standard internal/MIME |

vantaged users and disadvantaged users. From an implementation perspective, these mappings cost processing capacity and introduce latency and the possibility of exacerbating queuing problems. An architecture that reflects these needs of the protocol stacks will incorporate mapping tables that not only specify translations, but also specify the computational resource implications of invoking them.

The systems-level analysis lists the applications and identifies constraints among applications and waveforms. The results of this analysis then may be used to guide the software tradeoffs undertaken subsequently. A systems-level analysis of applications should identify the protocol-related issues as illustrated in Table 4-7.

The top row of this matrix lists the primary application, while the leftmost column indicates the interoperating application. The voice-voice entry shows that push-to-talk voice must interoperate with voice over IP, and that maintaining isochronism is the primary constraint. The design tradeoffs driven by this entry focus on ensuring that computational resources are provided to the IP protocol stack so that SDR internal delays do not contribute to delay spread. In addition, the waveform-related constraint is that this mode of linking two voice streams should code the IP voice at the highest data rate available on the physical channel. In other words, it is better to burst the IP packets so that they arrive early, and then time division multiplex the channel with other users, than to send the packets at a more leisurely pace, risking loss of isochronism. The implication is that the memory and processor facilities for making all this happen must be part of the architecture.

In the voice-VTC entry, a video-teleconference stream is to be joined by a voice-channel subscriber. The entry indicates that the voice subscriber may send an initial image (e.g., digital snapshot), but that in transferring the VTC stream to the voice-channel user, the voice is to be given priority over the image part of the stream. The voice-email entry indicates that a voice channel may be mated to an e-mail connection if the voice user leaves a voice message which is later appended to an e-mail as a voice annotation. (The e-mail may be sent on voice command by the voice user, for example.). The second row identifies mode-related protocol-level interactions among VTC as a secondary application in one channel and voice, VTC, or e-mail as the primary service. Its entries parallel the first row. In the last row, e-mail may be sent on a voice

| Communications Services | *Applications* | *Protocol Mappings* |
|---|---|---|
| | Voice, facsimile, data (e-mail, file transfer) | *WAP* |
| | Databases, voice mail, multimedia, VTC | *CORBA* |
| | *Related Services* | *SQL* |
| | Location finding | *X.400* |
| | Over-the-air downloads | *TCP/IP, MPLS* |
| | | *MIME ...* |
| **Radio Applications** | *Air interfaces (waveforms)* | *Protocol Mappings* |
| | State machines, modems, etc. | *Waveform/Air* |
| | *Synchronization algebra* | *AVD, DS0, Group IV...* |
| **Radio Infrastructure** | Data movement, memory management... | *Protocol Mappings* |
| | *Domain manager* | *CORBA* |
| | | *ODP/X.900* |
| **Hardware Platform** | Antenna(s), analog RF hardware, ASICS, FPGAs, DSPs, microprocessors, instruction set architecture, operating systems | |

**Figure 4-27**   Protocols in a layered software radio architecture.

or VTC circuit if the Alternating Voice and Data (AVD) protocol [168] is used. The e-mail-e-mail entry says that two distinct e-mail channels will be mapped to a standard internal format that supports MIME, the Multipurpose Internet Mail Extension [169]. A full-scale version of Table 4-7 provides a useful tool in both designing SDR systems and assessing candidate software-radio architectures.

An important contemporary focus of applications-layer development is Internet access via wireless devices (e.g., cell phones initially and PDAs later). Wireless Internet access is seen as a growth area demographically, since there are far more cell-phone users than PC users. According to Phone.com (formerly the Unwired Planet), there will be over 500 million wireless subscribers by 2001 [170]. They envision corporate applications in sales force automation and fleet dispatch. Real-time delivery of weather, news, sports, stock market quotes and traffic alerts could be customized to the user and location-aware. Banking and electronic commerce is increasingly using the Internet, and mobile applications could add a new dimension of tailored merchandizing. Cell phone applications are currently being fielded in the Wireless Markup Language (WML), an open-architecture standard defined by the Wireless Application Protocol (WAP) forum [171]. Compatibility of the software radio architecture with these rapidly emerging applications and protocol standards is therefore essential for the architecture to retain its commercial relevance. This can be achieved by incorporating WAP and other commercial protocols into the appropriate layer of the software-radio architecture, as illustrated in Figure 4-27.

In addition, interactions among applications must be accommodated. Some applications classes may mix readily with others. For example, location-aware-

ness can be readily established through the embedding of a satellite location service (e.g., GPS). Alternately (or in addition), the network may include
emitter location technology. Subscriber locations may then be used for other
services. For example, voice mail might be annotated with the location of the
user that left the mail. A high-quality software-radio architecture standardizes
the exchange of location information in a formal meta-data structure such as a
location API. In general, a high-quality architecture formalizes the exchange of
any data that is likely to be needed across multiple applications. A reasonable
framework for the exchange of such information is in the layering of the
software-radio architecture as illustrated in Figure 4-27.

Increasingly, applications provide end-to-end encryption. This constrains
the way in which lower layers can access applications data. Suppose, for
example, one would like to set up a secure video-teleconference. If the participants all have the same data rate, setup is straightforward. Suppose, however,
that some of the links are disadvantaged radio links with low data rates. Radio
access points (RAPs) cannot provide rate control by recoding the videostream
because they have no decryption capability. Instead, the network must set up
a secure rate-adaptation server, and the RAPs have to deliver the recoded
video within the delay tolerance of the source coder. The implication is that
secure services may constrain the network to connect through applications-
layer entities such as a secure rate adaptation server. Such a server may be
downloaded to an SDR RAP, simplifying network routing. The SDR therefore
expands the trade space, increasing the complexity of the protocol-mapping
process. If the SDR can host network services, then the flow of data through
the network may be simplified, resulting in reduced overhead and enhanced
reliability.

The above analysis provides a top-down framework for structuring the analysis of the lower layers of the protocol stack: the network, data link, and
physical layers.

## B. The Network Layer

The network layer contributes additional constraints on SDR design. First,
the well-established network analysis tools [172, 173] may be employed to
analyzing routing, queuing, and related aspects of the network. This level
of analysis establishes resource bounds, such as maximum buffer sizes and
processing latency in a node. These network-level parameters may then be
allocated to the resource budgets of the horizontal components of the SDR
(e.g., modem software, INFOSEC, etc.). Buffer space may be allocated to the
*service and network support function* (recall Figure 1-3), for example. In a
conventional digital radio, these resources are fixed at design time, but in a
software-defined implementation, they are defined during operations. In order
to allow the deferral of resource allocations until operations, the architecture
must include a facility to assign hardware resources (e.g., buffer space) to
software entities (e.g., the service and network support object). In a high-

quality software-radio architecture, this task, domain management, is allocated to the *evolution support function* (also Figure 1-3).

Commercial [174] and military [175, 176] radios may also support dynamic, ad-hoc networking in which the links in the network are constantly changing and varying in QoS due to the movement of nodes and the variability of radio propagation. The IETF has established a working group on mobile ad-hoc networks [177]. Ad-hoc networking is the subject of much current research that is outside the scope of this text. Since the performance of ad-hoc networks is difficult to assess, specialized simulation tools have been developed [178, 179]. From a software radio perspective, ad-hoc networking can induce additional constraints on time and buffer space within the software radio. For example, most algorithms require some kind of route cache [180], the information the router (algorithm) has discovered processing route-reply and route-error messages. Most require a route-request table as well. The routing algorithms have to handle numerous conditions including nonpropagating route requests, snooping, salvaging, and gratuitous replies. This algorithm complexity introduces uncertainty in the time required for networking algorithms to run to completion. Dynamic variability of data rates, FEC, and other air interface protocol parameters is an aspect of ad-hoc and dynamic networks that can also cause conflicts for processing capacity. The software-radio architecture therefore must have facilities for mediating dynamic conflicts for processing resources. This task also is part of domain management.

If these circuits are bearers for packet-switched traffic, the network layer of the protocol stack must be supported as illustrated in Figure 4-28. The data stream is segmented into packets, framed, and protected with FEC coding. Each packet is independently routed to its logical destination. Therefore, some packets may arrive out of order. The receiver then decodes the FEC, reorders the packets, and reassembles the original data structure (e.g., file of data being transferred). The software radio implementation of these layers of the protocol stack is generally straightforward. TCP/IP [6] is evolving to deal with the characteristics of radio as a physical layer medium. Initially, TCP was an extremely inefficient way of moving data over radio. TCP has two parameters that it adjusts to learn the expected time delay before receiving a block acknowledgment (ACK). When SNR is high and BER is low, TCP learns that the remote end ACKs the packets very quickly. As the radio enters a fade, the physical-layer mechanisms such as ARQ cause retransmissions which both get the data through and take more time than TCP would like. As a result, TCP's threshold is exceeded and it retransmits a block which has been roughly 80% transmitted successfully. Mechanisms have been defined that interact with TCP so that this does not continue to happen [429].

Mobile IP [181], another important protocol, deals with the fact that a mobile user has no fixed IP address. It employs tunneling, the encapsulation of address and related information so that packets are routed from a home

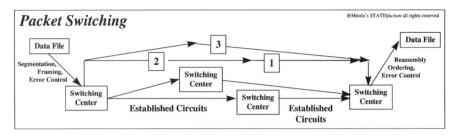

**Figure 4-28**   Packet switching independent of the physical network.

agent to a foreign agent where the packet is unencapsulated and delivered locally to the mobile subscriber. In most cases, the software implementations are limited to the data processing "back end" of the radio architecture. Since these bitstreams occur at relatively low data rates (e.g., less than 2 Mbps) and since the streams are relatively forgiving of delay spread, there are few implementation challenges to be addressed by the software-radio architecture.

*Handoff* is the transfer of data link continuity from one base station to another as an MS transitions across cell site boundaries. *Handover* is the transfer of continuity from one mode to another, possibly in response to a transition out of a preferred service area (e.g., PCS) with fallback to another mode (e.g., satellite mobile). Mode handover may also occur due to the failure of the current mode, e.g., due to lower-than-acceptable BER. Or it may occur in order to provide a specific service such as handover to a 155 Mbps EHF short-range (100 m) mode to receive a new terrain map in preparation for a military exercise. The data link layer of the ISO stack has to maintain continuity and indicate link status to the higher layers of the protocol. The software-radio architecture therefore has to accommodate band/mode bridging at the network layer of the protocol stack.

## C. The Data Link Layer

The data link layer of the ISO/OSI protocol stack is responsible for establishing point-to-point connectivity. The analysis of the data link layer includes an assessment of addressing, handoff/handover, framing, error control, and multiplexing as summarized in Figure 4-29. Addresses in cellular systems include the identity of the country, network, and node being used by a mobile subscriber. In GSM, for example, the International Mobile Subscriber Identity (IMSI) establishes a home location register (HLR) for a subscriber. Some link-layer modes used in a multimode path may not explicitly support such addressing (e.g., PTT voice). The information may need to be carried as side-information in the intervening SDR nodes in order to preserve that information (e.g., using AVD on a PTT circuit) for use in a downstream mode that requires it. Addressing, for the purposes of SDR mode-bridging architecture

| Function | Definition | *Data Link Layer* |
|----------|------------|-------------------|
| **Addressing** | Use of overhead bandwidth to establish the radio node to which traffic is to be passed | |
| **Handover** | Assignment of mobile subscriber to node | |
| **Framing** | Partitioning of traffic into frames for synchronization and control | |
| **Error Control** | Detection and correction of errors, typically by the addition of overhead bits | |
| **Multiplexing** | Means by which multiple subscribers share a single radio carrier | |
| **Multiple Access** | Means by which multiple subscribers share BW | |
| **Modulation** | Imparting signal structure to an (RF) carrier | |
| **Timing** | Identification of bit boundaries with precision necessary to reliably recover information | |
| *Physical Layer* | | |

**Figure 4-29**   Data link layer.

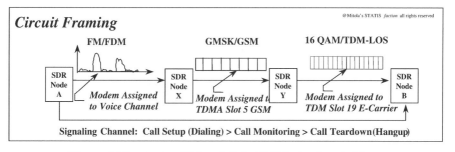

**Figure 4-30**   Analog FM and digital PCM circuit switching.

analysis, also includes data required by the signaling system such as dialed numbers.

Of these functions, modulation, framing, and timing warrant further attention. In particular, the software-radio architecture must bridge across significant differences between modes in order to support multimode operations. Figure 4-30 illustrates important differences among circuit-switched channels. In this example, four SDR nodes carry traffic from node A to node B using a sequence of mode bridges. Intervening SDR node X has to translate (bridge) the appropriate FM/FDM channel to GSM. Since FM/FDM is not a framed structure, but GSM uses 4.615 ms frames, the bridge must map the output of a point process, the FM/FDM demodulator, to a block process, the GSM burst modulator. FM/FDM demodulation is called a point process because it is memoryless. That is, the algorithm operates on a stream of digital samples. The FM demodulation process requires only a few digital samples of memory. Subsequent digital frequency conversion to baseband and filtering of the subscriber signal also takes typically fewer than two-dozen memory elements for the taps of the filter.

Finally, a second FM demodulation process recovers baseband audio. The digital FM/FDM demodulator produces an isochronous stream (e.g., a DS0 channel). The GSM's RPE-LTP voice coder transforms the baseband DS0 channel into a vocoded bitstream. Those bits then must be partitioned into groups of 60 bits, the payload of each GSM burst. These bits are then packaged with tail bits, a training sequence, etc. and modulated by a second point process, the GMSK modulator. A stream manager links the sampled signal (e.g., a UNIX stream) from the digital IF, to the FM/FDM demodulator, to the GSM vocoder. A block process accumulates 60 vocoded bits for framing. Then another stream process modulates the bits at GSM's 270.83 kbps rate. The SDR must maintain time-coherence among all these block and stream processes to within one audio sample period, 125 $\mu$sec.

SDR node Y has more difficulty synchronizing the GSM mode with the E-carrier (e.g., on a LOS 2 GHz radio-relay channel). E-carriers employ a 2.048 MHz data rate, divided into 32 channels, 30 of which support traffic. This SDR now must recode the GSM voice format into the PCM format of the E-carrier. This is the GSM TRAU function. E-carriers expect each traffic channel to deliver one voice sample every 125 microseconds. The SDR therefore has to translate the 270.83 kbps to 8 kbps, and must repack the bits from 60 bits in 4.615 ms to one 8-bit sample every 125 $\mu$sec. Since the GSM clock is defined by the GSM network, but the LOS clock is defined by the receiving end of the link, node Y has to allow both of these links to have independent clocks. In addition, both of these clocks have different drift rates than the node's internal clock. In order to do this, the SDR design must employ some method for establishing time independently on each channel. This may be accomplished using synchronization routines based on trigger events. In order for the events (e.g., start of frame) to be scheduled accurately, the synchronization package must include separate timing data sets, and time-correction factors (offset and drift rate) for each channel. A high-quality software-radio architecture will incorporate these facilities in its *radio infrastructure* layer (refer again to Figure 4-27).

Framing, bit-stuffing, interleaving, randomization, and error control are primary data link layer mechanisms for establishing link connections. The choice of specific techniques for multiplexing and error control can have a strong impact on the data quality. In this area, the tradeoff pits QoS against algorithm complexity, and hence processing resources that must be allocated for the advanced algorithms. A tight design for an SDR product would therefore minimize the complexity of link personalities. A robust software-radio architecture, on the other hand, would provide specific facilities that promote the evolution of algorithms as multiband-multimode applications increase. The layering of the software-radio architecture provides the mechanism for this growth.

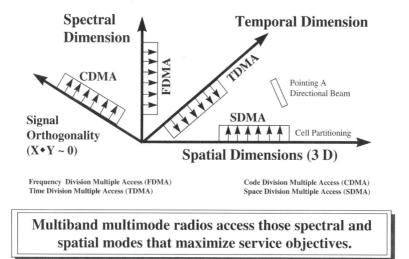

**Figure 4-31**   Physical layer: The radio dimensions.

## D. The Physical Layer Analysis

The analysis of the physical layer should identify the range of approaches to be supported by the architecture for delivering services in spite of the physical limitations of the individual bands and modes. From a software-radio perspective, the fundamental parameters of these limitations are propagation range, fading, interference, and usable bandwidth. In general, the architect has to consider any feasible sharing of radio resources that involve multiple access to signal epochs in time-space-frequency-parametric dimensions. To ensure access, one must have an appropriate multiple-access scheme. There must also be sufficient radiated power, antenna gain, and receiver processing gain to recover the transmitted signal(s) with QoS-defined SNR, carrier-to-interference ratio (CIR), or BER. Providing these mechanisms is in the domain of radio, network, and air interface design. To ensure viable evolution paths, the software-radio architecture must explicitly encourage the introduction of cross-band and cross-mode techniques, evolving over time. This is the domain of software-radio architecture.

As illustrated in Figure 4-31, the physical layer may be shared by any of the spectral, spatial, temporal, and code dimensions. Spectral sharing includes the allocation of bands and the assignment of modes to a given RF carrier. In first-generation cellular systems, each subscriber was allocated a dedicated RF carrier. The 2G systems improved the efficiency of spectrum use, measured in terms of the number of simultaneous subscribers per Hz. Efficiency may also be measured at a specific QoS profile, such as:

$$\text{BER} < 10^{-3}, \qquad \text{Packet Loss Rate} < 10^{-2},$$

$$\text{Delay Spread} < 100 \text{ ms}, \qquad \text{and} \qquad \text{GoS} > 95\%$$

In 2G, this was accomplished through timesharing a medium-bandwidth channel (e.g., GSM) or by digitally multiplexing more than one user in a channel previously occupied by a single user (e.g., digital AMPS). These time division multiple-access (TDMA) approaches require synchronization on each user burst. Spectral efficiency is reduced by the guard-time between bursts. Guard-time is a function of maximum user distance from the base station. The speed of propagation converts maximum propagation distance to equivalent guard time, nominally at 1 foot per nanosecond.

Code sharing depends on the mutual orthogonality of spreading codes in direct-sequence spread-spectrum (DSSS) systems. Although such codes are essentially orthogonal, the 20 to 30 dB of gain afforded by the codes is small compared to the near–far ratio. The 90 dB near–far ratio of terrestrial wireless dwarfs even the 30 dB spreading gain of DSSS. Imperfections of physical devices, the near–far ratio, and the imperfect orthogonality of practical codes leads to spectral efficiency of CDMA that is lower than researchers at first predicted. About 64 CDMA users occupy bandwidth previously occupied by 59 FDMA users, but with 1 : 1 frequency reuse. This yields an improvement of spectrum efficiency of nominally 7/3 and practically $> 200\%$ compared with 2G. Differences in cell deployment, power control, and voice coding make it difficult to compare FDMA, TDMA, and CDMA accurately.[22]

Space division multiple access (SDMA) includes cell partitioning and the use of sectorized and smart beamforming antennas to create spatial extents within which spectrum can be reused. Spatial sharing is also enhanced by the use of higher carrier frequencies which allow smaller beams and lower sidelobe power for a given antenna aperture. Such directional beams can be dynamically pointed to track a subscriber to maintain a high data rate in spite of mobility.

For fixed protocols, the guard-time is established in the standard. For future dynamically defined protocols, the guard-time may be decided among the few participants in an ad-hoc network, based on their ranges from each other. In many specific scenarios, the necessary closed-world assumption is valid. Although these protocols do not yet exist, some are being investigated [48]. Since today's research can be tomorrow's standard, a robust software-radio architecture must accommodate such growth paths.

Multiband, multimode radios access those spatial and spectral modes that maximize service objectives. In the past, dedicated hardware was deployed for each radio mode. With software radios, however, even a basic software radio has sufficient processing capacity and RF access to create hybrid FDMA, CDMA, and TDMA modes needed to optimize service delivery, provided the other radios in the network are also software radios. If not, then the network will include a subnetwork that is compatible with the legacy mode, but the software radio can also operate in enhanced modes with other software radios.

---

[22]A review of the IMT-2000 proposals UWC136, CDMAone, and W-CDMA reveals how challenging this is.

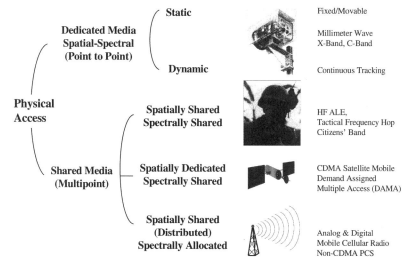

**Figure 4-32**   Physical layer modes.

SDMA requires multiple antennas with parallel RF and signal processing for beamforming, a significant increase in processing capacity over the GFLOP required for a nominal four-channel narrowband software radio.

The software radio presents alternatives for physical access as illustrated in Figure 4-32. The media may be dedicated such as with point-to-point digital radios or may be shared spatially and/or spectrally. Spatial/spectral sharing is accomplished using the FDMA, CDMA, and/or TDMA approaches discussed earlier. In addition, dedicated media may be used in a static employment such as a building-to-building link, or in a dynamic way such as timesharing a dedicated millimeter wave link by pointing it sequentially at first one user and then another. Dynamic spatial employment of dedicated media might be appropriate, for example, in the vicinity of a database server so that database updates are delivered at a very high data rate (e.g., 155 Mbps) for short periods (e.g., 100 ms). Delivery of battlefield maps, situation data, and operations orders might be provided from the operations building to soldiers in the vicinity using such a technique. Those soldiers might then download the same data at similar data rates during a face-to-face meeting. Even a very-low-power transmitter can deliver high data rate over a few meters. This somewhat speculative treatment is intended to suggest some of the many opportunities that software radio technology makes possible. Each such opportunity is earned at the price of increased radio complexity, primarily in terms of increased software complexity, but hardware phase shifters, antenna pointing, and other hardware are currently required for some of the physical modes such as continuous-tracking high-data-rate physical-layer modes. Thus, one must define a market niche and time-table with anticipated technology insertion opportunities in order to constrain one's enterprise architecture to feasible physical modes.

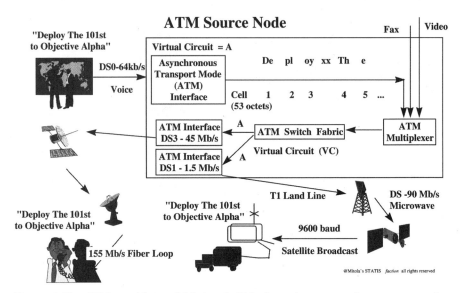

**Figure 4-33**   ATM provides variable bandwidth channels over synchronous networks.

## E. Alternate Protocol Stacks: Wireless ATM

Asynchronous Transfer Mode (ATM) was originally designed for highly reliable SDH circuits of the class offered by fiber-optic systems. With extremely low bit error rates, fiber provides an ideal mechanism for packing
voice and data channels more efficiently than is possible using the native
synchronous mode of SDH. As illustrated in Figure 4-33, ATM breaks subscriber traffic into 53-byte (octet) cells. Each cell has a 48-byte payload and
a 5-byte header. Although illustrated by short segments of text, the example of the figure is meant to convey the segmentation of a vocoded voice
stream.

A major advantage of ATM is its flexibility. As illustrated in Figure
4-33, an ATM packet-switched channel can be established over fiber, cable,
microwave, satellite, and mobile wireless radio channels. Wireless ATM (W-
ATM) is a growing technology for distributing telephony and Internet service
to the wireless local loop (WLL).

The U.S. DoD was a pioneer in the use of ATM over radio. In the early
and mid-1990s, the U.S. DoD conducted experiments in which ATM streams
were provided to tactical aircraft. One of the lessons learned in early experiments is that loss of frame synchronization is a major circuit fault for ATM.
Essentially all the ATM cells are lost until synchronization is restored.
This can take several seconds or longer, resulting in the loss of critical
data.

Li [182], however, noted that there is significant redundancy in the ATM
header. An unrestricted ATM header dedicates 5 bytes to addressing and FEC

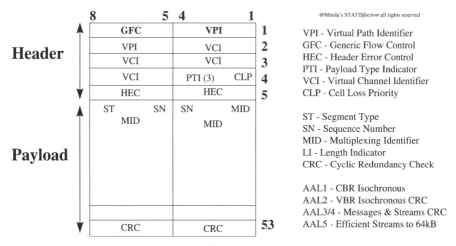

**Figure 4-34** ATM cell structure [6].

protection of the header. These bits can support up to $2^{40}$ combinations of address bits. He observed that there are rarely more than 100 actual users within radio LOS of a mobile ATM node. He therefore developed a technique for coding the 100 addresses that are needed into the 5 bytes that are available in a very robust and forgiving way. As a result, the header can be used for both synchronization and addressing with bit error rates down to worse than 1/100. The robustness of this frame synchronization yields a reliable wireless ATM link in spite of major channel impairments typical of radio applications. (See Figure 4.34.)

**F. Exercises**

1. Name the major components of a GSM network. List the critical parameters of the GSM air interface. What is the relevance of this network to the disaster-relief case study?

2. Name the major functional components of a military network. What classes of data rate need to be supported? What is the relevance of these classes of radio to the disaster-relief application?

3. What JTRS air interface has critical parameters similar to GSM? Which JTRS air interfaces present bandwidth challenges to an SDR implementation?

4. What is mode-parameter analysis? How would one apply this technique to develop a technology insertion roadmap? Provide a ten-year roadmap for the disaster-relief application.

TABLE 4-8  Systems-Level Architecture Parameters

| Parameter | Remarks |
| --- | --- |
| Number of Radio Units | Software radios have the potential to provide the same performance as discrete radios, but in a way that shares computational resources, reducing the number of radio units that must be provided to accomplish a task. |
| Radio Resource Management | Software radios have the potential to significantly enhance the use of spectrum through continuous spectrum monitoring with dynamic adaptation of power, antenna pattern, waveform, data rate, and quality of service. Spectrum rentals offer unique opportunities. |
| Transmission Parameter Agility | Although some discrete radios adapt power (e.g., IS-95 CDMA), the integrated adaptation of all of these parameters including air interface waveform to optimize resources for the communications task is the unique domain of the software radio. |
| Reception Parameter Agility | There is an interplay among the adaptation mechanisms that are available; modeling tools are necessary to estimate performance for dynamic optimization. |

## V. SYSTEMS-LEVEL ARCHITECTURE PARAMETERS

Development of software radio nodes requires management of the critical node parameters. Table 4-8 identifies the critical parameters with top-level remarks regarding how such parameters may be used in a software radio node architecture.

### A. Exercises

1. Determine the number of radio units in the disaster-relief case study. What approaches can be used to share computational resources among these radios? Can client-server, fat-client, and/or thin-client approaches [430, 431] be used between the disaster-relief vans? within a van? What is the single most significant architecture parameter associated with such an approach?

2. What functions must be included in an SDR architecture in order to support the concepts summarized in Table 4-8? What components are needed to support these functions that were not in prior-generation discrete/digital baseband radios? Which of these components are COTS? Which need breakthrough(s)?

3. Describe the functions of a domain manager that supports the concept of Table 4-8. Which functions pose large software development risk?

# 5 Node-Level Architecture Analysis

This chapter analyzes node-level software radio architecture. Attention turns to the internal functions, components, and design rules within a radio node. The canonical node architecture partitions software-radio functions into segments within which functions are functionally cohesive, and between which the segments are data-coupled. This approach conforms to well-established principles of *structured design* [183, 184]. SD has been superseded in contemporary practice by object-oriented technology (OOT) [185]. SDR precursor systems to which the author contributed were organized according to the SD principles of functional cohesion and data coupling. Message passing was a necessity for distributed processing among multiple minicomputers and microprocessors. These high-end military command-and-control systems employed federated parallel processing. Full-custom ASICs and special-purpose digital signal processing boards were integrated with a dozen minicomputers and over 100 Intel-8080-class microprocessors to create early cutting-edge signal processing capacity. This progenitor technology anticipated the emergence of commercial DSP chips and boards by about ten years. The design principles of functional cohesion, data coupling, and message passing developed then apply to software-radio architecture today. The military progenitor systems, however, prioritized mission effectiveness, maximizing technology insertion with cost as a relatively unconstrained variable. In the application of SDR technology to both military and commercial domains today, cost is a highly constrained input-variable. Therefore, the node-level architecture analysis treats cost and other externally imposed design constraints (e.g., standards) as explicit design rules.

OOT has matured the foundation principles of SD, adding features that promote software reuse. For example, UML, as an OOT computer-aided software engineering (CASE)-tool, integrates complementary views of a collection of objects. Use-case scenarios, logical structure, components, and deployment aspects may be developed independently, but UML assures that they are consistent for the set of objects being defined. This chapter develops the SDR node-level application of OOT and UML. This is not merely another treatment of object-oriented design. Instead, architecture emerges as an object-oriented framework for the peaceful coexistence of otherwise mutually incompatible (structured, object-oriented, and ad-hoc) designs. As DSP technology has

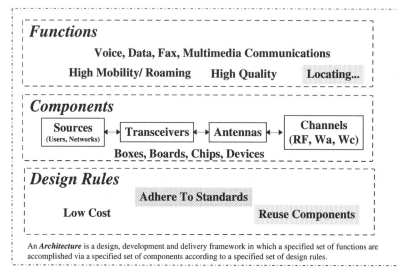

**Figure 5-1**   Aspects of architecture.

developed, available processing capacity has increased according to Moore's Law. Radio functions have been packaged into analog RF, digital hardware, and software many times. A slowly evolving set of functions has been repeatedly rehosted into a rapidly evolving set of (hardware-intensive) components. Node-level design principles have evolved through this process into the software-radio architecture strategy defined in this chapter.

## I. ARCHITECTURE REPRESENTATION

As illustrated in Figure 5-1, an architecture is a framework in which a specified family of functions may be accomplished via specified classes of components according to specified design rules. In this figure, architecture is represented as a collection of associated concepts with little obvious structure. The node-level analysis of software radio architecture begins with the organization of hardware components so as to maximize software flexibility. The hardware components include wideband antennas and RF/IF processing, wideband ADCs, DACs, parameter-controlled ASICs, FPGAs, DSPs and general-purpose processors. The software components consist of data structures and procedures organized into software objects. The analysis process begins with an examination of consistency and conflicts among functions, components, and design rules.

Consider the static hierarchies illustrated in Figure 5-2. These functions, components, and design rules characterize a top-level radio function. In this case, the function is to "transduce voice from [auditory] compression waves to radio waves." This function may be implemented via a processing thread that

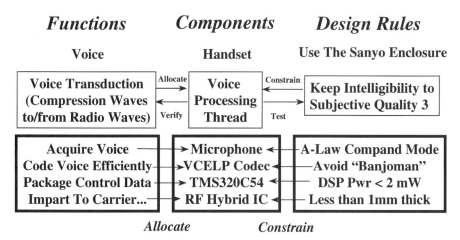

**Figure 5-2**   Radio architecture map.

begins with the microphone. The ADC component converts the signal from the microphone into a sampled baseband signal. A DSP algorithm vocodes the voice waveform efficiently, and another imparts the channel coding. A baseband DAC converts the signal to analog so that analog IF/RF stages can up-convert and impart it to a carrier.

In this representation of architecture, the mapping between the "voice transducer" function and the hardware and software components are more explicit than in Figure 5-1. In addition, the relationship between components and design rules has been made explicit. During the design stages of a project, the emphasis is on allocating functions to components. The components also are subject to the constraints of design rules. During the integration stages, the emphasis shifts to the verification that components implement functions while adhering to the design rules.

Business considerations create another dimension of design rules. A need to control costs may be met through the reuse of a (notional) "Sanyo enclosure." That module brings packaging constraints. These may violate thermal design rules. Many subsystems will be overspecified or overconstrained. One must therefore sacrifice some performance constraint or violate some design rules slightly to achieve a realizable, affordable design. During development, the components must be verified to accomplish the allocated functions. They must be tested to assure compliance to type-certification design-rules, such as radiated power and EMI limits.

Continuing with Figure 5-2, the middle series of boxes (e.g., "Acquire Voice," "Microphone," etc.) shows how one might hierarchically decompose the top-level functions. The resulting sequence of subordinate functions may be associated with other subordinate components. These are associated with other design rules. Reading across the figure, the subordinate function "acquire voice" is allocated to a microphone component. This notional component has a

## *Functions*

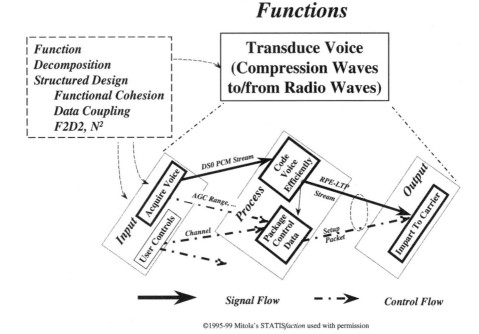

©1995-99 Mitola's STATIS*faction* used with permission

**Figure 5-3**   View of a traditional function hierarchy.

built-in analog compander. The notional VCELP codec complies with the design rule "avoid banjoman." There is a condition in GSM when the RPE-LTE algorithm resonates due to bit errors in a way that sounds to Americans like someone is playing the banjo, making a twanging sound. VCELP avoids this sound under severe BER conditions. The associated design rules encompass power, packaging, design, and performance issues. This yields an interlocking web of constraints that can easily become tangled unless subjected to rigorous engineering discipline. One simple but reliable engineering discipline suggested in the figure is the identification of one-to-one relationships between functions, components, and design rules. One-to-one (1 : 1) relationships localize the issues associated with tradeoffs. With 1 : 1 relationships, the steps needed to meet specific technical, performance, or schedule goals are clearly allocated a specific subsystem or component. This greatly simplifies the trade space, reducing the time and effort required to reach design decisions.

### A. Functional Design Hierarchies

The structure of a representative functional design hierarchy is illustrated in Figure 5-3. This is a three-dimensional view, with input on the left, output on the right, and higher levels of abstraction organized vertically. The top-level function "transduce voice" is hierarchically partitioned into the subordinate functions "acquire voice," "code voice efficiently," and "impart to carrier." These three functions accomplish the isochronous signal flow as shown in the bold arrows. The user interface and the packaging of control data is accomplished in the companion flow of control messages.

The structured design (SD) concepts of cohesion and coupling provide concrete criteria for partitioning the software into modules. Cohesion is the relationship among elements within a module. Functional cohesion is the tightest and hence most desirable kind, representative of the relationship between a function and its arguments. Yourdon defined other types of cohesion ranging from functional as the tightest to incidental as the loosest. In incidental cohesion, functions in a module share little or no relationship with each other. Modules may be coupled to each other through passing data (e.g., message passing), the loosest form of coupling. They may call each other, which is a tighter form of coupling. If they exercise internal control over each other's internal functions, or cause unanticipated side effects, they are maximally coupled. Data-coupled modules may be changed independently with no side effects. Data coupled, functionally cohesive modules are the ideal.

Techniques for getting to well-structured designs include the creation of *functional flow diagrams and descriptions* (F2D2). In Figure 5-3, the lowest level is represented in F2D2. A sequence of rigorous descriptions of the functions and interface are included in a formal F2D2 product. One also must examine the interfaces among the resulting modules (e.g., the coupling of the modules) by creating an $N^2$ diagram. This is a matrix with modules listed on the rows and columns. The values in the matrix represent the interfaces among the modules. The analysis process should yield interfaces that are well defined, complete, and consistent.

To go beyond design into architecture, one must consider a collection of different designs. The collection of designs should include designs produced by different teams at the same time, and designs that evolve over time. Consider the reuse of the software components that resulted from the functional decomposition above. A serious difficulty in the reuse of such software components becomes evident trying to map a functional component, the voice processing thread, into the hardware and software configuration hierarchies (Figure 5-4). The configuration hierarchies express packaging for configuration management. The figure shows a configuration hierarchy within which the notional handset's voice transducer could be configuration-managed. The Mobile Station (MS) hierarchy includes hardware and software components. As illustrated, the voice processing functional thread is distributed across codec, programmable RF IC, and TMS320 DSP chip. Which software functions are allocated to which components? This is not clear because the functional components are distributed in dedicated hardware and in software algorithms. A high-quality architecture framework therefore should help one express the functional components in an intuitively appealing, but implementation-independent way. That framework must also map these slowly evolving functional components to the rapidly evolving hardware and software components.

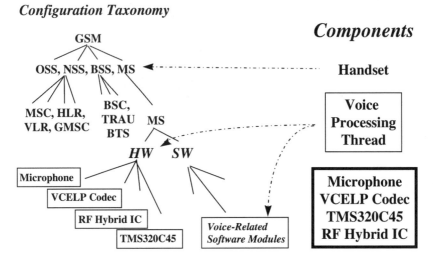

**Figure 5-4**  Configuration hierarchies obscure functional relationships, increasing the cost of component reuse.

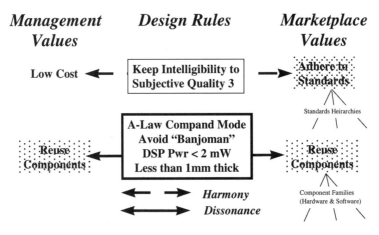

**Design rule conflicts are resolved through cost/benefit-over-time analyses.**

**Figure 5-5**  Design rule hierarchies further complicate design tradeoffs.

Design rule hierarchies complicate matters further as suggested in Figure 5-5. Design rules may be driven by customer-centric marketplace values including quality and aesthetics. Competition-centric values such as adherence to standards for compatibility and reuse of components for low cost may take on a central role in some design tradeoffs.

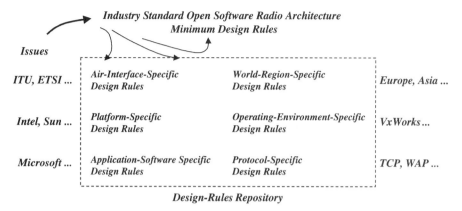

*Industry Standard Open Software Radio Architecture*
*Minimum Design Rules*

**Figure 5-6**   Design rules from complementary disciplines feed a design-rule repository.

When design rules are in harmony, the design tradeoffs are straightforward. Keeping subjective quality high, adherence to standards, and low cost harmonize in the narrow choice of a voice-coding algorithm. They are often not in harmony, however. There may be a conflict between reusing components and driving the form factor to a reduced thickness. Algorithm tradeoffs often influence size, weight, and power in a handset. The robust algorithms generally require more processing capacity and hence more powerful, power-hungry chips. It is thus helpful to find a design framework within which the entire set of issues related to functions, hardware and software components, and design rules are most manageable.

The open-architecture framework should impose the minimum design rules necessary to insure that architecture objectives are met. The objectives of open-architecture are often mutually contradictory. For example, open-architecture needs to both promote commonality and encourage competition. Commonality means buying components of one design, while competition requires one to buy components from multiple suppliers. The solution to this technical quandary invariably rests on a social process of letting competing interests pull in opposite directions until everyone is equally unhappy, but still willing to participate in the architecture. Architecture analysis facilitates that process by making the tradeoffs clear. It can also facilitate the process by providing mechanisms that allow one to hide or safely ignore contentious issues. For example, the banjoman phenomenon applies only to a particular GSM vocoder. The avoid-banjoman rule may be hidden as a vocoder-specific constraint in a multistandard open-architecture framework. Once the banjoman phenomenon becomes known, however, that knowledge should be retained for future use in appropriate circumstances.

An enterprise-architecture repository of design-rules can serve this purpose. The design-rules repository of Figure 5-6 illustrates such a knowledge repository. An enterprise may maintain its own repository of such rules, comprising part of its own intellectual property and competitive advantage. With the growth of open taxonomies on the World Wide Web, an open design-rule repository for software radio could emerge as a network of web- accessible domain-specific design rules. Industry bodies seeking to promote open software-radio architecture could adopt the minimum set of design rules necessary to insure the benefits of the open-architecture framework. It could defer the majority of the design rules to (centralized or distributed) design- rule repositories. Those developers with a vested interest in sustaining the knowledge required to successfully integrate hardware and software components into functioning systems that obey *all relevant* design rules would cooperate to maintain shared design rule repositories, e.g., in a users' group.

## B. Object-Oriented Approaches

Object-oriented design incorporates the core strengths of SD, while increasing the cohesion among data and algorithms. Objects encapsulate functionally cohesive components. SD criteria assist one to decide which software functions should be coalesced into an object. There are incidentally cohesive objects, as well as functionally cohesive objects. The incidentally cohesive objects built in C++ can be as hostile to reuse as spaghetti-coded FORTRAN. One may use OOT as a way to move forward, building on established design principles including SD and message-passing. One may then build on the principles of functional cohesion and data coupling in defining SDR object boundaries.

OOT makes significant progress beyond the basics, however, as illustrated in Figure 5-7. Objects encapsulate data and associated procedures, called methods, into a single software entity. Object classes are reusable by definition. Objects are instantiated from classes and new classes are defined in terms of existing ones. The figure illustrates the reuse case in which three existing object classes are merged to quickly create a vocoder. The assumption is that the developer has experience with voice coding, having developed a general vocoder class and other voice processing algorithms. The general voice-codec class, first of all, would define primitive digitized-signal flows. This could include double buffering and clock-synchronous delivery of speech data to the modem stream. SDR applications of OOT emphasize the isochronous operation of such multi-object processing threads.

A more specific class, the RPE-LTP algorithm of GSM, may be based on this general-purpose voice codec class. A notional Regular Pulse Exciter (RPE) vocal tract modeling algorithm is then combined with a Long-Term Predictor (LTP) analysis/synthesis and filtering algorithm. The new algorithm inherits the properties of each of these existing algorithms. This includes slots for data embedded in the objects. It also includes attached methods that define object behaviors. Newly created classes may override behavior from ancestor classes by declaring their own local methods.

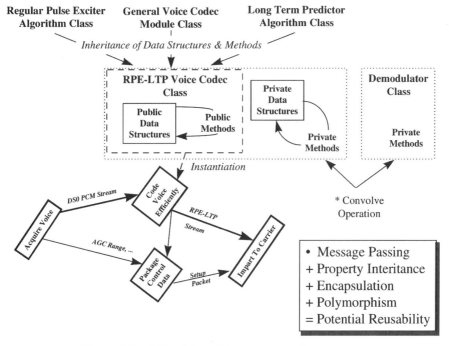

**Figure 5-7**   Object hierarchies promote software reuse.

Objects are coupled by message passing, institutionalizing as it were the good practice of data-coupling software modules. Message passing is a generalized form of remote procedure call since most messages need a response. The response may be regarded as a value returned from a procedure call. Message passing that requires the sending object to wait is, in fact, equivalent to a procedure call. Threaded message passing, however, does not require the sending object to wait for the response. Java's threading facility [186], for example, implements this aspect of OOT.

Messages may invoke public methods and may access public data structures, which the object makes available to the outside world. Objects may also use private data structures and methods, promoting reuse in another dimension. In the example, the "convolve" operation has been defined earlier. Convolution is a common operation that may be useful in both RPE-LTP and a channel demodulator. Convolve( )[23] is not a public method of RPE-LTP, but it is an internal operation from which the class is composed. The demodulator also employs Convolve( ). Thus the Convolve( ) operator may be drawn from a general-purpose library and used in both objects as a private method accessible only by members of the RPE-LTP (or demodulator) class itself. This approach effectively reuses the existing Convolve( ) code.

---

[23]Methods and function calls are indicated by the notation MethodName( ), by OOT convention.

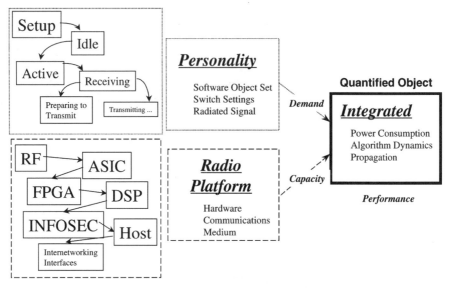

**Figure 5-8** Radio platform model integrates hardware and software characteristics.

There are several perspectives from which one may define objects like the vocoder or Modem. Some constraints on the vocoder may be evident from a use-case vignette that characterizes QoS. Working through such scenarios allows one to define objects in terms of their interactions with external entities. This is called *use-case analysis*. Other constraints may become evident in packaging the vocoder for deployment in an ASIC.

This brief introduction highlights aspects of OOT that are relevant to software-radio architecture. OOT provides the analysis and system design framework for software design developed in later chapters. The core notions of OOT include the encapsulation of objects by defining public slots and methods. Property inheritance allows specific objects to inherit slots and methods from more general classes of objects. OOT facilitates message passing as a generalized procedure call. It then couples object instances into a cooperative distributed processing framework, including support for multiple independent execution threads. Polymorphism also helps software developers gracefully extend existing functions to new data structures and behaviors. Related OOT literature includes introductory texts [31] and in-depth treatments of object-oriented design for real-time systems [187].

## C. Reference Platform Integration

OOT is readily extended from software design to hardware, software, and systems integration. Figure 5-8 illustrates an object-oriented way to define the relationship between hardware and software. The software provides the personality of the object, while the radio-platform provides the analog and digital processor hardware. Together, the software objects executing on the platform define a rich set of behaviors that are properties of the composite radio node. Some algorithms use more power than others. Power consumption is therefore a property of software and hardware. Standby power may be a hardware characteristic, but algorithm intensity will either use power conservatively or aggressively, changing the dynamic power consumption properties of the radio node.

Software is not the only thing that gives hardware personality. Physical switch settings may change the behavior of a chip, device, board, or other configurable hardware. The addition of daughter modules of different types may also change the properties of a platform. There are two ways to specify the radio platform. Traditionally, acquisition organizations wrote a specification that defined exactly what was required. Such specifications often constrain design, reducing competition and restraining the introduction of new technology. The more contemporary approach, introduced earlier is to specify only the essential features of the hardware necessary to support a wide family of software personalities. This statement of features is the reference platform. The end item may be modeled as an object that consists of two constituent objects: the hardware platform and the software personality. If the hardware fully complies with the reference platform, then the software personality will integrate successfully. By minimizing the feature set of the reference platform, compliant hardware has maximum degrees of freedom.

In Figure 5-8, a software-radio personality is suggested as a software state machine that includes Setup, Idle, Active, Receiving, etc. The host hardware is suggested as a mix of analog RF, ASIC, FPGA, DSP, programmable INFOSEC modules with a general-purpose host processor, plus some internetworking hardware interfaces. In planning the migration of existing services from current hardware to new hardware, one must define a reference-platform that captures the essential features needed to assure the integrated behavior. In particular, the desired features of the personality must be quantified in terms of demands that must be satisfied by the host radio platform. The set of quantified capacities of a radio platform is also called its "capability." Given a quantified capability, the required capacities may be summarized in the reference platform, and provided in the rehosting process. This assures the necessary level of integrated performance in the new hardware environment.

Figure 5-9 illustrates how the demands of personalities must be met in the capabilities of platforms in order for the desired performance to be achieved. The notional software radio system of the figure includes RF, modem, INFOSEC, and internetworking components orchestrated through some control component(s). The RF object needs suitable carrier frequencies, bandwidths, and dynamic range from the host analog RF components. Although there are many parameters that one may use to characterize the analog components, few parameters are as critical to SDR performance as bandwidth and linear dynamic range. These are carried through the architecture from antenna

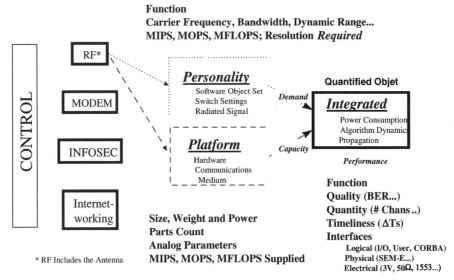

**Figure 5-9**   Capability quantification is essential to rehosting.

to wireline (e.g., DSO) or user interface (e.g., audio, video). Other parameters such as processing capacities are relevant to the configurable and programmable components including ASICs, FPGAs, DSPs, and general-purpose computers. These capacities are measured in *millions of instructions per second* (MIPS), *millions of operations per second* (MOPS), and or *millions of floating-point operations per second* (MFLOPS). They may also be measured with respect to an industry-standard instruction mix such as the SpecINT or SpecFP [46]. End-to-end performance characteristics such as quality (e.g., BER), quantity (e.g., number of channels, data rate), and timeliness (e.g., queuing delay) are the behaviors of the composite radio that are defined through balancing the demands of the personality against the capacities of the host components. ITU-T recommendation H.320 [188] for video-teleconferencing and Microsoft's TAPI [189] plug-and-play specifications both incorporate the capability-based approach for plug-and-play. Since the implementation of radio platform capabilities strongly influences software, the analysis of capability-based design is deferred to the software chapter.

## D. Using UML to Analyze Node Architectures

UML has evolved from early OOT approaches. Its current evolution includes increased emphasis on real-time and embedded applications [190]. UML models a system as a collection of objects with an associated set of relationships. At the systems level, these relationships are characterized intuitively in terms of views. The four UML views of a system are: *use-case view*, *logical view*, *component view*, and *deployment view*. Use cases are scenarios that express

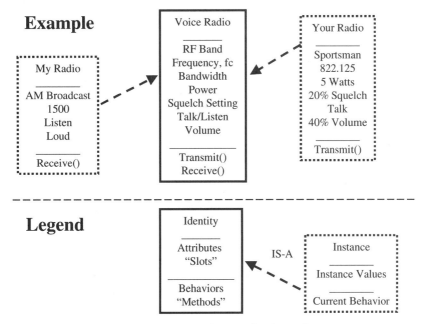

**Figure 5-10**   UML model of voice radios.

the behavior of a system in terms of its relationship with external actors. The logical view defines objects, classes, states, relationships, and interfaces along the lines developed above. The component view addresses the partitioning of the functionality of the logical view into (hardware and) software components with associated interfaces. Finally, the deployment view defines how components are related to physical entities. The analysis of radio node architecture is readily framed in terms of these four views.

***1. UML Objects***   A variety of graphical notations has arisen for describing the views and the objects of which a system is comprised. An object that implements the "voice transducer" function introduced above may be called a Voice Radio object. It is characterized in UML notation in the model of Figure 5-10.

The legend shows that the solid box represents a class, a generic object that embodies the concept of a voice radio. The dotted boxes, in this case, are instances. So My Radio "is a" voice radio, and Your Radio "is a" voice radio. The lines connecting the boxes represent relationships among objects. In this case, the IS-A relationship is useful. The arrow points from the more specific class to the more general class of object.

Each class and instance has an *identity*, shown at the top of the object-box. Each object has a set of *attributes* that are like *slots* into which values or other objects can be placed. The attributes of the instances have specific values, which define the *state* of that object. The procedures attached to software objects define *behaviors*. Some object-oriented software approaches call these *methods*. The physical properties of hardware devices may define their behaviors. In addition, UML allows one to assign *relationships* to objects. An object, then, is defined in terms of its set of attributes, behaviors, states, identities, and relationships.

***2. Regarding Object Notation***   Sometimes it is appropriate to be fanatical about notation. In this text, it is essential to be completely relaxed about notation. The motivation for this perhaps unusual treatment is that the software-radio architect has to be able to distinguish concept from notation. Defining architecture is about finding common ground among players who may never have spoken to each other before and who almost certainly do not use the same notation. Thus, the architect has to see beyond the notation to the underlying design frameworks. Within OOT, there are a variety of object notations, including Jacobson's [191], Booch's original approach [192], Coad-Yourdon [193], and real-time object-oriented systems analysis (RTOOSA) [194], to name a few. They have been integrated in UML, but one may find (documents or) practitioners that use the earlier approaches.

The use of multiple apparently ad-hoc notations in this text is not sloppy. The treatment is designed to give the reader practice needed to develop the skills to see through the notation. In addition, the notation used in this text is a simplified version of UML, tailored to convey the core concepts useful to the analysis of software-radio architecture, without making this a text on UML.

***3. Radio Classes, Subclasses, and Instances***   The example of Figure 5-10 shows the attributes of the class Voice Radio, as well as the specific values of those attributes at some point in time on My Radio and Your Radio. At some point, it will become clear that My Radio is just an AM/FM broadcast receiver, while Your Radio is a push-to-talk (PTT) transceiver. If that difference is important, two subclasses of Voice Radio could be created. The creation of subclasses makes it easier to recognize members of the class. For example, My Radio has no microphone, so its Transmit() function is null. One can represent My Radio as an instance of Voice Radio that has a null Transmit() function. If it is clearer or more helpful to the architecture-definition objective, a Broadcast Receiver class may be defined, as illustrated in Figure 5-11.

***4. Components in UML***   The structure of the components of the radio may also be expressed in UML, as in Figure 5-12. From one's knowledge of radio, it should be clear that the arrows represent the way in which one object is a component of another. The intervening classes are not populated, but the leaf nodes are those classes that are specific enough to make into instances (e.g., actual hardware and software components). The function "transduce voice" from above can readily be represented in this notation.

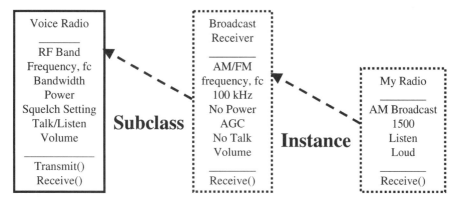

**Figure 5-11** The subclass establishes specific capabilities and default values of slots and behaviors.

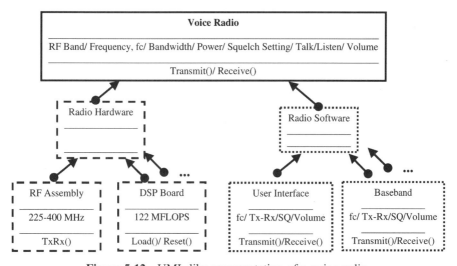

**Figure 5-12** UML-like representation of a voice radio.

## E. A Topological Model of Architecture[24]

The need to efficiently map radio functions to hardware configurations leads to the study of the mathematical foundations of system architecture. There are few. When two radio systems engineers with different backgrounds approach a systems-level design problem, they often come up with different approaches and solutions. The focus of this text is on architecture and hence on the common aspects of those endeavors, on systems engineering. How

---

[24]This section covers an advanced topic that may be safely deferred to a second reading.

does one go about deciding which approach is the best? Anyone who claims some "optimal" approach or general way of determining "best" in any universal sense is probably making an over-statement. Mathematical optimality may be formulated over mathematically tractable spaces. But systems engineering is defined over rugged, nonlinear spaces comprised of sets of overlapping functions to be deployed on somewhat inappropriate components according to mutually contradictory design rules. The complexity of the full trade-space thus defined is too large and complex to be represented as input for contemporary optimization tools. The term "optimal" may be accurately applied to carefully identified sub-problems such as carrier tracking in Gaussian noise. At present, it cannot be accurately applied to radio node architecture in the broad sense used in this text.

*1. In Pursuit of Commonality*   Systems design problems of any import deal with large, complex systems either in isolation or in combination. A national telecommunications system, for example, is a complex system; shipboard electronics for a battle group and the avionics suite of a new fighter jet also exemplify complex systems. In addition, today's large-scale telecommunications systems architectures are concerned with "systems of systems." That is, each component of a large-scale system is itself a very complex system. The national infrastructure (telecommunications, power, fuel, water, etc., all more or less interconnected) provides a good example of a system of [complex] systems. Researchers now categorize such systems as complex adaptive systems [195]. Some mathematical principles illuminate the quest for architecture. These principles take some study, the highlights of which are provided here and in [196].

What does an airborne avionics system consisting of communications, radar, navigation, IFF and a fly-by-wire control system have in common with a ground-based mobile military communications system? They both use RF, but avionics may prioritize remote sensing and navigation while the ground-based system may prioritize overcoming impairments of RF propagation in a battlefield environment. Both systems have signal generation, modulators, antennas, receivers, signal processors, embedded control and information systems, and user interfaces. Common use of the RF spectrum reveals much commonality. The radar and communications systems encompass radiated power, bandwidth, free space loss, and reflections. Some reflections result from a target, while others result from multipath reflectors. Both need noise suppression, and the correction of distortions introduced through propagation. Both use correlation gain to enhance the received signal, to correct errors, and to present results to a user. There are many shared abstractions and similar components. In addition, packing electronics into rugged enclosures applies disciplines of power distribution, thermal management, and control of electromagnetic interference (EMI). Thus, there are shared abstractions and technologies that differ widely in implementation. But the search for unifying *mathematical* principles that reflect the commonality is elusive.

**Radio Functions**

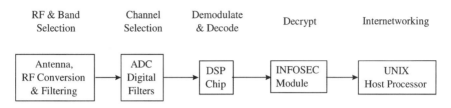

**System Design**

**Figure 5-13**   Radio functions may map onto system components.

*2. Toward Mathematical Structure*   Consider, however, Figure 5-13, which shows radio functions associated with system components. The figure gives an impression of a one-to-one relationship between functions and components. The figure, of course, is an abstraction that oversimplifies the situation. Nevertheless, figures like this give the (sometimes erroneous) impression of good design. Does this "nice" quality perceived when functions and components have a one-to-one (1 : 1) relationship have further support? That is, does this quality derive from mathematical foundations or can it be defended based on mathematical principles?

The topological model of software radios [196] shows that there is such a mathematical basis. Let $P$ be a set of primitive functions ("primitives") such as "RF Band Selection," "Channel Selection," etc. as illustrated in Figure 5-13. Each primitive function $P_{ij}$ operates over some domain, $D_i$, which is a *set* of inputs over which that primitive is defined. This primitive yields a *set* of results, the range, $R_j$, of that primitive. Some sets have rich structure. For example, filters, Fourier transforms, and wavelets are defined over vector spaces, which are metric spaces in which the additive group is commutative (Abelian), and the multiplicative group is Abelian [432]. Vector spaces obey distributive laws and the triangle inequality. This is a *lot* of mathematical structure. Most software functions have no such rich mathematical structure. A simple if-then-else statement, for example, might check the state of the receiver (e.g., the "carrier present" flag), wait a prescribed amount of time, and invoke a carrier fault-recovery process. Receiver state is a set of labels asserted by other algorithms. *Carrier fault recovery* is the name of the software process to be invoked. Such processes have mathematical structure [197], but there is no vector space in which a control state maps to a software process. Such a software process has a *point set topology*, however. This is a set (e.g., of state labels, process names, etc.) with a family of subsets (e.g., the ones over which the software operations are valid) that has topological properties.

*3. Topological Spaces*   A topological space is a set, $X$, and a family of subsets, $O_x$, which includes $X$ and the empty set. The family is closed under union and finite intersection [198, 199]. Point set topology is the study of families of subsets, functions over them, and mappings among them. Consider a set of states that a transmitter may assume such as "Transmit," "Receive," "Initialize," and "Ready." Let the set of such states be the set $X$. The control algorithm $C(x)$, may be defined such that exactly one state $x$ contained in $X$ (cf. $x \in X$) may be asserted at any time. The domain of $C$ is:

$$\text{Dom}(C) = \{\text{Transmit, Receive, Initialize, Ready}\} = X \qquad |X| = 4$$

But what is the topological structure of this domain? If exactly one $x \in X$ can be asserted at any one time, then $C$ is defined over $\{\{\text{Transmit}\}\{\text{Receive}\}$ $\{\text{Initialize}\}\{\text{Ready}\}\} = O_X$. That is, the state of the transmitter may not be $\{\text{Transmit}, \text{Receive}\}$, which would correspond to the situation that the radio is transmitting and receiving both at once. This might be acceptable for some radios, but not for one controlled by $C$. Each subset of $X$ that is allowed is in the control space. Topologically, the individual states are singleton subsets. $C$ is not defined over the empty set unless the state set contains the null set ($\phi$), the empty set. The rule $C(\phi) = C(\text{Initialize})$, for example, defines the default state. The extended topology is:

$$\{\{\text{Transmit}\}\{\text{Receive}\}\{\text{Initialize}\}\{\text{Ready}\}\phi\} = O_X$$

A topological space, however, must include $X$. Since $C(S)$ is undefined, $O_X$ is not a topological space. If one can induce a topological space on a software radio function, *all* of point set and algebraic topology becomes mathematical foundations for that function. If not, then the topologies could be embedded in a larger topological space that yields insights into the mathematical structure of software radio. For example, the product topology is the set of all subsets of $X$, $O_P$:

$$O_X \subset O_P$$

How could one define $C(X)$? $X$, $C$'s domain, is a property of the algorithm $C$. That is, $X$ is implicit in $C$. Historically, documentation may express that $C$ is defined over $X$. To explicate that implicit relationship, define $C(X,x)$, $x \in X$. $C(X,x)$ is the generalization of $C$ that "knows" that it is defined over $X$. The informal knowledge is formalized in $C(X,x)$. This leads one to define the ***Topologically Explicit Functions***: Functions defined over a domain $X$ with an explicit topology $(X, O_X)$ and range $Y$ with explicit topology $(Y, O_Y)$ are topologically explicit if $X$, $Y$, $O_X$ and $O_Y$ are computationally accessible. They are topologically well-structured if the topologies are topological spaces.

In the example,

$$X = \{\text{Transmit, Receive, Initialize, Ready}\} \qquad \text{and}$$

$$O_X = \{X, \{\text{Transmit}\}\{\text{Receive}\}\{\text{Initialize}\}\{\text{Receive}\}\phi\}$$

Is $O_X$ a topological space? To test this, consider unions and intersections of the members of $O_X$. Transmit $\cup$ Ready is in the set of unions over the members of $O_X$, but it is not in $O_X$. Therefore, this topology is not a topological space. It could be extended to define some value of $C$ for this condition. Since the behavior of $C$ specifies that such a condition is an error, the topology may be extended to capture this notion mathematically. In particular, define $\Phi$ as the distinguished symbol that represents an error condition. This is similar to the use of $\phi$ to represent the empty set. $O_X$ may be extended with $\Phi$:

$$O_X = \{X, \{\text{Transmit}\}, \{\text{Receive}\}, \{\text{Initialize}\}, \{\text{Receive}\}, \phi, \Phi\}$$

One writes $\{\text{Transmit} \cup \text{Ready}\}$ and all the other erroneous combinations, as members of the distinguished subset $\Phi$, which is shorthand for all those combinations. A convention says that anything that is not explicitly listed in $O_X$ as a valid member of the domain is one of the items that $C(X,x)$ will recognize as not valid, and to which it will raise an exception. This is an extension of $C$. It has some mechanism (e.g., a lookup table) for testing its input $x$ against $O_X$ (not just against $X$). If $x$ is not in $O_X$, then it asserts $\Phi$.

This shorthand $\Phi$ is not merely an editorial convenience. There are a huge number of possible topological spaces for any finite set X. The power set of $X$ is the set of all subsets of $X$. Since $X$ has four elements, there are $2^{|X|}$ or $2^4 = 16$ members of the power set ranging from $\phi$ to $X$ including the four singletons $\{\text{Transmit}\}$, etc., the six doubletons $\{\text{Transmit}, \text{Receive}\}, \dots \{\text{Initialize}, \text{Receive}\}$, the four triplets $\{\text{Transmit}, \text{Receive}, \text{Initialize}\} \dots \{\text{Receive}, \text{Initialize}, \text{Ready}\}$, $\phi$, and $X$ itself. A topological space may include just $\phi$ and $X$, which is called the *indiscrete topology*. Or it may include the power set, which is called the *discrete topology*. Or it may be any of the $2^{(2^{|X|}-2)}$ possible ways of taking subsets of the power set. For $|X| = 4$, there are $2^{(16-2)} = 2^{14}$ or over 16,000 ways to construct topologies—subsets of $X$—because the number of topologies is a double exponential. Instead of forcing one to pick through this intractable number of combinations, the $\Phi$ convention embeds $O_X$ in $O_P$. By defining the range of each primitive as a topological space, one may express all software radio primitives as maps over topological spaces.

**4. A Topological Framework for Architecture**   If a function is defined abstractly using UML, then its input and output spaces are defined, and the associated topological spaces may be inferred. In addition, since UML allows one to model hardware or to produce executable code, the topological properties may be extended to maps from abstract functions to hardware and software implementations. Such maps over topological spaces have useful mathematical properties including composability by the glueing lemma [199]. For example, if there is a map from an abstract function to a software module and thence to host hardware, the properties that have been demonstrated to be true of the abstract function are proved true of the hardware version. This can reduce type certification from an exhaustive testing process to a matter of checking

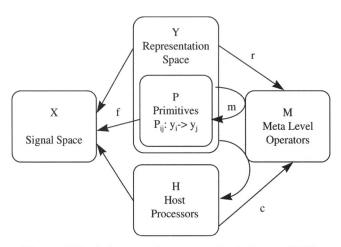

**Figure 5-14**   Software radio as a topological space [200].

that the topological maps defined in the (future) CASE tool are implemented in the hardware. This checking process is linear in the complexity of the hardware, whereas the testing of combinations of downloads, etc. is quadratic in the number of software modules.

In addition, one may compare the range of one primitive to the domain of another using a topological map called a *homeomorphism*, a topology-preserving mapping. Homeomorphisms are ONTO and 1 : 1 and they preserve topological properties (e.g., the inverse set-theoretic images of open sets $O_X$ are open in $O_Y$ under homeomorphism). So the desired relationship between a host processor and the function may be grounded in the theory of point set topology. One constructs the homeomorphism between the functional representation and the hardware representation. If the function is implemented in software, the homeomorphism is constructed between the functional abstraction in UML and the object implementation in C, Java, or FORTRAN and assembler.

This topological approach allows us to more precisely define other systems engineering issues such as the quality of the implementation and the resources required to host a software radio on a given hardware suite as illustrated in Figure 5-14. Defining the topological properties of the domain and range of each software radio primitive is an exercise in mathematical rigor. It helps determine the slots of the software objects. It can also help define the conditions over which complex functions such as control procedures are defined. Constructing the topological space may assist in identifying defaults and related behaviors of the objects.

The construction of topological spaces can be done by someone who knows topology theory using a knowledge-engineering approach. This is the way that the author has applied these principles except in rare cases where the designers had strong mathematical backgrounds. Alternatively, future CASE tools can be designed to guide the non-mathematician software-engineer through the potential topologies by a question-and-answer dialog. A data dictionary defines the dimensions of the topological spaces present in a software-radio system. The CASE tools could be extended to interpret the spaces over which the data elements are defined. Current CASE tools can check for consistency and completeness, but they cannot check even simple topological properties. RF, for example, is defined on a quantized subset of the real line. Valid RF values have to be properly quantized. That is, if RF is defined in the FM band, it should be quantized in 100 kHz steps, while in the AM broadcast band, it is quantized in 10 kHz steps. The FM band is the set $X$, and the values of $f_c \subseteq$ FM-Band are in $O_X$. Similarly, the CASE tool could process the UML to extract the discrete topology present in control variables, and could ask the user about combinations of conditions that the control algorithm does not explicitly check. It could thus induce $X$ and $O_X$ for each algorithm $C$.

In general, software radio objects defined on topological spaces may be defined with explicit spaces so that the functions may be more readily reusable. If $C(X, x)$ is defined explicitly, then $C$ can be accessed from a software reuse library based on $X$. If $X$ is RF, then the systems engineer has the ability to browse a reuse library based on RF. This may facilitate finding modules that may be modified to accomplish a new function. So there may be practical benefits to this somewhat abstract and unfamiliar treatment. This concludes the brief introduction to the topological model of software radios. The interested reader may pursue the topic further via [201 and 99, e.g., in 433].

### F. The Canonical Software Radio Node Architecture

Structured design principles admonish one to maximize cohesion within a component. Functional cohesion is evident in components whose elements share some common element. Applying this principle to SDR node architecture yields a grouping of functions that share common RF, IF, and baseband elements. Figure 5-15 shows the resulting segments of the software-radio node architecture. The power management and LNA elements of the RF conversion segment share the antenna, while the RF conversion elements share the RF frequency standard. The RF elements also share a need for proximity to the antenna. The LNA is placed near the antenna in order to set the system sensitivity. The power amplifier is near the antenna in order to deliver power efficiently to the antenna. The RF section may be placed remotely from IF processing (e.g., in diversity architectures). IF processing is therefore distinguished as a separate segment. The IF elements of a superheterodyne transceiver also share frequency standards. In PTT and FH radios, the transmitter and receiver IF are tightly coupled. In addition, IF processing in an SDR filters the wideband signal structure from the RF segment to yield the narrower baseband bandwidth. The transformation of bandwidth across the IF section therefore enhances its functional cohesion.

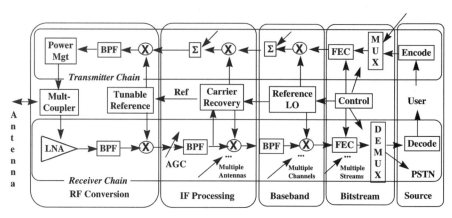

**Figure 5-15**   Canonical model of SDR node architecture.

ADCs may be inserted at the interface of IF to RF or IF to baseband, providing a basis for data-coupling between these segments. The baseband segment performs the modem functions, converting information between channel code and source code. This cohesive function is the basis for defining baseband as a segment. Soft-decision decoding delays the final transformation of channel symbols to baseband bits. It is therefore more cohesive with the baseband segment than with the bitstream segment.

The bitstream segment performs operations on bitstreams. This includes multiplexing, demultiplexing, interleaving, framing, bit-stuffing, protocol stack operations, and FEC. Turbocodes combine interleaving and FEC, exemplifying the functional cohesion of the bitstream segment. Control is included in the bitstream segment because of the digital nature of control messages. This may place the user-control interface in the bitstream segment. The source segment includes the user speech signal, the local source and sink of audio information. Source coding is the transformation of communications signals into bitstreams. This may occur locally (e.g., in a soundboard) or remotely, at the other end of the PSTN. This segment is coupled to the bitstream segment by standard bitstream interfaces such as DS0, T1/ E1, or a LAN. Although this formulation of the source segment permits a geographically distributed segment, the segment is functionally cohesive. Each of these segments is therefore functionally cohesive. Each accomplishes a single clearly identified function or closely related set of functions. In addition, the RF, IF, and baseband functions transform the data rate or bandwidth between input and output, typically by an order of magnitude or more.

These segments therefore comprise the canonical software-radio node architecture. This canonical form brings out aspects of architecture that are not highlighted in the functional architecture of Chapter 1. One could also formulate these segments as objects. Each segment is one object. The states of the segment are the slots of the object. The transformations of the segments are the behaviors of the objects. When simulated or implemented in software, each behavior corresponds to a method. When implemented in hardware, a behavior simulates a property of the hardware.

The primary signal flows of the canonical architecture are illustrated in Figure 5-15. There are two primary signal flows. The transmitter chain, first, transforms of source from its original analog waveform to a bitstream. The bitstream is further encoded and multiplexed. Channel coding is applied to the signal, which is then upconverted, amplified, and filtered for transmission at the antenna. The receiver chain, second, transforms the air interface waveform received at the antenna. Frequency selection, filtering, frequency translation, equalization, demodulation, error control, demultiplexing, and source coding yield the information signal to the user or to the PSTN interface.

The segments of this model may be mapped to the functional architecture and interfaces introduced in Chapter 1. The canonical model, however, explicitly represents properties of RF hardware (e.g., Local Oscillators) that are not explicit in the functional model. One of the goals of architecture is to facilitate the mapping of functions to hardware. Although there are many issues to be addressed in establishing such a mapping, the following three stand out:

1. Determining the node-level properties of antenna(s), RF conversion, and IF processing
2. Placing the ADCs and DACs at an appropriate interface point
3. Accommodating INFOSEC design criteria

Subsections consider these aspects of mapping to hardware in turn. The analysis of each of these areas refines the canonical model.

*1. Refining the RF Reference Platform*    The characteristics of the antenna(s), RF conversion, and IF processing determine the radio node's access to the RF environment. A radio design would specify the parameters of each of these segments. Architecture, on the other hand, specifies the absolute minimum necessary to assure that a specific radio function will be supported by the RF platform. It is convenient to encapsulate a consistent set of radio functions that implement an air interface with the associated physical data link, network, and applications interface functions as a waveform. To visualize the analysis, consider the generic block diagram of Figure 5-16.

Although the mix of antennas is intentionally unusual, the block diagram highlights RF platform issues. The physical placement of the antennas, for example, will determine the degree of coupling among them. If the antennas are sectorized, then each RF path has azimuth and elevation coverage. In a deployment, the antenna would be pointed in a specific direction. The value of this direction is not relevant to architecture, but controllability of direction is significant. Gimbaled antennas and phased arrays require tasking and/or control from some processor. These functions require control flow across segment interfaces, and any such flows are architecturally significant. Antenna

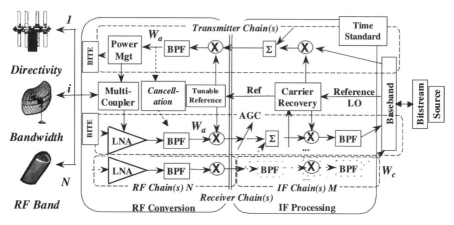

**Figure 5-16**   RF-related radio platform.

gains must be known for propagation modeling. Any other programmable parameters are also architecturally significant.

The RF conversion segment is matched to the antenna segment. Each antenna requires at least amplification and filtering, and this function is assigned to RF conversion on a 1 : 1 basis for each antenna element. RF conversion therefore sets the bandwidth of RF access for each antenna path. Since transmission on one band may interfere with reception on that band or on other bands, RF conversion may include active cancellation circuits. Alternatively, an RF cancellation path may be processed at IF or baseband, increasing the number of RF-IF chains. Internal calibration and/or built-in-test-elements (BITE) also increase the number of RF paths.

$$N_{\text{RFpaths}} = N_{\text{antennas}} + N_{\text{cancellation}} + N_{\text{BITE}}$$

If RF conversion includes a superheterodyne stage, then RF conversion yields analog IF at some reference frequency. The distribution of reference frequencies is a critical design issue for SDR. Raising the choice of IF into architecture imposes design constraints on the internal structure of the receiver. This choice may be encapsulated in a receiver component to protect intellectual property. Alternatively, it may be unencapsulated to allow antenna-RF modules to be acquired separately from the IF/baseband components. An architecture that forces all players to make the same design decision lacks flexibility but promotes competition for interchangeable antenna-RF modules. An architecture that allows both encourages the inclusion of a variety of transceiver designs, promoting competition at a different level of aggregation of components. The purpose of this text is not to propose specific solutions to such conflicts. Instead, it is to define architecture alternatives, pointing out the strengths and weaknesses of the alternatives so that the reader may make informed decisions. As one probes deeper into the mapping of functions to components, such conflicts are constantly arising.

There may be fewer IF processing paths than RF paths. A diversity combiner, for example, accepts multiple RF-segment inputs, producing a single IF output. IF filtering also defines the channel bandwidth, $Wc$, available to the baseband segment. There may be multiple programmable bandwidths. IF conversion defines the reference frequency, $f_0$, of the interface to the baseband segment. In addition to these primary segment-level parameters, other parameters may have architectural implications for the RF reference platform at some point in the future.

Finally, there are RF platform-level parameters. The RF and IF conversion stages often share common frequency control and timing references. Since many waveforms require a minimum level of frequency stability, this is an architecture-level parameter. In addition, many waveforms require knowledge of time to a specific accuracy and precision. If one thinks of the RF platform as encapsulating the analog aspects of the radio platform introduced in Chapter 1, then the RF platform parameterizes the timing capability of the radio platform.

The RF reference platform thus consists of that minimum set of features of antennas, RF conversion, and IF processing necessary to support a set of communications applications. The RF platform is further described in Table 5-1.

*2. Placement of the ADCs and DACs*   Several approaches are feasible for the conversion of analog RF signals to digital form. The first is to not convert signals from analog to digital at all, but to extract the information content directly. A direct-digital-RF receiver extracts baseband bitstreams from the (filtered) RF waveform without RF conversion. A direct-conversion receiver, similarly, may include one stage of RF conversion to IF equals zero at which the information-bearing bitstream may be extracted. A CDMA rake receiver, similarly, may consist of three matched-filter despreaders that yield information-bearing bitstreams, tracking Doppler and delay, and managing bitstream timing internally. These alternatives encapsulate the entire RF-IF-baseband process in one module within which bitstream extraction happens. A high-quality architecture allows for these possibilities.

Alternatively, wideband ADCs and DACs may be placed at the RF-IF interface or at the IF-baseband interface. For cell site applications, the placement of wideband conversion at the RF-IF interface of a superheterodyne conversion receiver enables digital IF processing. This choice is less likely for handsets.

*3. INFOSEC-based Partitioning*   INFOSEC imposes an additional partitioning on radios that seek to provide transmission security and/or encryption. Bitstreams may be either encrypted or unencrypted. INFOSEC practice designates the encrypted bits with the color black. Unencrypted bits are designated red bits. The bitstream segment, then, is partitioned into an encrypted black subsegment, an INFOSEC module, and an unencrypted red subsegment. The

**TABLE 5-1   RF Reference Platform**

| Segment | Parameter | Remarks |
|---------|-----------|---------|
| Antennas | Number of antennas (N; I) | Matched to RF; high cost impact |
|  | Antenna coupling | Matrix of losses (dB), which may be a function of antenna pointing |
|  | Directivity | Azimuth and elevation coverage |
|  | Pointing (fixed/controllable) | Determines control flow |
|  | Gain | Gt, Gr, or G/T for propagation modeling |
|  | Programmability | Architecturally significant |
| RF Conversion | $Wa^i$ (= $RF_{max} - RF_{min}$ per path) | RF access per path |
|  | $f_{IF}$ | Standard IF promotes competition for front-end hardware |
|  | BITE | Parameter set for control and diagnosis |
|  | Active cancellation | Capability by band, with control parameters |
| IF Processing | Diversity combining | N : M mapping defines IF paths |
|  | $f_0$ | IF of the baseband segment interface |
|  | $Wc^j$ | Bandwidth of the $j$th baseband segment interface |
| Platform | Frequency stability | Defines *minimum* quality of LOs |
|  | Timing accuracy | Defines *minimum* quality of timing standards |

INFOSEC module implements transmission security and/or communications security.

According to the best INFOSEC practice, there should be as much physical separation as possible between the black and red bitstreams. Therefore, many radios, including most military radios, divide the radio hardware into a black half and a red half, joined by the INFOSEC module (Figure 5-17). The antenna, RF, IF, baseband, and encrypted bits fall on the secure or black side. In addition, it is sometimes useful to process black bitstreams before decrypting them. This leads to the additional notion of a black-side data processing component. Classified source or object code may be encrypted and stored on the black side. The red side includes all nonencrypted bitstream processing, which may be called the internetworking subsegment. The red side also includes the user interface, system control, and encryption key management. In the canonical model, these functions are part of the source segment.

Encryption alone does not provide sufficient INFOSEC for emerging applications (e.g., e-cash). Encryption increases one's expectation of privacy, one INFOSEC function. But integrity is also essential. If an e-cash packet is garbled, changing the amount one is charged, the communication cannot be

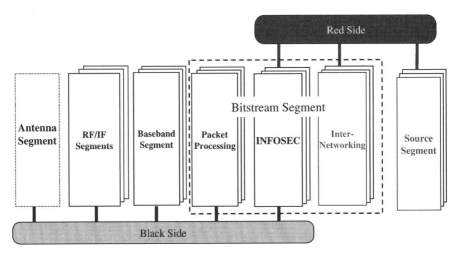

**Figure 5-17**  INFOSEC partitioning

regarded as secure. In addition, if someone steals your credit card, you would like the merchant to recognize that the user is not valid. Digital signatures [434] provide better protection of identity than personal identification numbers (PINs). Such authentication guards against spoofing. A robust INFOSEC capability further protects the radio against spoofing. Waveform design also provides robustness in the presence of jamming. Antijam is regarded by many as an important aspect of INFOSEC. With that view, CDMA spreading and hopping that suppress multiple-access interference in cellular systems are also INFOSEC measures. Commercial practitioners rarely view things this way, but some military practitioners do. In addition, the military often desires low probability of intercept (LPI). Thus FH and DSSS were initially pursued by the military for INFOSEC before their benefits to commercial communications were fully appreciated. INFOSEC issues may have implications from the air interface (e.g., synchronization) to the user interface (e.g., key insertion), and throughout the protocol stack. A robust architecture therefore partitions radio functions so that aggressive INFOSEC measures may be gracefully inserted with minimum impact on other radio functions.

*4. An Object-Oriented View of the Canonical Model*   The analysis of the canonical model has clarified some aspects of mapping the functional model of Chapter 1 to abstract (hardware and/or software) modules. The presence and placement of ADC and DACs determines whether IF processing components are digital or analog. INFOSEC considerations may partition bitstream processing into red and black components. These considerations may now be reflected in the emerging architecture.

UML helps visualize this architecture as shown in Figure 5-18. In this model, all radio nodes provide functions that are delivered via components

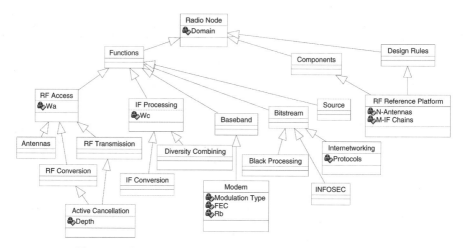

**Figure 5-18**   UML model of the canonical form (simplified).

that obey the appropriate design rules. The arrows all show inheritance relationships from the generic radio node class of object. The domain (e.g.,
handset, cell site, vehicular radio) property of the radio node is therefore inherited by the function, component, and design-rules objects. A few of the
architecture-level properties of the object classes are included in this class diagram to illustrate an architecture principle. The properties to be proscribed
in the architecture should be expressed as attributes of the object class that
is at the highest possible level in the inheritance hierarchy. This assures that
subordinate classes inherit this property. Some entities inherit from multiple
superior classes. Active cancellation, for example, inherits its power-handling
capability from RF Transmission, while its IF carrier is inherited from RF
conversion. All three of these classes inherit $Wa$ from the RF Access class
which has been created to aggregate parameters common to the RF-related
objects. The RF Reference Platform inherits from Components and Design
Rules, indicating that it reflects properties of components with the force of a
design rule.

The leaf nodes have no subordinates. In this diagram, as in all class hierarchies, instances of leaf-node classes provide the most specific attributes and
behaviors. They are therefore the most specific models of physical hardware or
software components. These subclasses will therefore be called radio objects.
This intentionally fails to differentiate a class from its instances. Instances
have exactly the attributes and behaviors specified by the class. A use-case
view of the radio objects can thus be used to define signal flows (Figure 5-19).
In this view, the arrows represent unidirectional relationships. Specifically, all
the arrows represent unidirectional signal flows.

The placement of the ADCs and DACs is not expressed in these views of
architecture. This is both a benefit and a shortcoming of the model. It is bene-

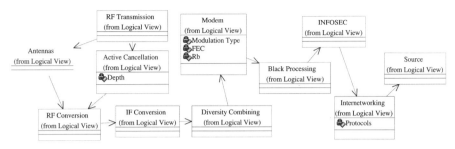

**Figure 5-19**   Use-case view of radio objects.

ficial because it admits any approach to ADCs and DACs. It is a shortcoming because it does not provide any basis for software-based representation of that aspect of the architecture. Thus, it cannot be used to formulate a capability profile of that aspect of a hardware platform. Therefore, control software would not know that there was a wideband ADC. It could not provide this information to an application that needs it (e.g., a spectrum-use mapping algorithm). Important internal details of these objects are hidden from view by the object model. This ability to hide details is essential to protecting intellectual property associated with novel implementations. More information hiding is possible. For example, one might define a superclass for the red objects and another for the black functions, yielding a radio composed only of red, black, and INFOSEC objects. Even with this degree of encapsulation, certain architecture parameters must be made visible, particularly those of the information flow.

## G. Digital Signal Processing Flow Parameters

The parameters of digital signal processing are critical to SDR performance. They must be addressed from an end-to-end perspective. First-order processing capacity requirements of the end-to-end digital signal processing flows may be computed from top-level parameters. The resource demands are computed from a top-level parametric model. Capacity estimates may then be derived from well-established rules of thumb. These initial estimates are appropriate when it is necessary to quickly assess the first-order feasibility of an SDR implementation. Subsequently, techniques will be described that accurately predict system performance.

*1. Mapping Functional Objects to Physical Objects*   An example of a mapping from functional objects (e.g., UML models) to physical objects (e.g., RF hardware, ASICs, DSP chips, and software load modules) proves helpful at this point. Figure 5-20 shows how the functional objects of the canonical model of a conventional cellular handset may be mapped to physical objects. In this case, RF conversion, power amplification, the ADC, and the DAC are implemented in a single ASIC. Similarly, the audio interface, including the

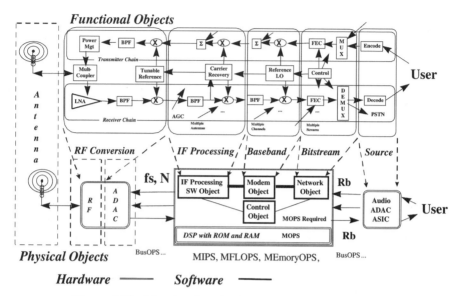

**Figure 5-20**   Mapping functional objects to physical objects.

vocoder, is implemented in an audio ASIC. ASIC designers are concerned with what goes on inside these chips. From the perspective of the SDR architecture, these details are encapsulated in the ASICs.

The structure and performance of the IF, baseband, and bitstream DSP hardware and software are architecturally significant at this level of abstraction. These components give the SDR its capability to be upgraded in the field (e.g., via software download). If this aspect of the architecture were a black box, then only complete memory maps could be downloaded. The component view of Figure 5-20 includes both the DSP platform and the software objects. This view supports incremental upgrade (e.g., by downloading a new modem software object. In addition, a component view at this level of abstraction supports reuse of the software objects shown. The software objects are shown in a signal-flow use-case view. The annotations show that the software objects require processing capacity (e.g., MOPS).

**2. Processing Resources**   The demand for processing capacity depends on the signal bandwidths and on the complexity of key operations within IF, baseband, bitstream and source segments as follows:

$$D = Dif + N*(Dbb + Dbs + Ds) + Do$$

Where   $Dif = Wa*(G1 + G2)*2.5,$

$Dbb = Wc*(Gm + Gd),$    and

$Dbs = Rb*G3*(1/r).$

**TABLE 5-2  Illustrative Processing Demand**

| Segment | Parameter | Illustrative Value | Demand |
|---|---|---|---|
| IF | $Ws$ | 25 MHz | |
| | $G1$ | 100 OPS/Hz | $Ws*G1*2.5 = 6.25$ GOPS[a] |
| | $G2$ | 100 OPS/Hz | $Dif = Ws*(G2 + G2)*2.5 = 12.5$ GOPS[a] |
| | $N$ | 30/cell site | |
| | $Wc$ | 30 kHz | |
| | $Gm$ | 20 OPS/Hz | $Wc*Gm = 0.6$ MOPS |
| Baseband | $Gd$ | 50 OPS/Hz | $Dbb = Wc*(Gm + Gd) = 2.1$ MOPS |
| | $R$ | 1 b/b | |
| | $Rb$ | 64 kbps | |
| Bitstream | $G3$ | 1/8 FLOPS/bps | $Dbs = G3*Rb/r = 0.32$ MOPS |
| Source | $Ds$ | 1.6 MIPS/user | $N*G4 = 4.02$ MIPS per user |
| | | | $N*(Wc*(Gm + Gd) + Rb*G3/r + G4)$ |
| | | | $= 120.6$ MOPS per cell site |
| | $Do$ | 2 MOPS | |
| Aggregate | $D$ | | 122.6 MOPS per cell site |

[a]Typically performed in digital hardware in contemporary implementations.

$D$ is aggregate demand (in standardized MOPS).

$Dif$ is the IF processing demand.

$Dbb$ is the baseband segment processing demand.

$Dbs$ is the bitstream segment processing demand.

$Ds$ is the source segment processing demand.

$Do$ is the management overhead processing demand.

$Wa$ is the bandwidth of the accessed service band.

$G1$ is the complexity of the service band isolation filter.

$G2$ is the complexity of subscriber channel isolation filtering.

$N$ is the number of subscribers.

$Wc$ is the bandwidth of a single RF channel.

$Gm$ is the complexity of modulation processing and filtering.

$Gd$ is the complexity of demodulation processing (carrier recovery, Doppler tracking, soft decoding and postprocessing for TCM, etc.).

$Rb$ is the data rate of the (nonredundant) bitstream.

$r$ is the code rate.

$G3$ is the complexity of bitstream processing per channel (e.g., FEC).

Table 5-2 shows how these parameters are related for an illustrative application. When designing a system, one must provide some excess capacity so that the processing demand given in the table can be met in spite of the statistical patterns of external and internal events. Simple radio nodes have only a single antenna and RF channel. The digital processing hardware consists of only one or two DSPs organized in simple linear flows. There are less than 10 k LOC performing a single fixed function. In this case one needs little spare capacity. If, however, there are multiple antennas and/or RF/IF chains, or the DSPs are organized in a pool to support hundreds of subscribers, the node is more complex. Such nodes generally have over 300 k LOC. Patterns of demand from the many users cause the pattern of demand on the processors to have a statistical distribution. In such cases, one must provide spare capacity. Too much spare capacity yields an unaffordable product. Too little yields an unreliable system that is constantly crashing due to the lack of sufficient processing resources. The performance management chapter of this book fully explains how to achieve the balance necessary for a reliable but cost-effective product.

If any of the above parameters are not known, then, under duress, one may employ the following simpler rule of thumb for receiver processing demand:

$$D \sim D_{\text{IF}} + D_{\text{DSP}}, \qquad \text{where}$$

$$D_{\text{IF}} = 100 * Wa * 2.5, \qquad \text{and}$$

$$D_{\text{DSP}} = 100 \text{ N} * (Wc + Rb) * \text{Uncertainty} * \text{Queuing}$$

Uncertainty is a multiplicative constant representing the degree to which processing is known or expected to increase. For a feasibility analysis of an architecture, Uncertainty = 2. Queuing is a multiplicative constant representing the excess capacity needed to manage internal queuing delays. Queuing = 2 provides 50% spare capacity. Unless a fine-grain performance-management model has been developed, Queuing should not be set less than 2.

Since $Wa$, $Wc$, and $Rb$ are known for all air interfaces, this formula can be applied essentially at any time. $D_{\text{IF}}$, the IF processing demand, is computed from the access bandwidth, the engineering version of the Nyquist factor, and a rough estimate of the processing complexity of digital IF conversion and filtering. If $Wa = 25$ MHz, $D_{\text{IF}}$ is 6250 MOPS (6.25 GOPS), where an operation is a multiply-add. This demand is generally allocated to an ASIC or FPGA. This front-end processor reduces the data rate to $2.5 * Wc$ for each of $N$ subscriber channels. For GSM, $Wc = 200$ kHz and Uncertainty = 1; then $D_{\text{DSP}} = 42.6$ MOPS per subscriber or 28.4 GOPS for a base station supporting 333 subscribers (receive only). Transmitter processing demand is typically $\frac{1}{4}$ that of the receiver, so the DSP pool would be sized at $1.25 D_{\text{DSP}}$.

When defining an architecture, one must ensure that it pays sufficient attention to digital processing performance management. If the architecture is to support plug-and-play as defined in Chapter 1, the architecture must proscribe the use of system management facilities that keep track of both software-driven processing demand and available hardware capacity. This may be accomplished using a constraint language to represent processing capacity and demand. An algorithm in the system manager may then compute whether

| Typical Application | RF fc | Wa | IF Wi | Channel Code | Baseband Wb | Bitstream States | FEC | Multiplexing | Privacy | Source Coding |
|---|---|---|---|---|---|---|---|---|---|---|
| FM Mobile | VHF | 30 kHz | 30 kHz | FM | 4 kHz | Inf | ---- | [PTT] | None | Compand |
| MCR Voice Control | UHF | 25 kHz | 25 kHz | FM | 4 kHz | Inf | ---- | [CA] | None | Compand |
| | " | " | | MSK | 10 kbps | 4 | Conv | TDM | None | Data |
| GSM | UHF | 25 MHz | 200 kHz | GMSK | 270.83 kbps 13 kbs | 4 | CPF | TDM | Encrypt | RPE-LTP |
| Mil FH | VHF | 60 MHz | 30 kHz Agile | FH-QPSK Agile | 16 kbs Digital | 4 | Conv ARQ | [PTT] | Encrypt | LPC/Delta |
| CDMA | UHF | 175 MHz | 50 MHz | M-PSK | 1.3 Mbs | M | Conv | [CD] | Privacy | LPC |
| JTIDS | L-Band | 200 MHz | 3 MHz | MSK | 150 kbps | 2 | Mult | TDM | Encrypt | LPC |
| Microwave | SHF | 20 MHz | 20 MHz | 64 QAM | 90 Mbps | 64 | Intlv | TDM | Bit Stuff | PCM |

[Implicit Multiplexing]: [PTT] = Push to Talk; [CA] = Circuit Assigned; [CD] = Code Division.

CPF: GSM specifies various modes of Convolutional, Parity and Fire codes [8].1es [8].

**Figure 5-21**  Processing-flow related parameters.

capacity will be sufficient for the personalities being instantiated on the radio [202]. Without such a constraint language the description of required capacities would be relatively ad-hoc and unstructured. Without such a system manager, the radio node would be rendered dysfunctional by the attempted execution of a combination of personalities or parameters that exceed the processing capacity of the hardware platform.

*3. Parameters of Illustrative Flows*  The parameters that most impact the digital processing flows are listed in Figure 5-21. $Wa$, the access bandwidth required to service all users, is derived from the air interface standards, adjusted for regional variations. For example, GSM specifies a set of frequencies in the 800–900 MHz band. In some countries, the GSM RF allocation to a specific operator may be a subset of the total allocation, limiting the required access bandwidth $Wa$ to half or $\frac{1}{3}$ that specified in the figure. Military, law enforcement, and others use frequency-hop agility, the entire bandwidth of which is reflected in $Wa$. JTIDS, for example, has an instantaneous direct spread bandwidth of only 3 MHz, but its FH agility bandwidth is the 200 MHz shown in the figure. An IF software radio for JTIDS requires $Wa = 200$. Alternatively, the fast-settling LO hardware of a dehopping transceiver may collapse the agility bandwidth to the spreading bandwidth of $W_{IF} = 3$ MHz. This approach reduces the processing demand by nearly 20 dB, but requires a fast-settling LO that is substantially waveform specific.

Baseband bandwidth, $Wb$, is the per-carrier analog bandwidth of the baseband channel modulation. First-generation FDMA systems have per-carrier bandwidths of 25 or 30 kHz. GSM has a carrier spacing of 200 kHz, related to the MSK data rate shown in Figure 5-21. CDMA systems based on IS-95 spread each subscriber's signal to about 1.25 MHz. The 3G W-CDMA proposals include bandwidths of 5, 10, and 20 MHz. Channel code, bitstream states, FEC, multiplexing, privacy, and source coding all contribute to the complexity of the software as illustrated in Table 5-2.

**TABLE 5-3   DSP Reference Platform**

| Parameter | Specification | Remarks |
|---|---|---|
| Pools | Identity, number, connectivity | Processor pools and their connections |
| Processor | ISA | Version and firmware level for DSPs and microprocessors; identity of ASICs and FPGAs |
| | $N_{PROCESSORS}$ | Defined per available pool |
| | Capacity | MIPS, MOPS, MFLOPS, SpecMark |
| | ROM | Read-only memory |
| | RAM | Random access memory |
| | Shared memory | With processor access matrix |
| | Environment | Operating system |
| Interconnect | Identity, capacity, connections | Realistic aggregate |
| Disk | Identity, capacity, transaction rate | Specify those on real-time threads |
| PSTN | Identity, SDH level, media (e.g., fiber) | Include traffic and SS-7 |
| LAN | Local interconnect | Specify net packets per second |
| Other | Other local interconnect | Required for growth |

*4. DSP Reference Platform*   A DSP reference platform parameterizes processing capacity and related capabilities (Table 5-3). This follows the strategy previously applied in defining the RF reference platform. In addition to processing-capacity, digital interconnect capacity and any other potential system bottlenecks must be defined in the digital reference platform. To be useful to resource management software, parameters necessary for plug-and-play must also be included. Binary load images, for example, that are specific to an instruction set architecture (ISA) must be defined in the reference platform. If capabilities are provided via ISA-independent methods (e.g., Java), then ISA-specific aspects need not be defined. An ASIC may be treated as a processor that has a proprietary ISA. FPGAs may be treated the same way. In this case, the FPGA's personality (e.g., load-image) is its ISA.

Processing capacity may be provided via a heterogeneous mix of processors. Each grouping, or pool, of processors should be identified so that the resource manager software knows where to install downloads. In addition, the connectivity among processor pools should be defined so that the system can compare the required connectivity to that provided in the hardware being loaded.

## H. Node-Level Architecture Capability Profile

The canonical model may be used to organize node-level architecture parameters by segment as illustrated in Figure 5-22. Some of these parameters apply to the radio as an end-item, while others are segment-specific. This view of the

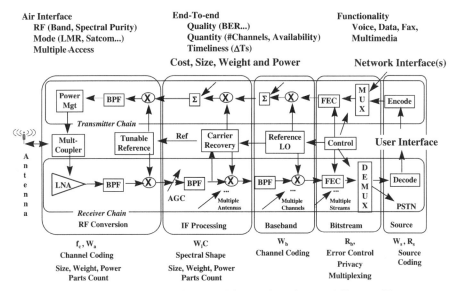

**Figure 5-22**  The canonical model organizes the capability profile.

canonical architecture focuses on the parameters necessary to support communications applications within the constraints of end-item packaging. The top-level parameters of a node may be aggregated into a capability profile. This profile becomes increasingly important as the number of potential personalities and hardware configurations grows over time.

*1. Capability Profile*  The set of air-interface or modes of which the node is capable is an architecture parameter at the node-level of abstraction. Air-interface parameters include the RF platform parameters. In addition, the spectral-purity required for type-certification and EMI control is relevant at this level. The node achieves spectral purity through a combination of hardware and software measures across multiple segments. Parameters defined by the effects of multiple segments are treated as node-level parameters.

Size, weight, and power consumption parameterize the node's packaging profile. Parts count contributes to manufacturing cost essentially linearly for mass production. Thus, an architecture model to be used by an enterprise to bring coherence to a product line must include this parameter. Similarly, acquisition organizations can use parts-count to estimate production learning curves in order to plan funding lines. Parts count is not as relevant for industry-standard open architecture hardware like PCI, however.

*2. Refined Radio Reference Platform*  The analysis of this section may be consolidated into a refined radio reference platform that extends the simple platform introduced in Chapter 1 (Table 5-4). This introductory treatment pro-

**TABLE 5-4   Refined Radio Reference Platform**

| Parameter | Specification | Remarks |
|-----------|---------------|---------|
| System Level | Size, weight, and power | Packaging-related |
| | Air Interfaces | Personality-defined |
| | Spectral purity | Depends on RF and algorithms |
| | Dynamic range | End-to-end |
| | Network timing | End-to-end accuracy possible |
| | Parts count | Mfg.-related, not open architecture |
| RF related | RF reference platform | Includes time standards |
| DSP related | Digital processing reference platform | Includes ASICs and FPGAs |

vides the tools for a critical assessment of architectures being proposed in industry forums.

## I. Exercises

1. Define node-level architecture analysis. How is node-level architecture analysis similar to system-level analysis? How are they different?

2. Use Figure 5-12 to represent the "transduce voice" radio node functions of Figures 5-2 through 5-7.

3. Describe the basic canonical model of SDR. What architecture-analysis objective does it serve?

4. Assess the completeness and consistency of the initial canonical model. What refinements were necessary to address the mapping of radio functions to hardware? What aspects of this mapping are made clearer in the UML object model?

5. Compare the use-case view of the radio objects to the canonical model of Figure 5-15. Which functions of the canonical model are absent from the use-case view? Insert these additional functions into the inheritance hierarchy, and insert the leaves of the new inheritance tree into the use-case view.

6. Complete the details of the object model of the canonical architecture sufficient to address the RF and DSP reference platforms. What aspects of the canonical model were left out?

7. What is the difference between computational demand and processing capacity?

8. What first-order parameters of the air interface determine the processing capacity required for IF processing? for baseband processing?

**9.** What additional parameters of SDR algorithms help determine the total computational demand?

**10.** Consider the disaster-recovery case study. Apply the techniques described in this section to that project. Include a UML model of the system.


## II. INDUSTRY-STANDARD NODE ARCHITECTURES

Industry-standard node architectures include open-architecture standards and de facto standards. The approach to defining open-architecture wireless taken by the SDR Forum is considered first. Subsequently, the reference architecture used by the ITU-T in its 3G deliberations is considered. These are both open-architecture standards. There are many PDRs and touted SDRs in existence, but there is as yet no single manufacturer that so dominates the industry that one could say a de facto standard exists in the year 2000.


### A. SDR Forum Architecture Framework

The SDR Forum published its initial technical report in July 1997. It defines an architecture framework for open-architecture plug-and-play software-defined radios. The SDR Forum appropriately differentiates architecture framework from architecture.

*1. Architecture Framework*   The SDR Forum architecture framework defines generic building blocks and interfaces. There are associated guidelines for tailoring the framework to create an architecture appropriate for a specific family of applications. The derived architectures embody subsets of the architecture framework appropriate to the application's functional needs, component constraints, and design rules. The focus of the Forum in 1999 and 2000 included the creation of an object-oriented architecture. This architecture is based on the earlier architecture framework, illustrated in Figure 5-23. The overall framework is based on the information transfer thread. It also envisions an API. API discussions included that of SPEAKeasy II as described in the literature [30], but no API was adopted. The heavy lines in the figure represent APIs mediated by a common bus (e.g., the separate red or black buses). The dotted lines show links between successive levels of expansion of the hierarchical decomposition of the information transfer thread. At the top level, the thread consists of a red front-end; an INFOSEC element; and a black back-end for message processing.

The third level of the SDR Forum framework separates the front end into RF and modem. INFOSEC is appropriately divided into TRANSEC and COMSEC. Message processing, according to the Forum, includes voice coding (vocoding), routing, signaling, input/output, and bridging. Bridging is the automatic conversion of a signal from one air interface format to another, such as from GSM to AMPS. Bridging generally signals a radio relay process in

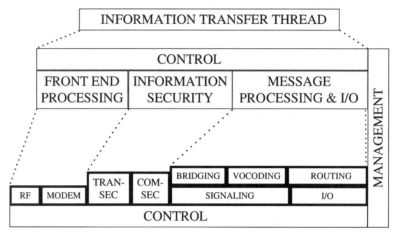

Figure © SDR Forum, 1999, reprinted with permission.

**Figure 5-23**   SDR Forum Information Transfer Thread architecture framework.

which the radio user is unaware that his unit is serving as a bridge, except
for a status indication. That is, no operator intervention need be required for
bridging.

One of the strengths of this framework is tailoring. Any of the building
blocks may be deleted (nulled) or replicated without violating the design integrity of the architecture framework. Some handsets, for example, have no
INFOSEC. The interfaces between front-end and back-end components will
work without the mediation of a physical INFOSEC module. Some applications require single-channel operation (e.g., an AMPS handset), while others
require four simultaneous channels. The modem, INFOSEC, and internetworking components may be replicated in a way that supports the four parallel
channels. These concepts were defined in 1996 and 1997, and have stood the
test of time. They have provided a basis for broad industry cooperation. The
details of these interfaces remain under development in 2000. The pace of this
evolution is slow, but the process retains the design philosophy in spite of the
many challenges of coming to industrywide consensus.

One aspect of this framework that warrants attention is the modem subthread. In this version of the SDR framework, IF processing was undefined.
Figure 5-24 suggests a revised partitioning with IF processing (IFP) and baseband processing (BBP). This conforms to the canonical model. The modem
functions are now explicitly baseband. The interfaces across these boundaries
conform to commercially available parts such as Intersil (formerly Harris) or
Graychip digital receivers.

*2. Entity Reference Model*   Figure 5-25 represents the next stage of SDR
architecture evolution. It encapsulates the elements of the IT thread into enti-

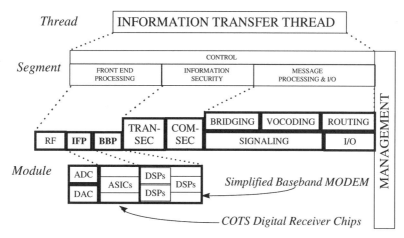

Figure © Mitola's STATIS*faction*, 1998, reprinted with permission.

IFP = IF processing; BBP = baseband processing.

**Figure 5-24**  Enhanced Information Transfer Thread.

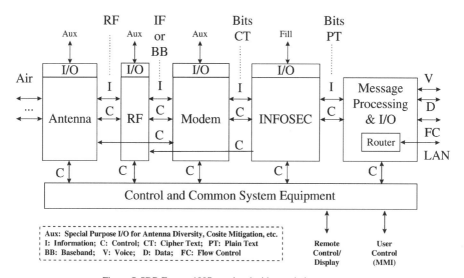

Figure © SDR Forum, 1997, reprinted with permission.

**Figure 5-25**  Module interfaces of the SDR Forum.

ties, object precursors. These promote the definition of interfaces among modules. Auxiliary interfaces ("Aux") are the special-purpose interfaces needed for input, output, antenna diversity control, cosite interference mitigation, and cryptographic key insertion (others are to be defined). The "I" or information

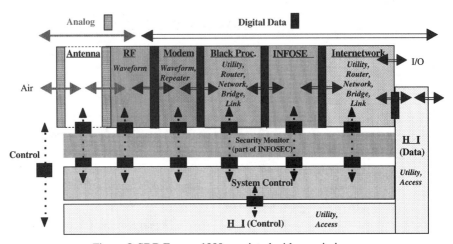

Figure © SDR Forum, 1999, reprinted with permission.

**Figure 5-26**   Notional entity reference model.

interfaces are paths for typically isochronous signal flows. "C" indicates exchange of control information among the modules. "CT" indicates cipher text
while "PT" indicates plain text. Other abbreviations include "BB" for baseband, "V" for voice, "D" for data and "FC" for flow control. Compare these
interfaces to the functional model of Chapter 1 and to the canonical model.

The interfaces among these SDR modules roughly correspond to the interfaces among segments of the canonical software radio. The RF interface
contemplated by the SDR Forum is an analog interface between the antenna
and the RF conversion module. This interface may be digital RF in the future
or analog as it is today. IF interfaces in this SDR model were also analog.
ADCs and DACs of the modem module provided the required conversions.
By 2000, this interface had become a digital IF interface in at least one successful commercial cell-site product. The bitstream interfaces among modem,
INFOSEC, and message processing & I/O are appropriate. Subsequently, the
forum added black-side data processing as a distinct entity. SDR module interfaces include local area networks (LANs) and an Internet Protocol (IP) router.
The "control and common system equipment" entity implies the existence of
logical and/or physical buses to link the other entities. Although INFOSEC is
present, there is no explicit red-black separation in this architecture.

A more recent proposal to the forum addressed INFOSEC in greater depth
as illustrated in Figure 5-26. The antenna, RF, modem, black processing,
INFOSEC, internetworking, and system control are entities in the Forum's
*entity reference model* (ERM).

The approach to these entities includes the object-oriented treatment of the
interfaces among the entities. In particular, the Forum has defined an inheritance hierarchy that allows objects to be controllable, initializable, activatable,

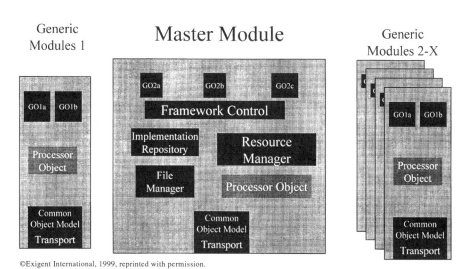

©Exigent International, 1999, reprinted with permission.

**Figure 5-27**   SDR Forum core framework.

and startable. The entities above are of the Transform class, which is activatable. Other classes in the hierarchy include Devices, Loaders, and Processors. Pentium, DSP, and PowerPC are illustrative examples of the Processor class. Classes for ASICs and FPGAs had not been defined as of June 2000. The members of the Transform class are loadable by a resource manager, which is part of the Forum's Domain Manager.

***3. CORBA Core Framework***   In addition, CORBA has been adopted by the Forum as the basis for integrating components that implement the entities. CORBA facilitates the definition of distributed processing middleware. This middleware is called the *core framework* (CF). It includes framework control, a repository of software resources, a file manager, and a resource manager as illustrated in Figure 5-27.

CORBA prescribes the definition of IDL interfaces to the ORB. Entities therefore conform to the Common Object Model, which links entities across distributed processors using the Processor Object abstraction. This corresponds to the infrastructure level of the layered virtual machine architecture introduced earlier in this text.

The architecture representation being deliberated in June 2000 by the Forum included an extensive object-oriented treatment. In order to appreciate the contributions of that architecture, one requires the more complete understanding of SDR software developed in Chapter 11 on software design tradeoffs.

***4. Handset Architecture Framework***   The SDR Forum also mapped its framework to the handset architecture model illustrated in Figure 5-28. The uppermost layer of this architecture accommodates ISO/OSI services. Layer 1 of the

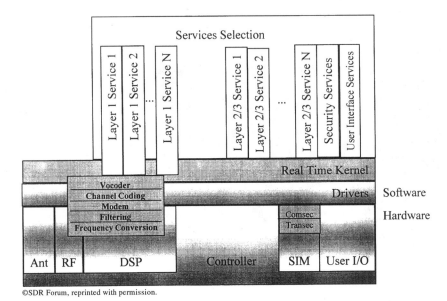

©SDR Forum, reprinted with permission.

**Figure 5-28**   SDR handset software architecture.

OSI model is the physical layer, the channel coding and related functions of the air interface. Layer 1 services are provided in the radio front end. The architecture accommodates multiple layer 1 services corresponding to the radio modes of the SDR. Thus, for example, a dual mode GSM/NMT handset would support the TDMA burst modem of GSM and the analog channels of NMT, as distinct layer 1 services. The Data Link (ISO/OSI layer 2) is also provided in the modem (or DSP as shown in Figure 5-28). As illustrated in the figure, the SDR architecture does not map easily to the canonical model. Vocoding, for example, is certainly a DSP function, but it is more appropriately aligned with the source coding of the bitstream segment. This is a back-end (source) function rather than a front-end (modem or black-side) function. This handset model therefore reflects engineering pragmatics. Comparing this to the canonical model yields design insights needed for product design-for-reuse.

The software in the handset model of Figure 5-28 is partitioned into drivers, a real-time kernel, and services. A rich set of drivers is required to implement the SDR modes. Although the SDR Forum has not yet fully defined these functions, subsequent drafts of its technical report may be expected to move in this direction. As the SDR Forum continues its deliberations, these interfaces may be refined and augmented with detail appropriate to the construction of hardware and software.

In addition to the technical contributions briefly highlighted above, the Forum has defined an open-architecture standard for downloading software to a mobile handset [203].

**TABLE 5-5    Parameter Alternatives for 3G Candidates (Circa 2000)**

| Parameters | ETSI UTRA & T1P1 | ARIB (W-CDMA) | TTA CDMA II | WIMS | Cdma2000 |
|---|---|---|---|---|---|
| Multiple Access | DS-CDMA | DS-CDMA | DS-CDMA | DS-CDMA | DS-CDMA |
| Bandwidth | 5 MHz 10, 20 | 5 MHz 1.25/10/20 | 5 MHz 1.25,10/20 | 5, 10, 20 MHz | 5 MHz 10/20 |
| Chip Rate | 4.096 Mcps (8.192/16.384) | 4.096 Mcps (1.024/8.192/ 16.384) | 4.096 Mcps (1.024/8.192/ 16.384) | 4.096, 8.192, 16.384 Mcps | 3.686 Mcps $(1.2288*n)$ |
| Carrier Spacing | Flexible/200 kHz carrier raster | Flexible/200 kHz carrier raster | | Flexible with carrier raster | $1.25*n$ MHz 30 or 50 kHz ras |
| Inter-BS Timing | Asynchronous | (Sync | … | Possible) | Synchronous |
| Cell Search Scheme | 3-step code acq nonscrambled | 3-step code acq nonscrambled | Two-pilot scheme | 3-step with search code | Pilot channel offsets |
| Frame Length | 10 ms | 10 ms | 10 ms | 10 ms | 20 ms (5 ms opt) |
| Data Modulation | QPSK | QPSK | QPSK | QPSK | QPSK |
| Spreading | BPSK | QPSK | QPSK | QPSK | QPSK |
| Scrambling | 10 ms | 10 ms | 10 ms | 625 $\mu$sec | Variable |
| Pilot per | Traffic channel | Traffic channel | Common | Common | Common |
| Pilot Structure | Time Mux | Time Mux | Code Mux | Time Mux | Walsh Mux |
| Pilot Coding | IQ/code Mux | IQ/code Mux | Code Mux | Time & code | IQ/code Mux |
| Power Control | Closed loop | based on | Channel | SIR | Dedicated |
| Scramble Length | 10 ms or 256 chp | 216 × 10 ms | 10 ms | 1 sym | $2^{42} - 1$ |
| Channel Code | Conv RS, Turbo | Conv, Turbo | Conv, Turbo | Conv, Turbo? | Conv, Turbo |
| Superframe | 720 ms | 720 ms | 640 ms | 720 ms | 80 ms |

Mux = multiplex; Conv = convolutional; RS = Reed Solomon Code.

## B. ITU-R IMT-2000 Device Architecture

The parameter alternatives presented by 3G candidates in Table 5-5 illustrate the similarities and differences among standards competing for the 3G standard. For a given infrastructure, these parameters must be fixed, but with SDR handsets, they could vary from one infrastructure to another, subject to deployment and/or handoff needs.

With current technology, a multiplicity of 3G standards would fragment the market unacceptably, escalating handset prices excessively. The prospects for SDR handsets expand the alternatives. Anticipating this technology, the ITU endorsed "SDR devices" as a mechanism for accommodating the differences [204]. The generic block diagram of an IMT-2000 device (Figure 5-29) defines some aspects of a reference platform. Its partitioning of functions carries strong design overtones because of the explicit placement of the ADC.

## C. Exercises

1. Describe the Information Transfer Thread model of the SDR Forum. What are its strengths? weaknesses? Compare the interfaces of Figure 5-24 to those of the functional model of Chapter 1.

2. Describe the entity reference model of the SDR Forum. What are its primary contributions? Which interfaces among entities tend to promote com-

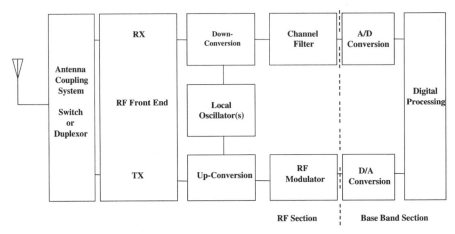

**Figure 5-29**   Generic block diagram of 3G device.

petition for entity-level components? What role could hardware interfaces play in this architecture?

3. Contrast the entity reference model of the SDR Forum to the (refined) canonical model. What is the contribution of the class hierarchy of the UML version of the canonical model? What role does the SDR Forum's entity reference model serve without something like the inheritance hierarchy of the UML canonical model (Hint: There is one, and it is important)?

4. What properties are addressed in the inheritance hierarchy of the canonical model? of the ERM of the SDR Forum? What SDR properties are best expressed some other way?

5. What function is served by the core framework of the SDR Forum? What other alternatives are potentially applicable to provide distributed processing middleware? How do these alternatives compare to CORBA?

6. Describe the 3G-reference platform proposed in the ITU. Compare and contrast it with the functional model of Chapter 1. Compare it to the UML canonical model.

7. The ITU's model could almost be the block diagram of a design, including the placement of the ADCs and DACs. What are the benefits of taking such a definitive model as a reference platform? How might such a reference platform shape or constrain competition?

8. What are the likely economic ramifications of the ITU's reference platforms? That is, is such a model likely to increase or decrease the cost of migration to 3G compared with the status quo where no such model exists?

9. Compare the handset model of Figure 5-28 to the canonical model. What end-to-end parameters of the canonical model need to be reflected in the handset model? Which are implementation-dependent? waveform-dependent?

10. Try to integrate the handset and waveform models into one model. Is yours better or worse than using two distinct models? Why?

## III. PROGRAMMABLE DIGITAL RADIO (PDR) CASE STUDIES

To develop the relationship between architecture and implementations, attention turns to a series of case studies of the progenitors and research implementations of the software radio. The first example is a simple SDR progenitor, a DSP baseband HF radio implementing analog voice and digital modem modes. This is the classic PDR. Progenitors of increasing complexity and sophistication are described, concluding with technology pathfinders. The material is organized according to the UML views to facilitate UML-based analysis of these radio nodes.

### A. A Basic Commercial PDR

HF radio uses narrow bandwidths (typically < 10 kHz) and moderate intermediate frequencies (IFs) (e.g., 550 kHz). As such it is an ideal domain for software radio applications. This section examines an SDR progenitor, a digital HF modem. This example proceeds in VML style, use-case first.

*1. Use-Case View*   The STR-2000 product line was developed by Standard Marine AB in 1992–93. Assume that the intended market included amateur radio enthusiasts, commercial shipping, and the military. To illustrate design issues, consider the feasibility-study phase of the project. At this point, functions are known, but the implementation decisions have not been made. Decisions at some point will focus on the partitioning of the functions into ASIC(s), FPGAs, DSP chips, and/or general-purpose CPUs. In this scenario, the focus is on the feasibility of a DSP platform. One goal is to determine either (a) which DSP chip would make an appropriate platform, or (b) how many such DSP chips would be required.

*2. Logical View*   The software radio functions shown in Figure 5-30 are to be implemented in the product. The design exercise for this example is to determine how much processing capacity is required for these functions by inspection. That is, one has to be able to look at the block diagram, identify the parameters that drive DSP capacity, and calculate MFLOPS with pencil and paper.

Figure 5-30 shows three of the radio's six functions. The modulator operating in J3E mode creates AM voice modulation on a 6 kHz IF carrier, which

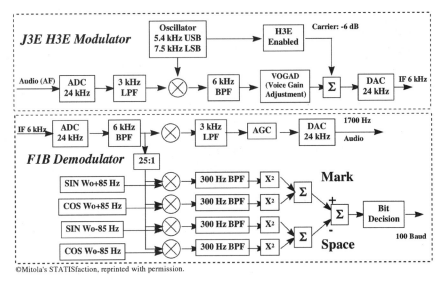

©Mitola's STATISfaction, reprinted with permission.

**Figure 5-30**    Standard Marine's HF digital IF software radio.

is translated up to HF frequencies using analog circuitry. In H3E mode, the vestigial sideband carrier is inserted after the voice gain adjustment (VOGAD algorithm). The receiver, similarly, expects a 6 kHz analog IF, which it digitizes at 24 kHz. Voice is translated to baseband and then filtered to a 1700 Hz audio band. The F1B frequency shift keyed mode is demodulated through the mark-space filter bank shown in the figure. Since the F1B mode has a low data rate, the algorithm decimates the IF samples by 25 : 1 before frequency translation and band pass filtering to extract the FSK energy at + or −85 Hz from the carrier.

*3. Component View*    One step in object-oriented SDR design maps the logical view to the component view. It consists of tracing the path of the ADC stream through the algorithm, estimating the number of instructions required per software task, and aggregating these to determine rough-order-of-magnitude (ROM) processing requirements. Generally, filtering requires more operations per sample of the signal waveform than most other operations. Bandpass filters are more computationally demanding than low-pass or high-pass filters. A typical 7-pole digital filter requires about 10 instructions per pole or 70 instructions per input point for a nominal filter. The process of rounding up from 70 instructions to 100 accommodates setup and control tasks not considered at this level of detail. At a sampling rate of 24 kHz, 100 instructions per point equals 2.4 million instructions per second (MIPS) for a simple filter. The 6 kHz bandpass filter might require more capacity while the low-pass filter might require somewhat less. Thus, the J3E or J3H modulator modes and the analog (digitized) voice mode should require somewhere around $2.4 \times 2 \sim 5$ MIPS.

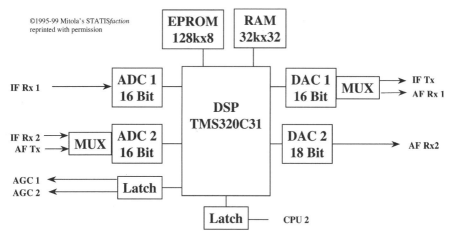

**Figure 5-31**   STR-2000 hardware block diagram.

But what about the F1B demodulator mode? Although there are four filters, the initial decimation rate of 25 : 1 reduces the sample rate from 24 kHz to about 1 kHz. Consequently, the filters and mark/space decision logic of the F1B algorithm can be 25 times more complex than the voice or modulator modes. Since there are only 4 filters in this mode, the complexity of filters is only twice that of the J3E mode. If the square-law or summing junction stages included complex bit-timing logic, then one would have to take a closer look. But from the (non-UML) logical view of the software functions, this is unlikely, so 2.5 MIPS can be used as a loose upper bound on F1B as well.

The accuracy of this estimate is plus or minus a binary order of magnitude (a factor of 2), for a range of from 2.5 to 10 MIPS. Compiler efficiency, the skills of the programmers, and the tradeoff between time-to-market and time to refine an implementation all contribute up to an order of magnitude in uncertainty of processing demand.

***4. Deployment View***   Processing demand of 2.5 to 10 MIPS is compatible with the TI TMS320 C30 (20 MIPS/40 MFLOPS). In fact, these algorithms were implemented on an early C30 with 10 MIPS capacity. It is usually not possible to support an application that needs 10 MIPS on a processor that delivers 10 MIPS. In general, one must supply a 50% processing-capacity margin for reliable performance. The physical packaging that results is shown in Figure 5-31. The link to the additional CPU provides interprocessor coordination so that the product could support two parallel DSP channels.

***5. Architecture Implications***   The example in Figure 5-31 shows how to assess the feasibility of a DSP implementation of a waveform that is defined in block diagram notation. This is a good step in determining SDR feasibility. That particular kind of logical view highlights the aspects of the application that are likely to have substantial impact on processing capacity. Block-diagram level of detail may be available from an air-interface specification, for example. This level of assessment is sufficient for sizing DSP hardware early in a development program.

The example also shows how the basic notions of SDR are present in digital radios. This radio has multiple DSP personalities. Its processing platform has been sized for the most computationally demanding personality. It has so few channels and such simple software that the complexities of large-scale SDR are not yet evident.

## B. Multimode Conventional Radios

The next case study examines a more complex radio system. This example proceeds using architecture-analysis style with functions, components, and design rules preceding UML analysis.

*1. Functions*   This example addresses medium-scale multiband radio systems. The services provided are voice circuits, with connectivity between about two-dozen air-interface modes and the T-Carrier PSTN. Customers include high-level military users. Specialized corporate communications networks also may use such technology. High availability and global connectivity are the driving requirements. A handful of base stations spread around the globe maintain communications connectivity at fixed locations. They also keep customers connected while on the move (usually via airplane).

*2. Design Rules*   User requirements dictate the use of discrete communications security devices such as KY58 and KG84 key generators. Cost may not be the major concern for such systems. During the early to late 1990s, several such systems were upgraded to SDR precursor technology.

*3. Baseline Components*   The baseline versions of this family of systems consisted of many discrete radio transmitters, receivers, power amplifiers, modems, and switches (Figure 5-32). Physical interfaces for data and control differed substantially from one device to another, resulting in hardware complexity of approximately ten to fifty 19-inch of racks of equipment in the radiotelephony base station. In addition, thousands of meters of inter-rack wire and analog cable were used. Much of the switching and equipment configuration was accomplished by patch panel. The analog lines were carefully matched for voltage standing wave ratio (VSWR), gain, and voltage level. Maintenance and repair was labor-intensive, requiring complex procedures involving signal generators, oscilloscopes, and other manual test equipment.

*4. Use-Case View*   Radiotelephone operators used minicomputers for monitoring and control. Customized displays, sometimes two or three per operator workstation, assisted telecommunications operators in setting up, tear-

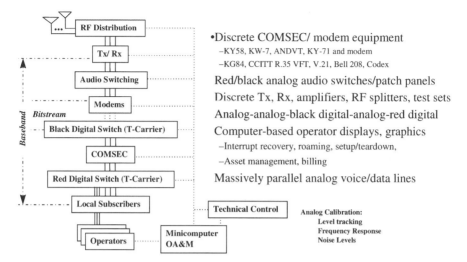

**Figure 5-32**   Conventional multiband-multimode radio system.

ing down, and monitoring quality of long-haul circuits. Alphanumeric interfaces displayed control information. Lessons learned in minicomputer-based OA&M laid the groundwork for the SDR-precursor implementations of the early 1990s. Some of these systems were replaced in the 1990s by modular digital systems.

*5. Digital Radio Components*   As illustrated in Figure 5-33, modem boards, computer-based switching, and open-architecture backplanes (VME and multibus) replaced patch panels. Most baseband communications and switching functions were implemented digitally in ASICs or DSP chips. The front-end radios became a hybrid of analog RF, programmable analog synthesizers, and programmable digital hardware. The use of programmable COMSEC units like the KG84 expanded, but compatibility with legacy COMSEC units was a requirement. Enhanced V-series modems emerged to raise voice channel modem data rates from 2400 baud to over 19 kbps. Antennas, RF conversion, IF processing, and filtering still were accomplished by hard-wired analog devices. Embedded software expanded the flexibility of baseband processing, switching, routing, and related services. Conference calls, group intercom, and database support were added.

*6. SDR-enhanced Use Case*   Operational improvements included the simplification of reductions in labor-intensive maintenance functions. In addition to avoiding some analog calibration and test, these systems began to employ instrumentation buses (e.g., GPIB) and computer-controlled BITE. This allowed a few personnel to maintain a large, distributed radio network. OA&M database servers assisted system operators with setting up communications channels and networks. This supported globally mobile users. Although the

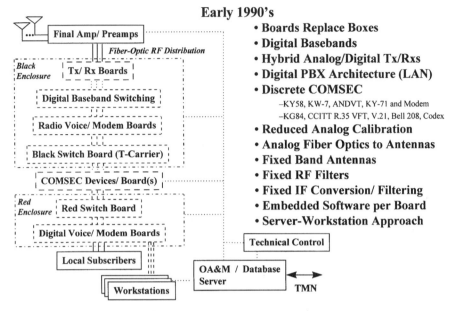

**Figure 5-33**   Radio system evolution.

markets were limited, the technology investments in part set the stage for the emergence of modular programmable digital radio products.

## C. GEC's Programmable Digital Radio

In the mid-1980s, a TIA panel advocated the development of programmable digital radios. GEC Marconi's programmable digital radio (PDR) was the first PDR product described in the literature [205, 206].

*1. Functions and Components*   GEC's PDR product [207] was organized into a black (encrypted) segment, a programmable message processor and a red, unencrypted segment as illustrated in Figure 5-34. The black segment employed a low-speed black-interconnect bus. Transceivers, all on the black side, consisted of an exciter module and a receiver module. The exciter fed a power amplifier, which in turn drove a multiband antenna subsystem.

The programmable message processor had two major functions: INFOSEC and the interconnection of the red and black segments of the radio. The red side included a CPU and I/O processor plus power supplies.

*2. Architecture Implications*   One advantage of the exciter/receiver modules is RFI/EMI control. Each module is encased in a shielded enclosure. High-power artifacts from the exciter modules are therefore readily controlled. Internally, the receiver module contains substantial digital signal processing capa-

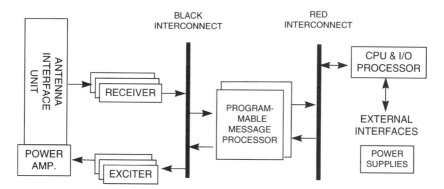

**Figure 5-34**  GEC programmable digital radio (1993).

bility in a programmable modem. High-speed clock transients from the DSPs were also isolated to the enclosure. Externally, however, the data rates are limited to voice and data rates of a few tens of kHz per channel. The low data rate of the red and black buses is also an advantage given the nature of the bus traffic. High-speed data (e.g., from the ADC) was connected directly to the DSP inside the receiver module. Interconnect is thus not a system bottleneck as it can be in other designs. This design was presented at the MMITS forum in 1996 as an architecture to consider in its deliberations.

## D. ITT Digital Radio

ITT Corporation also developed a digital radio in the early 1990s [208].

*1. Functions and Components*  The top-level components of the ITT radio are organized like the GEC PDR with black, red, and INFOSEC segments. The PC104 card-format and interconnect is used instead of a backplane bus, with separate PC104 card stacks in the red and black segments. As illustrated in Figure 5-35, there are three black hardware modules: RF, modem, and waveform processor. The RF module includes the power amplifier, transmit/receive switch, and transmit/receive modules. The modem module includes ASICs and FPGAs to perform hardware-intensive tasks. These tasks include a rake equalizer, Viterbi codec, signal acquisition, and synchronization circuits. The high equivalent computational capacity of the modem minimized the computational capacity of the waveform processor. Consequently, it could be implemented using a COTS Intel 486/66 MHz processor. This processor controlled the radio, and implemented the channel state machines. It also parsed the waveforms for transmission and reception. Other tasks included link-level bitstream processing of the air-interface protocol.

*2. Architecture Implications*  The software architecture associated with this design paralleled the hardware architecture as illustrated in Figure 5-35.

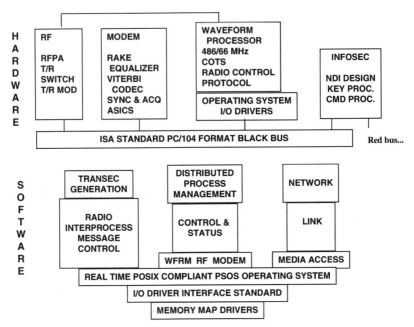

**Figure 5-35**  PC-104 digital radio architecture.

TRANSEC was generated in the software associated with the modem module. Since the radio included heterogeneous distributed DSP and general-purpose processors, it was controlled by interprocess messages. Such message passing is a precursor to imparting full object-oriented structure. The waveform processor managed the distributed processing including setup, status, and control of the black segment of the radio. This software is foundational to object-oriented resource managers. Due to the high degree of integration between the modem and waveform processor, these acted together as a programmable RF modem. Reprogramming required attention to software ripple effect to both of these components.

A real-time POSIX-compliant operating system (PSOS) was used in the front-end and back-end processors. Custom I/O drivers and memory maps were needed for interprocessor interfaces. As distributed processing technology continues to mature, these facilities should be provided in the COTS (e.g., CORBA) software tools. The partitioning of the software is informative. The POSIX kernel, I/O drivers, and memory map drivers define a core virtual machine. A subset of the radio and link control functions constitutes the radio-applications-level virtual machine built on this core. Higher-level radio functions and telecommunications services are built on this layer. These include control and traffic channel management. This partitioning implements many of the principles of object-oriented design. It helped set the direction for open-architecture software radio evolution.

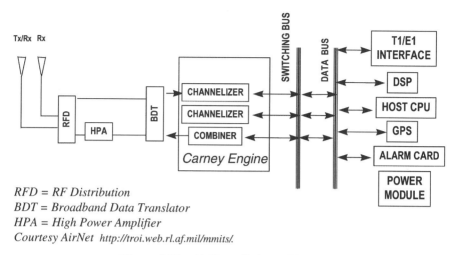

RFD = RF Distribution
BDT = Broadband Data Translator
HPA = High Power Amplifier
Courtesy AirNet  http://troi.web.rl.af.mil/mmits/.

**Figure 5-36**  AirNet cell site architecture.

## E. Commercial Progenitors: AirNet

AirNet, an early-1990s startup company, offered DSP-based wireless PBX to the commercial sector. It described its architecture to the MMITS forum in June 1996 [209].

*1. Functions and Components*  The architecture block diagram of Figure 5-36 identifies the organization of its main components. Since this radio node provides no information security, it lacks the black/red partitioning of the military radios. Its switching bus delivers isochronous streams between the T1/E1 interface, the COTS DSP elements, and the "Carney Engine." This DSP engine is the customized computational core of the PBX cell site. The data bus mediates the message-passing control interfaces of the system.

The Carney Engine [165] consists of ASIC channelizers and combiners. The channelizers are digital filters that extract subscriber signals from the broadband data stream created by the broadband data translator (BDT). These digital filters also preprocess the control and subscriber channels sufficiently that the COTS DSP modules can complete the modem functions within the allocated timing windows. The combiner aggregates many subscriber transmissions into a single digital stream. This stream is up-converted by the BDT and amplified by the high power amplifier (HPA) for transmission in the FDD output RF band.

*2. Architecture Implications*  AirNet's software-radio architecture may have been a little ahead of its time. At the first European Workshop on Software Radios in May 1996, Ericsson and Nokia presented technical papers describing the impediments to the deployment of software-radio cell sites. These included the lack of dynamic range in the ADCs necessary for near–far performance.

Although AirNet's product did not address the analog cellular standards, the full impact of these impediments may not have been fully appreciated by early pioneers. In addition, some early products were known to sometimes drop calls for no apparent reason. Power amplifier nonlinearities, lack of processing capacity, and software errors often cause such behavior in ways that can be difficult to diagnose and correct.

## IV. TECHNOLOGY PATHFINDERS

Technology pathfinders include thrusts in the military and commercial sectors. This section provides relevant highlights of these projects.

### A. COTS Research Pathfinders

The canonical model reflects implementation realities including compatibility with COTS hardware architectures. The Versa Multibus European (VME) form factor, for example, was one of the first widely successful COTS back-planes (e.g., [210]). Although the VME bus has a strong position in the DSP marketplace, SDR programs like SPEAKeasy II [211] and the ACTS FIRST software radio testbed [212, 213] chose the PCI bus and the PCMCIA form factor over VME. PCI's 132 M byte per second (MBps) data rate surpass VME's 40 MBps rate.

*1. Functions and Components*   As illustrated in Figure 5-37, the segments of the canonical software radio map readily to open-architecture VME or PCI hardware components. Commercial radio manufacturers offer COTS transmitter and/or receiver boards. Some of these integrate ADCs and DACs while others provide wideband analog IF outputs for COTS ADC boards as illustrated in the figure. The IF segment may be implemented digitally using COTS digital filter boards [34, 40, 42]. The wideband ADC is typically followed by programmable digital receiver chips. These chips digitally mix and filter subscriber channels to deliver the narrowband decimated digital channels to the DSP boards. These boards provide a pool of DSP resources that perform baseband and bitstream processing. The bitstream segment may include COTS T1/E1 interfaces to wireline networks.

*2. Architecture Implications*   These COTS DSP architectures map well to SDR functions. They therefore speed product development. If the deployment environment is benign (e.g., laboratory-like), and if the production quantities are small (e.g., < 10 systems), then the COTS architecture may be delivered essentially as-is. If the physical environment is more severe, then VXI may be more appropriate, or the software architecture may have to be rehosted to a rugged hardware environment. In addition, if the production quantities are large, greater integration of functions may yield lower production costs.

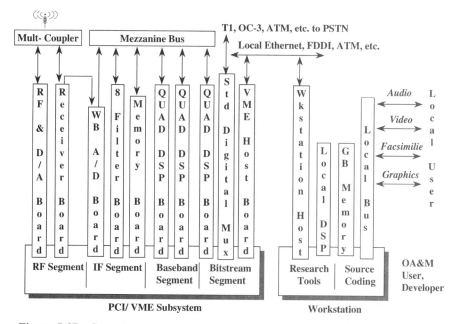

**Figure 5-37** Canonical architecture segments map to open-architecture hardware.

These systems help one understand the relationship between research-quality development platforms and production-quality systems. Production systems typically provide less processing capacity and limited arithmetic capability (e.g., 32-bit fixed point versus 64-bit floating point arithmetic).

## B. SPEAKeasy, the Military Technology Pathfinder

SPEAKeasy has been called the military software radio [3]. SPEAKeasy included two contract phases (SPEAKeasy I and II) in a military research and demonstration program in software radio technology. SPEAKeasy was to encompass 200 families of discrete military radios in a single product. It addressed the need to consolidate multiple discrete single-band radios into a single software-programmable unit with flexible RF access between 2 MHz and 2 GHz. A SPEAKeasy radio that can support four simultaneous modes in this frequency range replaces six to ten discrete radios in ground-based, shipboard, and airborne applications. Such consolidation simplifies logistics support, while the programmability enhances interoperability. This allows differently equipped military forces to work together. SPEAKeasy I was awarded to the Hazeltine Corp. (now GEC Marconi Hazeltine) in 1993. It was implemented as a six-foot rack of radio and open-architecture digital signal processing equipment. Although cumbersome, it proved the technical feasibility of the narrowband SDR. SPEAKeasy I also identified technology development issues. SPEAKeasy II was awarded to Motorola in 1995. It achieved a 4 : 1

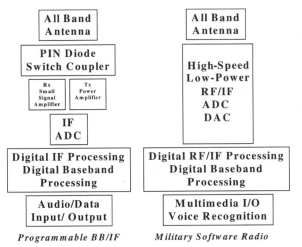

| All Band Antenna | All Band Antenna | • SPEAKeasy I |
| | | • Hazeltine (Prime), TRW, Rockwell (RF), Motorola |

Programmable BB/IF        Military Software Radio

**Figure 5-38**  SPEAKeasy migration from programmable baseband to software radio.

reduction in size and weight over the baseline military radios, while increasing computational capacity to over 1 GFLOP.

*1. SPEAKeasy I*  SPEAKeasy I was a joint program of the U.S. DARPA, the U.S. Army's Center for Electronics and Communications (CECOM), and the U.S. Air Force's Rome Laboratory (RL). An overview of the goals of SPEAKeasy I is provided in Figure 5-38. The original goal included the design of a single antenna that would span this entire band, directly accessing the RF digitally using a 5 GHz ADC and multimedia I/O as shown in the figure.

The original goal for waveform implementation included the 17 different modes listed in the figure.

*2. SPEAKeasy I Functions and Components*  SPEAKeasy I adopted the functional partitioning illustrated in Figure 5-39. Functional modules are partitioned such that each module makes significant translations in data structure and data rate. Consequently, the SPEAKeasy I modules are functionally cohesive and data coupled. The goal of a single wideband antenna was not met. The frequency plan included three antennas with associated RF and IF processing. The low band accessed HF, the middle band extended from 30 MHz to 400 MHz, and the high band was to extend from 400 MHz to 2 GHz. The antenna segment delivers analog RF to frequency conversion and filtering, the RF/IF processing segment of this radio. The analog IF at this point is broadband, appropriate for conversion to digital samples using a 40 MHz ADC. Digital filtering selects the narrowband channels.

Detection and demodulation processing yield a symbol stream. That is, a channel symbol (e.g., an MSK burst) is converted to the corresponding data symbol (e.g., 01, 10, etc.). These symbols are converted to bits through the

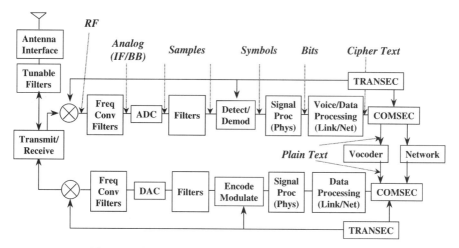

**Figure 5-39**   SPEAKeasy I functional block diagram.

appropriate bitstream processing such as differential decoding, bitstream de-interleaving, FEC decoding, etc. These initial segments handle the physical and data link layers of the ISO/OSI protocol stack. Data processing handles the higher levels of the protocol stack. The vocoder is an internal element of the SPEAKeasy I radio node, but the network block shown is an external element. INFOSEC is partitioned into COMSEC and TRANSEC. TRANSEC requires actions (e.g., FH) that are distributed throughout the front end of the radio.

AM double and single sideband modes were implemented for HF. SPEAKeasy I software has a set of modules named ALE that implemented the narrowband HF AM modulation mode, but never fully implemented ALE. In the ALE protocol both sides of the link choose a nearly optimal frequency for communicating through the ionosphere. This mode was dropped from the SPEAKeasy I program in part because of the complexity of the mode. Current ALE packages have 130 k LOC. HAVE QUICK I is a slow frequency hop mode which can be implemented with a relatively slow IF synthesizer. Due to the limitations of the wideband ADC technology (discussed in detail in the design chapters), the capabilities of the radio fell short of requirements for a production radio. Other waveforms were dropped due to the lack of software programming resources.

***3. SPEAKeasy I Use Case***   SPEAKeasy I demonstrated its capabilities in a series of experiments. In 1995 SPEAKeasy I participated in the Fort Franklin interoperability demonstration at Hanscom AFB. It bridged between a citizens' band (CB) radio network and a (nominally) frequency hopped HAVE QUICK network. In bridging, one transceiver channel was set to CB mode while a second simultaneous transceiver channel was set to HAVE QUICK mode. The network/vocoding functions translated CB audio to the vocoded

**TABLE 5-6   SPEAKeasy I Characteristics**

| Component | Remarks |
|---|---|
| Waveforms | AM, FM, VHF SINCGARS, UHF HAVE QUICK I & II bridging between voice waveforms |
| RF | Ad hoc discrete design per band, 1 mW power output, 4 MHz–400 MHz |
| Networking | None |
| User Interface | Sparc workstation with X Windows, hardcoded custom GUI, no remote display head |
| INFOSEC | TRANSEC in ASICs (No Cypris) |
| Modem | Quad C40 VME cards |
| I/O | Voice |
| Size | 6 ft rack (28 cu ft) $66 \times 26 \times 26''$ |
| Weight | 300 lb. (includes Sparc) |
| Power Consumption | 3300 W (15 A x 220 VAC) |
| Reliability | 20 hr. MTBF (Mean Time Between Failures) |
| Computer Bus | VME plus unique high-speed bus |
| Card Modules | Custom 12U VME plus unique hardware |

HAVE QUICK mode. This raised security concerns about bridging from an unencrypted network to a potentially encrypted network. It also illustrated the benefits of multiband, multimode radios with several simultaneous channels.

*4. SPEAKeasy I Architecture Implications*   One lesson established on this program was the difficulty of accurately predicting the amount of software resources required for implementing SDR waveforms. SPEAKeasy I also identified technology investment issues including the following:

- Programmable INFOSEC (Red/Black)—Resulted in the Cypress programmable INFOSEC module and the Motorola Advanced INFOSEC Module (AIM) chip.
- Software Cryptography & Certification—Identified this as a key issue for the U.S. National Security Agency (NSA), the INFOSEC authority for the U.S. government.
- Digital Frequency Conversion—Identified cosite interference as a key technology issue for multiband-multimode radios, with digital frequency conversion identified as one essential approach to interference mitigation.
- Cosite interference—Subsequently grew to become a more central issue in subsequent developments. Only JCIT produced hardware to address this challenge.
- Multiband RF—Multiband antennas and amplifiers; turnable preselectors; RF interference cancellation; low noise amplifiers (LNA), and wideband mixers.

- Air Interface Compliance—Digitally synthesized waveforms required the development of advanced techniques for ensuring spectral purity in the air interface including digital predistortion and dynamic envelope shaping.
- Spectral Purity for Interoperability—Direct digital synthesis (DDS) and spectrally pure DACs were identified as key product development challenges.
- Wideband Processing—SPEAKeasy I analyses showed that DSP chips fall far short of the processing capabilities required for wideband processing, resulting in a mix of DSPs and programmable ASICs (e.g., digital filter chips, sometimes called digital receivers). Lessons learned resulted in flexible ASICs from Harris (now Intersil), Graychip, and others.
- DSP Capacity—SPEAKeasy I DSP requirements were clearly beyond the capacity of fully programmable DSPs, and lessons learned which were fed back to device suppliers such as Texas Instruments led to advances in subsequent products.
- Low-power, Multi-Chip Modules (MCM)—Cost and power dissipation density remain the key challenges to packaging of sufficient ASICs and DSPs into small size, weight, and power required for military radios, particularly manpack and handsets.

*5. The SPEAKeasy II Program*   SPEAKeasy I laid the groundwork for SPEAKeasy II. It demonstrated the value of new services like integrated mode bridging. It also identified the technology issues to be addressed by SPEAKeasy II. While SPEAKeasy I was conceived as an applied research program, SPEAKeasy II was more of an advanced development program. Its emphasis included implementation pragmatics, notably the packaging of an SDR into open-architecture COTS hardware of reasonable size, weight, and power.

*6. SPEAKeasy II Components*   Figure 5-40 shows the main components of SPEAKeasy II. The black DSP modules include a control processor, which is the PCI bus host. Module partitioning falls into the prototypical pattern of the military radio. INFOSEC (in this case COMSEC) partitions the radio into red and black buses.

Speakeasy II originally contemplated four model-year releases, one each in 1996–1999. The first model year required significantly more resources than originally planned. Although software productivity was high, the system surprisingly included more than 300 k LOC. The balance of the program was restructured in January 1997. Model year 1 achieved the goals listed in Table 5-7 (*in italics*).

*7. SPEAKeasy II Use Case*   The SPEAKeasy II (Model 1) radio was demonstrated in the Task Force XXI experiment at the National Training Center at Ft. Irwin, CA. It replaced the GRC-206, shown in Figure 5-41. The GRC-206

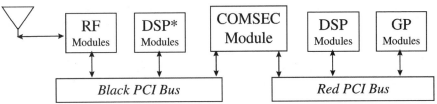

*DSP includes COTS, DSP chips, custom multiplier chips, FPGAs, and COTS bus host.

**Figure 5-40** SPEAKeasy II functional block diagram.

**TABLE 5-7 SPEAKeasy II Characteristics (Model 1, Task Force XXI)**

| | |
|---|---|
| Waveforms | *AM, FM, VHF SINCGARS (partial), UHF HAVE QUICK I & II*, UHF SATCOM, *GPS* |
| RF | "Ad hoc" discrete design per band, *2 W* power output, *4 MHz–400 MHz* |
| Networking | *Internet Protocol Suite (4) over Ethernet (IEEE 802.3), router capability* |
| User Interface | *Pentium laptop Win 95, modular object-oriented GUI, remote control w/stylus* |
| INFOSEC | TRANSEC in ASICs *with clear text bypass Cypris context switching (partial)* |
| Modem | Quad *C44 PCI COTS* cards *w/FPGAs, bridging among voice and data modes* |
| I/O | *Voice and data (802.3, RS-232)* |
| Size | *11 × 17 × 22″ (2.4 cu ft)* |
| Weight | *85 lb. (Includes laptop & RF amps)* |
| Power Consumption | *420 W (15 A × 28 VAC)* |
| Reliability | *500 hr. MTBF (estimated)* |
| Computer Bus | *PCI plus ISA* |
| Card Modules | *COTS (Pentium, PCI) except RF/control* |

weighs 368 lbs and requires space that measures 38 × 30 × 16 inches, over ten cubic feet. Its radio interfaces are limited to AM, VHF/UHF, HF single side band (SSB) with Automatic Link Establishment (ALE), and SINCGARS. It has KY99 Army/Navy Digital Voice Terminal (AN/DVT), KY65 and KY57 secure voice modes, plus T1 wireline compatibility. The remote radio set control may be placed in the front seat of a HWMMV while on the road, or it may be placed in a command shelter when in bivouac. The enclosure is rugged and supports the replacement of field-replaceable units while deployed. The SPEAKeasy II developmental unit weighed only 80 lbs. It required only 2.6 cu ft of space, a reduction of approximately 4 : 1. This was significant progress, especially considering the high COTS content, estimated by Motorola, the prime contractor, as 80% COTS—most of which was PCI general-purpose computer and DSP boards.

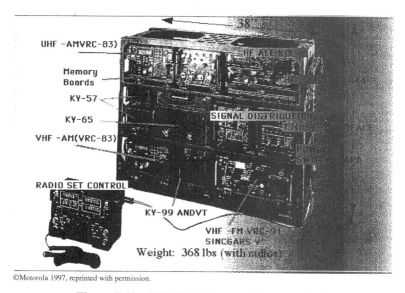

UHF -AMVRC-83)

Memory
Boards

KY-57

KY-65

VHF -AM(VRC-83)

RADIO SET CONTROL

KY-99 ANDVT

VHF -FM VRC-91
SINCGARS V
Weight: 368 lbs (with radios)

©Motorola 1997, reprinted with permission.

**Figure 5-41**   GRC-206 air-ground data terminal.

SPEAKeasy II was reprogrammed in the field to accommodate an unanticipated FM UHF mode. The new mode was needed for a Tactical Air Control Party (TACP). A 10 oz handheld commercial UHF radio was bridged to HAVE QUICK. This VHF FM mode was used by Close Air Support (CAS). The new mode was programmed in one week. The new software personality was downloaded to the field by modem. It was then loaded into the field-deployed units with essentially no interruption of the operational capability.

In addition, SPEAKeasy relied on COTS technology to make field-support easier in some ways. Again in Task Force XXI, the contractor repaired a failed system motherboard with one from a nearby computer store. The engineering approach called for "pure" COTS—no special modifications to the commercial motherboard. This approach enabled field repair from a local supplier, using simple procedures at low cost in this case. There are other positive and negative features to "pure COTS." The program demonstrated the advantage of commercial technology. For example, SPEAKeasy II's GFLOPS of processing capacity was provided by the latest COTS DSP chips. In this case, the benefits accrued as expected.

***8. SPEAKeasy II Architecture Implications***   The original evolution plan for the SPEAKeasy II program is illustrated in Figure 5-42. Unfortunately, the program was curtailed after only a single model year. A combination of minor hardware problems and significant software and integration cost overruns contributed to this restructuring. Others have discovered as well that it costs about two years and $25 M to transition from a world-class digital radio to one's first SDR [214]. In spite of the change of direction, this program

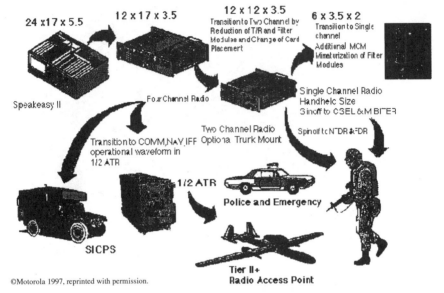

**Figure 5-42**   Initial SPEAKeasy II evolution concept.

clearly established narrowband software radio technology as viable for the
U.S. DoD. The technology is being applied to the JTRS, the ground-air data
modem (GADM), DMR, and other programs. GADM is to be a production
capability for the Air Force, replacing the older generation of GRC-206 radios.

Lessons learned from SPEAKeasy in many ways shaped the creation and
evolution of the SDR Forum. SPEAKeasy also laid the groundwork for the
PMCS and JTRS. In addition, SPEAKeasy lessons learned have helped to
shape the Future Modular Multi-Mode Tactical Radio (FM3TR) in NATO. And
the influence of the SDR Forum with its blend of SPEAKeasy, PDRs, JTRS,
commercial smart antennas, and multimode handsets is being felt in Europe
and Asia. Lessons learned from both SPEAKeasy programs are interleaved
into subsequent design chapters.

### C.  Joint Communications Interoperability Terminal

The Joint Combat Information Terminal (JCIT) is a multiband, multifunctional
terminal with programmable modulation and demodulation being developed
by the U.S. Navy Research Laboratory [215].

*1. JCIT Use Cases*   JCIT is a multiband, multimode SDR somewhat in the
spirit of SPEAKeasy. While SPEAKeasy was intended to be a technology
pathfinder from the outset, JCIT has been developed as a deployable product
for the U.S. Army aviation community. As illustrated in Figure 5-43, JCIT

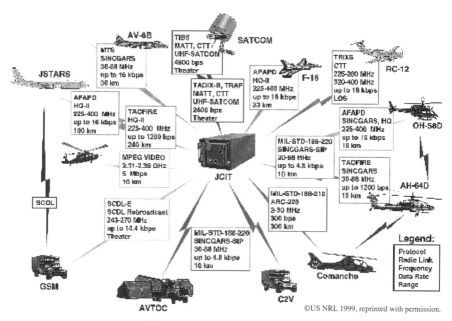

**Figure 5-43** JCIT applications.

will interoperate with numerous military radios, in RF bands from HF through 2.5 GHz. Initially, it will be used in helicopters to provide situation awareness and enhanced communications with supported forces.

***2. JCIT Functions and Components***   JCIT includes INFOSEC, link processing, message processing, and a user interface. The program has emphasized hardware capacity, including the use of FPGAs and DSPs for maximum processing capacity. The present system, for example, includes over 60 digital signal processors. The hardware is militarized, employing the rugged SEM-E form factor, as illustrated in Figure 5-44.

Its specifications are as follows. Eight multiband transceiver modules cover the integrated radio frequency range of 2–30, 30–88, 108–175, 225–530, 960 –1215, and 2400–2500 MHz. It provides the following modulation formats: AM, FM, SSB, BPSK, SBPSK, QPSK, SQPSK, SOQPSK, MSK, CPFSK, DSSS, FH, GPS, CDMA, TDMA. The transceiver output power is 0 dBm. Data rates range from 50 b/s to 5 Mb/s. Its I/O includes: RS-422, RS-232, MIL-STD-1553, RS-170, RS-330, RS-343, RGB, SCSI, and Ethernet. The protocol stack includes: ATM, TRAP, TADIXS-B, AFAPD, TACFIRE, TIBS, JTIDS, TRIXS, TADIL-A/B/J. Three integrated INFOSEC modules provide HQECCM, KY58A, KGV-10, KGV-11A, KGV-8A, KG-96, KOV-1; supports DS101/102 key loader. It is housed in a 3/4 ATR enclosure that consumes 300 watts power.

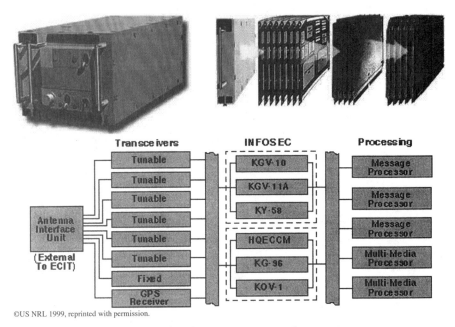

©US NRL 1999, reprinted with permission.

**Figure 5-44**   JCIT components and structure.

The *transceiver* function provides up to eight receive and transmit functions, along with control signals for external high-power transmit amplifiers. It provides amplification and filtering, frequency conversion, modulation and demodulation, and coding and decoding capabilities. It supports low probability of intercept (LPI) and low probability of detection (LPD) capabilities. It also supports Global Positioning System (GPS), frequency hopping (FH), and direct-sequence, spread-spectrum (DSSS) codes for transmission security (TRANSEC).

The *processing* function provides general-purpose and specialized processing capabilities for a variety of link formats (TRAP, TADIXS-B, TRIXS, EPLRS, AFAPD, TACFIRE, TIBS, JTIDS, DAMA, MIL-STD-188/220). JCIT supports user interfaces for a variety of digital I/O (RS-232, RS-422, MIL-STD-1553, Asynchronous Transfer Mode). JCIT supports graphics output processing and imagery data processing for user interfaces (RGB, RS-170, RS-330, and RS-343) and provides coding and decoding for secure voice interfaces (LPC-10, CVSD, and CELP). The software exceeds 500 k LOC, and has taken over four years to develop.

*3. JCIT Architecture Contributions*   By June 2000, the prototype units had achieved many of these specifications. The end-to-end data rate, however, was limited to 75 kbps (not 5 Mbps). This occurred because of the way that STel (TM, Stanford Telecom) digital filtering chip was tightly integrated into the front-end signal stream. Hardware platform upgrades to the front-end filtering digital architecture are needed to approach Mbps end-to-end.

JCIT's primary architecture contribution to date may be the support of the SDR Forum's domain manager. This software loads waveform objects onto processing resources using plug-and-play design principles of the Windows registry. In addition, a simple constraint language assures that sufficient computational resources are available for real-time performance.

## V. EXERCISES

1. What are the constituents of architecture? What design principles of architecture have been evolving since the 1970s. What new technology has assisted in the maturation of these design principles into architecture principles?

2. How does OOT contribute to architecture analysis? What conceptual tools are applied in performing an OO analysis of a software-radio architecture?

3. What type of architecture minimizes the bandwidth of digital interconnect between modules?

4. What type of architecture maximizes waveform flexitility?

5. What is the role of technology pathfinders? Describe lessons learned from one such project.

6. What methods are useful for measuring architecture complexity?

7. What contributions did SPEAKeasy I make to the understanding of SDR?

8. What contributions did SPEAKeasy II make to the understanding of SDR? JCIT?

9. What contribution did JCIT make?

10. Find ACTS FIRST on the world wide web. Compare its contributions to those of the technology pathfinders of this chapter. Why is this also an important technology pathfinder?

11. Apply the approach of Figure 5-33 to the disaster relief case study. Is this an affordable design? Are you sure that you have covered all the bands and modes needed using the discrete computer-controlled radios? Did you include TETRA? How about $6\frac{1}{4}$ kHz APCO 25? How easy is it to make changes?

# 6 Segment Design Tradeoffs

## I. OVERVIEW

The six steps in the systems-level design process associated with the software radio are illustrated in Figure 6-1. The tradeoffs proceed from front end to back end. The choice of antennas (step 1 in the figure) determines the number and bandwidth of RF channels (step 2). This, in turn, constrains the numbers and bandwidths of ADCs (step 3). Some waveforms may require dedicated ASICs (e.g., W-CDMA despreaders) in front of the ADCs. Additional parallel IF processing and ADC paths may be necessary to support multiple-service bands simultaneously. The ADCs provide high-speed streams for heterogeneous multiprocessing (step 4). Digital interconnect fans these streams out to digital-filter ASICs. The resulting narrowband streams then interleave among DSPs, medium-speed interconnect, and general-purpose processors yielding a multithreaded, multitasking, multiprocessing operating environment. Software objects must be organized into real-time objects (step 5). The effective hosting of these objects onto this complex operating environment requires a refined set of techniques unique to this text called *SDR performance management* (step 6). These six tradeoff steps are introduced in this section and discussed in depth in the subsequent chapters.

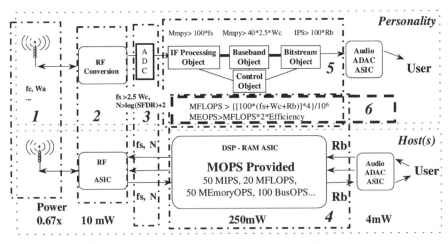

**Figure 6-1**   Six-step segment design-tradeoff process.

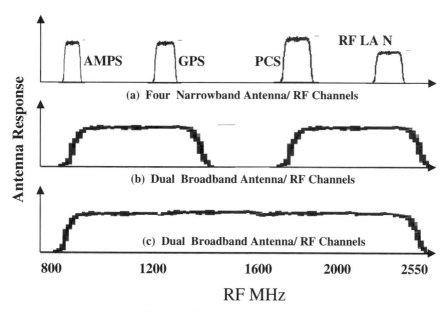

**Figure 6-2**   Antenna tradeoffs.

## II. ANTENNA TRADEOFFS

The first tradeoff defines the structure of the antenna segment and implicitly the RF conversion segment. Maximum system performance requires resonant narrowband antennas. As illustrated in Figure 6-2a, this approach typically results in multiple parallel antenna/RF conversion channels. In the specific case illustrated in the figure, an advanced PDA needs to operate in first-generation cellular (AMPS), and second- or third-generation digital cellular (PCS) bands. In addition, for location-aware services, the PDA has a GPS receiver. Finally, in order to operate on the corporate RF LAN, it supports a LAN band. One could fabricate such a PDA with parallel RF channels. The physical integration of the disparate RF devices presents challenges. Given a commodity GPS chip and a Bluetooth-class RF LAN, however, the parts costs could be low.

The broadband approach illustrated in Figure 6-2b simplifies the antenna and RF design to only two parallel channels. Finally, Figure 6-2c illustrates the spectral coverage of a single wideband antenna. Note that the antenna response is not uniform across such a broad range. The broad spectral coverage needed for SDR flexibility therefore can drive one toward many parallel overlapping narrowband channels. This can be an effective approach if cost is not a major consideration. Alternatively, the design may employ fewer channels with wider RF coverage per channel. Transmission efficiency and matching voltage standing wave ratios (VSWR) are more challenging as bandwidth increases. Below 100 MHz, multi-octave antennas and RF segments are well-established products. As frequency increases, wavelengths approach the physical dimensions of the RF devices, making it more difficult to exceed an octave of RF coverage. Since antennas, RF conversion, and IF processing through the first ADC can account for over 60% of the manufacturing cost of an SDR node, reducing the number of RF channels is a significant design tradeoff.

## III. RF AND IF PROCESSING TRADEOFFS

The second tradeoff concerns RF and IF conversion. Multichannel transceivers in TDD bands require an interference-suppression architecture that could include antenna isolation, programmable analog filters, and/or active cancellation. The transmitter may require both linear operation (e.g., for W-CDMA and QAM waveforms) and nonlinear operation (e.g., class-C amplifier for high-power efficiency with FSK or PSK waveforms). Single-channel receivers may resort to nonlinear distortion of the incoming waveform (e.g., for a narrowband direct-conversion architecture).

Multichannel receivers, on the other hand, must match the RF and IF conversion parameters to the ADC, digital filtering, and signal recovery algorithms of the back-end. The goal of this tradeoff is to balance the noise, spurious components, intermodulation products, and artifacts as illustrated in Figure 6-3. The receiver may include multiple RF/IF conversion stages (e.g., a superheterodyne—"superhet"—receiver). Alternatively, a single stage may convert the RF signal directly to baseband (the direct-conversion or "homodyne" receiver) [241]. Since the direct-conversion receiver has fewer parts, its manufacturing costs may be less than the superhet, which may have better performance. The thermal noise floor will be determined by the total bandwidth (e.g., in interference-limited bands below 400 MHz), or by the first LNA (e.g., in cellular and microwave bands). The thermal noise will be processed through the RF and IF conversion stages, resulting in noise shaping across the passband. Spurious responses ("spurs") and LO leakage can mask subscriber signals if ineffectively filtered. LO leakage can be particularly problematic in homodyne receivers [245]. A well-balanced architecture keeps the peak energy of all noise, spurs, and artifacts at about half of the least significant bit (LSB) of the wideband ADC.

## IV. ADC TRADEOFFS

The third tradeoff is the design of the ADC segment. Maximum sampling rate is obtained for a given clock technology in a quadrature-sampling ADC architecture. Such ADCs can introduce nonlinearities due to mismatching between the in-phase (real) and quadrature (imaginary) conversion channels. Real oversampling with digital quadrature provides a lower-complexity alter-

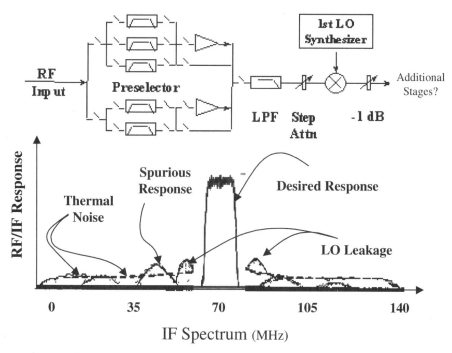

**Figure 6-3**  RF tradeoffs include staging, conversion frequencies, and filtering.

native. In addition, one must match the ADC architecture to the structure of the service bands being supported. If two or three 25 MHz bands spaced hundreds of MHz apart are to be supported, one may have more total dynamic range using multiple medium-bandwidth ADCs instead of one superwideband (e.g., 500 MHz) ADC. Each medium-bandwidth ADCs access band may be programmable by tuning the final LO. Such an approach therefore complicates the RF architecture, but reduces the interconnect bandwidth and processing capacity of the next stage. The set of ADC architecture alternatives is similar to the RF access alternatives illustrated in Figure 6-2.

## V. DIGITAL ARCHITECTURE TRADEOFFS

The fourth tradeoff concerns the mix of parallelism and pipelining of the digital signal processing hardware from ASICs and FPGAs to DSPs and general-purpose processors. Figure 6-4 illustrates the processing flow among wideband ADCs, DACs, and reconfigurable ASICs or FPGAs. High-speed (gigabyte-per-second) digital interconnect is necessary to fan wideband ADC streams out to digital filtering FPGAs or ASICs. Reconfigurable processors and despreader ASICs may reduce or eliminate the need for wideband dig-

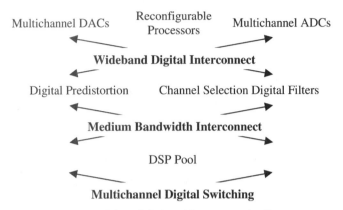

**Figure 6-4**   Digital architecture tradeoffs.

ital interconnect by either embedding the interconnect on-chip or producing baseband streams directly.

Digital filtering of high-data-rate ADC streams yields much lower data rate subscriber baseband channels. Medium bandwidth digital interconnect (hundreds of megabytes per second) then provides flexible paths among DSPs and general-purpose processors. The architecture of local and global memory among the processors also can be a significant contributor to algorithm performance. Balancing these high-speed data flows and bandwidth reduction steps against clusters of processing capacity and memory is a central concern of the digital architecture tradeoffs. The selection of an appropriate ISA is part of this step that determines the availability and maturity of software tools.

## VI. SOFTWARE ARCHITECTURE TRADEOFFS

The fifth tradeoff concerns the organization of the radio software into appropriately packaged data structures and real-time algorithms. Figure 6-5 summarizes software design tradeoffs as a function of the level of abstraction of the capability. At the highest level, interfaces among applications and services need to be radio-aware so that the radio's low data rates, high variability in data transfer times, and occasional outages do not severely curtail user satisfaction with the services. In the radio applications layer, object-oriented design techniques help group related data structures with appropriate algorithm methods. This simplifies detailed design, development, testing, deployment, and evolution of the software architecture. The terminology and approaches of object-oriented design may be applied to all the functions of the radio. This facilitates the realization of those functions in hardware, firmware, or software, as a function of technology and project needs.

| Communications Services | *Applications and Services* - Defining Interfaces Among Services |
| --- | --- |
| Radio Applications | ***Air Interfaces ("Waveforms")***<br>Partitioning radio software into reusable objects (e.g. Modem)<br>Defining Radio-level Interfaces (e.g. DS0 Voice Streams) |
| Radio Infrastructure | Achieving Real-time signal flows using CORBA<br>Assigning low-level objects to processors |
| Hardware Platform | Defining the Operating Systems Profile and Tuning it for Performance<br>Selecting Real-Time Kernel and Libraries (e.g. VxWorks, PSOS...) |

**Figure 6-5**   Overview of software architecture tradeoffs.

Projects may be implemented using conventional software techniques. With such approaches, the radio applications and infrastructure software elements are intricately interwoven. Open-architecture approaches now favor the use of the industry-standard CORBA [216] in radio infrastructure middleware. Such middleware reduces the coupling between radio functions and distributed processor hardware. This adds flexibility but requires processing capacity above that which is needed for a closed architecture. Accurate characterization of the processing requirements of modular collections of open-architecture software can be challenging. At the lowest level of abstraction is the real-time interface to the hardware platform. This includes not just the processors, but the many computer-controlled features of the analog radio platform. Software design tradeoffs, then, include both top-down application of open-architecture principles and bottom-up integration of existing hardware, firmware, and software components.

## VII. PERFORMANCE MANAGEMENT TRADEOFFS

The final major tradeoff concerns the management of processing demand offered by the software against the resources provided by the hardware platform. Accurate characterization of processing demand requires benchmarking. Accurate prediction (e.g., at proposal time), can be accomplished using queuing theory techniques that have been refined and reduced to the structured method described in the corresponding chapter of this text. A sustained measurement and instrumentation campaign to monitor performance implications of development decisions reduces development risk. Performance prediction and management steps add cost to a software radio development program. One therefore must balance the cost of performance management against the benefits of reduced development risk. This text shows how to manage performance affordably.

## VIII. END-TO-END TRADEOFFS

Other important tradeoffs include end-to-end tradeoffs. One of the most important in the software radio is the allocation of dynamic range among RF, IF, ADC, DSP hardware (e.g., FPGAs and ASICs), and algorithms. Another is the allocation of software objects to hardware components. After considering the antenna segment, the RF, ADC/ DAC, DSP, software, and integration aspects are considered in turn.

## IX. EXERCISES

The following questions review the material in the first part of the text. They also motivate the subsequent chapters.

1. Review the disaster-recovery case study. List the RF bands that need to be addressed. What antennas are needed? Search the web for COTS antenna products. List antenna products that you would employ in a contemporary design. Which of these antennas are programmable? Is there any flexibility that could be software-defined? What about a next-generation product?

2. Continuing with the case study, list the RF conversion components that are needed in a conventional design. What is the modularity of these components? What fraction of these components are integrated transceivers, and what fraction are modular at some other level of granularity? Is the data processing in devices physically separate from the RF access? Could the baseband streams (e.g., voice and data) be switched in software on LANs and workstations? How much work is it to determine the answer to this question? Did you use a systematic method to address this question? Do you know of one that you could have used?

3. Also in the case study, from the bandwidths of the RF access bands derive the bandwidths of the ADCs necessary to access these bands digitally. What are the output bandwidths of these ADCs in MBps?

4. Suppose narrowband channels (e.g., AM and FM push-to-talk) require 10 MFLOPS and second-generation standards like GSM require 30 MFLOPS per channel. How much digital signal processing capacity is required for the disaster-recovery application? How many vans are in your system? If you do not know, assume there are five. Can the channel capacity be partitioned among these vans? How many of what kinds of processors do you need? You could multiply the number of channels times the processing requirements per channel and then divide by the capacity of a processor. You could also apply some factors for realism, such as multiplying the number of processors by two so that each has 50% spare capacity. Does this yield an adequate sizing of the processors? Are they dedicated or pooled?

**5.** What software is needed for the disaster-relief case study? Of this software, which modules are in the front-end, which are in the back-end, and which are distributed across the system? How many lines of code will be in the system? How can you estimate this parameter?

**6.** What are the end-to-end issues in this case study?

# 7 Antenna Segment Tradeoffs

The antenna segment establishes the available RF bands. Although much research has been applied toward creating an "all-band" antenna, multiband radios generally require at least one antenna per decade of RF band (e.g., HF, VHF, UHF, SHF, etc.). In addition, the antenna determines the directional properties of the receiving system. Sectorized antennas, static beamforming arrays, and adaptive beamforming arrays (smart antennas) each have different spatial and temporal properties, the most significant of which is the pattern of transmit and/or receive gain. The antenna may also constrain the phase noise of the overall system. In addition, the interface between the antenna and the RF conversion stage determines VSWR, insertion loss, and other miscellaneous losses. In bands above 100 MHz, this interface can determine the overall system noise floor. This chapter characterizes the systems-level antenna segment tradeoffs relevant to SDR architecture.

## I. RF ACCESS

From a SDR perspective, the enabling RF-access parameters of the antenna segment are RF band and bandwidth as illustrated in Figure 7-1. Antenna-type in the figure lists the mechanical structure and the physical principle on which the antenna is based. Bandwidth is expressed either as a percent of carrier frequency or as a ratio of lowest RF to highest RF over which the antenna efficiency, VSWR, etc. are workable. Narrowband antennas have only a few percent relative bandwidth. Frequency limits are typically defined in terms of the 3 dB bandwidth of the antenna. An HF antenna, for example, that is operable between 2 MHz and 20 MHz has a relative bandwidth of 20/2 or 10 : 1. An antenna that operates effectively between 2 and 4 GHz, on the other hand, has a relative bandwidth of only 2 : 1. This ratio is one octave. Wideband antennas such as log periodic and equiangular spirals require a large number of resonant elements and therefore have a relatively high cost compared to narrowband resonant antennas. Helical antennas may be wound into whip or stub mechanical configurations for PCS applications [217].

For the ideal software radio, one needs a single antenna element that spans all bands. Requirements of the JTRS program are illustrated in Figure 7-2a. More than forty bands and modes must be supported in that program. With conventional technology, nine or ten antenna bands would be required as shown in the figure. Anticipating the JTRS program, SPEAKeasy attempted to

244

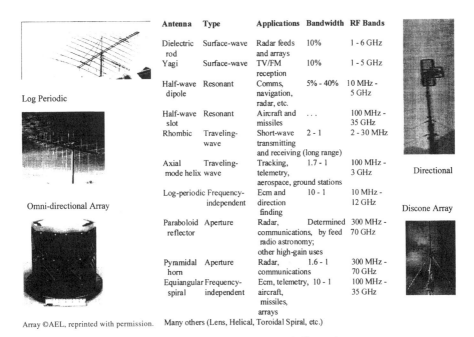

| Antenna | Type | Applications | Bandwidth | RF Bands |
|---|---|---|---|---|
| Dielectric rod | Surface-wave | Radar feeds and arrays | 10% | 1 - 6 GHz |
| Yagi | Surface-wave | TV/FM reception | 10% | 1 - 5 GHz |
| Half-wave dipole | Resonant | Comms, navigation, radar, etc. | 5% - 40% | 10 MHz - 5 GHz |
| Half-wave slot | Resonant | Aircraft and missiles | . . . | 100 MHz - 35 GHz |
| Rhombic | Traveling-wave | Short-wave transmitting and receiving (long range) | 2 - 1 | 2 - 30 MHz |
| Axial mode helix | Traveling-wave | Tracking, telemetry, aerospace, ground stations | 1.7 - 1 | 100 MHz - 3 GHz |
| Log-periodic | Frequency-independent | Ecm and direction finding | 10 - 1 | 10 MHz - 12 GHz |
| Paraboloid reflector | Aperture | Radar, communications, radio astronomy; other high-gain uses | Determined by feed | 300 MHz - 70 GHz |
| Pyramidal horn | Aperture | Radar, communications | 1.6 - 1 | 300 MHz - 70 GHz |
| Equiangular spiral | Frequency-independent | Ecm, telemetry, aircraft, missiles, arrays | 10 - 1 | 100 MHz - 35 GHz |

Log Periodic

Omni-directional Array

Array ©AEL, reprinted with permission.    Many others (Lens, Helical, Toroidal Spiral, etc.)

Directional

Discone Array

**Figure 7-1** Candidate antenna configurations.

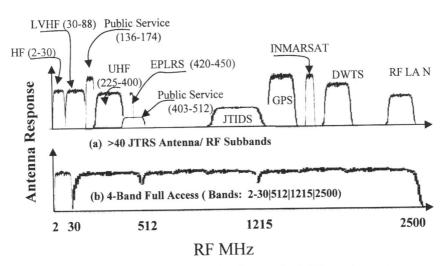

**Figure 7-2** Four software radio bands span the JTRS requirements.

realize a full-band antenna. The RF range extended from 2 MHz to 2000 MHz, a ratio of 1000 : 1 or 3 decades. Figure 7-1 shows that this requires a technology breakthrough, since the maximum relative bandwidth of the well-established designs is 10 : 1, or one decade. Through in-depth antenna studies conducted by Rockwell, Hazeltine, and others, it was determined that at least 3 bands are needed for this range. In fact, SPEAKeasy employed three bands as follows: (a) 2–30 MHz; (b) 30–400 MHz; and (c) 400–2000 MHz. To be precise, only band b was fully implemented in SPEAKeasy I and only bands a and b were implemented in SPEAKeasy II. For the foreseeable future, affordable RF access will probably be limited to octave coverage in the bands above 100 MHz. One configuration of antenna coverage that employs four conservatively designed bands is illustrated in Figure 7-2b.

## II. PARAMETER CONTROL

From a systems-engineering perspective, one must allocate end-to-end performance to parameters of the appropriate segment. The use of wideband antennas that enable SDR levels of performance complicates the control of SNR, timing, and phase parameters as follows.

### A. Linearity and Phase Noise

Wide bandwidth is sufficient for detection, but high SNR is necessary for good SDR algorithm performance. As the antenna bandwidth is increased, the thermal noise power increases linearly. Thus, the antenna channels must be filtered to select only those subsets of the band required to service subscriber signals. This is accomplished in the RF conversion and digital IF processing segments.

Low phase noise is also critical for phase-sensitive channel modulations such as high-order QAM (> 16 states). Phased array antennas that form beams through the switching of delay elements can have high phase noise induced by switching transients, making high-order QAM impractical.

### B. Parameters for Emitter Locations

In addition, precision timing or RF phase control may be necessary. For example, the commercial sector now has requirements for the location of mobile stations from which emergency calls are placed. The US E911 service requires location to within 125 meters. Network-based emitter location techniques include time-difference of arrival (TDOA) and angle of arrival (AOA) estimation using phase interferometry. TDOA [218] requires timing precision on the order of 100 ns, systemwide, to meet E911 requirements. Similarly, AOA [219, 220] requires phase measurements equivalent to a few degrees of angle uncertainty, which is equivalent to a few electrical degrees of phase error.

Smart antennas generally derive some estimate of the direction-manifold of the received signals. This information can be translated into AOA. In addition, TDOA techniques may be used alone or in conjunction with smart antennas to estimate the location of mobile subscribers. TDOA is particularly

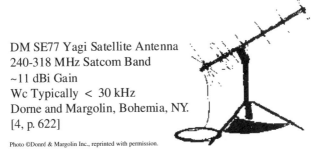

DM SE77 Yagi Satellite Antenna
240-318 MHz Satcom Band
~11 dBi Gain
Wc Typically < 30 kHz
Dorne and Margolin, Bohemia, NY.
[4, p. 622]

Photo ©Dorné & Margolin Inc., reprinted with permission.

**Figure 7-3**   Yagi illustrates mechanical configuration issues.

relevant to CDMA systems because they continuously estimate time of arrival (TOA) in order to recover the direct-sequence, spread-spectrum waveform. The conventional rake receiver may be augmented with, for example, extended multipath tracking Kalman filters in order to improve the TDOA measurement [221]. The presence of multipath can degrade both AOA and TDOA measurements.

## III. PACKAGING, INSTALLATION, AND OPERATIONAL CHALLENGES

Challenges facing the SDR systems engineer include the packaging of antennas with the desired capabilities into suitable hardware formats. For precision applications like emitter location, antenna arrays must be calibrated periodically. In addition, the influence of the human body on the antenna patterns of hand-held units should be understood. Software techniques may mitigate some of these effects to yield a corrected, more idealized antenna response.

### A. Gain versus Packaging

A typical UHF satellite antenna has a fractional bandwidth of less than one octave, but relatively high gain, as illustrated in Figure 7-3. This specific antenna from Dorne & Margolin uses crossed grounded elements for a ground plane, with a relatively complex Yagi array of receiving elements that enhance the gain. Since this antenna operates only over the satellite band between 240 and 318 MHz, the narrow relative bandwidth is not a limiting factor. The high gain is available only within about 20 degrees of the direction in which the antenna is pointing. In addition, such narrow bandwidths and beamwidths seriously limit RF access, or increase overall system cost. If, for example, the Yagi were the standard antenna for the 240–318 MHz band, the node would not be able to receive other communications in that band from any direction other than that in which the Yagi is pointing. Alternatively, one could provide six to ten parallel Yagi's for omnidirectional coverage, but this increases cost

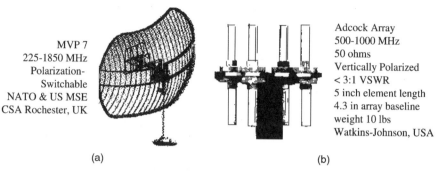

(a)                                                                (b)

**Figure 7-4**  Wideband antennas degrade over time. (a) Highly directional dish antenna; (b) Omnidirectional phased array.

and is not needed because of limited satellite geometries. As the SDR engineer increases band coverage to satisfy the need for agile RF access, the likelihood of needing to point the antenna's gain in more than one direction increases. Other antenna configurations that provide wide relative bandwidths with omnidirectional coverage include the Adcock array shown in Figure 7-4b. This array provides 10 : 1 relative bandwidth. The parabolic dish also shown in the figure provides a decade of bandwidth.

An alternative is to accept lower directional gain, using an antenna with greater relative bandwidth. This may not be physically possible in come cases. For example, the satellite link budget requires the 11 dBi of antenna gain for acceptable outage probability.

## B. Bandwidth versus Packaging

The microstrip [222] patch antenna illustrated in Figure 7-5 provides a much more convenient physical structure, but with only moderate relative bandwidth. Such patch antennas might easily be embedded in a PDA or soldier radio. Several such antennas could be combined using an analog received signal strength indicator (RSSI) circuit to yield reasonable gain in most directions. Using a lower gain antenna reduces the link margin and therefore increases the outage probability proportionally. However, the SDR design process must entertain the use of such suboptimum antennas. That is, the SDR antenna may be suboptimal for a specific band, but may be optimal in terms of aggregate cost and quality of information services across the *combination of bands and modes* over which the radio operates.

## C. Antenna Calibration

Commercial cellular systems historically have not required extensive antenna calibration. The narrow bandwidth of first- and second-generation air interfaces allowed one to ignore the minimal distortion introduced by the antenna

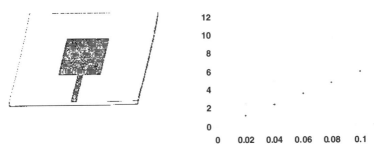

**Figure 7-5**  Microstrip and patch antennas provide small fractional bandwidth.

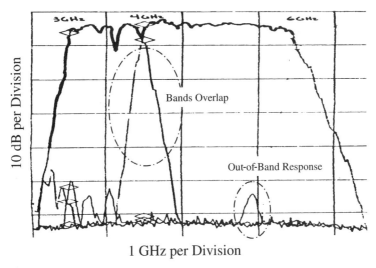

**Figure 7-6**  Amplitude vs. frequency response of antenna in the field.

response. Third-generation bandwidths of 20 MHz at 900 MHz carrier frequencies benefit from element calibration and real-time normalization. In addition, smart antennas require normalization of both amplitude and phase responses in order to form accurate beams and/or nulls that enhance CIR. This section therefore provides a systems-level introduction to the antenna-calibration process.

As the test data in Figure 7-6 shows, antennas are vulnerable to divergence from ideal responses, and to degradation over time. The scale of the figure is 10 dB per vertical division. Marks are provided at 3, 4, and 6 GHz in the horizontal dimension. The relatively deep notches in the amplitude response result in phase and amplitude distortion to the degree that subscriber signals span those artifacts. In the band-overlap region, one must select the subscriber signal from the appropriate channel. If each band has its own antenna, RF conversion, and wideband ADC, the choice of band in the overlap region may be made digitally. In addition, the spurious out-of-band response shows that a high-powered out-of-band signal can create distortion within the operating band of the antenna, degrading communications capability. The out-of-band energy can alias back into the passband through the digital sampling process.

These variations from the ideal response may be compensated for through calibration of the antenna system. To correct the amplitude response, one first establishes a reference amplitude (e.g., 0 dB). The amplitude versus frequency response is then measured by tuning the calibration signal, noting the difference from the reference amplitude. A narrowband calibration table is then created by stepping the known frequency-amplitude source by a small increment, $\delta f$. If $Wa$ is the bandwidth accessed by the antenna, then $N = Wa/\delta f$ is the number of points in the narrowband calibration table. For the notional antenna response of Figure 7-6, $\delta f$ of 100 MHz appears reasonable. The narrowband calibration table is indexed by the input frequency. The values in the table are the constants by which to multiply the observed amplitude in order to recover the reference amplitude. Narrowband signals are those for which a single amplitude calibration constant normalizes the signal. A single constant is a good approximation to the frequency response if the bandwidth of the signal is much smaller than the bandwidth of the deepest/narrowest notch.

If the bandwidth of the signal spans multiple $\delta f$ points, then these wideband signals should be normalized or "prewhitened." The normalization process attempts to drive the normalized components to equal amplitudes across the band. Since signals that are uniform in the frequency domain are called "white," the normalization process is sometimes called prewhitening. This may be accomplished by transforming the signal to the frequency domain (e.g., by an FFT), multiplying the signal by the calibration table values, and transforming the signal to the time domain. Alternatively, the calibration table may be transformed to the time-domain and the signal may be convolved with impulse-responses from the wideband table. If the subscriber signal spans $2n + 1$ values of the narrowband calibration table, then each entry of the wideband table should have $2n + 1$ time-domain impulse response coefficients. The Fourier transform of the calibration table yields the impulse response stored in each entry of the wideband calibration table:

$$y(t;f) = F(C_{f-n}, C_{f-n-1}, \ldots C_f, C_{f+1}, \ldots C_{f+n}) * x(t;f)$$

where $*$ is the convolution operator.

The antenna signal $x(t;f)$ must be indexed into the wideband calibration table at point $f = k\delta f$, which could be the frequency on which the subscriber signal is supposed to be transmitted. Doppler and frequency errors could introduce distortion errors. Generally, Doppler spread is much smaller than $\delta f$, so these errors may be neglected.

Phase may be calibrated using an analogous approach. Let $z = Cv + n$ be an ideal data model, where $v$ is the ideal array response, $n$ is the noise component, and $z$ is a (complex) measurement. The structure of the calibration algorithm is given by:

$$\min_C \sum_k \|z_i - a_i Cv(\theta_i)\|^2$$

In this equation, $(a_i, \theta_i)$ are the known amplitude and phase angle of the source for the $i$th measurement, $z_i$. The calibration table $C$, in this case, is a matrix, is constructed to minimize the total square error. Each element in an array antenna system must be calibrated and corrected using the calibration tables in real-time. Since the values in the calibration tables change only when the antenna is recalibrated, and since the size of the tables is not large and is well known and fixed, antenna calibration can be allocated to an FPGA or programmable ASIC. If well-known signals are present in the deployment environment, then antennas may be recalibrated in the field. Usually, however, the system must be moved to a facility in which the antenna pattern may be recalibrated using precision sources and test equipment. This process should generally be undertaken when the antenna subsystem undergoes configuration changes. Movement of a large antenna to a new site may necessitate recalibration using portable test equipment. Structural changes to a vehicle on which the antenna(s) are mounted may also necessitate recalibration.

### D. Antenna Separation

The physical separation of antennas can substantially control self-generated interference. Local oscillators from one band, for example, can leak into other bands. This can be particularly problematic for a SDR in a low band (e.g., SINCGARS) on a platform in which a fast-tuning LO is operating in a high band (e.g., JTIDS). If these two antennas are located in the same antenna enclosure or on the same mast, the JTIDS LO leaking through the antenna could cause interference in the SINCGARS band or on another low band. The benefits of physical separation may be estimated using a link budget spreadsheet. Consider, for example, the placement of an HF antenna with respect to a UHF antenna in a vehicular application. If these antennas are separated by 10 ft instead of 1 ft, the path loss of out-of-band spurs increases by 20 dB to $-11$ dB. Near-field effects and local reflections may reduce this to 5 to 10 dB. Skin currents in metal structures can also contribute to coupling and can cause passive intermodulation. Mounting the antennas as much as possible on opposite sides of the vehicle tends to suppress these effects.

Separation among multiple vehicles can also be a problem for military vehicles. A military operations center, for example, may contain a half-dozen or more vehicles with a dozen or more radios operating in the 30 to 500 MHz RF bands. Using typical military radios such as SINCGARS, these radios will jam each other. Operational steps may reduce the number of networks to only the highest-priority few that do not interfere with each other. SDRs may be programmed to search the mode-parameter space of power, available hop sets and network activity to automatically identify the combination of modes and networks that maximizes an objective function (e.g., network throughput) subject to constraints (e.g., "must have the primary command network"). Alternatively, the SDR equipped with propagation prediction and measurement software can recommend redeployment of command center vehicles that would optimize the communications goals subject to operational constraints.

## E. Human Body Interactions

Human body interactions are also important to the SDR handset engineer. These interactions include the distortions of antenna pattern induced by the human body, and the health risks of radiation. The body influences antennas very much like a cylinder of salt water [223]. The most popular antenna configurations studied for handheld devices are wire antennas and planar arrays, although many new configurations are under study [224]. Handheld units that tilt a wire antenna away from the body and shield the head with the structure of the handset or PDA absorb least into the body and radiate with greater efficiency. Dual frequency antennas (900/1800 MHz) have also been studied, but at present the kind of wideband, multiband antennas needed for advanced SDR PDAs do not appear to have been reported in the literature. The evolution of the SDR antenna platform, then, should include further attention to the biological-interaction properties of wideband, multiband antennas.

Since antennas radiate energy, one has to consider the possibility that this energy may have harmful interactions with the human body. These effects have been studied extensively [225, 226]. Communications emissions interact with the body by raising its temperature, and perhaps by changing other fine-scale medical features of the organism [227]. Internationally recognized limits on exposure to radio energy are given in terms of *specific absorption rate* (SAR), defined as $SAR = dP/dm = \sigma/\rho E^2 = c \, dT/dt$, where $m$ is mass, $\sigma$ is dielectric conductivity, $\rho$ is the tissue density, and $c$ is the specific heat capacity. The exposure recommendations of leading regulators are summarized in Table 7-1. Due to the software radio's ability to concentrate energy, software constraints may be required to preclude unacceptable exposure levels. For example, a four-channel radio might be permitted to operate at peak power on only two of its channels. Alternatively, the software could take into account the absorption coefficients for the specific antenna configuration to conform to both whole body average and spatial peaks. Using CDMA impedance-matching techniques, the radio may be able to measure its proximity to the body [228] to dynamically fine-tune its radiation properties.

**TABLE 7-1  Recommended Maximum Radiation Exposure Levels**

| Regulator | US FCC | CENELEC | ARIB STD-T56 |
|---|---|---|---|
| Whole body average SAR (mW/kg) | 0.08 | 0.08 | 0.08 |
| Spatial peak SAR (W/kg) | 1.6 | 2 | 2 |
| Averaging time (minutes) | 30 | 6 | 6 |
| Averaging mass (g) | 1 | 10 | 10 |

## IV. ANTENNA DIVERSITY

Since propagation channels introduce multipath fading, the reception system must be designed to overcome fading in some way. The available alternatives include:

- Reduced channel symbol rates to reduce intersymbol interference (ISI)
- Structuring the data to be resilient to the effects of fading
- Diversity transmission and/or reception
- Slow FH
- Increased instantaneous bandwidth for multipath resolution and equalization

Reducing channel symbol rates may be necessary if other measures are ineffective. One prefers, however, to support larger data rates if possible. Interleaving and FEC reduce the impact of erasures introduced through fading, but one would also like to reduce the probability of a nonrecoverable fade depth. Diversity transmission and reception reduces this probability as follows. Suppose that one has established the channel symbol rate and forward error control and empirically determined that the probability of a nonrecoverable fade depth is $P_e$. The question arises whether the addition of an additional receiving antenna at a place distant from the primary antenna will be faded as well. If the signal strength that causes a nonrecoverable fade is $S_{min}$, then

$$P_e = P\{S < S_{min}\}$$

The spatial distribution of $S$ is given by the spatial structure of the multipath. If $P_r$ is the probability that the signal is also faded to $S < S_{min}$ at range $r$, then diversity reception that chooses the strongest received signal strength yields an improved error floor, $P_e' = P_e P_r$, the product which is ideally the probability of the joint event. The strength of correlation of $S$ at two such antenna elements as a function of mutual displacement is called the spatial coherence of the signal. Experimentation and in-depth analysis of spatial coherence yields insights for diversity antenna architecture tradeoffs.

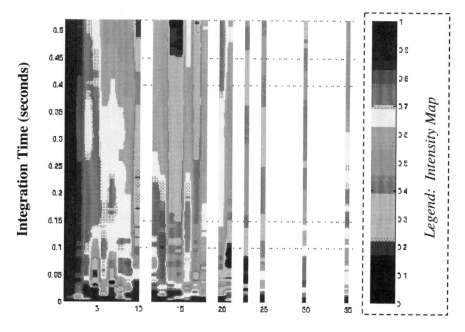

**Separation (wavelengths)**

**Figure 7-7**   Signal coherence simulation.

## A. Spatial Coherence Analysis

Let $\mathbf{r}_i(t)$ be the $i$th received signal component. The mutual *coherence* between the $m$th and $n$th received components is given in the following equation:

$$\rho_{mn} = \frac{\left| \int_0^{\approx 0.3 \text{ sec}} r_m(t) \times r_n^*(t)\, dt \right|^2}{\int_0^{\approx 0.3 \text{ sec}} |r_m(t)|^2 |r_n(t)|^2\, dt}$$

This equation represents the inner product of the two path components, normalized by the total power in the corresponding interval. Since the signal strength varies as a function of time for realistic fading models (Rayleigh, Log-Normal, etc.), one must also select a meaningful integration period. Figure 7-7 shows how this correlation varies as a function of both antenna separation (in wavelengths) and integration period. Integration for 0.3 seconds yields substantial decorrelation at 10 wavelengths of separation. The simulation of this figure was tested in an experiment on terrestrial fading [229], yielding the empirical result of Figure 7-8.

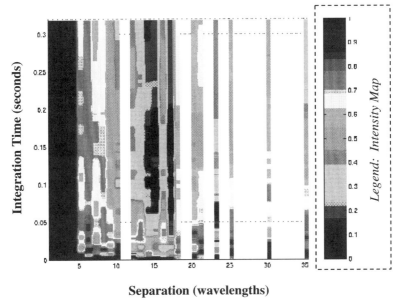

**Figure 7-8** Empirical verification of coherence model.

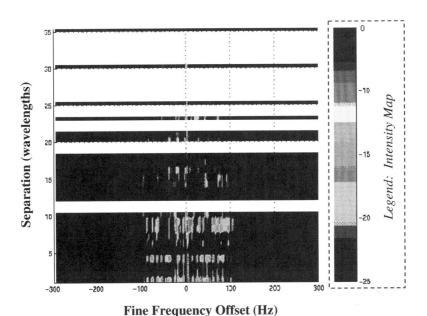

**Figure 7-9** Doppler-spread induces decorrelation.

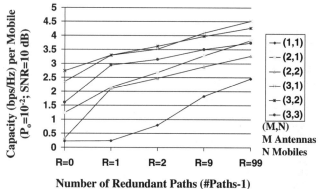

**Figure 7-10**   Spatial diversity simulation characterizes benefits.

In addition to the spatial structure of the reflectors, Doppler changes the inner product of the two received components. Figure 7-9 illustrates the relationship. Doppler spread is proportional to the carrier frequency times the ratio of the maximum velocity of a transmitter divided by the propagation velocity (approximately $c$, the speed of light). This introduces decorrelation as a function of antenna separation as well.

The trend of decorrelation at 10 wavelengths and 0.3 seconds of integration time establishes a rule of thumb for antenna diversity. Given an antenna separation of 10 wavelengths or more, there is a significant probability ($> 60\%$) that the diversity signal will be substantially decorrelated from the primary signal. At 900 MHz, a wavelength is about 333 centimeters, so the rule of thumb can be met with a separation of about $3\frac{1}{3}$ meters (11 ft), which is practical for most cell sites.

## B. Potential Benefits of Spatial Diversity

In most bands from VHF to EHF, spatial and/or polarization diversity provides substantial fade protection. Cellular antenna systems are now routinely deployed with three-way (120-degree) sectorization. The sectors may be assigned separate RF channels, separating them into the functionally distinct sectors required for high subscriber densities. Figure 7-10 illustrates the potential benefits of spatial diversity characterized through diversity simulation [230]. In this simulation, there are $M$ antennas and $N$ mobile units. The number of redundant paths, $R$, is a function of the multipath, which depends on the details of the propagation. With one antenna and one mobile receiver, the capacity available in bits per second per Hz is relatively low as shown in the lowest curve in the figure. As the number of antennas increases without increasing the number of mobiles sharing the channels, the capacity increases to the upper curve $(3, 1)$ in the figure. As the number of mobiles increases to 3, the capacity decreases somewhat (see [230] for details).

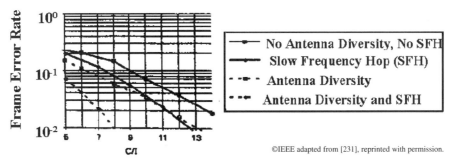

**Figure 7-11**   Joint spatial and frequency (hopping) diversity.

©IEEE adapted from [231], reprinted with permission.

## C. Spatial and Spectral Diversity

FH also provides fade resistance for slow-moving mobiles. If one is stopped at a traffic signal in a deep fade with $f_c$ of 850 MHz, the fade will with high probability be less severe if the frequency is switched to 860 MHz. As shown in Figure 7-11, slow FH improves radio performance. GSM's slow-FH plan effectively averages out deep fades, enhancing SNR. The research reported in [231] compares slow FH to antenna diversity and to combined slow-hopping and diversity. The measure of effectiveness of the techniques is the frame error rate. With no diversity or hopping, about 15 dB of carrier-to-interference ratio (CIR) are required to achieve a bit error rate of $10^{-2}$. With either diversity or FH, the required CIR is reduced to about 13 dB. The combination of both techniques, however, reduces the required CIR to only 8 dB.

Research into the instantaneous value of a received signal strength indicator (RSSI) as the criteria for diversity combining [232] reveals the high degree of variability of RSSI as a function of distance between transmitter and receiver. This research reports success in modeling the value of RSSI, subject to variances of 20 dB or more as shown in Figure 7-12.

These variations in received signal strength are accommodated by the AGC function, provided the received CIR supports demodulation (e.g., 7–12 dB for discrete channel symbols). The result improves signal quality, as a result of spatial and/or spectral diversity. The primary tradeoff, then, is to provide diversity in the architecture in a way that balances benefit against cost. CDMA's inherently wide bandwidth is robust in multipath, but also benefits from diversity combining, subject to receiver complexity constraints [233].

## D. Diversity Architecture Tradeoffs

A canonical model of diversity antenna system is shown in Figure 7-13. As illustrated, diversity combining typically occurs in an IF stage. Analog diversity combiners may simply pick the diversity channel with the largest received signal strength [234]. Digital combiners may insert a variable time-delay and linearly add the signals to yield a stronger, more coherent and more noise-free

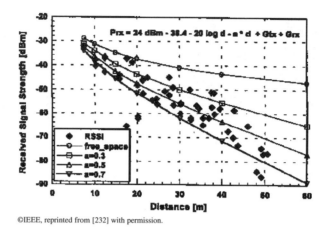

**Figure 7-12**   Received signal strength indicator (RSSI) measurements.

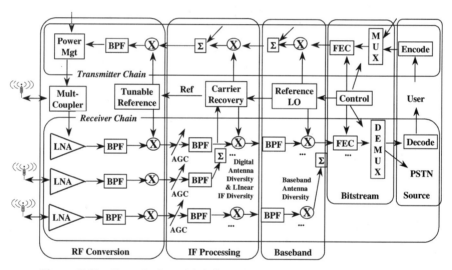

**Figure 7-13**   Canonical model defines diversity antenna insertion points.

resultant signal. Digital combiners are easiest to implement at baseband, but IF combiners are also feasible, e.g., using FPCA's.

The impact of including diversity in an SDR includes both technical and economic challenges [235]. Diversity antennas require parallel RF/IF conversion and ADC channels, increasing the cost of the system. They make it possible to delay and combine diversity paths more precisely and adaptively than is possible with analog approaches, enhancing CIR by 5 to 15 dB or more. In addition, any of the diversity and slow-FH techniques described above may be implemented using the pooled DSP resources in an SDR architecture as

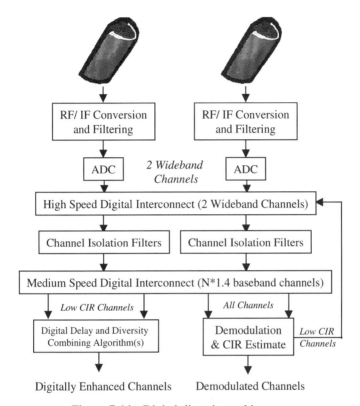

**Figure 7-14**   Digital diversity architecture.

illustrated in Figure 7-14. The economic challenges center on minimizing the cost of such parallelism. The antenna, RF/IF processing, and ADC path can account for upwards of 60% of the procurement cost of a base transmission station.

Figure 7-14 shows the diversity-processing path including antenna elements; RF/IF amplification, filtering, and conversion; ADC; digital channel isolation filtering; and the diversity integration algorithms. These algorithms are typically hosted on FPGAs or DSPs with sufficient memory to introduce relative delay of a few microseconds. The cost of the additional DSP resources can double or treble the cost of the digital back-end. On the other hand, not all subscribers need spatial diversity combining at once. Depending on the geometry, 20 to 40% or fewer subscribers need this enhancement. Although all subscribers require channel filters, not all require the diversity combining. In the figure, a low CIR estimate in a conventional channel results in a command to the high-speed interconnect to create a path from the diversity channel through an additional channel isolation filter and on to the digital diversity combining algorithm. This requires 20 to 40% more isolation filters than subscribers in order to process the diversity paths. In addition, the paths

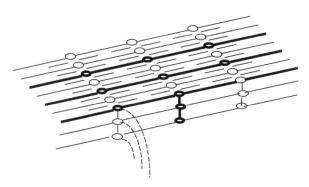

**Figure 7-15**   Optically-fed reconfigurable array antenna.

from the ADC to the channel isolation filters may not be hard-wired. Thus, dynamically-pooled DSP resources may enhance those subscriber channels with low CIR. The fundamental parameters of this tradeoff, then, are the cost of these digital resources versus the increased revenue stream provided by the enhanced QoS and the reduced dropping of faded calls.

## V. PROGRAMMABLE ANTENNAS

Military applications of software radios require RF access from 2 MHz to 2 or 3 GHz while commercial applications outlined in BellSouth's RFI addressed frequencies from 40 MHz to 60 GHz. Such wide frequency ranges cannot be met using conventional resonant RF structures. One of the interesting research areas that offers promise is the optically fed reconfigurable antenna array [236] as illustrated in Figure 7-15. The array consists of resonant elements, micro-electromechanical system (MEMS) optoelectronic switches, optical fiber, and a control subsystem. The resonant elements are arranged in a three-dimensional array embedded in layers of dielectric material. The elements are individually resonant at specific frequencies. In addition, optically controlled switches connect elements in the same row. When the switches are open, each element resonates at its own characteristic frequency. If it is connected (e.g., electrically or by a field mechanism) to a balun (matching network or "feed"), the individual resonant element establishes the resonance of the array. But if a switch is closed, then the total length of the interconnected elements defines the resonance, which may be several multiples of the length of the individual elements. If an individual element resonates at frequency $f_o$, then $N$ elements in series resonate at approximately $f_o/N$. Selection of $f_o$ and $N$ defines a programmable frequency range in one dimension from $f_o/N$ to $f_o$. This frequency range may be called the *agility band* of the antenna. Although it will not access all radio frequencies in this band simultaneously, it may be programmed to a resonance band (typically an octave) within this overall agility band.

The MEMS switches are typically bistatic, with the state changed by pulses of light. Alternatively, the presence of light may cause the switch to assume one state (e.g., open), while the absence of light causes the other state (e.g., closed). These switches are controlled by light delivered through nonmetallically shielded fiber optic cable such as graded index of refraction (GRIN) fiber. GRIN fiber passively channels the light toward the center of the fiber. Any metal in the fiber-optic control subsystem would distort the antenna pattern. Therefore, electrically controlled antenna arrays pose technical challenges in the isolation of control wires from the antenna elements. In discrete phased-array applications, the control switches may be situated behind a ground plane, essentially eliminating interaction with the antenna pattern. However, in a programmable array, the number of switches and their proximity to the resonant elements and their presence in the dielectric material would distort the array pattern. GRIN optical control cables, however, do not interact significantly with the RF waves, even at relatively high frequencies such as EHF.

Three-dimensional structures allow one to adjust the location of the ground plane by grounding elements and interconnecting grounded elements into a mesh that acts as a ground plane. Researchers have demonstrated the use of MEMS switches on dipoles, and have postulated designs for distributing micromachined MEMS switches on waveguide [237]. California Microwave [236] has implemented a prototype array. At present, such arrays have significant drawbacks. The VSWR, first of all, is difficult to control. In the future, electronically programmable analog circuits (EPACs) may be adapted to program the balun so that VSWR is maintained across an operating subband which is programmed within the overall agility band of operation. In addition, the antenna patterns lack the uniformity of antennas with shaped resonators and ground planes. Finally, one might expect the phase stability of such structures to be less than that of conventional antennas because of the unavoidable reflections that occur at the switch points. Tests on research antennas confirm raggedness in the antenna patterns, inconsistent VSWR, and greater phase noise than with conventional antennas. On the other hand, no conventional antenna has such a large agility band as the optically fed programmable resonant array.

SDR applications could benefit from such arrays. Obviously, the programmability of the array extends the notion of programmability of the radio to the antenna itself. In addition, the DSP capacity of SDR architectures may be applied to compensating the amplitude, directional, and phase errors of the programmable antenna arrays. One should not anticipate the use of such antennas in operational environments until the research and engineering issues have been successfully addressed.

## VI. COST TRADEOFFS

Since cost of production electronics is nearly a linear function of parts count, the number of antennas and related RF/IF paths is critical. For each antenna, there must be at least some minimum amount of RF circuitry. And in most multiband, multimode radio designs, the parallelism of analog RF equipment extends to the ADC. As a result, the antenna and RF subsystems can account for upward of 60% of the reprocurement costs of a radio node. SPEAKeasy I and II, therefore, put considerable effort into developing all-band antennas, but with little success. The antenna therefore remains one of the most challenging aspects of SDR platform technology development.

## VII. SUMMARY AND CONCLUSIONS

The antenna is the most challenging subsystem of the software radio in many respects. It is not possible to synthesize a single antenna that provides acceptable performance from 2 MHz to 2 GHz for military applications. A single antenna with tolerable performance from 400 MHz to 2500 MHz is possible using, for example, helical, spiral, Yagi, or other broadband antenna structures. In addition, the physical location of transmit and receive antennas (e.g., on a command vehicle or aircraft) has a considerable impact on self-generated EMI, which is also called *cosite interference*.

Due to the lack of bandwidth and programmability, military users are generally driven to *channelized* architectures in which there is a dedicated antenna and RF conversion subsystem for each subband accessible by a resonant antenna. Octave antennas with good VSWR ($> 2 : 1$), uniform patterns, and acceptable phase performance are practicable in most bands. Consequently, a channelized SPEAKeasy architecture for high performance could have as many as eight subbands: 2–20 MHz; 20–40 MHz; 40–80 MHz; 80–160 MHz; 160–300 MHz; 320–600 MHz; 600–1200 MHz; and 1200–2400 MHz. SPEAKeasy I used just three bands: 2–20 MHz; 20–400 MHz; and 400–2000 MHz. This approach followed intensive study by Hazeltine and Rockwell as summarized in Figure 7-16.

The antenna characteristics determine not only the gain due to aperture effects, but also several critical characteristics of the SDR, including:

- The number of antenna channels required to support multiband multimode operation
- Usually, the number of parallel RF conversion chains
- Often, the number of ADCs and DACs required

Parallelism is a major cost driver for software radios. Higher-gain antennas achieve this gain over relatively small segments of RF (e.g., 5% of the carrier). As a result, it might take 20 such narrowband antenna elements to provide high gain across an entire operating band. Wideband antennas provide wider bandwidth at increased cost and manufacturing difficulty. A wideband antenna architecture allows one to span an entire operating band with far fewer antennas. Performance of such wideband architectures can be high, provided one

- **Hazeltine RF Study** (2–2000 MHz)
  - – Multiband antenna technology needed
- **Physical Limits Are Challenging**
  - – Antennas/parallelism is a major size, weight, and cost driver
  - – Size/aperture function of platform (huge retrofit costs)
- *Cosite Is Critical:*
  - – Multiband, multimode antenna interactions
  - – Eddy currents & near field effects
  - – Electrical isolation, structure, age
  - – Multicarrier passive intermodulation (PIM) [119]
  - – Problematic on wearables, vehicles, large C2 assemblages
- **Programmability/configurability has potential**
  - – Optically fed; biomimicry: high risk
  - – Intelligent antennas (appliques: range, node/subscriber density)

**Figure 7-16**   Significant challenges of the antenna segment.

compensates for amplitude and phase errors across the bands. In the future, further advances in RF MEMS may permit the introduction of reconfigurable antennas.

## VIII. EXERCISES

1. Identify the parameters of the antenna segment that define the fundamental operating constraints of an SDR.

2. What RF access bandwidths are readily attainable in compact resonant antennas such as microstrips? What bandwidths are feasible with readily available wideband antennas?

3. Describe the impact of mechanical packaging constraints on antenna segment parameters. Suppose the host vehicle is an aircraft? An HMMWV? A luxury automobile?

4. What antenna effects have to be taken into account in the design of a handheld radio product? Describe the potential impact of this aspect of the radio on the control software in an aggressive SDR application.

5. What functional contributions are provided by diversity antennas? Are there other ways of obtaining these benefit(s)? How much can the combination of diversity and related techniques improve reception quality? Translate this benefit to a fraction of additional subscribers supported at a given QoS.

6. Describe the tradeoffs associated with multiple antenna elements and related RF chains.

7. What differentiates an antenna segment design from an antenna segment architecture?

8. Hypothesize a timeline over which antenna technology in an SDR PDA could advance from narrowband to wideband antenna technology. What market forces tend to drive one toward the wideband technology? Develop a simple model of cost per band for narrowband and wideband antennas. For example, one might assume that the antenna and RF and IF processing (including ADC or despreader ASIC) account for 50% of the cost of a dual-channel PDA. One could then assume that a wideband antenna costs $x\%$ more than a narrowband antenna. Write the equation of such a cost model. Use it to answer the next question.

9. What configurations of services and/or spectrum allocations drive one toward the selection of wideband antennas over the narrowband antenna(s) based on your cost model? When will that evolution occur on your timeline? If there is no point at which wideband technology makes economic sense, what is it about your market niche that causes this to be the case? Consider other markets (e.g., military) in which the case for wideband antennas is stronger. What circumstance would tend to drive this market niche toward a single integrated market?

10. Consider the disaster-relief scenario. What antennas would be necessary to support the bands and modes you think are required? Suppose now that you must include HF, LVHF, VHF-UHF, cellular (900 MHz), PCS (1800 MHz), and 2.5 MHz RF LAN? How does the concept of operations influence antenna design? How many vans (nodes) are necessary to cover a disaster area that is $40 \times 40$ miles in extent? Suppose the terrain is highly populated (e.g., New York City)? Suppose the terrain is mountainous?

11. Recall the object-oriented analysis of the RF platform. Develop an object-oriented model of the antenna segment. Define classes of antenna, in which more-specific antenna classes inherit general properties from higher-level classes. How many levels are in your inheritance hierarchy? Could the hierarchy be defined with fewer layers? With more? What are the advantages of fewer layers?

12. Recall the tabular RF reference platform. Instantiate the antenna aspect of the reference platform based on the material from this chapter. Would hardware built in accordance with your reference platform meet the needs of the disaster-relief case study?

# 8 RF/IF Conversion Segment Tradeoffs

This chapter introduces the system-level design tradeoffs of the RF conversion segment. Software radios require wideband RF/IF conversion, large dynamic range, and programmable analog signal processing parameters. In addition, a high-quality SDR architecture includes specific measures to mitigate the interference readily generated by SDR operation.

## I. RF CONVERSION ARCHITECTURES

The RF conversion segment of the canonical software radio is illustrated in Figure 8-1. The antenna segment may provide a single element for both transmission and reception. In this case, a multicoupler, circulator, or diplexer protects the receiver from the high-power transmission path. In other cases, the transmit and receive antennas may be physically separate and may be separated in frequency. First-generation cellular radio and GSM systems separate downlink and uplink bands by typically 45 MHz to limit interference.

The transmission subsystem intersects the RF conversion segment as shown in Figure 8-1. This includes a final stage of up-conversion from an IF, bandpass filtering to suppress adjacent channel interference, and final power am-

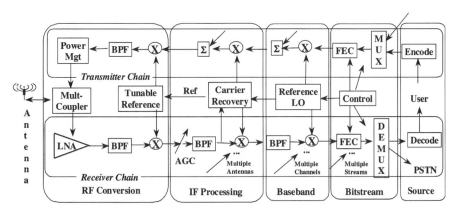

**Figure 8-1** The canonical model characterizes RF/IF segment interfaces.

265

plification. First-generation cellular systems did not employ power control to any significant degree. CDMA systems, including third-generation (3G) W-CDMA, require power control on each frame (50 to 100 times per second). SDRs may be implemented with a DAC as the interface between IF up-conversion and the RF segment. Alternatively, a high-speed DAC may directly feed the final power amplifier.

Power amplifiers have less-than-ideal performance, including amplitude ripple and phase distortion. Although these effects may be relatively small, failure to address them may have serious consequences on SDR performance. Amplitude ripple, for example, degrades the transmitted power across the band, particularly near the band edges. IF processing may compensate by preemphasizing the IF signal with the inverse of the power amplifier's band-edge ripple. Feher [238] describes techniques for compensating a sequence of channel symbols, shaping the transmitted waveform in the time domain to yield better spectral purity in the frequency domain. The concept behind Feher's patented design is straightforward. Sequential symbols may have the same relative phase, yet the channel-symbol window in which the sinusoids are generated modulates the amplitude at the symbol boundaries. When adjacent symbols have different phase, this symbol weighting reduces frequency domain sidelobes and hence adjacent-channel interference. Feher suppresses the modulation further with an extended symbol that includes the sequential symbols of the same phase generated with constant amplitude, thus without the weighting-induced amplitude modulation. The result is that energy that normally is redirected into the adjacent channels by the phase discontinuities remains within the channel because the discontinuities have been suppressed.

The receiver subsystem intersection with the RF conversion segment is shown in Figure 8-1 also. This includes the low noise amplifier (LNA), one or more stages of bandpass filtering (BPF), and the translation of the RF to an IF. In conventional radios, a tunable-reference local oscillator (LO) may be shared between the transmitter and receiver subsystems. FH radios often share a fast-tuning LO between the transmitter and receiver. In military applications, the LO executes a frequency-hopping plan defined by a transmission security (TRANSEC) module. In commercial systems (e.g., GSM), a fixed frequency-hopping plan that suppresses fades may be used instead of a complex TRANSEC plan. The radio then either transmits or receives on the frequency to which the LO is tuned. Any radio which employs a physically distinct programmable LO may be a programmable digital radio (PDR), a type of SDR, but it is not a software radio. Software radios use lookup tables to define the instantaneous hop frequencies, not physical LOs. This approach, of course, requires a wideband DAC. One advantage of using such a DAC is that the hop frequency settles in the time between DAC samples, typically $Wa/2.5$—hundreds of nanoseconds. The hop frequency is pure and stable instantly, subject to minor distortions introduced by the final power amplifier.

Since the receiver must overcome channel impairments, it may be more complex and technically demanding than the transmitter. Thus, this chapter focuses on receiver design.

Again referring to Figure 8-1, IF processing may be null, as may baseband processing. The direct conversion receiver, for example, modulates a reference signal against the received RF (or IF) signal to yield a baseband binary analog waveform in the in-phase and quadrature (I&Q) channels. Although this kind of RF conversion has nonlinear characteristics, it is particularly effective for single-user applications such as handsets. It may not work well for multiuser applications, however.

This chapter examines the SDR implications of the RF conversion segment. The following section describes receiver architectures. Programmable component technology including MEMS and EPACs is described. RF subsystem specifications are then analyzed. The chapter concludes with an assessment of RF/IF conversion architecture tradeoffs.

## II. RECEIVER ARCHITECTURES

This section describes the superheterodyne architecture used in base station applications, the direct conversion receiver used in handsets, and related research.

### A. The Superheterodyne Receiver

The Watkins-Johnson company [239] publishes the frequency plans of its receivers, an example of which is shown in Figure 8-2. This superheterodyne receiver [240] consists of a preselector and two conversion stages. The preselector consists of a matrix of bandpass filters and amplifiers that are switched as defined by the frequency plan for the specific frequency to which the receiver is tuned. The preselector filters cascade with a low-pass filter and step attenuator that keep the total power of the signal into the first conversion stage within its linear range.

Each conversion stage includes one LO and additional filtering and amplification. The first local oscillator is tuned in relatively coarse steps (e.g., 2.5 MHz in Figure 8-2). The first conversion stage converts the RF to 3733.75 MHz. Higher IF frequencies minimize the physical size of the inductors and capacitors used in the filters. The modulator that converts the RF into the initial IF generates sum and difference frequencies in addition to the desired frequency. The bandpass filter then suppresses these intermodulation products. The low-pass filter further suppresses out-of-band energy. An amplifier and pads with variable gain determine the power into the second conversion stage. The operation of the second stage is similar to the first except that it down-converts the 3733.75 MHz to a standard wideband IF, in this case, 21.4 MHz. In addition, this stage has fine-tuning steps of 1 kHz.

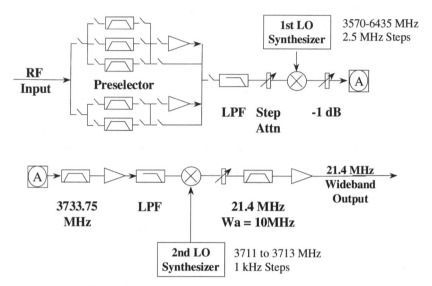

**Figure 8-2**   Superheterodyne receiver architecture.

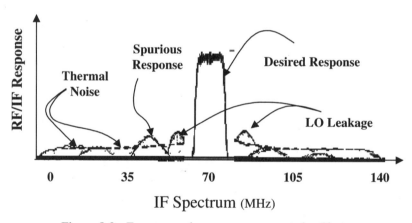

**Figure 8-3**   Frequency plan suppresses spectral artifacts.

Artifacts must be controlled in the conversion process [241, 242]. In addition to the desired sideband, the conversion process introduces thermal noise, undesired sidebands, and LO leakage into the IF signal as shown in Figure 8-3. Thermal noise is shaped by the cascade of bandpass and low-pass filters. Depending on the RF background environment, thermal noise in the receiver may dominate or thermal-like noise or interference from the environment may dominate the noise power.

Superconducting IF filters suppress noiselike interference generated in one cellular half-band from a second, immediately adjacent half-band (e.g.,

12.5 MHz of active signals). See [243] for superconducting filter test results that show a 30 dB suppression of such noise. Undesired sidebands are always present at some very low level because filtering operations suppress sideband energy but do not completely eliminate it. LO leakage occurs because a modulator acts in some ways as a transmission line with imperfect matching. Consequently, part of the power of the LO is transmitted through the modulator to the output.

When the IF is processed digitally, these artifacts can be characterized. Long-term averaging using an FFT, for example, will reveal shape of the noise and the degree of suppression of the LO leakage and of the undesired sidebands. When designing a PDR, one is concerned that these artifacts not distort the baseband enough to degrade the output SNR or BER unacceptably. When designing an SDR, none of these artifacts should degrade any of the subscriber channels by more than the degradation of the least significant bit (LSB) of the ADC. To accomplish this, the in-band artifacts need to be as uniform as possible and the maximum level anywhere in the operating band (e.g., in the cell channels) cannot exceed half of the LSB of the ADC. As shown below, this constraint implies that the ADC, postprocessing algorithms, and RF plan must be designed to mutually support each other. Algorithm designers who employ floating-point precision at design time may not be familiar with the noise, spurs, and other analog artifacts of the analog RF circuits that limit useful dynamic range constraints. These effects limit the digital dynamic range, and thus reduce the requirements for arithmetic precision in the digital hardware and software. Thus, the effects of each of the disparate analog, digital, and software-signal processing stages have an effect on the sampled signal.

When these effects are properly balanced, the wideband superheterodyne receiver yields hundreds of analog subscriber channels that have been structured for the ADC. As a result, the ideal software radio base station replaces hundreds of parallel narrowband analog channels with one wideband channel digitized by a wideband ADC, followed by hundreds of parallel digital channels. Since the digital channels inherently cost less than analog channels, the software radio base station may be more cost-effective than the baseband digital design. Yet most base stations deployed up to 1999 had a baseband digital architecture, not an SDR architecture. The inadequacy of the prior generation of ADC technology explains this situation as discussed in the sequel. Wideband ADCs were within about 6 to 10 dB of the performance required to effectively compete with baseband architectures in the base station. By June 2000, digital IF base stations began shipping, but manufacturers did not publically disclose this fact in order to protect this competitive advantage.

Tsurumi's discussion of zero-IF filtering with up-conversion in a handset architecture provides an innovative approach to multiple-conversion receivers for handsets [244]. By heterodyning multiple bands to zero-IF, Tsurumi prefilters any of the commercial standards using a simple programmable low-pass

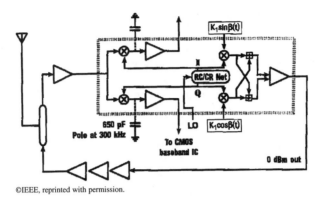

**Figure 8-4**   Alcatel direct conversion receiver.


filter. Subsequent up-conversion before digitizing yields a standard digital IF for multiple commercial standards.

## B. Direct Conversion Receiver

The superheterodyne receiver is relatively complex. Its wideband performance is appropriate for base station applications where hundreds of subscriber channels are to be processed at once. But suppose there is only one channel of interest as in the handset receiver. In this case, there is little benefit to the wideband performance of the superheterodyne receiver.

Instead, a direct conversion receiver may be more appropriate [245]. The homodyne receiver translates RF to baseband, with the center frequency tuned to zero Hz in one step. The direct conversion receiver is a homodyne receiver that may use nonzero baseband center frequency and may also demodulate the signal into baseband bitstreams in the same circuit. LO leakage and DC bias can be significant problems with such an approach is used for wideband digital signal processing. On the other hand, Alcatel's direct conversion GSM receiver represents the kind of approach taken in a viable commercial product. It selects channels via switched capacitor filters in a mixed-signal integrated circuit (IC) as shown in Figure 8-4. The RC-CR network generates quadrature phases [246]. The feedback loop at the output of the modulators is filtered for the GSM's 280 kHz RF channel bandwidth in such a way that the I&Q amplifiers yield level-shift analog baseband signals. This analog signal has two nominal states, corresponding to the two channel symbol states of the MSK waveform. Siemens [247], Philips, and numerous other manufacturers make similar chip sets [248]. See [249] for a direct-conversion GPS receiver.

In the past, gallium arsenide (GaAs) circuit technology was necessary for RF circuits, precluding one from implementing the RF circuitry and the microcontroller of a handset with the same circuit technology. Differences in power supply, thermal properties, and bonding between CMOS and GaAs complicated handset design. Recently, however, CMOS silicon RF 50 W to 40 GHz has been reported. One of the .18 micron CMOS chips [250] supports 2.4 GHz RF at 1.8 Volts. CMOS devices that have been demonstrated include low-noise amplifiers, mixers, differential oscillators, IF strips, and RF power amplifiers with 1 W output and 40 to 50% efficiency at 1 to 2 GHz [251].

## C. Digital-RF Receivers

PhillipsVision [252] created some excitement by announcing a software-radio on a chip. The interesting aspect of their product announcement is that the demodulator is said to "operate at RF." Due to the necessarily vague nature of the statements, it is impossible to determine the exact nature of the demodulation process. This announcement plus the recent interest in digital demodulation at RF makes it useful to address this alternative. The comments below may not be representative of the PhillipsVision product, but they reflect research approaches to digital demodulation at RF.

Since GHz clocks can be fabricated in single ASICs, one may employ such a clock to demodulate certain modulation types at RF. One approach is the one-bit direct conversion digital receiver, which may be called the RF zero-crossing demodulator. With this approach, the RF is amplified until it is hard-limited into a square wave. Reference square waves are synthesized for each channel-symbol state. An MSK waveform, for example, has two square waves. One corresponds to the mark, say, the lower of the two frequencies. By generating digital streams at mark and space frequencies and counting the number of coincidences between mark and space streams in the incoming RF signal, one can estimate the state of the RF waveform. A bit-timing logic state-machine can then determine bit timing to produce the baseband bitstream. All this can be implemented for a single channel-modulation type in an FPGA using less than ten thousand gates. One advantage of this architecture is that the bit patterns for the channel states may be stored in a lookup table. Different waveforms at different frequencies correspond to different lookup tables. By using clever data-compression techniques, the lookup tables may be kept compact in spite of the large number of entries in the table.

A similar approach simply counts zero-crossings of the RF. Once the variance of $N$ zero-crossing counts becomes small, a signal is present. The strongest of two or more cochannel signals will be reflected in the subsequent counts for CIR $> 7$ dB. This phenomenon is the digital equivalent of FM capture [245, p. 497]. Random noise generates zero-crossings with large variance, but a sinusoid has a tight variance. Frequency modulations like GMSK exhibit two different zero-crossing rates, one corresponding to mark, and the other to space. The output of a zero-crossing counter, then, can be gated and reset at the expected channel-symbol rate. A threshold determines whether the channel symbol was mark or space, yielding the baseband bitstream. Timing logic can also estimate and track symbol timing. Although the logic has to operate at the GHz rates of the RF zero-crossings, the counter logic is simplicity itself.

However, certain problems have precluded this receiver architecture from being widely used. First, low-power, high-speed logic has not been available until recently. Thus, the architecture seems timely. In addition, however, the incoming signal cannot be equalized using this type of receiver. The BER floor therefore is worse than that of an equalized receiver. In addition, the recovery of a timing reference is difficult in fading, again raising the BER floor. There does not appear to be much in-depth discussion in the literature of such receiver architectures.

## D. Interference Suppression

The first line of defense in suppressing interference is in the antenna and RF conversion segment of the receiver. Physical antenna separation, frequency separation, programmable analog notch filters, and active cancellation are steps that help control interference at RF. In addition, the software of a well-conceived SDR will include mutual constraints among air interfaces that could be invoked simultaneously so that self-generated interference is avoided or minimized.

*1. Frequency Separation*   Interference introduced into a receiver from out-of-band energy created by a nonideal transmitter is the convolution of the out-of-band signal with the bandpass characteristic of the receiver [245]. Although out-of-band interference of high-performance transmitters rolls off to less than $-100$ dB within 20 MHz of the transmitted frequency, radios with less EMI control may present more like $-70$ or $-60$ dB of rolloff. The presence of a half dozen signals within the overall operating band can then cause substantial interference. FDD standards separate the uplink and downlink to minimize this kind of interference. SDRs operating in TDD bands can create dynamic FDD nets by a protocol that dynamically define uplink, downlink, and frequency separation. This is a novel approach to interference suppression.

*2. Programmable Filters*   The application of a programmable interference suppression filter is illustrated in Figure 8-5. The filter may be called a roofing filter because the interference captures the dynamic range, establishing a maximum (roof) and minimum (floor) linearly processable signal level. It is also called a cosite filter in military jargon because the interference may be generated by the colocation of two transmitters in the same locale (site). Prior to the application of the roofing filter, the roof of the dynamic range is so high that weak signals fall below the floor, resulting in dropped calls. After the application of the filter, the roof has been lowered such that the dynamic range is now on the noise floor. Although the interference is still present, it has been suppressed enough to control the available dynamic range. In order for this approach to be effective, the filters have to have low insertion loss, programmable center frequency, and programmable bandwidth. Amplitude and phase ripple across the band has to be kept to near zero to avoid distorting the other subscriber signals.

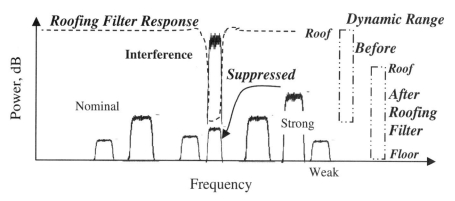

**Figure 8-5**  Workable situation for roofing filter.

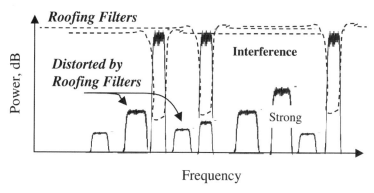

**Figure 8-6**  Roofing filters distort subscriber signals.

Not all situations can be addressed effectively using roofing filters, however. If there are more than a few strong interference signals in the passband, the roofing filters may introduce excessive distortion into the subscriber signals. This situation is illustrated in Figure 8-6.

Factors that determine the number and characteristics of allowed roofing filters include the modulation of the subscriber signals, and the bandwidth of the interference relative to the overall passband. If the subscriber signals are robust to phase and amplitude distortion (e.g., FSK), then more filters or filters that introduce more severe distortion may be used. If the subscriber signals are phase-sensitive (e.g., 16 QAM proposed in many of the 3G alternatives), no more than one analog roofing filter is likely to be workable.

***3. Active Cancellation***   Active cancellation is the process of introducing a replica of the transmitted signal into the receiver so that it may be somehow subtracted from the input signal. A detailed treatment of cancellation techniques is beyond the scope of this text, but the following introduces the essential notions.

Active blanking of radar signals from the input to communications systems on the same platform is an example of active cancellation. In this case, the radar transmitter provides a control line that is active a few microseconds before it transmits so that the communications system can activate a grounding circuit. The RF stage passes no signal at all to the rest of the communications system until the control line is inactive [245].

Active communications cancellation circuits may delay the transmitted signal and attenuate it in such a way that the transmitted and received signals are exactly out of phase, shifted by $\pi$ radians (at RF or IF) with respect to each other. In principle, such a circuit should cause the transmitted signal to be completely removed from the received signal. In practice, the cancellation is not ideal. In part, this is due to the inexactness of fabrication of analog circuits. In part, modulation of the transmitted signal distorts each IF sinusoid slightly, and the filtering-induced distortion through the transmitting antenna and into the receiving antenna (or through the circulator) differs slightly from the distortion of the cancellation circuit. The result is that simple linear techniques can achieve only about 10 to 20 dB of cancellation. Complex phase-tracking circuits can improve performance, but nonlinear techniques are required to approach 30 to 40 dB. Few of the nonlinear techniques are in the public domain.

The cancellation that is needed is the difference between the maximum nondistorting input signal and the radiation level that reaches the receiving antenna.

Required-Cancellation

$$= \frac{(Peak\ energy\ at\ the\ output\ of\ the\ receiver\ antenna\ terminals)}{(Maximum\ linear\ energy)}$$

If this power is not suppressed or dissipated, it will capture the roof of the dynamic range and cause either intermodulation distortion or lost subscribers or both.

Not all cancellation has to be accomplished using analog circuits. Any cancellation that occurs in the early stages of RF amplification and filtering also improves system linearity and contributes to dynamic range improvement just like roofing filters. Residual components may be further suppressed using digital techniques.

*4. Software-based Interference Mitigation*   SDR architecture exacerbates interference mitigation by driving the radio platforms toward the use of wideband antennas and RF. It also can contribute to interference suppres-

**TABLE 8-1  Mode Constraint Table (Minimal)**

| Mode/Constraint | PTTj | EPLRSj | GSMj |
|---|---|---|---|
| **PTTi** | $i,j < \text{PTTmax}$; Pmax $f_{\text{PTTi}} - f_{\text{PTTj}} < F_{\text{PTT}}\min$ | $N_{\text{PTT}} + N_{\text{EPLRS}} <$ $N\max f_{\text{PTT}}$ $-f_{\text{EPLRS}} < F\min$ | $N_{\text{PTT}} + N_{\text{GSM}} < N\max$ $f_{\text{PTT}} - f_{\text{GSM}} < F\min$ |
| **EPLRSi** | $R_{\text{PTT}} + R_{\text{EPLRS}} < R\max$ | $i,j < \text{EPLRS}\max$; $P\max$ | $N_{\text{EPLRS}} + N_{\text{GSM}} <$ $N\max$ |
| **GSMi** | $R_{\text{PTT}} + R_{\text{GSM}} < R\max$ | $R_{\text{GSM}} + R_{\text{EPLRS}} <$ $R\max$ | $i,j < \text{GSM}\max$; $P\max$ |
| **PTT, EPLRS, GSM** | $i + j + k < N\max$ | $Ri + Rj + Rk <$ $R\max$ | $Pi + Pj + Pk <$ $P\max$ |

sion in several ways. A well-designed SDR has a table of constraints among combinations of waveforms that can operate simultaneously on the platform. The entries of the constraint table specify parametric limits on power, frequency, data rate, and number of simultaneous channels supported, as a minimum.

Table 8-1 provides a minimal example of a constraint table. In this case a notional dual-use military-commercial PDA has three possible waveforms: push-to-talk (PTT) AM/FM voice, EPLRS, and GSM. The entries on the diagonal limit the number of channels that can be used in each mode to less than ⟨mode⟩ max, where ⟨mode⟩ is PTT, EPLRS, or GSM. If the radio has four channels, it may be capable of supporting all four as push-to-talk channels, but it may have some capacity limit to only one EPLRS channel and only two GSM channels. When used in combination, however, the number of PTT and GSM channels may not be the sum of the individual limits. The entries "$N\langle\text{mode1}\rangle + N\langle\text{mode2}\rangle < N\max$" specify the limits when two modes are used in combination. In addition, the PTT row has been augmented with limits on the frequencies of the modes. The first column specifies that any two PTT channels must have the minimum frequency separation $F_{\text{PTT}}\min$. The other entries specify limits on the separation of combinations of modes. Additional entries specify joint limits on data rate ($R\langle\text{mode}\rangle$) when modes are used jointly. One may specify a total data rate for all subscribers that cannot be exceeded. Other constraints to be included in such a table are the presence and status of an active cancellation circuit, or the measured distance from the transmitting node to the nearest colocated node. This distance may be estimated using round-trip leading-edge delay techniques similar to the way radio distance measuring equipment (DME) operates [399]. An SDR with a 100 MHz ADC/DAC channel and an FPGA with access to the digital IF signal could send a DME signal to be transponded by nearby radios. The internal delays can be calibrated so that the distance can be estimated to within 100 feet or so.

The constraints in such a table must be checked before initializing a mode. An entry may not be available for a mode to be loaded (e.g., because of a download). If so, then the system must warn the user or the network that an uncontrolled mode is about to be used (e.g., at one's own risk). Alternatively, the network might specify that if constraints are not known the mode may not be instantiated.

The combinatorial complexity of such a table deserves attention. Suppose there are $N$ waveform families available in the waveform library. Let the radio platform support up to $C$ simultaneous RF channels. Assume that power, $P$; aggregate data rate, $R$; frequency separation, $\Delta F$; and number of channels of family $i$ in configuration $j$, $Nij$, must be constrained, for a total of four basic constraints ($k = 4$). For each waveform family, there will be four constraints for the waveform used alone (e.g., no other waveforms are instantiated). In addition, each pair of waveform families must be mutually constrained. There are $N - 1$ pairs, yielding an additional $4(N - 1)$ constraints. There are only $N - 2$ triples, yielding another $4(N - 2)$ constraints, and so forth, to one final constraint when all families are instantiated. This yields a formula for the number of constraints as follows:

$$M = k \sum_{j=0}^{N-1} (N - j) = kN(N + 1)/2$$

This number of entries in the constraints table grows like $k/2$ times $N^2$. If there are 30 waveform families, then there are $k(465)$ constraints, or 1869. Forty families yields 3280 constraints. The number of channels, $C$, limits the number of families that may be initialized (e.g., for operational use). But it does not necessarily limit the number that could be instantiated (e.g., loaded into memory, among which a user may choose a subset for operational use). Therefore, $C$ provides no practical limit on the number of constraints that have to be known to the SDR. These constraints may be organized into a constraints database. The challenging aspect of such large numbers of constraints is the labor-intensive process of analyzing each combination of waveforms to determine their potential for generating mutual interference. Whenever a new waveform is to be introduced into an existing family of $N$ waveforms, $N$ new combinations must be analyzed for interference-generation potential. In addition, not all mutual constraints are as simple as those of the minimalistic type shown above. This notion of mutual constraints among waveform families in the context of some host radio platform is a theme that will be further developed in subsequent chapters as more types of potentially problematic interactions are examined. The combinatorial growth of mutual constraints is one of the aspects of SDR that causes unpleasant surprises during the integration process. The analysis, testing, and management of such mutual constraints therefore emerges as a central theme of the design and implementation of software radios.

## III. RF COMPONENT TECHNOLOGY

This section provides highlights of RF component technology relevant to the development of SDR platforms. One objective is to characterize RF technology in terms of its potential to support the increasingly wider bandwidths needed by SDR platforms. The primary objective is to identify those aspects of the analog RF platform that are or may become programmable in the future.

### A. RF MEMS

RF integrated circuits (ICs) generally require off-chip resonators, inductors, and capacitors. Each discrete device increases the cost of production manufacturing, which is nearly a linear function of the number of parts (cost per part is not the first-order driver of manufacturing cost). In addition to replacing discrete devices, MEMS RF switches provide an electromechanical alternative to electronic switching circuits, in some cases substantially reducing size, weight, and power while improving performance. MEMS RF devices are beginning to emerge as an alternative to both discrete devices and switching circuits. Initial academic demonstrations have been sufficiently promising to attract substantial military, academic, and industrial investment. The bandwidths and programmability of RF MEMS foreshadow substantial increases in the capability and reprogrammability of RF platforms for SDR.

*1. Resonant Structures* MIT's Microsystems Technology Laboratories (Joseph Lutsky) reported at the International Electron Devices Meeting in December 1996 the development of VLSI-compatible, sealed-cavity, thin-film resonator (TFR) devices that use sputtered piezoelectric films. The resultant devices are freestanding structures that exhibit a 1.36 GHz fundamental longitudinal resonance with a 3.5 dB insertion loss [253]. This technology can achieve quality factors (Q) of 70 to 80,000 in 250 square microns. This is one of the first filters referred to as a MEMS device. The size of the device is six orders of magnitude less than discrete component LRC circuits. RF products that take advantage of such device technologies historically have been introduced about five years after the introduction of the core technology. This leads to the expectation of wideband RF MEMS by 2001–2003.

Resonators include designs that suspend nanoscale I-beams above cavities in the silicon. The mechanical frequencies of these I-beams depend on the size and stiffness of the I-beam and the distance between the beam and the bottom of the cavity. Using bulk, acoustic, or piezoelectric effects, these devices have sharp resonance. Qs of over 10,000 have been measured on some of these devices. Unfortunately, the best performing devices to date have operating

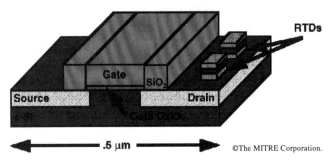

**Figure 8-7**   RF MEMS employs 3D mixed technology devices.

frequencies that are either below 70 MHz or above 2 GHz. An ideal high-Q
filter for cellular applications would have an operating frequency in the 800–
2000 MHz range. A MEMS resonant structure is illustrated in Figure 8-7. This
is a resonant tunneling diode (RTD) circuit. In addition to the conventional
source, gate, and drain, the RTD requires a freestanding three-dimensional
stack of active material. Conventional manufacturing processes are incapable
of depositing these 3D components due to the relatively shallow slope of the
sidewalls of conventional etched structures. MEMS deposits new material using novel techniques such as LIGA machining to achieve true 3D as illustrated
in the figure. MEMS devices have been fabricated in nickel at low temper-
atures (250°C) [254]. This allows the MEMS components to be added to a
prefabricated silicon chip without melting the chip in the process. Dow and
Intarsia are integrating passive components using novel process technology
[255]. New substrates are also appearing [256].

DARPA Electronics Technology Program MEMS electronic filters are to
be used in the detection and suppression of jamming signals for GPS by the
year 2000 [257]. These filters condition the RF signals electroacoustically in
an analog manner. Circuits with these filters have higher Q, lower dissipated
power, and smaller size than equivalent discrete circuits. MEMS contractors
are the University of Michigan and the Naval Surface Warfare Center, China
Lake (DARPA/Air Force Contract Number F30602-97-2-0101).

MEMS capacitors and inductors have been fabricated in laboratory settings
[258]. A 12.5 turn inductor was characterized at 24 nH [259]. In addition,
variable-geometry capacitors have programmable capacitance between 1 and
4 pF [260]. A variable plate geometry capacitor can have a Q of 20,000.
These developments will have a positive impact on SDR RF platforms during
the next five to ten years. The micron scales of the devices should permit the
fabrication of arrays of narrowband filters that may be selected under software
control [261]. In addition, the programmable capacitors permit the software
to set the exact parameters of RF circuits. This will constitute a significant
breakthrough for the flexibility of SDR platforms with palmtop-class size,
weight, and power.

*2. RF Switches*    Many military communications systems operate on vehicles that place a premium on the size, weight, and power consumption of electronic systems, such as tactical aircraft. MEMS switches and tunable capacitors were demonstrated in FY98 to function for radio frequencies up to 40 GHz. They were to be inserted into antenna interface units for the Comanche Helicopter and the F-22 Fighter, targeting a frequency range of 30 MHz to 400 MHz in FY 00 [262].

An industry-standard figure of merit for an RF switch is R1C0, the product of the ON-resistance and the OFF-capacitance. This product is measured in femtoseconds, fs, $10^{-15}$ second. Typical MESFETs attain R1C0 of 500 fs with a 10 ns switching time, while PIN diodes achieve 250 to 100 fs, depending on power dissipation [263]. MEMS switches have been measured with R1C0 of from 2.5 to 12 fs. DARPA expected R1C0 to be reduced by another order of magnitude during 1999–2000 [264].

One airborne application replaced 1044 components of a PIN-diode switch array with 36 MEMS components, reducing size by a factor of over 10,000. Each PIN diode requires 15 components (2 diodes, 2 transistors, 3 capacitors, and one inductor plus resistors). Each 15-component diode was replaced by a single capacitive MEMS RF switch. Essentially a microhinge, the switch-state is controlled by an applied voltage pulse that switches the local charge. Consequently, the circuit draws no power unless it is switching (contrast to the PIN diode). Thus, the MEMS assembly was 1/10,000 the size and consumed 1/1000 the power of the PIN diode. Instead of performance degradation, which is often a tradeoff in miniaturization, the MEMS switches have over 100 dB of off-isolation. This is 30 dB better than the PIN diode. There is continuing research into the fabrication of MEMS switch arrays including a 1 Gbps data rate reconfigurable in 100 ns [265], a prototype of which is illustrated in Figure 8-8.

Other MEMS devices include accelerometers (e.g., in automotive airbags), piezoelectric motors, micromirrors, and flow meters [266]. Due to the broad base of commercial applications, new design tools have been introduced. Tanner, for example, introduced a "system-level" MEMS tool called MEMS Pro. The company says MEMS Pro is the first tool suite for both device-level and system-level design [267]. Based on Tanner's previous IC layout and Spice simulation tools, MEMS Pro lays the groundwork for a tool suite that targets both device-level and system-level design. It lets users create systems that integrate MEMS devices with analog and digital circuitry. It also may facilitate designs such as Analog Devices' widely used air-bag controller, one of the first examples of MEMS integrated onto a single chip.

*3. SDR Applications*    MEMS components enhance the possibilities for the programmability needed by software radios in at least two ways. Initially, MEMS switches may select among arrays of analog circuits and components so that the RF conversion segment has more degrees of freedom. This is a direct application of MEMS to conventional designs. The size, weight, and

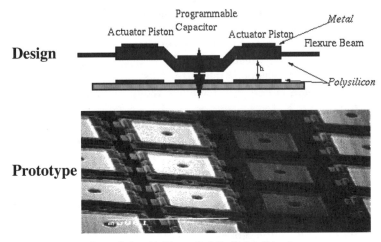

**Design**

**Prototype**

Courtesy Professor Paul Franzon, North Carolina State University.

**Figure 8-8**   High performance MEMS switch fabric.


power saved through MEMS would be reallocated in part to additional degrees of freedom. The second stage of programmability is the reengineering of the RF subsystems. A piezoelectric MEMS motor might move a nanoscale I-beam in a second-generation MEMS resonator. Such motor drives could reconfigure the frequency plan of a superheterodyne receiver so that one device could operate effectively in VHF and UHF as an array of specially tuned switched VHF and UHF resonators.


## B. Superconducting Filters

The wide bandwidth of SDR architecture accepts more noise and interference into the RF stages than equivalent narrowband RF architectures. Thus, applying this architecture to cellular base stations benefits from the reduction of broadband noise. In particular, adjacent service providers mutually interfere with each other. In first-generation systems, for example, a 25 MHz AMPS allocation would be split equally among two service providers. Consequently, a hundred users in the 12.5 MHz band of one service provider are supplying adjacent-channel interference (ACI) into the band of the other service provider. Although the absolute level of the ACI power is low (e.g., −80 to −100 dB below peak power), 100 subscribers increase this ACI by 20 dB. Superconducting analog filters reduce the total noise and interference by up to 30 dB [268], compared to conventional analog filters. Superconductus and Illinois Superconductor both offer products for commercial cellular applications. These products use high-reliability closed cooling systems or thermoelectric coolers (TECs) to maintain the high-temperature superconductors at the required operating temperature near 70 K. By combining superconducting

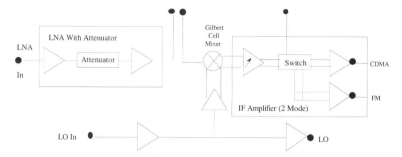

**Figure 8-9**   Multimode amplifiers.

filters and conventional antialiasing filters, one may achieve better spectral purity of the wideband-digitized signals of the SDR.

## C.  Dual-Mode Amplifiers

Dual-mode handsets require RF devices of limited programmability. For example, there is a conflict in design approaches between linear RF and power-efficient RF signal generation. The most efficient RF amplifiers operate in a saturated mode (Class C), which nonlinearly distorts the output waveform. This characteristic is acceptable for amplitude-insensitive modes such as FM and QPSK. Modes in which the instantaneous amplitude envelope contains information, such as QAM, are degraded by the collapsing of amplitude states in such power amplifiers. Dual-mode amplifier chip sets such as the one shown in Figure 8-9 emerged (e.g., for dual-mode satellite mobile and PCS applications [269, 270]). This device includes a Gilbert cell mixer and two different final power amplifier circuits. Note from the numerous external connections that this chip set requires discrete external tuning circuits. These components and the internal switches are candidates for MEMS technology insertion.

Voltage requirements for power amplifiers and low-noise receiver amplifiers continue to drop. Phillips, for example, offers a GaAs low-noise amplifier (LNA) that operates from a single 3.6 V power supply [271]. Dual-mode and low-power RF MEMS components are enablers for SDR approaches into handsets.

## D.  Electronically Programmable Analog Components

Programmable RF requires programmable analog components. The electronically programmable analog circuit (EPAC) provides a specific architecture for the programmability of such analog components (See Figure 8-10). EPACs, also called Field-Programmable Analog Arrays (FPAAs), combine traditional analog circuits such as amplifiers and filters with programmable interconnect. In addition, the operating parameters of the analog circuits are also digitally programmable. These circuits guarantee performance over wide temperature

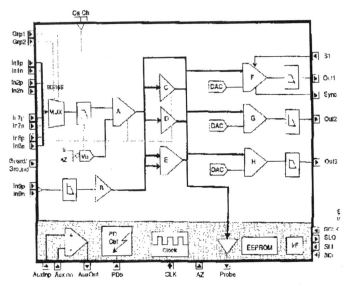

**Figure 8-10**   Electronically programmable analog circuit (EPAC).

ranges. Support software assists in the design and programming of the circuits. Typical programmable functions include amplifiers, comparitors, multiplexers, DACs, track-and-hold circuits, filters, power supplies, and interconnect. Circuits that provide gain are also feedback-stable over the temperature range. Some devices allow group switching of gains and offsets. Devices on the market in 1998 could switch in 4 $\mu$sec and reconfigure in 200 msec [272]. Motorola's FPAA [273] has a clock of 1 MHz and an effective bandwidth of 200 kHz. These narrow bandwidths limit the circuits to baseband at present. But the marriage of multimode RF MEMS devices with EPAC control technology may usher in a new generation of RF programmability. The dual-mode amplifier mentioned above, for example, could be extended to a multiband, multimode base station transmitter using EPAC technology, for example. Although there was no significant demand for multimode base stations in June 2000, incremental deployment of IMT-2000 increases the demand for such technology.

## IV.  RF SUBSYSTEM PERFORMANCE

Critical parameters of the RF segment are illustrated in Figure 8-11. In this example, the RF band ranges from 20 to 500 MHz with a 4 MHz IF bandwidth. The receiver adds 13 dB of noise to the input signal, but it maintains a spurious free dynamic range (SFDR) of 70 dB. This dynamic range establishes the range of input power for which there are no sinusoidal RF conversion artifacts in the output. Such artifacts can mask a weak signal. Large dynamic

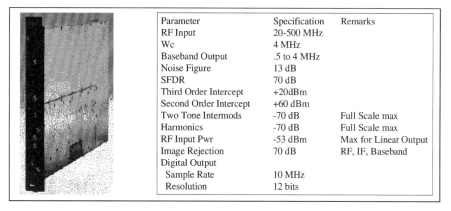

| Parameter | Specification | Remarks |
|---|---|---|
| RF Input | 20-500 MHz | |
| Wc | 4 MHz | |
| Baseband Output | .5 to 4 MHz | |
| Noise Figure | 13 dB | |
| SFDR | 70 dB | |
| Third Order Intercept | +20dBm | |
| Second Order Intercept | +60 dBm | |
| Two Tone Intermods | -70 dB | Full Scale max |
| Harmonics | -70 dB | Full Scale max |
| RF Input Pwr | -53 dBm | Max for Linear Output |
| Image Rejection | 70 dB | RF, IF, Baseband |
| Digital Output | | |
| Sample Rate | 10 MHz | |
| Resolution | 12 bits | |

**Figure 8-11** Digital receiver subsystem performance. Figure derived from ©Watkins–Johnson photographs.

range is more challenging to achieve across wide bandwidths than in narrow bandwidths. The second- and third-order intercept points also characterize receiver linearity. Two-tone intermodulation products can be induced by device nonlinearities when two sinusoids are present at different frequencies at the same time. Harmonics of the fundamental (carrier) frequency may also be present [245].

In the receiver described in Figure 8-11, the SFDR, intermodulation products, and harmonics are controlled to a consistent level of $-70$ dB with respect to full-scale input. The RF conversion process will yield unwanted images of the desired bands. The maximum power of these images is also controlled to 70 dB below the RF, IF, and baseband power. Consequently, this receiver has a useful dynamic range of 70 dB. At nominally 6 dB per bit, 70 dB is equivalent to 11 2/3 bits of ADC resolution. The receiver provides 12 bits of resolution, which is consistent with the dynamic range. In addition, the sampling rate of 10 MHz oversamples the 4 MHz IF passband by a ratio of $10/4 = 2.5 : 1$. The Nyquist criterion specifies that one must sample a band-limited analog waveform by at least twice its maximum frequency component in order to reconstruct the signal unambiguously. This ratio of 2.5 : 1 represents good engineering practice, slightly oversampled with respect to the Nyquist criterion.

Useable dynamic range is arguably the most critical parameter of the analog processing stages of an SDR. These processing stages include the antenna, RF/IF conversion, and the analog circuits of the ADC. Figure 8-12 illustrates linear dynamic range in more detail.

The horizontal axis (abscissa) represents the input power level. Consider the point at which the output power of a single input sinusoid is equal to thermal noise. Call this point $P_{min}$. As the input power of that sinusoid increases, the output power also increases linearly. At some point, the power of the third-order intermodulation product is equal to the power of the output noise. Call

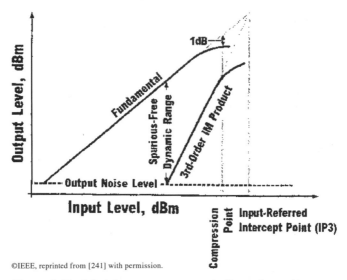

**Figure 8-12**  Modulator stages constrain linear dynamic range.

this point $P_{max}$. The useful dynamic range (DNR) is a dimensionless quantity, which represents the ratio of the largest processable signal to the smallest detectable signal. If power is measured in dB, then:

$$DNR = P_{max} - P_{min}$$

which applies when testing the receiver, so

$$P_{min} = P_{max} - DNR$$

which is used to determine $P_{min}$ when $P_{max}$ is defined by the roofing filter and DNR is the specification of the wideband receiver.

In order to process a signal effectively, it must have a positive SNR, S. Thus, one must differentiate a receiver's specified DNR from DNR-S, which is the dynamic range of processable signals. This is the near–far ratio that a receiver with a given DNR can support if the class of modulation requires an SNR of S dB for the required BER. Suppose, for example, that the near–far ratio in a GSM system is 90 dB. Suppose further that S = 9 dB is required to process a GSM signal with minimum acceptable BER. Then an SDR must offer 90 dB of near–far range plus 9 dB to process the signal of minimum energy, or 99 dB of useable dynamic range. An equivalent way of stating this condition is that given a 99 dB DNR and a 9 dB SNR for the minimum processable signal, an SDR has a near–far capability of DNR – S = 90 dB.

The processable dynamic range is also important in analyzing the effects of RFI and EMI. RFI originates with high-power radio sources that are external to the software radio and its host platform. EMI originates within the host hardware or host platform. Otherwise, the two are very similar. Consider RFI

- **Commercial:**
  - Single RF/IF up- and down-conversion (one ADC/multiple users)
  - Multimode RF power ICs increasingly available, multimode handsets soon
  - Superconducting filters ( Conductus , Illinois Superconductor [196])
- **Key Decisions:**
  - Direct conversion, zero-IF (challenging), superhet tradeoffs
  - Wideband/digital frequency plans with power amplifier and filter matching
- **Military Applications:**
  - Tunable preselectors for cosite rejection
  - ADCs need narrowband roofing filters
  - Separate RF/IF chains for each band drive costs
  - Low phase noise/fast tuning synthesizers for JTIDS, EPLRS
- **Key Parameters:**
  - Size, weight, power; RF, Wc, SFDR, AGC range, ripple, phase noise

**Figure 8-13**   Critical issues in RF conversion.

in a military context. A SPEAKeasy-like radio may have to operate within a kilometer of a high-power (90 dBm) troposcatter communications system. Metal structures may reflect some of the radio energy with nonlinear intermodulation products that inject appreciable narrowband power levels into the SDR receiver. Received power of 0 to $-10$ dBm narrowband artifacts is not impossible. If the narrowband sensitivity of the receiver is, say, $-110$ dBm, then a distant FM signal at $-100$ dBm received power level should be processable. Given a $-10$ dBm troposcatter harmonic, the system must have a useable dynamic range of $((-10) - (-110))$ or 100 dB. The RF system described above has only 70 dB dynamic range, so it would be incapable of detecting the FM signal in the presence of the high-power RFI. EMI could originate from the fundamental of a processor clock at, say, 266 MHz that is leaking into the RF in the UHF band. Local oscillator (LO) radiation from one band may leak via skin currents or near-field reflections on the vehicle platform into other bands. A cosite filter could be tuned to suppress such an interference source. If the filter suppresses the interference by 40 dB, then the total dynamic range from minimum detectable signal to highest power interference is reduced from 100 dB to just 60 dB. SPEAKeasy I, for example, used cosite filters as roofing filters to limit total dynamic range in this way.

## V. RF/IF CONVERSION ISSUES

Critical issues in RF conversion are summarized in Figure 8-13. Commercial applications of wideband RF are feasible in part because of the FDD design of cellular standards. This promotes an architecture focused on migration toward the single wideband digital channel as the wideband RF technology matures.

Multimode RF ICs are leading in the near term to reconfigurable RF. Further advances in this technology plus the introduction of RF MEMS resonators and switches will expand the programmability of the RF conversion segment. In addition, superconducting filters can extend the dynamic range of cellular applications. All these developments bode well for the introduction of SDR base stations and for the migration of those products toward the ideal software radio during the next ten to twenty years.

In handset applications, multimode RF-power ICs increase the flexibility of the handset to support multiple modes. The pivotal decisions for RF conversion in the handset center on the choice of an efficient integrated RF conversion subsystem. The direct conversion receiver may distort adjacent channels, so it may not be effective in base stations, but the receiver architecture is among the most efficient for handsets. Zero-IF receivers minimize receiver parts-count, but the possibility of leaking LO and DC into the signal passband renders this architecture very challenging to implement. With the introduction of 3G, some wideband CDMA mode will also be required. The central challenge in RF conversion will be the integration of multiband RF conversion with common despreading.

Military SDRs require cosite roofing filters and/or active cancellation for typical command-and-control applications where high-power external RFI will be present. It would be possible to dynamically restructure the transmission and reception patterns of the military bands to reduce cosite interference substantially if all users had SDRs. Consider a simple approach to the 60 MHz LVHF band. One could allocate the lower 15 MHz to downlink and the upper 15 MHz to uplink from a mobile command center. If all command centers adopt this strategy, then all mobiles may transmit only in the uplink band. The result would be the separation of transmit and receive frequencies by an average of 15 MHz. Of course, legacy radios could not operate this way. But SDRs could.

Additional constraints on the RF are implicit in the ADC. In particular, the total system dynamic range is determined by the combination of RF, ADC, and algorithm dynamic range. RF dynamic range is a function of the analog design. ADC dynamic range is discussed in the next chapter.

## VI. EXERCISES

These exercises include a mix of review of topics covered in this chapter and questions designed to focus on architecture issues to be addressed later in the text.

1. What functions are accomplished in the RF conversion and IF processing segments of the canonical software-radio architecture? Which of these functions are amenable to enhancement through the introduction of digital techniques?

2. Consider a wideband radio that implements frequency hopping with a fast-hopping LO. What steps are necessary to migrate that implementation toward a software-radio architecture? What benefits would accrue through such migration? What new functions would be readily implementable on such a platform that were not implementable with the LO-based design?

3. What does it mean for a function to be null in an architecture? Suppose one wants an architecture that will accommodate diversity combining but does not require diversity combining. How would you abstract the design of diversity combining to accommodate that? How do you do this in such a way as to enable analog, digital, and hybrid implementations of diversity combining?

4. Identify receiver architectures appropriate for base station and for handset applications. Describe an architecture that would permit one to implement the receiver hardware with either design, yet would not require radio applications to have specific provisions for different receiver designs.

5. List the types of noise and interference that may be present in the passband of a receiver. How does one estimate the level of thermal noise in a receiver? How does one estimate the magnitude of spurious responses that may be produced by intermodulation products? What design principle should be followed in the RF section to control the noise and interference in the signal delivered by wideband RF/IF stages to an ADC?

6. Define the digital-RF receiver. What are its primary benefits? What challenges need to be met in order for such architectures to be widely deployed in the future?

7. List the techniques that may be used to suppress interference in wideband communications systems in the RF/IF stages. List the primary constraints on the use of these techniques. Describe how constraints on the use of these techniques could be incorporated into a constraints-table. Should this table be a function of waveform?

8. What are RF MEMS? What functions implementable with RF MEMS support the migration of narrowband fixed-function RF/IF circuits toward SDR or software-radio architectures?

9. What other RF component technologies facilitate migration toward the SDR? For each technology, identify its relevance. Define a notional timeline for the insertion of RF MEMS and these technologies into commercial base stations, handsets, and military systems. What economic factors will determine whether this timeline will be accelerated or delayed? What technology breakthroughs are needed to accelerate the migration?

10. What performance parameters best characterize the RF platform in terms of its support to software-based definition of radio functions? Describe the relevance of these parameters to the evaluation of whether an RF platform

will support a given air-interface standard. Apply your approach to GSM. To WCDMA.

11. Consider the disaster-relief application. What RF bands should be supported? How much must be specified about the modes in use in these bands in order to design an RF platform that will support the needs of an SDR operating in these bands? What operational constraints shape the design parameters of the RF segment? What RF architecture maximizes flexibility? What RF architecture maximizes performance within a fixed, limited size/weight/power envelope (e.g., a handset)?

12. Recall the object-oriented approach to architecture analysis. Apply that approach to the RF segment, given the characteristics of RF developed in this chapter. Define an object hierarchy. Assess the capability of this hierarchy for the disaster-relief case study.

13. Recall the tabular RF reference platform developed previously. Define the RF portion of this reference platform for the disaster-relief case-study. How are the tradeoffs of this chapter reflected in that tabular treatment? Is the insertion of new technology stimulated or suppressed by your profile? Would a programmable aperture with FPAA-based RF comply with your RF reference platform?

# 9 ADC and DAC Tradeoffs

This chapter introduces the relationship between ADCs, DACs, and software radios. Uniform sampling is the process of estimating signal amplitude once each $T_s$ seconds, sampling at a consistent frequency of $f_s = 1/T_s$ Hz. Although there are other types of sampling, SDRs employ uniform-sampling ADCs.

## I. REVIEW OF ADC FUNDAMENTALS

Since the wideband ADC is one of the fundamental components of the software radio, this chapter begins with a review of relevant results from sampling theory. The analog signal to be converted must be compatible with the capabilities of the ADC or DAC. In particular, the bandwidths and linear dynamic range of the two must be compatible. Figure 9-1 shows a mismatch between an analog signal and the ADC. For uniform sampling rate $f_s$, the maximum frequency for which the analog signal can be unambiguously reconstructed is the Nyquist rate, $f_s/2$. The wideband analog signal extends beyond the Nyquist frequency in the figure. Because of the periodicity of the sampled spectrum, those components that extend beyond the Nyquist frequency fold back into the sampled spectrum as shown in the shaded parts of the figure (thus the term *folding frequency*). This is well known as aliasing [274, 275]. Although some aliasing is unavoidable, an ADC designed for software-radios must keep the total power in the aliased components below the minimum level that will not unacceptably distort the weakest subscriber signal.

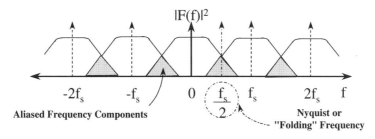

**Figure 9-1**   Aliasing distorts signals in the Nyquist passband.

## A. Dynamic Range (DNR) Budget

If acceptable distortion is defined in terms of the BER, then dynamic range (DNR) may be set by the following procedure:

1. Set $BER_{THRESHOLD}$ from QoS considerations
2. $BER = f(MODULATION, CIR, FEC)$
3. $BER < BER_{THRESHOLD} \rightarrow CIR > CIR_{THRESHOLD}$, from $f(\ )$
4. $DNR = DNR_{ADC} + DNR_{RF-IF} + DNR_{OVERSAMPLING} + DNR_{ALGORITHMS}$
5. $P_{ALIASING+RFIF+NOISE} < \frac{1}{2}(DNR_{ADC} + CIR_{THRESHOLD})$

Consider the situation where the channel symbol modulation, MODULA-TION, is fixed (e.g., BPSK). BER is a function of the CIR. The first step in establishing the acceptable aliasing power is to set the $BER_{THRESHOLD}$ by considering the QoS requirements of the waveform (e.g., voice). The $BER_{THRESHOLD}$ for PCM voice is about $10^{-3}$. The next step is to characterize the relationship between BER and CIR. In the simplest case, this relationship is defined in the BER-SNR (CIR or Eb/No) curve for MODULATION (e.g., from [275]). In other cases, FEC reduces the net BER for a given raw BER from the modem. In such cases, net BER has to be translated into modem BER using the properties of the FEC code(s) [276, 277]. $BER_{THRESHOLD}$ is then translated to $CIR_{THRESHOLD}$ using $f$ (e.g., 11 dB). Finally, one must incorporate the instantaneous dynamic range requirements of the ADC. Total dynamic range must be partitioned into dynamic range that the AGC, ADC, and algorithms must supply. In the simplest case, the total dynamic range is just the near–far ratio plus $CIR_{THRESHOLD}$. If the RF and/or IF stages contain roofing filters or AGCs, then some of the total system DNR is allocated to these stages. In addition, since the wideband ADC of the SDR oversamples all subscriber signals, digital filtering can yield oversampling-gain. Other postprocessing algorithms such as digital interference cancellation can yield further DNR gains. Each such source of DNR reduces the allocation to the ADC. From these relationships, one establishes $DNR_{ADC}$. The power of aliasing, spurious responses introduced in RF and/or IF processing, and noise should be kept to less than half of the LSB of the ADC.

If the total power is less than the power represented by $\frac{1}{2}$ of the least significant bit (LSB) of the ADC, then all of the ADC bits represent processable signal power. If the power exceeds $\frac{1}{2}$ LSB, then this extra precision presents a computational burden that has to be justified. For example, the extra bits may result from rounding up from a 14-bit ADC to the more convenient 16 bits in order to transfer data efficiently. When this is done, the difference between accuracy and precision should be kept clear.

## B. Anti-Aliasing Filters

When the aliased components are below the minimum acceptable power level (e.g., $\frac{1}{2}$ LSB) the sampled signal is a faithful representation of the analog sig-

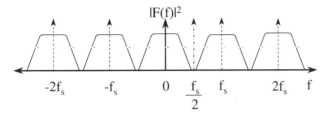

$|F(f)|^2$

$-2f_s$          $-f_s$          $0$     $\dfrac{f_s}{2}$   $f_s$              $2f_s$     $f$

**Figure 9-2**   Anti-aliasing filters suppress aliased components.

| Resolution (bits) | LSB Weight (fraction) | (dBc) | Stop Band Attenuation (dB for < 1/2 LSB) |
|---|---|---|---|
| 8 | 1/256 | -48 | -54 |
| 10 | 1/1024 | -60 | -66 |
| 12 | 1/4096 | -72 | -78 |
| 14 | 1/16384 | -84 | -90 |
| 16 | 1/65536 | -96 | -102 |

**Figure 9-3**   High resolution requires high stop band attenuation.

nal, as illustrated in Figure 9-2. The wideband ADC, therefore, is preceded by anti-aliasing filter(s) that shape the analog spectrum to avoid aliasing. This requires anti-aliasing filters with sufficient stop-band attenuation. Figure 9-3 shows the stop-band attenuation required for a given number of bits of dynamic range. Since the instantaneous dynamic range cannot exceed the resolution of the ADC, the number of bits of resolution is a limiting measure of the dynamic range. High dynamic range requires high stop-band attenuation. To reduce the power of out-of-band energy to less than $\frac{1}{2}$ LSB, the stop-band attenuation of the anti-aliasing filter of a 16-bit ADC must be $-102$ dB. This includes the contributions of all cascaded filters including the final anti-aliasing filter.

To suppress frequency components that are close to the upper band-edge of the ADC passband, the anti-aliasing filters require a large shape factor. The shape factor is the ratio of the frequency at which $-80$ dB attenuation is achieved versus the frequency of the $-3$ dB point. Bessel filters have high shape factors and thus slow rolloff, but they are monotonic. Monotonic filters exhibit increased attenuation as frequency increases. Nonmonotonic filters have decreased-attenuation zones. These admit increased out-of-band energy and distort phase. Those filters with fastest rolloff also have high amplitude ripple and distort phase more than filters with more modest rolloff. Filter design has received much attention in the signal-processing literature [278]. (See Figure 9-4.)

| Transition and Stop Band Performance (-80 dB) | | | | |
|---|---|---|---|---|
| Filter | Poles | Shape Factor $(f_{@-80}/f_{@-3})$ | Rolloff | Monotonic |
| Bessel | 8 | 6.068 | Slowest | Yes |
| Butterworth | 8 | 3.162 | Faster | Yes |
| Chebyshev | 8 (.1 dB) | 2.183 | Next Fastest | Yes |
| Cauer | 7 (6 Zero) (0.1 dB) | 1.661 | Fastest | No |

**Figure 9-4** Attenuation rolloff, amplitude ripple, and shape factor determine antialiasing filter suitability.

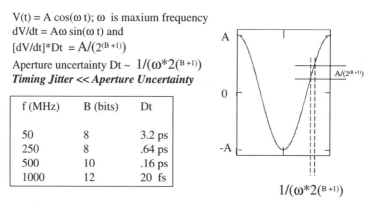

$V(t) = A \cos(\omega t); \omega$ is maxium frequency
$dV/dt = A\omega \sin(\omega t)$ and
$[dV/dt]*Dt = A/(2^{(B+1)})$
Aperture uncertainty $Dt \sim 1/(\omega * 2^{(B+1)})$
***Timing Jitter << Aperture Uncertainty***

| f (MHz) | B (bits) | Dt |
|---|---|---|
| 50 | 8 | 3.2 ps |
| 250 | 8 | .64 ps |
| 500 | 10 | .16 ps |
| 1000 | 12 | 20 fs |

$1/(\omega * 2^{(B+1)})$

**Figure 9-5** Sample-and-hold circuits limit ADC performance.

## C. Clipping Distortion

In most applications, one cannot control the energy level of the maximum signal to be exactly equal to the most significant bit. One must therefore allow for some AGC or for some peak power mismatch. Clipping of the peak energy level introduces frequency domain sidelobes of the high power signal. These sidelobes have the general structure of the convolution of the signal's sinusoidal components with the Fourier transform of a square wave, which has the form of a $\sin(x)/x$ function. Frequency domain sidelobes have a power level of $-11$ dB, which is clearly unacceptable interference with other signals in a wideband passband. In practice, avoiding clipping may occupy the entire most significant bit (MSB). Usable dynamic range may therefore be one or two bits less than the ADCs resolution.

## D. Aperture Jitter

Sample-and-hold circuits also limit ADC performance as illustrated in Figure 9-5. Consider a sinusoidal input signal, $V(t) = A\cos(\omega t)$, where $\omega$ is the maximum frequency. The rate of change of voltage is as shown, yielding a maximum rate of change of $2A/(2^B)$ or $A/(2^{(B+1)})$. The time duration of this differential interval is inversely proportional to the frequency and the exponential of the number of bits in the ADC. This period is the aperture uncertainty, the shortest time taken for a maximal-frequency sine wave to traverse the LSB. The timing jitter must be a small fraction of the aperture uncertainty to keep the total error to less than $\frac{1}{2}$ LSB. Therefore, the timing jitter should be 10% or less of the uncertainty shown in the figure. An 8-bit ADC sampling at 50 MHz requires aperture jitter that is less than a picosecond (ps).

This stability must be maintained for a period of time that is inversely proportional to the frequency stability that one requires. If, for example, the minimum resolvable frequency component for the signal processing algorithms should be 1 kHz, then the timing accuracy over a 1 ms interval should be less than the aperture uncertainty. Short-term jitter can be controlled to less than 1 ps for 1 ms with current technology. If the spectral components should be accurate to 1 Hz, then the stability must be maintained for 1 second. Due to drift of timing circuits, such performance may be maintained for $10^9$ to $10^{11}$ aperture periods, or on the order of 1 to 100 ms. Stability beyond these relatively short intervals is problematic due to drift induced by thermal changes, among other things. A sampling rate of 1 GHz with 12 bits of resolution requires about 2 fs of aperture jitter or less. This stability is beyond the current state of the art, which corresponds to 6.5 to 8 bits of resolution at these sampling rates.

### E. Quantization and Dynamic Range

Quantization step size is related to power according to [279]:

$$P_q = q^2/12R$$

where $q$ is the quantization step size, and $R$ is the input resistance. The SNR at the output of the ADC is

$$SNR = 6.02\,B + 1.76 + 10\log(f_s/2f_{max})$$

where $B$ is the number of bits in the ADC, $f_s$ is the sampling frequency, and $f_{max}$ is the maximum frequency component of the signal.

For Nyquist sampling, $f_s = 2f_{max}$, so the ratio of these quantities is unity. Since the log of unity is zero, the third term of the equation for SNR above is eliminated. The approximation for Nyquist sampling, then, is that the dynamic range with respect to noise equals 6 times the number of bits. This equation suggests that the SNR may be increased by increasing the sampling rate beyond the Nyquist rate. This is the principle behind the *sigma-delta/delta-sigma* ADC.

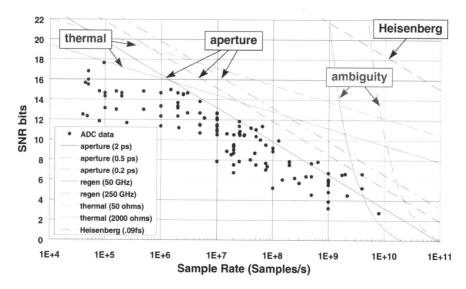

Figure © R. Walden and Hughes Research Laboratories, reprinted with permission.

**Figure 9-6**   Walden's analysis of ADC technology.

## F. Technology Limits

The relationship between ADC performance and technology parameters has
been studied in depth by Walden [280, 281]. His analysis addresses the electronic parameters, aperture jitter, thermal effects, and conversion-ambiguity.
These are related to specific devices in Figure 9-6. The physical limits of
ADCs are bounded by Heisenberg's uncertainty principle. This core physical limit suggests that one could implement a 1 GHz ADC with 20 bits
(120 dB) of dynamic range. To accomplish this, one must overcome thermal,
aperture jitter, and conversion ambiguity limits. Thermal limits may yield to
research in Josephson Junction or high-temperature superconductivity (HTSC)
research. For example, Hypress has demonstrated a 500 Msa/sec (200 MHz)
ADC with dynamic range of 80 dB operating at 4K [435]. Walden notes
that advances in ADC technology have been limited. During the last eight
years, SNR has improved only 1.5 bits. Substantial investments are required
for continued progress. DARPA's Ultracomm program, for example, funded
research to realize a 16-bit × 100 MHz ADC by 2002 [282]. Commercial research continues as well, with Analog Devices' announcement of the AD6644,
a 14-bit × 72 MHz ADC consuming only 1.2 W [282].

## II. ADC AND DAC TRADEOFFS

The previous section characterized the Nyquist ADC. This section provides
an overview of important alternatives to the Nyquist ADC, emphasizing

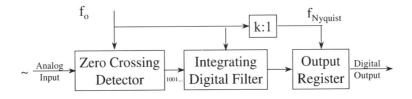

$$f_o, \text{ the oversampled rate} = k*f_{Nyquist}$$

**Figure 9-7**  Oversampling ADCs leverage digital technology.

the tradeoffs for SDRs. It also includes a brief introduction to the use of DACs.

## A. Sigma-Delta (Delta-Sigma) ADCs

The sigma-delta ADC is also referred to in the literature as the delta-sigma ADC. The principle is understood by considering an analogous situation in visual signal (e.g., image) processing. The spatial frequency of a signal is inversely proportional to its spatial dimension. A large object in a picture has low spatial frequency while a small object has high spatial frequency. Spatial dynamic range is the number of levels of grayscale. A black-and-white image has one bit of dynamic range, 6 dB. But consider a picture in a typical newspaper. From reading distance, the eye perceives levels of grayscale, from which shapes of objects, faces, etc. are evident. But under a magnifying glass, typical black-and-white newsprint has no grayscale. Instead, the picture is composed of black dots on a white background. These dots are one-bit digitized versions of the original picture. The choice between white and black is also called zero-crossing. The dots are placed so close together that they oversample the image. The eye integrates across this 1-bit oversampled image. It thus perceives the low-frequency objects with much higher dynamic range than 6 dB. The gain in dynamic range is the log of the number of zero-crossings over which the eye integrates. Zakhor and Oppenheim [283] explore this phenomenon in detail, with applications to signal and image processing. Thao and Vetterli [284] derive the projection filter to optimally extract maximum dynamic range from oversampled signals. Candy and Temes offer a definitive text [285].

*1. Principles*  The fundamentals of an oversampling ADC for SDR applications are illustrated in Figure 9-7. A low-resolution ADC such as a zero-crossing detector oversamples the signal, which is then integrated linearly. The integrated result has greater dynamic range and smaller bandwidth than the oversampled signal. The amount of oversampling is the ratio of the sampling frequency of the analog input to the Nyquist frequency, shown as $k$ in the figure. This follows

$$\text{SNR} \cong 6\,\text{B} + 10\log(f_s/2f_{\text{max}}) = 6\,\text{B} + 10\log(kf_{\text{Nyquist}}/2f_{\text{max}})$$

Since $f_{\text{Nyquist}} = 2f_{\text{max}}$, the oversampling rate must be at least $2kf_{\text{max}}$. With continuous $1:k$ integration of the zero-crossing values, the output register contains a Nyquist approximation of the input signal.

Since the integrated output has an information bandwidth that is not more than the Nyquist bandwidth, the integrated values may be decimated without loss of information. Decimation is the process of selecting only a subset of available digital samples. Uniform decimation is the selection of only one sample from the output register for every $k$ samples of the undecimated stream. If the signal bandwidth is 0.5 MHz, its Nyquist sampling rate is 1 MHz. A zero-crossing detector with a sampling frequency of 100 MHz has an oversampling gain of ten times the log of the oversampling ratio (100 MHz/1 MHz), 20 dB. The single-bit digitized values may be integrated in a counter that counts up to at least 100. Although this is the absolute minimum requirement, real signals may exhibit DC bias. A counter with only a capacity of 100 could tolerate no DC bias. A counter with range that is a power of two, e.g., 128, tolerates up to $\log(28)$ bits or 4.7 of DC bias. For a range of 128, a signed binary counter requires $\log_2(128)$ bits or 7 bits plus a sign bit. The counter treats each zero-crossing as a sign bit, $+1$ or $-1$. The decimator takes every 100th sample of this 8-bit counter, with an output-sampling rate to 1 MHz as required for Nyquist sampling.

Zero-crossing detectors do not work properly, however, if there are insufficient crossings to represent the signal. For example, if DC bias drifts beyond the full-scale range of the detector, then there will be no zero-crossings and no signal. A signal may be up-converted, amplified, and clipped to force the required zero-crossings. A similar effect can be realized in linear oversampling ADCs through the addition of dither. A dither signal is a pseudorandomly generated train of positive and negative analog step-functions. The dither is added to the input of the ADC before conversion (but after anti-alias filtering). The corresponding binary stream is subtracted from the oversampled stream. Alternatively, an integrated digitized replica of the dither signal may be subtracted from the integrated output stream. This forces zero-crossings, enhancing the SNR. One may view dithering as a way of forcing spurs generated by sample-and-hold nonlinearities to average across multiple spectral components, enhancing SNR.

In addition, high power out-of-band components will be sampled directly by the zero-crossing detector. These components will then be integrated, subject to the bandwidth limitations imposed by the integrator-decimator. The anti-aliasing filter therefore must control total oversampled power so that it conforms to the criteria for Nyquist ADCs.

*2. Tradeoffs*   There are several advantages to oversampling ADCs. First, sample-and-hold requirements are minimized. There is no sample-and-hold circuit in a zero-crossing detector. Simple threshold logic, possibly in con-junction with a clamping amplifier, yields the single-bit ADC.

Aperture jitter remains an issue, but the jitter is a function of the number of bits, which is 1 at the oversampling rate. This minimizes aperture jitter requirements for a given sample rate. As the oversampled values are integrated, the jitter averages out. In order to support large dynamic range for narrowband signals, the timing drift (the integration of aperture jitter) should contribute negligibly to the frequency components of the narrowband signal. This means that integrated jitter should be less than 10% of the inverse of the narrowband signal's bandwidth, for the corresponding integration time.

In addition, the anti-aliasing filter requirements of a sigma-delta ADC are not as severe as for a Nyquist ADC. The transfer-function of the anti-aliasing filter is convolved with the picket-fence transfer-function of the decimator. Thus, the anti-aliasing filter's shape factor may be $1/k$ that of a linear ADC for equivalent performance. Many commercial products use oversampling and decimation within an ADC chip to achieve the best combination of bandwidth and dynamic range.

Oversampled ADCs work well if the power of the out-of-band spectral components is low. In cell site applications, Q must be very high in the filter that rejects adjacent band interference. Superconducting filters [286] may be appropriate for such applications.

## B. Quadrature Techniques

Nyquist ADC samples signals that are mathematically represented on the real line. Quadrature sampling uses complex numbers to double the bandwidth accessible with a given sampling rate.

*1. Principles*   Real signals may be projected onto the cosine signal of an LO and onto the sine reference derived from the same LO. This yields an in-phase (I) signal and a quadrature (Q) signal, an I&Q pair. The in-phase signal is the inner product of the signal with a reference cosine, while the quadrature signal is the inner product with the corresponding sine wave. In the complex plane, the in-phase component lies on the real axis, while the quadrature component lies on the imaginary axis. If the underlying technology limits the clock rate to $f_c$, then the real sampling rate is also limited to $f_c$. The Nyquist bandwidth is limited to $f_c/2$. On the other hand, if the signal is projected into I&Q components, each channel may be sampled independently at rate $f_c$. The Nyquist bandwidth is then the same as the sampling rate as illustrated in Figure 9-8. This doubles the Nyquist rate for a given maximum ADC sampling rate.

Quadrature sampling is the simplest of the polyphase filters. The concept may be extended to multirate filter banks [287]. These advanced techniques include the parallel extraction of independent information streams from real signals.

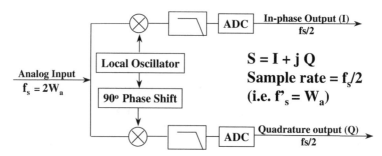

**Figure 9-8**  In-Phase and quadrature (I&Q) conversion reduces sampling clocks.

*2. Tradeoffs*  Although theoretically interesting, analog implementations of quadrature ADCs are challenging. Refer again to Figure 9-8. The modulators, signal paths, and low-pass filters in each I&Q path must be matched exactly in order for the resulting complex digital stream to be a faithful representation of the input signal. Any mismatches in the amplitude or group delay of the filters yields distortion of complex signal.

Historically, it has been difficult to obtain more than 30 dB of fidelity from quadrature ADCs. Military temperature ranges exacerbate the problems of matching the analog paths. Integrated circuit paths are more readily matched than lumped components. Short lengths of signal paths are easier to match, as are resistors and other passive components on IC substrates. Since the components are very close together, the thermal difference between the filters is less than in lumped-circuit implementations. IC implementations of I&Q ADCs can be effective.

To date, the best results for research-quality ADCs have been obtained using real-sampling wideband ADCs in conjunction with digital quadrature and IF filtering. This was the approach used in SPEAKeasy I, for example.

### C. Bandpass Sampling (Digital Down-Conversion)

Nyquist sampling is also called low-pass sampling because the ADC recovers all frequency components from DC up to the Nyquist frequency. Bandpass sampling digitally down-converts a band of frequencies having the Nyquist bandwidth but translated up in frequency by some multiple of $f_s/2$.

*1. Principles*  When frequency components are recovered from a Nyquist ADC stream, the maximum recoverable frequency component is $f_s/2 = W_{\text{Nyquist}}$. The minimum resolvable frequency is inversely proportional to the duration of the observation interval. The observation interval is defined by the number of time-domain points in that observation. The time-delay elements in a digital filter constitute an observation interval. A fast Fourier transform (FFT) is an observation interval of $N$ real samples. If $N = 1024$ and $f_s = 1.024$ MHz, then

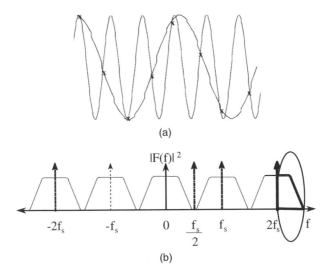

**Figure 9-9**  Bandpass sampling converts channels directly to baseband. (a) time domain; (b) frequency domain.

the minimum recoverable frequency and the resolution of each cell are both $f_s/1024$, or 1 kHz. The FFT has a DC component that is the average value of the signal over the observation interval $T_s * 1024$, which is 1 ms. The first $N/2$ or 512 FFT bins are not redundant. They represent the frequencies from $f_s/N$ to $f_s/2$, 512 kHz. Thus, the low-pass nature of the Nyquist sampling process defines frequency components from DC to $f_s/2$.[25]

The principle of bandpass sampling is to sample a passband of bandwidth $W_{\text{Nyquist}}$ centered at frequency $kfs$ ($k \geq 2$, $k$ is even), at the Nyquist rate $fs$. The high-frequency components are translated to baseband by the frequency-translation property of subsampling. Figure 9-9a illustrates the subsampling process in the time domain. The high-frequency sinusoid represents the upper cutoff frequency of a bandpass signal, occurring at an integer multiple of the Nyquist frequency. Sampling this frequency at the Nyquist rate creates a beat-frequency, which translates the signal to baseband, the low-frequency sinusoid of the figure. The frequency-domain representation (Figure 9-9b) shows how a passband centered at $2f_s$ (circled) is translated to baseband below $f_s/2$.

One advantage of this approach is that the subband of interest is translated in frequency without the use of a mixer stage, and with no LO, either analog or digital. The primary disadvantage is that all of the power in the frequency components between the selected subband and DC are aliased into the baseband. Therefore any residual energy in the bands centered at $kfs$ is integrated

---

[25]This analysis employs sinusoids as the basis functions used in the observation. Wavelet-basis functions yield different observations.

into the baseband. Bandpass-filtering requirements for this approach therefore must keep the total power in the intervening bands to less than $\frac{1}{2}$ LSB.

*2. Tradeoffs*  SDR RF bands generally have bandpass characteristics, not low-pass characteristics. A cellular uplink, for example, might consist of 25 MHz from 824 to 849 MHz. The ideal software radio would convert directly from RF at a sampling rate of say 2.3 GHz. The Nyquist frequency defines a low-pass digital spectrum from DC to 1 GHz. Bandpass sampling of the same cellular band requires a bandpass sampling rate of only $2 * 25 = 50$ MHz, not $2 * 849 = 1698$ MHz. In terms of ADC rates, bandpass sampling presents an attractive alternative to Nyquist sampling.

In order to translate the passband without distortion, the intervening spectra between 3 $fs/2$ and 5 $fs/2$ must be suppressed. A high-$Q$ analog bandpass filter or cascade of filters suppresses the unwanted parts of the spectrum. Such filters have historically not been available. Consequently, the superheterodyne receiver translates the bandpass signal to an IF where Nyquist sampling techniques suffice. High-$Q$ filters such as those emerging from the MEMS program may facilitate bandpass sampling. At present, one cannot obtain equivalent signal quality and dynamic range from bandpass sampling as from the superheterodyne receiver. Bandpass sampling will no doubt continue to attract research and development interest [288].

### D. DAC Tradeoffs

DACs convert digital signals to analog waveforms. Good DAC design incorporates not just level conversion but also high linearity (low intermodulation products), integrated filtering, grounding, and isolation of the digital clock from the analog output waveform. In addition, DACs for cell site applications require oversampling for improved smoothness. This reduces out-of-band artifacts. The design principles of DACs are similar to ADCs. DAC setup and hold corresponds to the sample and hold of the ADC. Setup-and-hold time therefore determines the fidelity of signal reconstruction in a way that corresponds to the effects of aperture jitter in ADCs.

Harris Corporation's 12-bit 100 MHz DAC (HI5731) has spurious-free dynamic range of $-70$ to $-85$ dBc (depending on windowing and oversampling). Its integral linearity error is 1.5 LSB. Full-scale gain error is 10% maximum [289]. For cell site applications, this DAC will generate 12.5 to 25 MHz of total output bandwidth. Amplifiers used in the cable TV industry have 1 GHz output bandwidth from a few MHz to 1 GHz with flat amplitude and phase response. Such amplifiers are appropriate for the amplification of analog IF in cell site applications.

Phase coherence of the multiple parallel IF waveforms combined into one DAC stream can cause the output amplifiers to saturate at peak power. There is a 20 dB difference in peak-to-average power ratio between a single sine wave and a base station application with 100 phase coherent IF sine waves. The

| Software Radio Application | Sampling Rate ($f_s$) | Dynamic Range (dB) |
|---|---|---|
| HF Baseband | .5–8 kHz | 24–64 |
| Modems | 8–32 kHz | 48–64 |
| Music | 20–100 kHz | 60–96 |
| VHF-UHF BB | 50–150 kHz | 20–60 |
| UHF-SHF FDM | .1–25 MHz | 48–96 |
| HF-IF | .2–10 MHz | 72–120 |
| Cellular Radio | .2–75 MHz | 48–90 |
| UHF Air Nav | 2–25 MHz | 48–90 |
| SHF QAM | 12–100 MHz | 30–72 |
| LVHF-IF (FH) | 12–200 MHz | 66–108 |
| VHF/UHF-IF | 25–500 MHz | 60–96 |
| HF RF | 75 MHz | 130 |
| SHF CDMA | 125–500 MHz | 60–90 |
| VHF RF | 650 MHz | 96 |
| SHF Agility | .5–3 GHz | 48–72 |
| SHF-IF | 1–8 GHz | 48–90 |
| UHF RF | 5.4–10 GHz | 48–90 |

**Figure 9-10**  Sampling rate depends on the application.

random phasing of these digital signals reduces the peak-to-average power ratio proportionally. This improves the efficiency of the amplifier and reduces the likelihood of saturation. One should therefore assure that the RF modem software randomizes phase to distribute output power uniformly in the time domain. This is another example of a way in which the design of the digital processing algorithms and hardware can yield benefits (or cause problems) for the analog parts of the software radio.

## III.  SDR APPLICATIONS

ADC and DAC applications are constrained by sampling rate and dynamic range. The pace of product insertion into wireless devices is also determined by power dissipation. Infrastructure applications that are not power-constrained may evolve toward digital RF. This section highlights these aspects of ADC and DAC applications.

### A.  Conversion Rate, Dynamic Range, and Applications

ADC sampling rates and dynamic range requirements depend on the application. Figure 9-10 shows how increasingly wideband applications require increasingly large instantaneous dynamic range. Analog filtering and AGC achieve 90 to 100 dB or more of total dynamic range. As one increases the instantaneous bandwidth, one must also increase the instantaneous DNR as shown in Figure 9-10. It differentiates baseband (BB), IF, and RF ADC requirements. Baseband refers to the bandwidth of modulation of a single RF

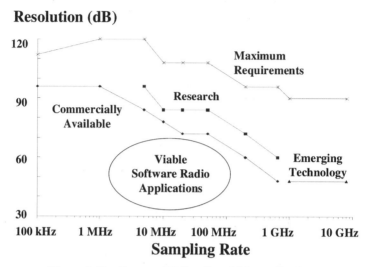

**Figure 9-11**   Present ADCs offer viable applications.

carrier. Thus, HF baseband consists of typically 5 Hz to 3 kHz of modulated
RF carrier. HF automatic link establishment (ALE) may employ linear FM
(chirp) waveforms that use more bandwidth, increasing sampling rate requirements accordingly. Voice channel modems and music require only a few tens
of kHz of bandwidth, but with appreciable DNR for high fidelity applications.
Baseband ADC is the technology of classical programmable digital radios.

Frequency division multiplexed (FDM) signals have a few MHz IF-band-
width, while PCM, cellular band allocations, 3G, and air navigation signals
require tens of MHz. IF-ADC is the technology of SDR. Miller [290] derives
the RF DNR requirements of HF as 120 dB, consistent with [291]. CDMA
bands are not as demanding of DNR because they are power managed. The
RF-ADC is the technology of the ideal software radio. As the bandwidth
increases from BB to IF to RF, the instantaneous DNR increases by about
30 dB per change.


**B. ADC Product Evolution**

Figure 9-11 shows the relationship of commercially available ADC performance to research devices, emerging technology, and maximum requirements
from Figure 9-10. Many viable SDR applications are workable with currently
available technology. Fielded applications include baseband digital signal processing in programmable digital radios. Emerging applications include SDRs
that use IF conversion and parallel ADC channels to obtain high dynamic
range. SPEAKeasy I and II, for example, both employed IF conversion with
moderate (1 MHz) and wideband (70 MHz) ADC channels. The dynamic

# Resolution (bits, nominal)

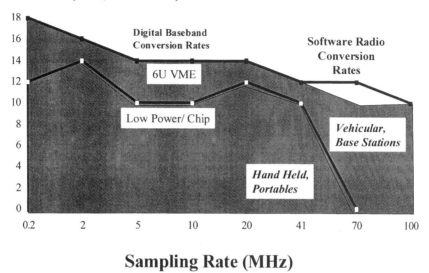

## Sampling Rate (MHz)

**Figure 9-12** Low-power ADCs driven by wireless marketplace.

range of these implementations did not fully address the maximum requirements for radio applications. But they established the feasibility of the technology, allowing developers to gain experience with SDR architecture.

### C. Low-Power Wireless Applications

The recent evolution of ADC product has been driven significantly by the wireless marketplace. Handheld commercial audio devices motivate investment in devices with less than 1 MHz sampling rates but more than 100 dB SNR. Wireless handset applications provide much of the impetus behind low-power wideband ADC chips. Figure 9-12 shows the difference in sampling rate and dynamic range between low-power ADCs and ADCs for board-level products (e.g., for research and laboratory instrumentation markets). The 10- and 12-bit 70 MHz ADCs are rapidly evolving to 14-bit products.

### D. Digital RF

As ADCs continue to evolve, they will enable the digital RF architecture illustrated in Figure 9-13. Traditional RF subsystems include preamplifiers, LNAs, filters, RF distribution, and frequency translation and filtering stages that translate RF to usable IF signals. Such RF subsystems may comprise upward of 60% of the manufacturing cost of a radio node. These subsystems require large amounts of expensive touch-labor to assemble waveguide, coaxial cable,

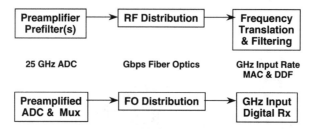

**Figure 9-13**   Digital RF replaces analog waveguide/coax with digital fiber.

**Dynamic Range (DNR) Improves With Oversampling**

| (Over)Sampling Rate MHz | ADCs SFDR | Cell Site Bandwidth MHz | Gain In DNR | Cell DNR | Subscriber Bandwidth kHz | Subscriber DNR |
|---|---|---|---|---|---|---|
| 6000 | 30 | 25 | 21 | 51 | 30 | 80 |

**Figure 9-14**   Digital RF could provide 80 dB of dynamic range.

| Sample Rate | Resolution | Manufacturer | Sample Rate | Resolution | Manufacturer |
|---|---|---|---|---|---|
| 4000 | 6 | Rockwell International | 750 | 8 | Signal Proc. Technology |
| 2000 | 8 | Hewlett-Packard | 3000 | 8 | ARPA (US DoD) |
| 70 | 10 | Pentek | 50 | 12 | Hughes Aircraft |
| 100 | 12 | ARPA (US DoD) | 24 | 14 | Hughes Aircraft |
| 10 | 18 | Hewlett-Packard | 6000 | 6.5 | TI |

**Figure 9-15**   High-performance ADCs have been demonstrated.

and other discrete components. The digital RF alternative, also shown in Figure 9-13, uses a preamplified ADC and multiplexer at the antenna to create a Gbps fiber optic signal [292]. Digital RF distribution via gigabit fiber optics weighs less and costs less per meter than RF distribution via coax or waveguide. In addition, fiber optics costs less to install and maintain than coax and waveguide. Lack of dynamic range, digital-RF's major shortfall at this time, can be enhanced using digital filtering techniques.

To see this, consider the use of a 6 GHz ADC [280] as illustrated in Figure 9-14. Although the RF ADC has a limited dynamic range, its high sampling rate oversamples the bandpass bandwidth of an AMPS signal. The oversampling gain increases the dynamic range through integrating digital filters as discussed above. The 25 MHz bandwidth of the cell site is 21 dB less than the RF sampling rate, yielding 51 dB of dynamic range within the cellular band. The subscriber bandwidth of 30 kHz offers an additional 29 dB of gain, yielding an aggregate DNR of 80 dB. Thus, the power in the digitally integrated baseband signals may range linearly over 80 dB. This results from the 30 dB of DNR at RF and the integrating digital filters that follow. Figure 9-15 shows some recent high-performance ADC products with sponsor or manufacturer. Any of the products with 2, 3, 4, and 6 GHz sampling rates could be the

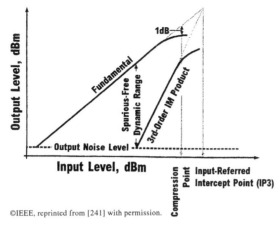

**Figure 9-16**   Nonlinearities characterized by compression and intercept points.

basis for digital RF. The Hypress supercooled ADC may accelerate progress towards digital RF [435].

## IV.  ADC DESIGN RULES

The ADC determines the quality of the digital signal available for subsequent digital signal processing. The parameters that most shape SDR applications are linearity and dynamic range. Dynamic range can be established empirically by the measurement of SNR. Several methods are available for making such measurements, and some are more relevant to SDR than others. Such measurements allow one to establish SNR and DNR budgets from the antenna through the product delivered to the user by matching SNR at each SDR interface. In addition to the related design rules, the parallelism of ADCs and DACs has a first-order impact on SDR architecture. This section provides an overview of these technical issues and the associated design rules.

### A.  Linearity

ADCs exhibit nonlinear behavior characterized by the compression and intercept points illustrated in Figure 9-16. Just like a mixer stage in a receiver, as the input power is increased, the signal output power increases. It reaches the output noise floor at a level defined by the equivalent thermal noise temperature of the device. Continued increase in the input power yields a continued increase in the output power of the fundamental.

The point at which the power of the third-order intermodulation product of the ADC is tangential to the output noise level determines the spurious-free dynamic range of the ADC. As the third-order product increases, its power eventually intersects the fundamental. This point is called the input-referenced

| Specification | Definition | Application |
|---|---|---|
| Signal to Noise Ratio (SNR) | Signal Power* Residual Error Power* | Desired signal bandwidth equals Nyquist bandwidth |
| Spurious Free Dynamic Range (SFDR) | Signal Power* Power of Largest Spurious Product* | Desired signal bandwidth is less than Nyquist bandwidth |
| Noise Power Ratio (NPR) | Spectral Density Outside of Notch Spectral Density Inside of Notch | Desired signal spectrum contains many narrow-band channels |
| Full Power Analog Input Bandwidth | Upper 3dB Frequency minus Lower 3dB Frequency | Bandpass sampling |

* Indicates use mean-squared value.

**Figure 9-17**   ADC specifications depend on applications.

third-order intercept point (IP3). The output power of the fundamental satu-
rates well before IP3, however. The point at which the output power of the
fundamental differs from the ideal output power by 1 dB is the 1dB compression point. If two tones are present in the input, the spurious-free dynamic
range (SFDR) is termed the two-tone SFDR (2-SFDR). The maximum two-
tone spur may appear when the tones are separated by an amount that is a
harmonic of the sampling rate, for example. Generally, it is difficult to predict
the combination of tones that yields the maximum spur. The search for tone
combinations is combinatorially explosive. Therefore, it is important that the
ADC supplier characterize the two-tone spurious-free dynamic range at critical points, including integer multiples and halftones of the sampling rates.
Tone separations at integer multiples and harmonics should also be tested.

## B. Measuring SNR

In addition to SFDR and 2-SFDR, SNR measurements are useful in specifying ADC performance. The SNR of an ADC is the ratio of signal power to
nonsignal power. Nonsignal power includes thermal noise and other residual
errors of the converter (Figure 9-17). This metric is most appropriate when the
bandwidth of the signal of interest is approximately the Nyquist bandwidth of
the ADC. Radar-matched filtering exemplifies such applications. Radar pulses
are typically wideband square waves. The matched-filter receiver is optimal for
the square wave when the bandwidth of the receiver is the Nyquist bandwidth
of the ADC.

The SFDR is a more appropriate metric when the bandwidth of the signal of
interest is much less than the Nyquist bandwidth of the ADC. First-generation
cellular base stations exemplify this situation. A 30 kHz AMPS carrier is
more than two orders of magnitude smaller in bandwidth than the 12.5 MHz

spectrum-allocation accessed by a cell-site ADC. The seven-cell frequen[c]y reuse pattern of first-generation systems reduces the maximum density of narrowband signals to 1 : 7, not considering interference from adjacent sites. GSM's 1 : 3 frequency reuse pattern is also well characterized by SFDR.

The density of narrowband carriers may be high, as in an analog FM-FDM with 100% channel occupancy or CDMA, with 1 : 1 frequency reuse. In such cases, the noise power ratio (NPR) is a more appropriate metric. The NPR is the ratio of the spectral density outside of a notch filter to the maximum spectral density inside the notch filter. The measurement is taken when the Nyquist bandwidth is flooded with white noise. The notch filter must be deeper than the noise power inside the notch so that the measurement defines the leakage that the ADC causes from the adjacent channels into the channel of interest. By sweeping the notch filter across the band, the point of maximum spectral density inside the notch is readily identified. When all but one channel are occupied, the total power that leaks into a single unoccupied channel defines the dynamic range available to the unoccupied channel. In addition, the full-power analog input bandwidth is relevant to bandpass sampling. Since bandpass sampling converts signals directly to baseband, the full-power bandwidth specifies the maximum RF spectrum that may be thus converted.

### C. Noise Floor Matching

One approach to the allocation of SNR and DNR through an SDR is to match the radio noise floor to the ADC input noise level. The noise power from a noise-limited receiver may be matched to the power of the ADCs LSB using [279]:

$$P_m = -174 \text{ dBm} + 10\log(W_a) + \text{NF}$$

where

$P_m$ is the noise power of a noise limited receiver,

$-174$ dBm is $kT_oB$, Boltzmann's constant, temperature, and unit bandwidth,

$W_a$ is the receiver (access) bandwidth in Hz, and

NF is the receiver noise figure in dB.

This creates a design rule that total system noise should be less than $\frac{1}{2}$ LSB. This rule applies to $kTB$ bands in upper UHF and SHF.

ADC error noise should always be less than $\frac{1}{2}$ LSB, but receiver noise need not be so matched. At first, it appears inefficient to sample noise power with many bits. But in the HF bands, for example, the noise consists of the additive effects of large numbers of distant emitters and natural phenomena (e.g., lightning strikes). Consequently, the differentiation of noise from subscriber signal depends on differentiating impulsive noise from a subscriber signal such as an HF-ALE probe. Since the noise background may shift by 10 dB in a few milliseconds at HF, the allocation of 2 or 3 bits of ADC dynamic range to

## of Merit: Sampling Rate x Full

ment for No/Cmax

2000 MHz no AGC: –210 dBc/Hz

n no AGC, 0 dBm preselector: –180 dBc/Hz

with AGC, cosite filter: –150 dBc/Hz

## Sampling    fs x Dynamic Range

- 100,000 (50 dB/Hz) x 16 bits (96 dBc) = –146 dBc/Hz
- 30,000,000 (73 dB/Hz) x 12 bits (72 dBc) = –145 dBc/Hz
- 500,000,000 (86 dB/Hz) x 8 bits (48 dBc) = –134 dBc/Hz
- 5,000,000,000 (96 dB/Hz) x 19 bits (114 dBc) = –210 dBc/Hz

## Contemporary

- Contemporary monolithic ADC: –135 to –146 dBc/Hz
- Hybrids: –150 to –160 dBc/Hz

**Figure 9-18**   SPEAKeasy I ADC study–defined figure of merit.

noise characterization may be appropriate. In interference-limited bands, one may apply many bits of ADC DNR to the characterization of the interference. This technique allows one to apply algorithms that subtract an idealized replica of the demodulated interferer from the passband, enhancing the subscriber's CIR. An appropriate formulation of a design rule for ADC DNR is to allocate sufficient bits to the noise or interference to support the needs of the CIR enhancement algorithms.

### D.  Figure of Merit

A figure of merit that characterizes the level of ADC technology is the product of sampling rate times the full-scale SFDR as summarized in Figure 9-18. "Net" SFDR reduces full-scale SFDR by 2 bits or about 10 dB. One bit assures that the noise power is less than $\frac{1}{2}$ LSB. The other bit assures that there is sufficient dynamic range for the input AGC to avoid saturation.

SPEAKeasy I sought to access from 2 MHz to 2 GHz in a single RF channel with a single ADC. This feat would require an ADC with at least a 5 GHz sampling rate and 19 bits of SFDR for a total figure of merit of $-210$ dBc/Hz. Contemporary ADCs reach the values shown in Figure 9-18. The widest practical bandwidth for SDR applications is about 65 Msps (25 MHz) at 12 to 14 bits of SFDR (72 to 84 dB full scale or about 74 dB net). This performance is marginal for cell site applications.

### E.  Technology Insertion

ADCs shape the SDR architecture. In the handset, there may be no ADC because the extremely low-power budgets drive one away from the high dissipated-power of wideband ADCs. The lower total power of direct conversion receivers is more appropriate. The nonlinear aspects of direct conversion

| Application | Specification | Performance Issues |
|---|---|---|
| Spread Spectrum | SNR<br>SFDR<br>NPR | SNR for small signals in an<br>environment with strong interference<br>SFDR for spatial filtering<br>NPR for interchannel crosstalk |
| Wideband Digital<br>Receivers | SFDR | Reliable detection of weak signals<br>in strong in-band interference |
| Mobile Cellular | SNR<br>SFDR<br>NPR | SNR and SFDR for wideband<br>channelized receivers<br>NPR for interchannel crosstalk |
| Spectrum Analysis | SNR<br>SFDR | SNR and SFDR for high-fidelity<br>instrumentation |

**Figure 9-19**   ADC technology insertion issues.

receivers do not degrade the reception of the single channel-per-band of a handset.

In base station and mobile-node designs where large numbers of subscribers are supported digitally, the power dissipation of a wideband ADC is acceptable. The technical advantages of the wideband ADC-based IF architecture then apply. Oversampling may be appropriate to enhance the effectiveness of multiple-access interference-cancellation algorithms.

Once the ADC sampling rate is established, one may employ either real or complex sampling. In many cases, the ADC may employ oversampling to enhance dynamic range (e.g., sigma-delta ADC). ADC technology continues to advance, driven by wireless and radar markets.

Figure 9-19 summarizes performance issues for ADC applications. SDR technology has the potential to reduce the cost and complexity of first-generation mobile cellular base stations. One approach is to reduce the number of analog IF and baseband channels to one, from the maximum number of active subscribers (e.g., approximately 100). Alternatively, a dual-mode cell site capable of both GSM and first-generation standards could have five parallel 5 MHz digital channels instead of the twenty-five 200 kHz channels of a conventional GSM cell site. ADCs supporting this approach require 80 to 90 dB of SFDR. As the cellular standards migrate from single-subscriber-per-carrier to TDMA and CDMA, the density of occupancy of the wider-bandwidth RF carriers increases. In order to support 3G deployment with an SDR architecture, one requires an ADC capable of supporting the despreading of 20 MHz W-CDMA channels.

This alternative is unlikely to gain wide commercial acceptance due to the computationally intensive nature of digital despreading. A more reasonable alternative would be to add additional circuits to a W-CDMA rake receiver that permits it to digitize GSM carriers. The 2G waveform may then be processed using the 3G baseband DSP. This would be a dual-mode despreader-digitizer chip. Such chips might also be used in military applications to despread wideband waveforms, and to extract narrowband waveforms for SDR processing.

### F. Architecture Implications

The system DNR must be sustained from antenna through the information stream delivered to the wireline network. Consequently, the software-radio systems engineer must allocate DNR to RF conversion, ADC, and digital filtering to maintain the required system DNR. RF conversion, in particular, may employ AGC, which increases total dynamic range. If AGCs are incorporated in RF, ADC, digital filtering, and demodulation, then the interactions among these stages is complex. Consequently, SDR architecture should include facilities for the allocation and management of DNR.

## V. EXERCISES

1. Does the conversion of an RF signal to digital form require the use of an ADC? If not, what are the alternative ways of obtaining a digital representation of the RF signal?

2. Define uniform sampling. Define Nyquist sampling. Define aliasing.

3. What QoS metric should one use to determine anti-aliasing requirements? (e.g., time delay? cell loss rate? other?)

4. Differentiate among the Nyquist ADC, the sigma-delta ADC, and the quadrature ADC. What are the advantage(s) of the latter two over the Nyquist ADC?

5. Consider the disaster-relief scenario. What ADCs are applicable to operation with police and rescue aircraft in the 108 to 400 MHz band? What is the sampling rate of an ideal ADC to access this entire band at once? What are the performance ramifications of implementing a system using an ADC with this data rate? (Hint: How much dynamic range can be provided to a 25 kHz narrowband user?) What alternative ADC approaches are possible? How will they effect the cost of the system?

6. Considering the situation of question 5, what operational constraints could be imposed on users of legacy radios in this band to operate with the disaster-relief system? How can this reduce the requirements on the overall ADC suite?

7. How much has ADC technology improved in the last eight years? What are the purely theoretical limits of ADC technology? What three technological features of ADC technology now limit progress toward the theoretical limits? How much should the technology improve over the next five years?

**8.** Recall the object-oriented approach to architecture analysis. Define an inheritance hierarchy for digital processing including ADCs, ASICs, FPGAs, DSPs, and GP processors. What slots are needed on the ADC object class? What criteria apply to selecting such slots? Differentiate slots needed for an industry-standard open architecture from an enterprise-oriented architecture intended to reflect proprietary product plans.

# 10 Digital Processing Tradeoffs

This chapter addresses digital hardware architectures for SDRs. A digital hardware design is a configuration of digital building blocks. These include ASICs, FPGAs, ADCs, DACs, digital interconnect, digital filters, DSPs, memory, bulk storage, I/O channels, and/or general-purpose processors. A digital hardware architecture may be characterized via a reference platform, the minimum set of characteristics necessary to define a consistent family of designs of SDR hardware. This chapter develops the core technical aspects of digital hardware architecture by considering the digital building blocks. These insights permit one to characterize the architecture tradeoffs. From those tradeoffs, one may derive a digital reference platform capable of embracing the necessary range of digital hardware designs. The chapter begins with an overview of digital processing metrics and then describes each of the digital building blocks from the perspective of its SDR architecture implications.

## I. METRICS

Processors deliver processing capacity to the radio software. The measurement of processing capacity is problematic. Candidate metrics for processing capacity are shown in Table 10-1. Each metric has strengths and limitations. One goal of architecture analysis is to define the relationship between these metrics and achievable performance of the SDR. The point of view employed is that one must predict the performance of an unimplemented software suite on an unimplemented hardware platform. One must then manage the computational demands of the software against the benchmarked capacities of the hardware as the product is implemented. Finally, one must determine whether an existing software personality is compatible with an existing hardware suite.

**TABLE 10-1   Processing Metrics**

| | |
|---|---|
| MIPS | Millions of Instructions per Second |
| MOPS | Millions of Operations per Second |
| MFLOPS | Millions of Floating Point Operations per Second |
| Whetstone | Supercomputing MFLOPS Benchmark |
| Dhrystone | Supercomputing MIPS Benchmark |
| SPECmark | SpecINT, SpecFP Instruction Mix Benchmarks (92 and 95) |

Consistent use of appropriate metrics assures that these tasks can be accomplished without unpleasant surprises.

*1. Differentiating the Metrics*    MIPS, MOPS, and MFLOPS are differentiated by logical scope. An operation (OP) is a logical transformation of the data in a designated element of hardware in one clock cycle. Processor architectures typically include hardware elements such as arithmetic and logic units (ALUs), multipliers, address generators, data caches, instruction caches, all operating in parallel at a synchronous clock rate. MOPS are obtained by multiplying the number of parallel hardware elements times the clock speed. If multiple operations are required to complete a machine instruction (e.g., a floating-point multiply), then

$$\text{MIPS} = \alpha\text{MOPS}, \qquad \alpha < 1$$

If, on the other hand, the processor has a *very long instruction word* (VLIW), $\alpha$ may be greater than 1. Suppose, for example, that a processor includes a "smart" cache, an ALU, and two parallel multiplier units with a 250 MHz system clock. One could characterize this processor in terms of the operations of the ALU and multipliers. If $\alpha = 1$, then it can deliver $250 \times 3$ or 750 MIPS, maximum. If the multipliers accomplish one 32-bit floating-point multiply on every clock cycle, then the processor provides 500 MFLOPS. Thus, one may characterize such a device as capable of a peak of 750 MIPS/500 MFLOPS. This notation means "750 MIPS of which up to 500 may be MFLOPS." Digital filtering takes more floating-point operations than, say, protocol processing, or FEC algorithms. If the SDR application uses a mix of 50% ALU and 50% floating point operations, then the processor delivers a maximum of $0.5 \times 250$ ALU operations plus $0.5 \times 500$ MFLOPS for a total of $125 + 250 = 375$ MIPS. Clearly, processing capacity realized is a function of instruction mix.

Alternatively, one could consider just the memory cache operations, attributing 250 MOPS of memory operations (MEOPS). If the memory cache operates fast enough so that the ALU and multipliers are never waiting for data or instructions, then the memory cache is not a bottleneck. If, however, there are states in which it must wait, then the potential 750 MIPS will not be realized. In this case, since MEOPS < MIPS, then the peak of 750 MIPS cannot be sustained beyond the capacity of the cache. For extremely computationally intensive operations like digital filtering, one may in fact realize the maximum capacity because all the data is resident in cache. Cache-misses then degrade performance.

*2. Processor-Memory Interplay*    The execution of an instruction requires accessing memory for instructions and data or accessing local registers. Processors that are more complex may fill a pipeline with instructions to be executed concurrently. Pipelines produce no results until the pipeline is full. Thereafter, pipelines produce a result per clock cycle. Newer architectures may employ set-associative cache coherency and other schemes to yield a higher number of instruction executions for a given clock speed. In addition, there is statistical structure to the application, which will determine whether the data and instruction necessary at the next step will be in the cache (cache hit) or not (cache miss). Statistical structure is also present in the mix of input/output, data movement in memory, logical (e.g., masking and finding patterns), and arithmetic needed by an application. Some applications like FFTs are very computationally intensive, requiring a high proportion of arithmetic instructions. Others such as supporting display windows require more copying of data from one part of memory to another. And support of virtual memory requires the copying of pages of physical memory to hard disk or other large-capacity primary storage. This gives the programmer the illusion that physical memory is relatively unlimited (e.g., 32 gigabytes) within a physically confined space of, say, 128 Mbytes of physical memory.

*3. Standard Benchmarks*    Consequently, MIPS are hard to define. Often, the popular literature attributes MIPS based on a nonstatistical transformation of MOPS into instructions that *could be executed in an ideal instruction mix*. This approach makes the chip look as fast as it possibly could be. Since most manufacturers do this, the SDR engineer learns that achievable performance on the given application will be significantly less than the nominal MIPS rating. The manufacturer's MIPS estimate is useful because it defines an upper bound to realizable performance. Most chips deliver 30 to 60% of such nominal MIPS as usable processing capacity in a realistic SDR mix.

In the 1970s, scientists and engineers concerned with quantifying the effectiveness of supercomputers developed the Whetstone, Dhrystone, and other benchmarks consisting of standard problem sets against which each new generation of supercomputer could be assessed. These benchmarks focused on the central processor unit (CPU) and on the match between the CPU and the memory architecture in keeping data available for the CPU. But they did not address many of the aspects of computing that became important to prospective buyers of workstations and PCs. The speed with which the display is updated is a key parameter of graphics applications, for example. The SPECmarks evolved during the 1990s to better address the concerns of the early-adopter buying public. Consequently, SPECmarks are informative but these also are not the ideal SDR metric in that they do not generally reflect the mix of instructions employed by SDR applications. Turletti [293], however, has benchmarked a complete GSM base station using SPECmarks, as discussed further below.

*4. SDR Benchmarks*    At this point, the reader may be expecting some new "SDR benchmark" to be presented as the ultimate weapon in choosing among new DSP chips. Unfortunately, one cannot define such a benchmark. First

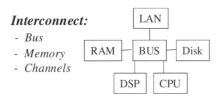

**Figure 10-1** Identify processing resources.

of all, the radio performance depends on the interaction among the ASICs, DSP, digital interconnect, memory, mass storage, and the data-use structure of the radio application. These interactions are more fully addressed in Chapter 13 on performance management. It is indeed possible to reliably estimate the performance that will be achieved on the never-before-implemented SDR application. But the way to do this is not to blindly rely on a benchmark. Instead, one must analyze the hardware and software architecture (using the tools described later). One may then accurately capture the functional and statistical structure of the interactions among hardware and software. This systems analysis proceeds in the following steps:

1. Identify the processing resources.
2. Characterize the processing capacity of each class of digital hardware.
3. Characterize the processing demands of the software objects.
4. Determine how the capacity of the hardware supports the processing demands of the software by mapping the software objects onto the significant hardware partitions.

There is a trap in identifying the hardware processor classes. ASICs and DSPs are easily identified as processing modules. But one must traverse each signal processing path through the system to identify buses, shared memory, disks, general-purpose CPUs, and any other component that is on the path from source to destination (outside the system). Each such path is a processing thread. Each such processor has its own processing demand and priority structure against which the needs of the thread will be met. One then abstracts the block diagram into a set of critical resources, as illustrated in Figure 10-1. This chapter begins the process of characterizing the capacity of SDR hardware. It summarizes the tradeoffs among classes of processor, functional architecture, and special instruction sets. Other source material describes how to program them for typical DSP applications [294]. The extensive literature available on the web pursues detailed aspects of processors further [295–298]. The popular press provides product highlights (e.g., [299–303]). This text, on the other hand, focuses on characterizing the processors with respect to the support of SDR applications. This is accomplished by the derivation of a digital processing platform model that complements the RF platform developed previously.

**TABLE 10-2  Mapping of Segments to Hardware Classes**

| Segment | Module | Typical Performance | Illustrative Manufacturers |
|---------|--------|---------------------|----------------------------|
| RF | RF/IF | HF, VHF, UHF | Watkins Johnson, Steinbrecher |
| IF | ADC | 1 to 70 Msa/sec | Analog Devices (AD), Pentek |
| IF | Digital Rx | 30.72 Mz Filters | Harris Semiconductor, Graychip, Sharp |
| IF | Memory | 64 MB at 40 MHz | Harris, TRW |
| IF, BB | DSP | $4 \times 400$ MFLOPS | TI, AMD, Intel, Mercury, AD, Sky |
| BS, SC | Bus Host | M68k, Pentium | Motorola, Force, Intel |
| SC | Workstation | 50–100 SPECmark 92 | Sun, HP, DEC, Intel |

Legend: BB = baseband; BS = bitstream; SC = source coding.

## II. HETEROGENEOUS MULTIPROCESSING HARDWARE

Segment boundaries among antennas, RF, IF, baseband, bitstream, and source segments defined in the earlier chapters make it easy to map multiband, multimode, multiuser SDR personalities to parallel, pipelined, heterogeneous multiprocessing hardware.

### A. Hardware Classes

Some design strategies map radio functions to affordable open-architecture COTS hardware. In one example, the VME or PCI chassis hosts the RF, IF, baseband, and bitstream segments as illustrated in Table 10-2. The workstation hosts the OA&M, systems management, or research tools including the user interface, development tools, networking, and source coding/decoding. Each module shown in the table represents a class of hardware. The parameters of these modules that assure that a software personality will work properly are defined in the digital processing reference platform.

Consider the roles of these hardware classes. The bus host serves as systems control processor. The DSPs support the real-time channel-processing stream, sometimes configured as one DSP per $N$ subscriber channels, where $N$ typically ranges from 1 to 16. The path from the ADC to the first filtering/decimation stage may use a dedicated point-to-point mezzanine interconnect such as DT Connect™, Data Translation. Customized FibreChannel and Transputer links have also been used. Synchronization of the block-by-block transfers across this bus with the point-by-point operations of the first filtering and decimation stage introduces inefficiencies that reduce throughput. Fan-out from IF processing to multiple baseband-processing DSPs also may be accomplished via a dedicated point-to-point path such as a mezzanine bus. Alternatively, an open-architecture high-data-rate bus might be used.

Instead of configuring such a heterogeneous multiprocessor at the board level, one might use a preconfigured system. Mercury™, for example, has offered a mix of SHARC 21060 [304] (Analog Devices), PowerPC RISC, and

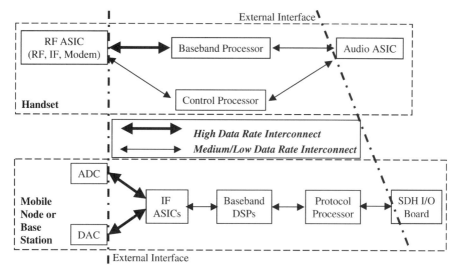

**Figure 10-2** Alternative processing modules and interconnect.

Intel i860 chips with Raceway interconnect [305–307]. Raceway I had nominally three paths at 160 MByte/sec interconnect capacity. Arrays of WE32's were used in AT&T's DSP-3 system. Arrays of i860's were available from Sky Computer [308], CSPI [309], and others. Of particular note is UNISYS' militarized TOUCHSTONE processor, which was also based on the i860 [310]. Although the i860 is no longer a supported Intel product, the architectures are illustrative.

System-on-a-chip level architectures also employ ASIC functions, shared memory, programmable logic arrays, and/or DSP cores. The physical packaging of these functions may be organized in point-to-point connections, buses, pipelines, or meshes. In each case, digital interconnect intervenes between functional building blocks and memory. Threads are traced from RF stimuli to analog and digital responses. Often in handsets, there is no ADC or DAC. Instead, RF ASICs perform channel modem functions to yield an alternative functional flow.

Figure 10-2 contrasts these complementary views of interconnect and other hardware classes. The boundaries of the digital flow are the external interface components. These include the display drivers, audio ASICs, and I/O boards that access the PSTN. Tradeoffs among internal interconnect are addressed in the next section.

## B. Digital Interconnect

Digital interconnect in systems-on-a-chip architectures is an emerging area. Over time, standards may emerge because of the need to integrate IP from a mix of suppliers on a single chip. Macroscale digital interconnect has a longer

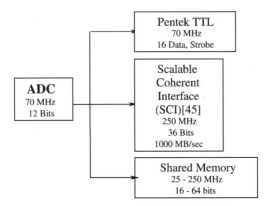

**Figure 10-3**   Illustrative classes of digital interconnect.

history of product evolution, and that is the focus of this discussion. These macroscale architectures may serve as precursors to future nanoscale on-chip interconnect.

Illustrative approaches to digital interconnect for open-architecture processing nodes are the dedicated interconnect, wideband bus, and shared memory (Figure 10-3).

*1. Dedicated Interconnect*   Dedicated interconnect is typically available from subsystem suppliers like Pentek [311]. Pentek provides 70 MHz ADC boards and Harris or Graychip digital receiver boards. Its MIX$^{TM}$ bus interconnects these cards efficiently. In addition, if the set of boards and interconnect does not work, the vendor resolves the issues. This approach leverages COTS products, with low cost and low risk. For applications with relatively small numbers of IF channels, it represents a solid engineering approach.

*2. Wideband Bus*   The next step up in technical sophistication is the wideband bus. The SCI bus [312], for example, has been used in supercomputer systems for several years. It is becoming available in turnkey formats including interface chip sets. The gigabyte-per-second capacity of the SCI bus could continue to increase with the underlying device technology. In addition, the design scales up easily to 8 × 140 MBps channels. The MIX bus, DT Connect, Raceway, SkyChannel [313], and other lower-capacity designs may be configured in parallel to attain high aggregate rates. This requires the hardware components to be appropriately partitioned. Other high-speed bus technologies are emerging, such as Vertical Laser at 115 GHz [314, 315].

*3. Shared Memory*   Shared memory can deliver the ultimate in interconnect bandwidth. Bulk memory of 64 MBytes easily has 16- to 64-bit paths. Scaling to 128 or 256 bits is feasible. Clock rates of 25 to 250 MHz are within reach. Thus, aggregate throughput of 3.2 to 64 gigabytes per second are becoming

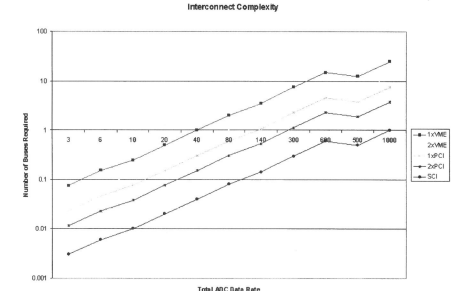

**Figure 10-4**   Wideband ADC rate versus interconnect complexity.

practicable with 4 ported shared memory. As the number of ports increases above 4, clock contention drives throughput down. But the switching, blocking and routing of data streams need not degrade throughput if the shared memory is supported by programmable direct memory access (DMA) or equivalent hardware. If only two very wideband input streams and two output streams need to be interconnected simultaneously (possibly out of a choice of 4 or 8), the shared memory architecture may be the best choice. Shared memory historically has the greatest performance, design/development cost, and risk of these approaches to digital interconnect.

***4. SDR Applications***   As illustrated in Figure 10-4, the ADC drives the digital interconnect architecture. Considering only the ADC's output data rate (in millions of bytes per second) and the nominal capacity of typical buses, the figure shows the relationship between aggregate ADC rate and number of buses. One 40 MByte per second VME bus can support a 3 MByte per second ADC stream using less than 1/10 of its capacity. As data rates increase, multiple buses and/or buses of greater bandwidth must be used to support the data rate. The 600 MByte per second ADC rate represents two bytes of resolution at 300 MHz, while the 500 MHz ADC has only one byte of resolution in this example. Interconnect efficiency is usually a function of the size of the data blocks being transferred. DMA transfers require setup, an overhead task that detracts from overall throughput. Buses also have bus-associated handshaking that constitutes overhead.

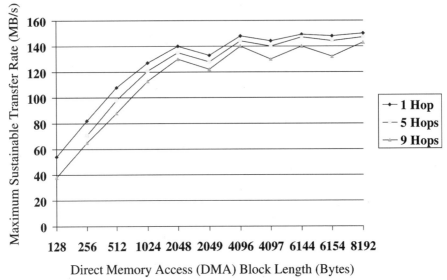

**Figure 10-5**   Interconnect efficiency.

Most buses experience low throughput for small block sizes. Mercury characterizes the performance of its products thoroughly. The maximum sustainable transfer rate of Raceway I varies as a function of DMA block length as illustrated in Figure 10-5. Although the peak rate of 160 MB/sec is not sustainable, it is approached with block sizes above 4096 bytes. Some devices (e.g., ADCs) may have short on-board buffers, constraining blocks to smaller sizes. In addition, algorithm constraints may proscribe smaller block sizes. A 0.5 ms GSM frame, digitized at 500 k samples per second, for example, may be processed with a block size of 250 samples (500 Bytes). If presented to Raceway in that format, the sustainable throughput would fall between 80 and 120 MB/sec as shown in the figure. If this is understood, then a constraint can be established between the algorithm and Raceway as an interconnect module. Constraint-management software can then assure that the capacity of the interconnect is not exceeded when instantiating a waveform into such hardware. In a more representative example, the entire bandwidth of the GSM allocation could be sampled at 50 M samples/sec, yielding 25.5 k samples per GSM frame, or over 50 kBytes. This data could be efficiently transferred to digital filter ASICs in 8 kByte blocks.

**5. Architecture Implications**   The physical format of digital interconnect (e.g., PCI, VME, etc.) need not be incorporated into an open-architecture standard for SDR. The less specific standard encourages competition and technology insertion by not unnecessarily constraining the implementations. On the other hand, such an architecture must recognize the fact that each class of physical interconnect entails implementation-specific constraints. An open architecture that supports multivendor product integration therefore must characterize those constraints to assure that software is installed on hardware with the necessary interconnect capabilities. Otherwise, interconnect capacity may become the system bottleneck that causes the node to fail or degrade unexpectedly.

An architecture standard used by a large enterprise to establish product migration paths, on the other hand, should specify the digital interconnect (e.g., PCI) and its migration from one physical realization to others as technology matures.

## III. APPLICATIONS-SPECIFIC INTEGRATED CIRCUITS (ASICs)

The next step in the digital flow from the ADC to the back-end processors in a base station is typically a pool of ASICs. ASICs particularly suited to software radios include digital filters, FEC, and hybrid analog-digital RF-transceiver modules with programmable capabilities. Waveform-specific ASICs are exhibiting increased programmability, mixing the capabilities of digital filters, FEC, and general-purpose processors for new classes of waveform (e.g., W-CDMA). In addition, DSP cores with custom on-chip capabilities are ASICs, but for clarity, they are addressed in the section on DSP architectures.

### A. Digital Filter ASICs

Base station architectures need digital frequency translation and filtering for hundreds of simultaneous users. Minimum distortion and nonlinearities are required in the base-station receiver architecture to meet near–far requirements. Digital-filter ASICs therefore extract weak signals in the presence of strong signals. The architecture for such ASICs is illustrated in Figure 10-6. The frequency and phase of the ASIC is set so that the complex multiply-accumulator chip (CMAC) translates the wideband input to a programmable baseband. For first-generation cellular applications, the decimating digital filters (DDFs) yielded 25 or 30 kHz narrowband voice channels through computationally intensive filtering.

Hogenaur realized that adjustment of the integrator, comb, and decimator parameters reduces aliasing as illustrated in Figure 10-7 [316]. Aliasing bands are folded into baseband at the complex sampling frequency. Choice of decimation rate and comb filter parameters places a deep null in the band of interest, achieving 90 dB of dynamic range using limited-precision integer arithmetic. The Hogenaur filter thus facilitated the efficient realization of the Harris ASICs. The product-line evolved to the HSP series now owned by Intersil.

Oh [317] has proposed the use of *interpolated second-order polynomials* as an improvement over the Hogenaur filter. Graychip has also been develop-

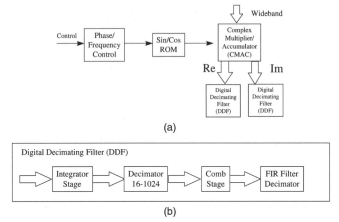

(a)

(b)

©Harris Corporation, reprinted with permission.

**Figure 10-6**   Digital filter ASIC architecture. (a) top-level ASIC architecture; (b) digital decimating filter architecture.

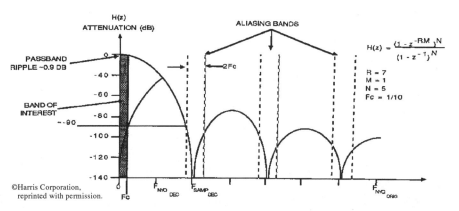

**Figure 10-7**   Hogenaur filter reduces aliasing.

ing filtering ASICs since the late 1980s. In addition, Zangi [318] describes a transmultiplexer architecture that yields all channels in a cell site using a Discrete Fourier Transform (DFT) stage. Zangi's transmultiplexer offers advantages for ASIC implementations. For example, with 1800 points per filter in a Digital AMPS application, $Fs = 34.02$ MHz, and decimation of 350, the DFT requires 1134 points for a complexity of 826 M multiplies per second. Such ASICs would simplify cell-site designs.

The complexity of frequency conversion and filtering is the first-order determinant of the digital signal processing demand of the IF segment. In a typical application, a 12.5 MHz mobile cellular band is sampled at 30.72 MHz (M samples per second). Frequency translation, filtering, and decimation requiring

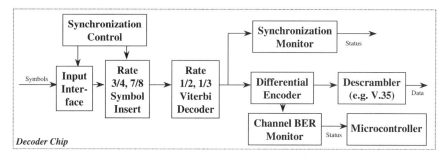

**Figure 10-8**　FEC ASIC architecture.

200 operations per sample equates to over 6000 MIPS of processing demand. Although GFLOPS microprocessors are now available, one may offload this computationally intensive demand to dedicated ASICs chips such as the Intersil or Gray digital receiver chip. Spreading and despreading of CDMA, also an IF processing function, creates demand that is proportional to the bandwidth of the spreading waveform (typically the chip rate) times the baseband signal bandwidth. This function also may be so computationally intensive that with current technology limitations, it is typically allocated to ASIC chips as well.

## B. Forward Error Control (FEC) ASICs

Forward error control ASICs offload computationally intensive aspects of error control coding onto dedicated hardware. As shown in Figure 10-8, the FEC decoder synchronizes the input bitstream, reverses symbol puncturing, and computes the majority logic best-estimate of the transmitted bits (e.g., using a Viterbi decoder). It then differentially decodes the stream and descrambles the resulting bitstream by adding the scrambling bitstream (e.g., V.35) synchronously to the output stream.

FEC operations are bit-serial, usually involving register lengths that are prime numbers like 11, 13, 17, etc. These bits operations do not pack and unpack efficiently into 8-, 16-, and 32-bit arithmetic offered by the typical DSP. Consequently, there is significant bit-masking and other nonessential steps to implement the FEC functions. When implemented in a conventional DSP, the FEC operations consume considerable power. An FEC chip, on the other hand, consists of exactly the right bitstream structure (e.g., an 11-bit register), with only those interconnects among bits required by the FEC algorithm. As a result, FEC ASICs dissipate the absolute minimum power for a given data rate. Some FEC chips are programmable across a range of FEC functions, without much loss of power efficiency. The issue of power efficiency is central to tradeoffs in the handsets where power is at a premium.

Turbocodes have been shown to improve error protection by interleaving two systematic concatenated codes. Since fading is generally correlated, it can have an impact on the success of turbocoding in CDMA systems [319]. The

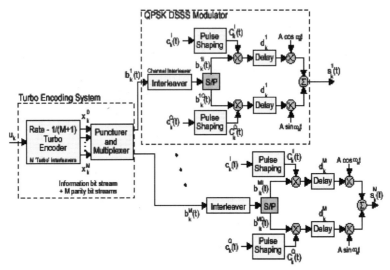

**Figure 10-9**  Turbocoded CDMA system.

complexity of the turbo encoding subsystem is such that it is a strong candidate for ASIC or FPGA implementation. In addition, the interleaver, pulse shaping, delay, and combining circuits may be included on the same FPGA or ASIC. The decoder has a somewhat higher level of complexity, as illustrated in Figure 10-10.

## C. Transceiver ASICs

Alcatel, Siemens, Motorola, Ericsson, Nokia and others employ direct conversion transceiver ASICs in handsets as presented in Chapter 8. Other RF ASICs integrate dual-mode amplifiers, matching circuits, and related RF and RF conversion modules in a single package. GaAs has been a popular device technology for these circuits, but RF CMOS is making progress for handset applications. Handset ASICs may nonlinearly distort the RF, provided the subscriber's signal is not distorted beyond recovery. Some digital ASICs include RF/IF functions.

The STEL-2000, for example, is a highly programmable ASIC with functions similar to the digital filter ASICs, but with additional transceiver functions as illustrated in Figure 10-11. The numerically controlled oscillator (NCO) and clock feed the CPSK modulator. The NCO's I&Q (SIN, COS) channels provide the reference signal for the down conversion stage. Differential encoding and decoding pairs are provided. The receiver clock generator, PN code generator, matched filter, power detector, and symbol tracking processor may function as a despreader. Control and interface logic permit an external microprocessor to integrate this ASIC into a spread-spectrum class

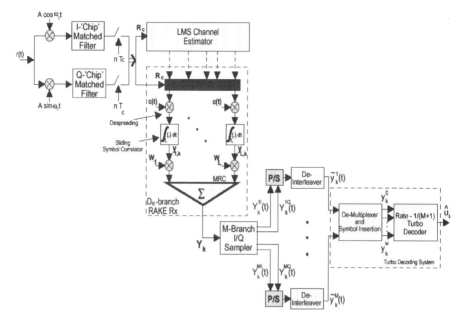

**Figure 10-10**   Turbocoded CDMA receiver archiecture.

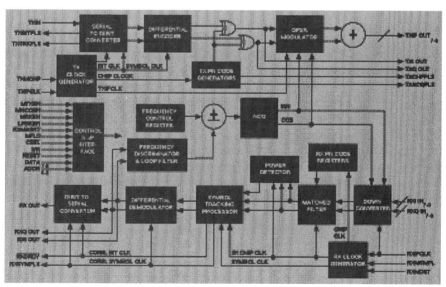

**Figure 10-11**   STEL-2000A block diagram.

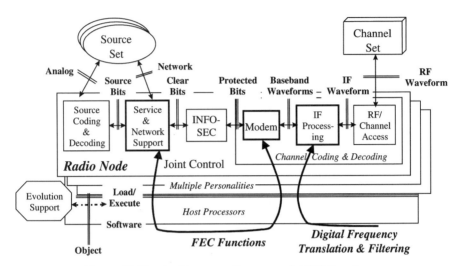

**Figure 10-12**   Architecture alignment of ASIC functions.

SDR. The Bitspreader-2000 SDR transceiver [320] integrates the STEL-2000, a synthesized sampling clock generator, and an FEC ASIC under the control of an 89C51 microcontroller. As gate densities continue to increase, such ASIC functions may be integrated around a DSP-core for volume production.

## D. Architecture Implications

Digital filtering ASICs contribute to both base-station and handset architectures. Since there is continuing research in this area, one can expect further development of associated intellectual property and related products. The same applies to FEC. The advantages of ASIC implementations include reduced size, weight, and power of the target devices. In addition, these devices reduce parts count, reducing manufacturing costs proportionally.

These ASICs represent a category of optimization of SDR products that must be addressed in SDR architecture. One approach is to encapsulate such devices within the modem entity. This blurs the distinction between modem and IF processing. FEC may be encapsulated within some modems, but digital filter ASICs are better represented as digital IF processing since they perform IF-to-baseband frequency translation and related filtering. This alignment of ASIC functions to architecture-level functions is illustrated in Figure 10-12. Clearly, the Modem function has been generalized to include some FEC aspects of bitstream processing. In addition, the service and network support function includes many aspects of protocol stack processing besides FEC.

If an SDR architecture is to facilitate the integration of such power-efficient devices as ASICs, then the architecture has to include a mechanism for passing control and data to these facilities. Efficient access from architecture-level

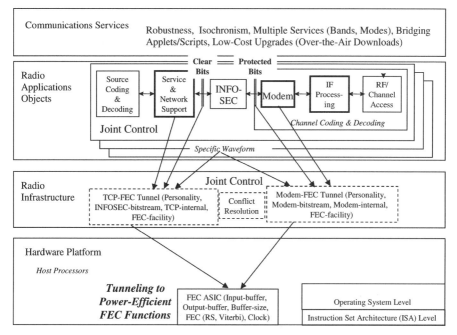

**Figure 10-13**   Tunneling provides open-architecture access to proprietary IP.

functions to component-level building blocks may be called *tunneling*. It requires the refinement of the layered virtual machine architecture illustrated in Figure 10-13.

Several aspects of the tunneling facility need to be pointed out. These include the definition of interface points, the use of the tunneled component, the identification of constraints, and the resolution of conflicts. These aspects are supported by Tunnel( ) functions that tell the radio infrastructure about the interfaces to the applications objects and the capabilities of the ASIC objects as follows.

First, the tunneling points are anchored to architecture-level functional components by the ⟨function⟩⟨ASIC⟩Tunnel( ) expression. In this format, the name of the tunnel includes the function requesting the tunneling service and the name of the object that is the target of the tunnel. In the figure, both the Modem and the TCP protocol tunnel to the FEC ASIC. The interface from the Modem function is specified independently of the interface from the protocol stack to the FEC ASIC. If the interface to the ASIC class conforms to the architecture-level interfaces, then the resource-management function of the radio infrastructure has the information it needs to establish streams between the software objects and the ASIC.

This may not always be the case. In the example, the TCP software for a specific waveform personality may use the ASIC to provide some additional block coding. The Modem function may apply further FEC, such as a convolutional encoder, to the bits prior to converting them to channel symbols. If the INFOSEC function is null, then the clear-bits and protected-bits interfaces are identical. Furthermore, these interfaces may be implemented inside of the FEC ASIC. Although the interface is known to the resource manager, tunneling makes it impossible for other software to access this interface unless the FEC ASIC provides access to its clear-bits interface. In order for the ASIC-enhanced personality to be compliant with the architecture, it would have to provide access to that radio-application level interface. Personalities with noncompliant interfaces may be acceptable for some reason (e.g., because it supplies the highest data rate the implementation technology will allow, within some power constraint). Flagging personalities as noncompliant allows third-party software suppliers to know that only a limited subset of standard streams are available in that SDR environment.

If INFOSEC is not null, then TCP bits first may be scrambled and then passed to the modem to add error-protecting redundancy. The FEC ASIC could allow buffers to be used independently by networking and modem functions via its FEC( ) method. In this case, the radio-applications-level software objects execute FEC(buffer) to block-encode the data in the FEC's input buffer. The driver associated with the ASIC converts this call to a signal on an appropriate hardware control line. This is similar to the Hayes AT language for modems. Instead of expressing commands as a sequence of ASCII strings, commands are expressed by passing a message to the FEC ASIC to execute one of its public methods.

An FEC ASIC has some maximum input buffer size and maximum throughput or FEC conversion rate. These parameters define constraints under which tunneling will yield specific levels of performance. Such constraints are typical for optimized devices. In order for tunneling to be effective, these constraints need to be represented in the architecture for the use of a constraint-manager. Architecture compliance, then, should entail a design rule that "constraints on ASICs are defined." The constraint manager must be capable of processing these constraints. Constraint-violation responses should be defined and the users should have an easy way of understanding the error conditions. Internal constraints might include clocking the bits through the ASIC at a certain data rate. Other constraints may include a limit on the number of input-output buffer pairs. There may be a limit on the size of a specific input buffer (e.g., Reed–Solomon coding occurs on blocks of specific integer multiples), or on initialization (e.g., convolutional codes remember the internal states of the shift register). All of the constraints may be enforced without user intervention if the computational demands of the radio application are compatible with the resources of the hardware platform. But the satisfaction of such constraints is only the first step in addressing potential conflicts between the personality and the platform.

Some INFOSEC design rules, for example, preclude the use of one ASIC to process both the clear bits and the protected bits. If so, then the FEC ASIC

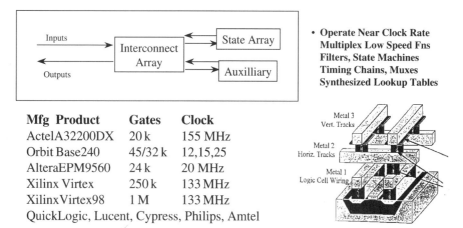

| Mfg Product | Gates | Clock |
|---|---|---|
| ActelA32200DX | 20 k | 155 MHz |
| Orbit Base240 | 45/32 k | 12,15,25 |
| AlteraEPM9560 | 24 k | 20 MHz |
| Xilinx Virtex | 250 k | 133 MHz |
| XilinxVirtex98 | 1 M | 133 MHz |

QuickLogic, Lucent, Cypress, Philips, Amtel

**Figure 10-14**  Overview of FPGA devices.

violates an architecture design rule. This conflict should be detected at the time the hardware platform is initialized, so that such INFOSEC is not instantiated. This design-rule conflict has to be detected during waveform instantiation before operational use. As a minimum, the resource manager should identify the design-rule conflict to the user (in user terms) so that the user may decide not to use the mode, or to use it in an appropriate way.

## IV.  FIELD-PROGRAMMABLE GATE ARRAYS (FPGAs)

A compromise between the cost of a unique ASIC and the high power dissipation per function of DSPs is the FPGA.

### A.  Introduction to FPGAs

FPGAs are high-speed configurable logic circuits packaged as high-density commodity chips (Figure 10-14). The physical and logical layout is designed for rapid implementation of state machines and sequential logic. A state machine is an automaton that can process a finite state language [321]. State machines consist of a memory that represents a finite number of states, an ability to detect and parse inputs, a set of state transition maps, and an ability to generate outputs as a function of state transition [322]. A state transition map is a correspondence between a current state and an input that determines the next state. The output map selects an appropriate output or side effect to be produced during a state transition.

FPGAs therefore are organized into sequential logic that detects the inputs and generates the outputs plus lookup tables for state memories and transition maps. Combinatorial "glue" logic such as buffer registers, decoders, and multiplexers may be implemented efficiently in FPGAs. Most commercial

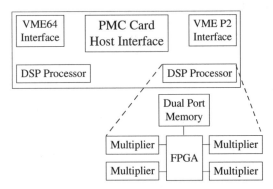

**Figure 10-15**   Reconfigurable FPGA processor.


chips also include ancillary timer circuits [323, 324]. FPGAs may be used for complex processes such as convolution, correlation [325], and filtering [326]. Because of their flexibility and ability to reduce parts count, FPGAs have attracted continued investment and research interest [327]. Consequently, clock rates continue to increase and gate densities per chip continue to increase as illustrated in Figure 10-14 [328].


## B. Reconfigurable Hardware Platforms

FPGAs provide a strong platform for specialized digital signal processing tasks for SDRs. They have been used with success in wireless research environments [329]. C. Dick, for example, describes FPGA-based FIR filters, extended precision arithmetic, and a CORDIC carrier recovery loop for a run-time reconfigurable digital receiver [330].

S. Srikanteswara et al. [331] implemented a single-user CDMA receiver with LMS equalizer using FPGAs. Their platform was a Giga Ops G900 board containing Xilinx XC4028EX processors operating at 1.25 MHz. The digital IF was converted and filtered by a Harris digital filter ASIC. The Giga Ops board then implemented a packet-driven, software-defined CDMA demodulator and equalizer. In this research, the packet headers define the hardware personality used in processing the packet payloads. The packet format defines four layers of abstraction. These are the application layer, the soft radio interface layer, configuration layer, and processor layer. The current research addresses the synthesis and testing of these four layers on a wormhole architecture [332].

Reeves et al. [333] describe a reconfigurable hardware accelerator. Their processor includes a high-gate-count FPGA, four floating point multipliers, a dual-port memory for signal streams, static coefficient memories, and a port for a configuration bitstream (Figure 10-15). The processing logic can be reconfigured in 100 microseconds.

The dual reconfigurable processor board includes two such processors, IO and a PMC mezzanine card. The data memory consists of 256 kBytes of dual-port static RAM with simultaneous access by the processor and the external input/output stream. This memory is optionally organized as either:

1. 16-bit real integers (128 k deep),
2. 16-bit complex integers (64 k deep),
3. 32-bit real floating point numbers (64 k deep), or
4. 32-bit complex floating point numbers (32 k deep).

The memory access controller's personality is customized to each application through a dedicated memory access/IO processor FPGA. In addition, the IO processing accommodates VME64, PCI mezzanine card (PMC), VME P2 connector, and a user-configurable front panel port. A radix-2 fast Fourier transform (FFT) with eight independent signal streams was implemented on the two processors. Four real multipliers and six real adders were required for the complex butterfly operation. The real multiplies are performed in the dedicated multipliers while the six real adders are mapped to the flexible FPGA core. At a clock speed of 36 MHz with ten such floating point operations in parallel, four multiplies and two adds yields 360 MFLOPS of 16-bit fixed point processing capability per reconfigurable processor. This is a 720 MFLOPS peak capacity for the full board. This results in a 68-microsecond average benchmark for a 1024-point FFT. Since the input and output occur in parallel, double buffering the signal stream in the dual-ported memory, this throughput is sustainable. By comparison, it would take approximately fifty-two TMS320C40's in parallel operating at 50 MHz on 16 VME boards to do the same thing. Alternatively, one or two C62s can be configured for the same throughput.

To probe the FPGA-DSP tradeoff further, consider Reeves' implementation of a *lattice filter*. The filter requires 12 stages with eight lattices per stage, but the data rate is reduced by $\frac{1}{2}$ between successive stages. Each lattice requires two multipliers and two adders, so two such stages can be implemented in parallel in each of the two processors (4 : 1 parallelism potential). Since all but the first stage is decimated by multiples of $\frac{1}{2}$, the last seven stages can be hosted on a single pair of multiply-accumulator resources in a processor. With an input rate of 7 Mword/sec ($\times 16$ bits per word) and a total of 112 million multiplies per second total, the seven subsequent lattice stages are reconfigured on the fly (with 100 usec per reconfiguration). Continuous throughput is nominally 120 MFLOPS. In this case, a Quad C40 board could implement the lattice filter in the same board area, consuming more power.

## C. FPGA-DSP Architecture Tradeoffs

These comparisons between FPGAs and DSPs support the assertion that FPGAs are more computationally efficient than DSPs. This may be true for specific algorithms like the FFT, convolution, digital filtering, and FEC. Such algorithms have what may be called limited data-scope. The data needed by the FFT is limited to the data points in the input block. The data needed for filtering is the set of delay taps and weights. Convolution may be accomplished either on signal blocks (using FFTs), or on streams using the pole-zero formulation of the transfer-function. In such cases, the scope of the data is extremely limited. The topology of such algorithms has been studied [334]. Algorithms with limited scope have ISA-like topologies. Digital filters, FFTs, etc. are topologically like hardware instruction sets, and therefore are amenable to FPGA implementations.

A database algorithm, on the other hand, accesses any data in bulk storage. Thus, an FPGA configured for database retrieval wastes most of its time waiting for the disk to return the requested data. This reflects the behavior of an algorithm with moderate data scope. To reconfigure the FPGA to do other tasks while waiting for the disk is possible. This approach runs the risk that the processor will be configured the wrong way when the disk returns the data. A radio control algorithm, for example, could access any data in the system. The uncertainty about the arrival of control tasks puts a premium on processing interrupts efficiently. General-purpose processors include interrupt hardware stacks that may be more efficient at handling these events than the FPGA. Therefore, as the algorithm mix expands to include functions with increasingly broad data scope, the FPGA's advantage is diminished. System control and protocol stack processing, for example, could force the repeated reconfiguration of the FPGA. Topologically, these algorithms have long sequences of different data-instruction combinations before repeating. Such algorithms therefore place high reconfiguration demands on FPGAs. A research breakthrough seems necessary to change this situation.

In a limited-scope digital radio application, one could reconfigure the processor to filter the signal, then again to demodulate it and then again to perform FEC. An FPGA should perform well within such a limited scope of processing throughput versus functionality. As the complexity of the total algorithm suite increases, the amount of hand-tuning required to pack incremental functionality into the FPGA goes up significantly. Rapid reconfiguration provides additional headroom as it were for algorithm growth, but the algorithms may outgrow the FPGAs. When this happens, the entire hardware design may need to be redefined—not a graceful evolution path.

## D. Table-Driven Signal Generation

FPGAs are also applicable to the generation of digitally preemphasized waveforms. The sampled waveform is typically stored in random access memory either off or on chip. The FPGA implements a state machine that reads the sampled waveform according to a precise clock sequence, delivering the sampled waveform to a DAC. The state machine may contain logic that modifies the contents of the waveform-memory in a data-driven way. Feher [238], for

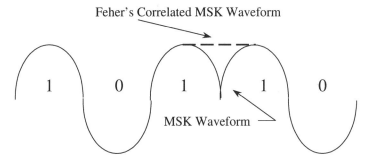

**Figure 10-16**   FPGA modification of waveform conditioned on data stream.

example, holds patents on adjusting the sampled waveform to the bitstream correlation. In particular, his conditional waveform bridges similar signal states to minimize the amplitude and/or phase discontinuities that would otherwise result. The concept is illustrated in Figure 10-16. Feher's adjustment to the envelope of a transmitted waveform may be made at baseband or at an IF. An IF lookup table would adjust the amplitude and/or phase of the current symbol based on a sequence of symbols, yielding the corrected time-domain waveform with sharper Adjacent Channel Interference (ACI) control. For IF sampling rates, such lookup tables may require the speed of an FPGA or ASIC.

### E.  Evolutionary Design of FPGA Functions

In the early 1990s, Hugo deGaris [336, 337] introduced evolvable hardware. Evolvable hardware ("E-Hard") controls the definition of FPGA functional blocks with genetic algorithms. FPGA-defining bitstrings are treated as artificial chromosomes by a *genetic algorithm* (GA) [338, 339]. DeGaris et al. in Kyoto (Advanced Telecommunications Research Institute) developed robotic-control system designs that evolve on their own using the Xilinx XC6264 family of FPGAs [340]. Although the early research envisioned communications applications, these have yet to be reported in the literature.

The implications of such an approach are worth considering for the future. E-Hard could permit a pool of alternate modem personalities to be represented by different FPGA bitstring-chromosomes. Modem performance could then be measured on incoming data, and the worse performers could be pruned from the population. After sufficient training, the survivors could be robust and nearly optimal. One advantage of this approach is that it substitutes machine learning for labor-intensive design, potentially saving time and cost. One disadvantage is the large number of data sets that must be processed by a large community of competing modems before the winner(s) emerge. If that disadvantage can be overcome, one would be faced with a high-performance modem that is an opaque black box. There would be no associated source code and no documentation per se. If the GA were also included in the modem, this modem could evolve to address the specific communications environment. In other words, what is encapsulated as a modem object could have complex, adaptive internal structure. It might consist of a GA and a small population of alternate modems, pruning and evolving during operations. It might have unknown desired and undesired properties, to be discovered during operations. A similar approach may be applied to protocol evolution [67, 361].

One architecture implication concerns the inclusion of such self-adaptive systems in SDR architecture. SDR downloads can be frozen and type-certified prior to use, but how is one to certify the type-acceptance of a modem that can adapt its performance as a function of its environment? At present this question is just entering the public debate about SDR type-certification [436]. Researchers might consider constraints under which the products of evolutionary processes can be constrained to a chromosome-space within which any defined behavior is type-certifiable. This is an open research question at this time.

## F. Architecture Implications

FPGAs have grown from hundreds of kilo-gates into the million-gate range. This increases the applicability of this technology to SDR. Low-power FPGAs are needed in handsets for reduced parts count. They may grow to assume an increasingly larger share of the processing within the scope of those tasks that have appropriate data-access topologies. A well-conceived software-radio architecture therefore must support the insertion of FPGA technology as opportunities emerge.

FPGAs may be accessed via tunneling as described above. In addition, however, SDR architecture must include FPGAs with multiple personalities. Srikanteswara et al., [331, 332] envision such soft radios as structured into the four layers illustrated in Figure 10-17. The processing layer contains the reconfigurable FPGA hardware. The configuration layer translates processing needs expressed in the packet headers into configuration commands. These are obtained from the soft radio interface layer. This layer also returns processed data and error messages to the applications layer. This uppermost layer controls the architecture parameters, provides data from the ADC, and delivers results to the host processor, user, etc. This stack forms a subset of the layered virtual machine architecture as illustrated in Figure 10-18.

In this model of architecture, the radio applications layer requests services from the lower layers. A CORBA ORB may be used in the infrastructure middleware to dispatch processes to processors. CORBA IDL is a suitable mechanism for translating an applications-layer request to filter a digital input stream into a request to the FPGA-specific soft radio interface (SRI) layer. The SRI then behaves according to the packet-driven layering described by Srikanteswara et al. The SRI translates the Digital_Filter( ) call into a bit pattern

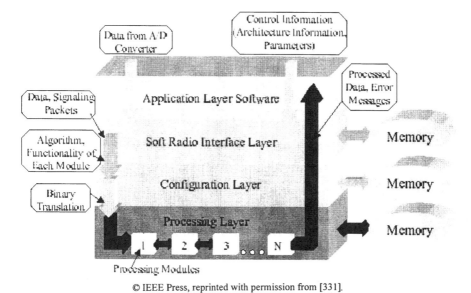

© IEEE Press, reprinted with permission from [331].

**Figure 10-17**  Layered architecture for FPGA-based "soft radios."

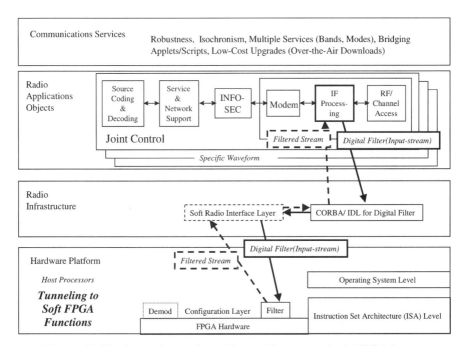

**Figure 10-18**  Layered virtual machine architecture embeds FPGA layers.

that reconfigures the FPGA hardware into its filter personality. At that instant, its Demod personality is not available (dotted in Figure 10-18). The filtered stream is then passed back up the layers to the radio applications object that initiated the request (e.g., by pointer manipulation). In a high-performance system, most of this layering is accomplished by activating pointers set up at initialization time, minimizing run-time overhead. In addition, tunneling constraints apply, as with ASICs. Conflicts for FPGA personalities also must be detected and resolved.


## V. DSP ARCHITECTURES

DSP chips are designed for efficient execution of computationally intense functions like filtering and fast Fourier transforms (FFTs). The early DSP chips such as Texas Instruments' (TI) TMS320C30 emphasized raw multiply-add computational power. Subsequent designs included greater on-chip parallelism and more capable input/output for multiprocessing. This section begins with a discussion of DSP "cores," DSP instruction sets embedded into wireless ASICs. The discussion then follows the evolution of DSP chips, emphasizing the ways in which the chips support the needs of SDR.


### A. DSP Cores for Wireless

The number of gates per chip is approaching 1 to 5 million. The opportunities for combining a standard digital signal processor with applications specific on-chip capabilities have led to a series of DSP ASICs for the wireless marketplace. These include, for example, the Motorola DSP56304, built on the DSP56300 core and illustrated in Figure 10-19. It includes additional interfaces and memory around the 80 MHz 24-bit DSP56300 core. With about 90 k × 24-bit words of on-chip RAM and ROM, the processor has sufficient capability for handset and fine-grain parallel processing applications. The SCI in the figure is a Serial Communications Interface. Enhanced synchronous serial interfaces (ESSI), and H108 host interfaces provide flexible IO. The 3V device was offered in a 144-pin thin quad flat pack, with commodity pricing on the order of $20.00 each in 1999.

The 24-bit integer arithmetic provides over 120 dB of arithmetic dynamic range, sufficient for wireless applications. Its modem applications require 50 to 100 kBytes of memory. This includes RF modem and voice-channel codec applications. This DSP is therefore useful without off-chip memory in many applications. In considering alternative DSP cores, one must characterize deliverable processing power versus notional values. With an 80 MHz clock and parallel multiply accumulator functions, this chip may reach over 100 MIPS instantaneously. But the sustainable throughput depends strongly on the mix of bus accesses and on-chip versus off-chip accesses required by the application [341].

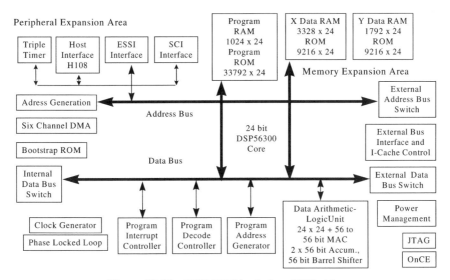

**Figure 10-19**    DSP 56304 wireless DSP chip.

Other popular DSP cores include the ADSP 21xx SHARC (16-bit, 15–20 ns cycle time). The TI TMS320/C54x is also a 16-bit processor with a 16.6 to 20 ns cycle time [342, 298]. Motorola's 56300 has 24-bit arithmetic and shorter (12–15 ns) cycle time. In addition, Star*Core [343], Advanced RISC Machines, the Oak/DSP Group, and others offer DSP cores. Integrating DSP cores with other proprietary circuitry raises intellectual property (IP) issues [344].

## B. Basic DSP: The TMS320C30

The Texas Instruments (TI) TMS320C30 (Figure 10-20) may be credited with starting the DSP chip COTS marketplace. AT&T developed the Western Electric (WE) DSPs contemporaneously with TI, but TI fostered commercial marketplace for such chips. These early devices included both instruction set extensions and data format extensions over contemporary microprocessors. These extensions maximize throughput by saving instruction cycles and by increasing the functionality of a sequence of cycles.

Instruction set extensions included register, direct, indirect, and immediate addressing modes. The immediate mode provides operands in the instruction itself, saving memory access. Bit-reversed addressing allows one to reorder the product of an in-place FFT without performing a separate address-arithmetic loop to move and reorder the data, saving on the order of $N$ instructions for an $N$-point FFT. Circular (Modulo $N$) addressing avoids the testing of buffer-end and resetting the buffer pointer to zero; this is all accomplished in hardware. Rapid context switching is available from hard-

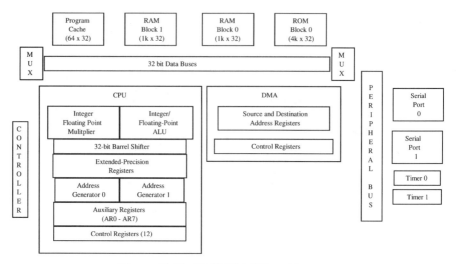

**Figure 10-20**   TMS 320 C30 architecture.

ware push/pop of register stacks. Hardware semaphores ensure glitch-free sharing of resources such as data buffers. And Repeat-N (No Loop Overhead) loops avoid software cycles by offloading the loop control logic to hardware. Of course, parallel multiply-accumulate stages are central to efficient digital filtering. Load, Multiply, Add, Increment, and Iterate instructions may be combined with parallel Multiply-Add to further optimize filtering.

Data format extensions include fix, float, double-, and, in some cases, triple-precision integer arithmetic. Precise timing may be derived from high-speed clocks with clock periods on the order of nanoseconds. Keeping track of, for example, channel symbol timing over long transmissions can cause double-precision integers to overflow, mandating triple-precision integer arithmetic for some algorithms.

## C. Increasing Interconnect Capacity: The C40 and SHARC

Although the C30 enjoyed immense popularity, it had serious shortcomings, the most significant of which was its paucity of independent input/output ports. The C40 and Analog Devices SHARC, on the other hand, were designed for multiprocessing. The C40 offered six 20 MB/s I/O Ports. Global shared and local memory enabled the partitioning of algorithms for efficient use of the separate local and global buses. One could, for example, filter signals on the local bus while moving new data on the global bus. As these products have increased in popularity, direct-connect products have emerged, including ADCs, DACs, FFT chips [345], and I/O spooler chips. The C40's rating of peak 250 M operations/sec and peak 50 MFLOPS led the TI family until the introduction of the C62 and C67 chips in 1998. Board-level products include Quad and Octal VME Boards, Quad PCI boards and daughter modules.

## D. Size-Power Tradeoffs: The c54x, and Motorola Chips

The TI TMS320C54x series of low-power chips were designed for cell phone and similar applications. Its 40-bit adder and two 40-bit accumulators support parallel operations. It can normalize and exponentiate in a single-cycle. Four internal buses and dual address generators enable multiple operand operations and reduce memory bottlenecks. The 40-bit ALU in dual 16-bit configuration supports dual one-cycle operations. A $17 \times 17$ multiplier allows 16-bit signed and unsigned multiplication. A version of the C54 provides baseband DSP, for example, in Ericsson ASM phones.

## E. Toward Greater Parallelism: The c80 and c6xx

TI's C80 was announced for the set-top video market. Its quad on-chip DSPs are managed by a RISC microcontroller. A full crossbar-switch links twenty-six 2 kByte shared RAMs. On-board clocks, video controller, joint test port (JTAG), and 64-bit MVP bus transfer controller round out the on-chip capabilities. The chip was, however, expensive. As a general-purpose video DSP, it was used in the development of HDTV algorithms. But any general-purpose chip is less efficient in the use of chip area than a dedicated chip. Thus, ASICs won the design competitions for set-top boxes, in some cases coupled with general-purpose microcontrollers.

The C80 was in some sense a precursor to C6x Series. The TMS320C62 advertised 1600 MIPS (200 MHz clock) with 512 k program and 512 k data memory (32-bit) plus a very long instruction word (VLIW) architecture. Its four DMA channels and two serial ports provide adequate I/O, particularly when coupled with shared memory. With 8 ALUs, the C62 compiler is the key to achieving the advertised performance. The C67 advertises 1000 FLOPS using the same 32-bit VLIW architecture but with 6 floating point ALUs and two fixed-point processors. This chip is essentially a C62 in which 6 of the 8 ALUs have floating point capability.

## F. Summary and Comparison of Contemporary Chips

Table 10-3 provides additional data on contemporary chips. The ADSP HARC and TI processors are particularly competitive for SDR designs, for example in base station. Analog Devices has made significant investments in wireless software including GSM base station, for example. Analog Devices offers this software with its chips and related support packages.

Motorola's 56 k series of DSPs find application in voice channel and radio products. These chips may provide the DSP core for a specialized chip. Alternatively, they may be packaged in board-level packages. AT&T's DSP products were originally developed by Bell Telephone Laboratories and man-

**TABLE 10-3  Comparison of Contemporary DSP Chips**

| Part | Mfgr | Speed @ Clock | On-chip Memory | Data Width | Dedicated I/O | Price[b] | Comments |
|---|---|---|---|---|---|---|---|
| ADSP21062 SHARC | ADI | 40 MFLOP 120 MIPS @ 40 MHz clock | 32 Kwords RAM | 32 bit float | 240 MB/s, 2 × 40 serial 6 × 40 MB/s | $120 | 2nd-generation multiproc 40 MHz clock |
| TMS320C32 | TI | 40 MFLOP 20 MIPS 40 MHz | 512 word RAM | 32 bit float | Two 10 MHz serial ports. | $25 | Low cost; Broad commercial base |
| TMS320C40 | TI | 50 MFLOP 25 MIPS 50 MHz | 2 Kwords 32 bit RAM | Six link ports (20 MBytes/sec) | | | First- generation multiprocessing |
| TMS320C80 | TI | 1600@200 MHz | 12 Kwords | 32 bit | Crossbar I/O, 400 MB/s | | On-chip MIMD 4 DSP + host |
| TMS320C62 | TI | 1600 MIPS @200 MHz | 512 k Pgm 512 k Data | 32/VLIW | 4 DMA, 2 Serial | $50? | 8 ALUs, compiler is key |
| TMS320C67 | TI | 1000 FLOPS @167 MHz | TBD | 32/VLIW | 4 DMA, 2 Serial | $50? | 6 FP ALUs in C62 |
| DSP96002 | Motorola | 60 MFLOP, 20 MIPS @40 MHz | 2 K wd pgm 2*512 w data RAM | 32 bit float | 2 ext bus ports, used for data & program as well as I/O | | Multiprocessing interconnect limited; det by ext ckts |
| DSP56690 | Motorola | 104 MIPS | 84 k × 24 FM | 24, 16, 32 | RT clocks, numerous I/O | | AMPS accelerator [361] |
| (56 K + Mcore 210) | | @104 MHz | > 256 k total | | (PIG, DSC, GPIO, RAM…) | | 1.8v part op at 2.2v |
| DSP32C | AT&T | 25 MFLOP 12.5 MIPS @50 MHz | 1.5 Kword RAM | 32 bits | Serial port, general-purpose buses | | Not fully commercialized |
| E1-32 | Hyperstone | 97 Dhrystone 60 MHz | 4 kB | 32 bits | Single port/limited I/O | | RISC DSP [100, 101] 1 k FFT ex 0.81 ms; 3.3 V |
| DDMP | Sharp | Async 3.8 BOP | Yes | Video | Data, PCI, memory | | 8 parallel data-driven cores |
| Alpha 21164 | DEC | 4-644/17-20[a] @667 MHz | 32kB Inst 64 MPY | 64 128 L3RAM | Memory- based SCSI, add-on chips | | 0.28 νm, 22 W, 2.5 V 2X Pentium II [250–252] |

RISC MPU's (5–8 FPUs), IA-64 MERCED (8 MPUs) and MMX (50 instruction types) 500 MHz clocks, 1 GFLOP performance. [a]400–644 SPECfp92; 17 SpecInt92; 20 SpecFP (95). [b]Price at introduction.

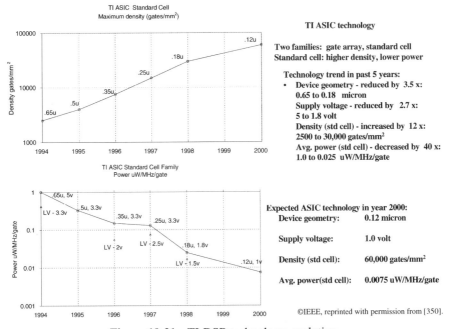

**Figure 10-21**   TI DSP technology evolution.

ufactured by Western Electric. Bellcore (Bell Communications Research) retains some DSP expertise, while Lucent has the large-scale manufacturing capability for DSP chips ("AT&T" in the table). The table also lists other DSP-capable processors with small market shares from Hyperstone and DEC.

RISC processors continue to encroach on signal processing applications that had historically been the domain of DSPs [346]. Competition among RISC machines and the Intel IA-64 Merced continue to propel DSPs on a collision course with general purpose processors [347]. Recent processor research thrusts took only 2 to 3 years to appear in commercial products (compare [348] with [349]).

## G. Potential Technology Limits

DSP technology continues to move forward, propelled by submicron-device technology. Figure 10-21 illustrates the evolution of TI's ASIC technology [350]. The most significant trends are the reduced power and reduced minimum device geometry. At a projected 0.12 microns and 1 volt in the year 2000, the average power of a standard cell drops from 1 microwatt per MHz per gate to .0075. These order-of-magnitude improvements will help make SDR practical for handsets.

## VI. INFOSEC PROCESSOR ARCHITECTURES

Information Security (INFOSEC) processors have increased in importance with the increased demand for privacy in the commercial sector. The U.S. National Security Agency (NSA) has the responsibility for INFOSEC for the DoD, while the U.S. National Institute for Standards and Technology (NIST) has the responsibility for promoting INFOSEC practices and technology for the commercial sector. This section provides a brief summary of INFOSEC technology and developments relevant to SDR.

### A. The Clipper Chip—Key Escrow Approach

NSA introduced the Clipper chip in the early 1990s with the concept of key escrow agents. The agents are a pair of government agencies (such as the FBI and NSA) who would each have part of the information required to recover or reconstruct Clipper chip keys. The assumption is that two agencies would not collaborate to break the law by recovering the key without a court order. This, the argument goes, assures Clipper chip users of the confidentiality of their communications unless they were suspected of criminal wrongdoing and probable cause sufficient for a warrant were presented to a judge. That approach was not well received by the commercial sector. NSA, however, remains publicly adamant that hardware encryption is much more difficult to compromise than software encryption. They assert that key escrow provides a good compromise between privacy and the needs of the law enforcement community to wiretap for evidence gathering in criminal cases.

### B. Programmable INFOSEC Modules

Early INFOSEC devices were limited by the technology to dedicated (usually digital) hardware key generators (the KG series). Other encryption functions included analog voice (KY series). The KG84 was one of the first programmable INFOSEC devices designed to include externally invoked modes by which it could implement the authentication and stream or block cipher behaviors of other discrete INFOSEC hardware, allowing some degree of software-defined INFOSEC. The CYPRIS programmable INFOSEC module was developed to meet user needs for reprogrammability across military modes such as SINCGARS, HAVE QUICK I and II, and other narrowband modes. The SPEAKeasy I program was instrumental in bringing this generation of programmable INFOSEC to the market. By early 1997, however, the CYPRIS technology was abandoned, to be replaced by the Advanced INFOSEC Module (AIM) developed by Motorola. In addition, a VME INFOSEC Module (VIM) was conceived as a delivery platform for the AIM chip into open-architecture applications. But by early 1997, the PCI bus had pushed ahead as the open-architecture backplane of choice for many applications, including

SPEAKeasy II, leaving a vacuum in standard PCI board-level products for open-architecture INFOSEC.

Finally, the JRS architecture includes programmable INFOSEC [437].

## VII. HOST PROCESSORS

UNIX VME hosts have been used in SPEAKeasy and other software radio technology pathfinders. In addition, the Motorola 601 and 604 series of general-purpose processors have been used for both DSP and general-purpose radio applications. In E-Systems' CellTap™ series of first-generation cellular law-enforcement products, the IBM PC Industry Standard Architecture (ISA) backplane supported multiple DSPs with a PC host. In SPEAKeasy II, the user interface was hosted on a COTS embedded Pentium PC. Others have used DEC Alpha chips and PowerPC chips for digital signal processing, unifying the ISA mix. This allows one to use the same operating system, debugging tools, and other utilities on all processors. It can be argued that such approaches fail to maximize the computational capacity for a given applications mix. It can also be argued that the reduced time-to-market and relaxed programmer training more than compensate for any loss of computational capacity.

## VIII. ARCHITECTURE IMPLICATIONS

SPEAKEASY I offers important lessons in the development of software radios that have stood the test of time. Programmable DSPs have the required high-order language (HOL) programmability, but they are inappropriate for front-end filtering tasks. The SPEAKeasy I DSP-based front-end filter design called for 100 massively parallel DSPs. Although not out of the question, these DSPs would have consumed much higher power than the Harris ASICs that were eventually chosen. The DSPs were more appropriate for channel modem and baseband signal processing tasks. FPGAs were also considered for front-end filtering. VHDL programmability provided flexibility, but these chips were also less power efficient for finite impulse response (FIR) filters and FFTs than dedicated ASICs. ASICs are clearly the most power efficient, but the mix of ASICs, FPGAs, and DSPs proliferates different types of modules. It also reduces the number of backup modes.

The net result is a hybrid mix of ASICs for front-end filtering; FPGAs for state machines and glue logic; and DSPs for baseband signal processing. General-purpose host processors still give the best overall combination of flexibility with manageable power dissipation. Processing therefore requires a mix of these technologies. Bit, byte, memory access, FLOPS, and I/O need to be characterized for each processor family to assure processing capacity needs are met. In addition, on-chip, on-board, in-chassis, and in-rack timing penalties have to be taken into account in this characterization process.

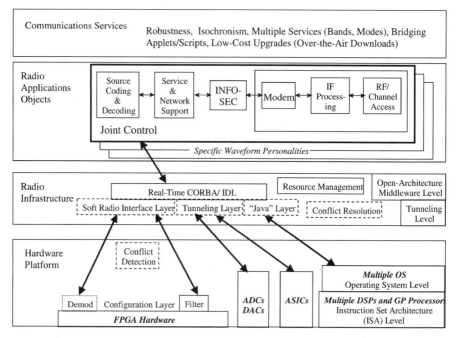

**Figure 10-22**  Layered architecture for heterogeneous multiprocessing.

This processor mix also places a premium on the integration of a heterogeneous mix of processors in a distributed, multiprocessing environment. This mix may occur in the IP of systems-on-a-chip, or it may occur at the macroscale in a multiband-multimode base station. Binding these entities together requires an architecture model with the abstractions illustrated in Figure 10-22. This view integrates the facilities developed in the analysis of digital processing alternatives presented above.

The parameters of specific processors highlighted in this chapter are not the central issue in architecture. What is important is design evolution within the framework of architecture. The continued validity of Moore's Law will continue to transform radios into computers with antennas. Although the layering represented in Figure 10-22 is excruciatingly computationally intensive, MIPS are becoming cheap at an exponential rate. The layered architecture presented in the figure therefore insulates radio applications from the rapid hardware evolution. In the near-term, these layers may be used during product development to facilitate the reuse of functional components (e.g., high-level software objects). The software factory will have to precompile and optimize these personalities to operate on computationally challenged low-power platforms. Over time, however, as these limits recede, the delivery platforms will increase the use of real-time middleware. Middleware will enable the services-driven restructuring of SDRs in the field. In addition, the processing capabilities of

SDR nodes will unleash innovation at every level of abstraction from the physical layer of the air interface through networking and applications. This places a premium on architecture flexibility.

## IX. EXERCISES

**1.** What metrics may be used to characterize digital processing and interconnect? Which are appropriate to characterizing a disk array? Suppose the disk array is to be used for voice mail. Would the same characteristics be used? others? Suppose the disk array is to be used for streaming video. What metrics now apply? Suppose the disk array is to be a part of a distributed database. What metrics now apply?

**2.** What extended instructions are provided by DSPs versus RISC processors? What classes of algorithm are accelerated by these additional instructions? Compare the Intel MMX instruction set extension to your favorite commercial DSP. What additional steps are needed for the MMX class of processor to usurp the unique features of DSP chips in this comparison? Are these additions likely to happen? What applications would cause them to be adopted? What geopolitical events could cause overlap to increase? decrease?

**3.** Compare the interconnect capabilities of the major DSPs with the high-performance GP processors. What classes of algorithm/application tend to drive one toward higher-capacity interconnect? What classes of algorithm need greater computational capacity?

**4.** Summarize the features of a programmable INFOSEC module. Familiarize yourself with the GSM SIM capability (e.g., using [18]). Is this an INFOSEC module? Is it programmable? What would be the advantages of a more programmable SIM card? the disadvantages?

**5.** Develop an object-oriented model of a heterogeneous digital hardware configuration discussed in this chapter. What class of object is at the root (top) of the inference hierarchy? Does this object correspond to any specific realizable device? Are there separate branches in the hierarchy for low-speed interconnect and high-speed interconnect? Is this a reasonable split? Why or why not? What criteria did you use to define processor classes? Subclasses? Instances? What additional hardware classes are needed to make your hierarchy general and re-usable for any SDR?

**6.** Define the concept of a digital processing hardware reference platform. What is the motivation for this concept?

**7.** Generate a digital processing reference platform for the DSP-core-based ASIC presented early in Figure 10-19. Does this reference platform apply to a PCI-based implementation of a cellular base station? Does it apply to a system-on-a-chip architecture in which memory and other devices that are external to the core-based ASIC are now internal? If so, what is the advantage of this feature; and if not, what is the disadvantage of this lack of generality? What are the benefits of your approach?

**8.** Express the layered architecture of Figure 10-22 as a digital processing reference platform. Express it as an OO model with classes and an inheritance hierarchy. Explain the advantages of each representation.

# 11 Software Architecture Tradeoffs

This chapter addresses software design for SDR nodes. This includes software functions, hardware-software interactions, object-oriented design, and software architecture. It also addresses the evolution of the software components of SDR designs. Architecture tradeoffs addressed include the partitioning of software into objects. The boundaries of functional-interfaces and levels of abstraction determine the potential for reuse. These boundaries also determine the ease with which software products from different development teams will integrate into a multi-mode SDR. Over time, the use of specialized hardware facilities may be first encouraged and later discouraged. In addition, there are continuing tradeoffs between system performance and algorithm complexity. The more computationally demanding algorithms often offer better QoS than the less computationally demanding algorithms. This puts pressure on algorithm designers, software architects, and configuration managers. The analysis presented in this chapter provides insights into how these tradeoffs shape SDR architecture.

The technical focus remains on the internal structure of SDR nodes. Network-level software architectures are as important as internal structure, but are beyond the scope of this text. Texts on specific air interface standards address networking issues [351–353], as do general wireless communications texts [354]. This text focuses on the process of structuring the software of high-performance SDRs, and on the architecture implications of the resulting software components.

## I. THE SOFTWARE DESIGN PROCESS

The tradeoffs of this chapter are set in the context of Figure 11-1. A specific SDR implements a subset of the radio functions shown. A top-down software requirements-statement should include services, numbers of channels, radio bands, and modes. In addition, a hardware platform may be specified, or its characteristics may be defined in a reference platform.

It is possible to design SDR software top-down from requirements and radio functions to be targeted to a class of radio platforms (e.g., base stations, mobile nodes, etc.). It is not wise to embark on a purely top-down design process, however. SDR technology includes many existing components, with more

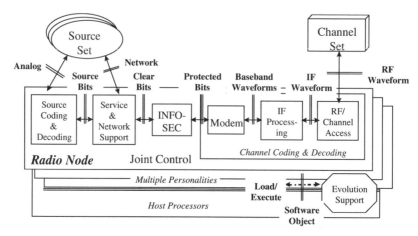

© Mitola's STATIS*faction*, used by permission.

**Figure 11-1**   Top-level radio node architecture.

added daily. The current commercial emphasis on wireless mobile computing and Internet access is producing software components for reuse. Corporate experience invariably includes one or more baseline systems with components that management envisions as potentially reusable whether they were designed for such reuse or not. Therefore, software follows a hybrid of top-down and bottom-up design. The top-down aspect identifies the behaviors and top-level partitioning of SDR software into objects. The bottom-up aspect identifies existing software components that may be encapsulated into objects at some appropriate level of abstraction to avoid the work of designing, coding, and testing those components again.

The functional model of Figure 11-1 is the basis for the partitioning of software into components. This model was derived by examining hardware-software partitions of SDR technology-pathfinders and precursor systems. Software functional entities and associated top-level interfaces have exhibited strong consistency over time and across implementations by different teams. The functional model is therefore the top-down framework. The process iterates between top-down and bottom-up aspects to yield software objects organized into a class hierarchy.

## II. TOP-DOWN, OBJECT-ORIENTED DESIGN

Top-down SDR software design is presented in this section.

### A. Object-Oriented Design for SDR

This section further explains OOT principles introduced earlier. The aspects of OO design that support top-down software development in-

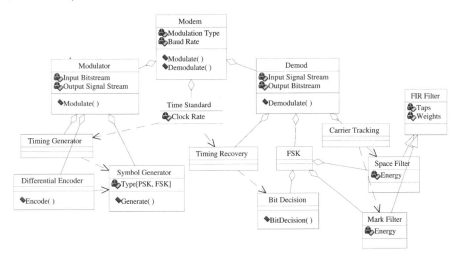

**Figure 11-2**   Partial object model of a simple modem.

clude encapsulation, message passing, property inheritance, and polymorphism.

*1. Encapsulation*   The first step in the development of object-oriented models, whether for modeling, simulation, or software development, is the identification of the object classes. This is accomplished by drawing a conceptual circle around a cohesive set of data and functions to define an object. Initially, one treats the entire radio node as an object in order to define its behaviors when stimulated by the external world. This encapsulates the entire system including software as one object. Subsequently, one defines the constituent software objects of which the radio node is comprised. A consistent set of software objects constitutes one of the radio's personalities. These lower-level objects provide the well-known radio functions of filtering, modulation, demodulation, timing and control, as well as objects that handle protocol stacks and user interfaces. The software objects should encapsulate groups of functions in ways that make sense to radio engineers, to promote object reuse and technology insertion.

The Modem class illustrated in Figure 11-2 provides a convenient example. Encapsulated object classes within the modem interact with each other to accomplish modem tasks by exchanging messages.

*2. Message Passing*   When a radio application sends a message to the Modem object to "modulate a baseband bitstream," it is effectively executing a procedure call of the Modem object's Modulate( ) method. To do this, the calling object executes Modem.Modulate( ), sometimes noted as Modem → Modulate( ). Early OO languages like LISP Machine LISP employed explicit message passing using syntax like Send(Modem, Modulate(bits)). This sent the Modem object a request to modulate bits. Contemporary OO languages like Java employ the more concise message Modem.Modulate(bits).

Message passing permits conceptually simple integration of software components. It also facilitates interconnections across physical boundaries of ASICs, FPGAs, DSPs, and GP processors. Layering includes the process of translating messages from one format to another. Tunneling includes the process of setting up a software path to a hardware entity. The driver software encapsulates the hardware by making public methods available to other objects in the system.

In addition, interrupt service routines and interprocess communication typically is based on message passing using the distributed processing approach. This has nothing to do with object-oriented design. On the other hand, the historical use of message passing in distributed processing makes it easy to encapsulate a processor as a software object. One may thus jointly address the needs of distributed multithreaded multiprocessing and object-oriented software by message passing. The path of transformations of a message defines a thread, in this framework.

The clear, "public" definition of the messages—syntax and semantics—is necessary for the successful integration of software and hardware. Types of messages useful in object interfaces include:

- Setup and control,
- State and state transitions,
- Information streams (an infinite sequence of messages of a specific format), with specified timing constraints,
- Timing and frequency-standard information,
- INFOSEC information such as the current level of protection on each channel,
- Operational parameters like hardware and software signal gain (which impacts linearity), and
- Resource needs and capabilities (e.g., for plug-and-play).

A data dictionary that includes the format (syntax) and meaning (semantics) of the messages should provide examples of when to and when not to use a given message. A comprehensive data dictionary also includes names for at least the public slots and methods of all objects in the system. The degree of "publicity" required is determined by the scope of the software components. If only new, locally designed software objects are to be used, then teamwide agreement on slots and messages suffices. If objects from multiple teams are to be used, then the agreement has to embrace all the teams. In particular, if the purpose of an SDR architecture is to engage all of industry in the creation of third-party products, then the data dictionary of messages should be part of the open architecture standard.

***3. Property Inheritance*** When a new object class is synthesized from existing object classes, that new class inherits data slots and behaviors. One may create an inheritance hierarchy with a generic Modem class at the top and with subclasses such as FSK-Modem and PSK-Modem. Alternatively, one may define Modem in terms of constituent objects, Modulator and Demodulator, with states that determine whether the object operates in PSK or FSK mode. In the example of Figure 11-2, the latter approach is taken. The modulation type and baud rate of the constituent objects are inherited from the Modem class in which they are defined.

Property inheritance allows one to define reusable classes of generic software objects like FIR filters, timing recovery, packet multiplex objects, etc. From these, one may synthesize task-specific objects like mark and space filters. The Filter inheritance hierarchy might begin with Filter at the root. Properties of the Filter might include pole-zero structure (e.g., FIR, IIR, etc.), for example. FIR Filter components could include time-delay storage elements and feed-forward weights.

The simple modem of Figure 11-2 recovers the carrier, extracts bit timing, estimates signal parameters (e.g., to estimate whether a mark or space is present), and makes bit decisions. It demodulates FSK using mark and space filters. The bit-decision object compares the energy in the mark and space filter at the appropriate time, deciding for mark or space based on the strongest filter energy. Since the mark and space filters are FIR filters, they inherit properties from the FIR filter class. The object model shows the constituent components of the modem. That is, the modem is constructed from software components like the mark and space filters that inherit their properties from other object classes. The Modem object dispatches Modulate( ) and Demodulate( ) tasks to the Modulator or Demodulator objects. It might perform only error checking and time synchronization internally, delegating essentially all the behaviors to its constituent objects. These objects inherit Modulation Type and Baud Rate from the aggregator class, the Modem.

***4. Polymorphism and Operator Overloading*** Polymorphism is the ability of a software object to assume different behaviors as the context dictates. The classic case is operator overloading in which the same operator, say "+", behaves differently for different data types. The + operator can do conventional addition on two scalars. Instead of being undefined for two vectors of unequal length, for example, an overloaded + could do an element-by-element addition starting at the first element of both vectors. To do this, the definition of + is augmented with a method that is invoked when both operands are vectors. Similarly, the addition of a scalar to a vector could be accomplished with the same + operator with a different method that adds the scalar, say, to every element of the vector. The + operator dynamically examines the types of its operands to invoke the appropriate method.

Operator overloading makes it much easier to write readable code. It also makes the code assume a degree of *independence of the underlying data struc-*

*tures.* That is, operator overloading allows a given algorithm to operate on a range of different data structures.

The Modem object could overload Modulate( ) if the input bitstream were packetized. A packetized stream includes a header containing control information and a body containing the signal [e.g., 331]. Modulate( ) could check the packet header and apply the type of modulation defined in that packet. Similarly, Demodulate( ) could be overloaded. In this case, it would need a modulation-class recognition algorithm in order to know whether to apply FSK, PSK, etc. to the signal. In traditional radio architectures, the channel modulation is rigidly defined by the air interface. In 3G, however, the channel modulation may range across several types as a function of QoS and SNR. The channel could use BPSK in low-SNR and 16 QAM in high-SNR conditions. In lieu of mode-change commands that waste channel capacity, the Demodulate( ) function could be overloaded, applying the appropriate demodulation algorithm for whatever signal is present.

A Filter-class's behavior could be overloaded as well to be either block oriented or stream oriented. When a filter with $N$ taps is presented with a data block of length $M$, it could load $N$ internal delay states from the prior block, filter the $M$ samples, and save the $N$ internal states for use in the next block. When the same filter is presented with a stream, it could pop the first element from the stream, apply the filter weights to the current $N$ values in its delay line, and return one filtered sample. Loading and saving filter values must be efficient for such software to operate in real-time. Objects facilitate efficiency through access to data structures in the slots. The Filter object, for example, can allocate a new delay line with different taps to each stream. The object then applies its (presumably highly optimized) multiply-accumulate algorithm to the appropriate slot(s) yielding the results without physically loading or storing the data.

Thus, the principles of encapsulation, message-passing, property-inheritance, and polymorphism are useful in SDR contexts.

## B. Defining Software Objects

One may apply the principles of object-oriented design to the design of an SDR node in a top-down way, as outlined in this section.

*1. Context Diagrams*    The context diagram treats an entire system as if it were a single object. Given the system as an encapsulated object, one must answer the question: What "messages"—in the most generic sense—will be exchanged with the outside world?

Figure 11-3 illustrates this process for a notional mobile telephone switching office (MTSO), including the base stations, transmitters, etc. This particular MTSO includes BTS and BSC functions to simplify the discussion. Thus defined, the MTSO includes radio and nonradio telecommunications functions necessary to make a mobile SDR node work with the larger PSTN.

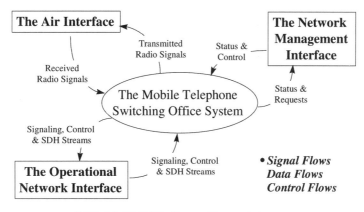

**Figure 11-3** Illustrative context diagram.

The air interface, network management interface, and operational network interface provide access to external objects like subscribers and networks. Abstract external objects like callers and networks are called *actors* in UML terminology. Actors have properties and/or behaviors that shape the system design. Between the MTSO and each external system there are two arcs, one for each direction of stimulus and response, which are modeled as message-passing. The air interface represents the MTSO's connection to the mobile subscribers. In this case messages include traffic channels and control channels. In contemporary digital air interfaces such as GSM and IS-95 (CDMA), virtual channels are multiplexed over physical channels. There also may be channels for establishing timing (e.g., CDMA pilot channels) in a complex array of streams. From these, isochronous traffic channels (message streams) and formatted control packets (messages) must be recovered.

The context diagram identifies all signal flows, data flows, and control flows with external entities. Signal flows are isochronous streams. Data flows contain near-real-time data packets. Control flows shift processor use among software objects. These flows may be defined in part by air-interface standards. Specific designs invariably introduce nuances, such as the application-specific use of air-interface bits that the air-interface standards leave unspecified.

*2. Event Lists* The context diagram is examined for external events that may stimulate the system. Applicable messages from the air interface, status requests from the telecommunications management network (TMN), and calls placed through Signaling System 7 (SS7) from the PSTN are examples of external events. Each must elicit an appropriate response from the software. For each external event, there may be more than one system response. These are collected into a comprehensive event-response list.

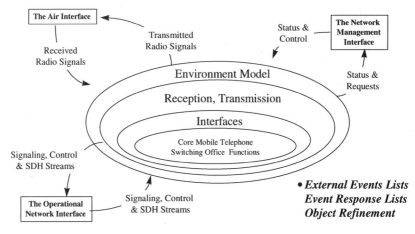

**Figure 11-4**   Layered context with event lists.

From the usual object-oriented design perspective, one builds the software objects that recognize the external events and generate the required responses. For SDR applications, this enumeration of external events and system responses must be tempered by considering the interface layers. These are mechanisms through which the outside world imposes on the radio constraints of external events and responses that give rise to events unanticipated in the initial analysis. The layered context illustrated in Figure 11-4 includes the radio propagation environment, which adds noise and interference. Interference may create false messages and may mask legitimate messages from subscribers. This interaction is taken into account by expanding the external-events list. One may establish positive acknowledgment across the air interface with timeouts and back-off mechanisms to ensure that a masked message cannot cause a permanent suspended state of a critical resource like a traffic channel.

Reception and transmission events might include pointing a smart antenna to maintain high CIR on a specific subscriber. In addition, the reception/transmission layer will constrain some interfaces to observe the demanding timing requirements of the Synchronous Digital Hierarchy (SDH) or SS7. On the other hand, the Interfaces layer may provide some hardware relief to the software challenges. SDH products, for example, include T and E carrier chip and board-level interfaces. These meet many of the SDH requirements provided the SDR fills buffers fast enough (but not so fast as to overflow).

Having examined these context layers for events that may not have been present in the initial context diagram, one defines an initial set of software objects.

*3. Use-Case Scenarios*   The event lists contain stimulus response pairs among actors. It is instructive to trace the path of such pairs through the system.

**Figure 11-5**   Object interaction threads.

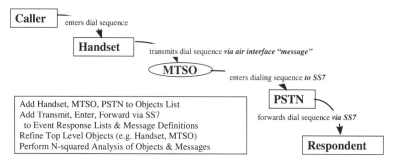

**Figure 11-6**   Sequence of objects with related actions.

Such a trace may be called a thread. The top-level trace of a caller dialing a respondent, for example, is shown in Figure 11-5. If Caller and Respondent are not on the objects list as actors, they are added. In addition, the dialing event is placed on the events list if it is not already there. Tracing threads provides a good check on the events lists while providing a natural basis for encapsulating objects and defining message flows. As illustrated in Figure 11-6, the trace reveals the existence of Caller, handset, MTSO, PSTN and Respondent objects with associated message flows. In UML terminology, the tracing of the interactions among external actors and the encapsulated system is called *use-case analysis.*

The effects of these top-level objects on the internal structure of the MTSO are shown in Figure 11-7. The thread extends from stimulus to response inside the MTSO. The dialed number is presented in an appropriate signaling structure of the air interface. In first-generation cellular systems, all dialing is expressed in a time-shared control channel accessed via a physically distinct analog receiver. In the SDR, this channel is one of many accessed in the wideband RF, IF, and ADC streams. In a first-generation scenario, this channel is accessed at the air interface by wideband analog antenna. It is translated in frequency and filtered to a wideband IF where it is converted from analog to digital via the ADC. In Figure 11-7, these operations are performed by the RF/ADC segment, yielding a wideband digital stream. The RF/ADC segment encapsulates the antenna, RF, and IF processing segments of the canonical model and the ADC of the hardware reference platform. Those details are hidden in this encapsulation because the focus is on the top-level objects. Since objects employ message passing for interobject communications, one may model the wideband digital stream as an infinite sequence of single-sample

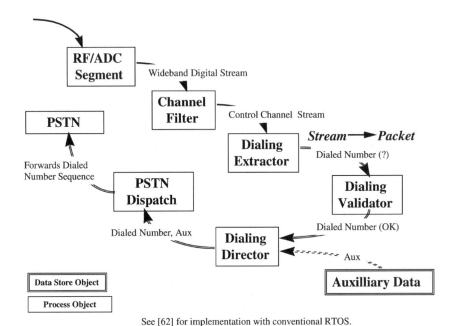

See [62] for implementation with conventional RTOS.

**Figure 11-7**   Sequence of internal objects and message flows.

messages. Digital filtering and down-conversion is then accomplished by an
IF processing object called the Channel Filter. It produces a control-channel
stream. In a first-generation system, this would be a 25 or 30 kHz bandwidth
analog stream sampled discretely at perhaps 50 k samples per second. In a 2G
air interface, there are multiple types of virtual control channels, multiplexed
onto physical data bursts. Object-oriented modeling of 2G protocols segregates
the physical representation from the logical representation. The Channel Filter
object has transformed the wideband digital stream "message" structure to a
narrowband stream.

   The next data transformation in the figure is to extract the dialing information from this narrowband channel. Figure 11-7 shows the Dialing Extractor object performing this transformation of the narrowband control-channel
stream into a Dialed Number packet. The narrowband control-channel stream
from the IF Processing object has not yet been demodulated from the sampled
(Manchester coded) waveform into information bits, so the dialing extractor
accomplishes at least two things. First, it demodulates the control channel
into a 20 kbps data stream. Next, it processes the protocol of the bitstream to
extract the dialing information. This behavior is acceptable in a process of object refinement. Abstractions allow one to consolidate functions that appear in
more than one radio object into abstract objects. Ultimately the Demodulator
object, which supports this Dialing Extractor object, would be aggregated into
the control-channel Modem object. Dialing information is a packet of format-

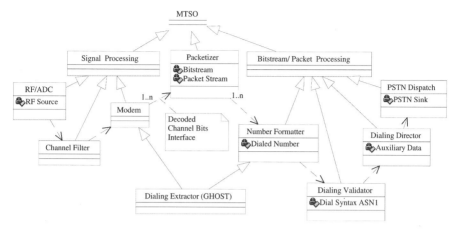

**Figure 11-8**   Illustrative MTSO inheritance hierarchy.

ted data, so the Extractor object has transformed both the format and data rate of the data stream. These dialing data packets must meet SS7 timelines.

Next consider the internal data stores. The formal object design methods have different nomenclature and formats for many different kinds of classes, objects, and relationships. These may be useful in rigorous object-oriented modeling. In this text, a variety of notations provides practice in interpreting alternative notations for the OO concepts. In Figure 11-7, data stores are differentiated from transforms. Signal transformations are evident in the different notation for signal streams versus packet streams. Thus, the notation is tailored to express the concept being studied. Contemporary OO technology often does not allow alternative representations that reflect different analytical objectives. At this stage of top-down analysis, alternative representations can be helpful.

Finally, the dialed number is validated. Ancillary data (e.g., the MTSO's identifying data, SS7 message type, etc.), is looked up in the data store. The Dialing Director appends it to the dialed number for presentation to the PSTN. The PSTN Dispatch object handles the details of the interface to SS7. Behaviors of the objects thus are defined by the needs of the thread.

***4. SDR software object representation***   This process continues with the analysis of additional threads until all stimulus-response pairs have been analyzed to determine data transformations and object behaviors. One result of this process is the definition of an MTSO object class diagram, as illustrated in Figure 11-8. In this example, the system consists of front-end signal processing classes and back-end packet processing classes. The tentative objects created earlier have been allocated appropriate roles. All of the front-end objects process streams of sampled signals at specific sampling rates and arithmetic precision. The signal-processing class contains the code that efficiently moves streams around, but does nothing else, leaving that to the subordinate objects. The back-end objects expect packets of bits as input data structures, transforming the packets and passing them on through the isochronous stream to the PSTN. The relationships (dashed arrows) show the signal flow paths. The Decoded Channel Bits Interface arises naturally as the interface between signal processing and bitstream processing. The Dialing Extractor object is no longer explicit since its functions have been subsumed into other objects. It can be maintained as a *ghost object* that checks the behavior of the modem, packetizer, and number formatter. Such objects may be implemented as abstract classes that check the behavior of the objects that are supposed to be doing the work. Such objects are useful in ensuring that downloaded objects have not violated constraints. Alternatively, the Dialing Extractor may be used as a waveform-independent object that implements dialing behavior by calling the waveform-dependent modem, protocol stack, etc.

The information flows among the objects are threads. Only one isochronous thread is illustrated in Figure 11-8. Threads are classified as isochronous, near real time (NRT), or noncritical. Isochronous threads must be accomplished within short timing windows. In OO software environments that support multiple inheritance (e.g., C++), the isochronous thread may be a class that checks the timing constraints. These constraints may be slots in the objects that are on such threads. The thread-object could then aggregate the timing budgets of each constituent object, keeping track of the probability that constraints can be met and detecting conflicts. In other OO environments (e.g., Java), the timing constraints may be expressed as relationships among objects (e.g., Java "Interfaces"). Timing constraints of NRT threads are not severe, but timing budgets and constraint checking can be helpful. NRT constraint violations can be expressed to users so they expect slower performance, (e.g., of the user interface). Other NRT constraint violations can be used to create back-pressure flow control into the network to reduce the demands on the node.

## C. Architecture Implications

Each SDR design team creates the threads, objects, slots, and methods that are tailored to the needs of the project. Thus, two SDR object structures are rarely identical. SDR architecture may embrace this diversity in two ways. First, industry-standard open architecture should specify only the minimum, highest-level aggregated classes necessary to define plug-and-play interfaces among the objects. The simple front-end/back-end partition at the top level of Figure 11-8, suitably augmented, would be a minimalistic approach. The subordinate object classes of that figure may not be entirely appropriate. The front-end objects of channel filter and modem map well to radio subsystems, so they might be the basis for a more fine-grained architecture. But the behaviors of dialing director and PSTN dispatch may not be as appropriate since they are subsets of more complex protocol stack functions not shown in the hierarchy. The reusability of software objects, then, is determined by the structure of

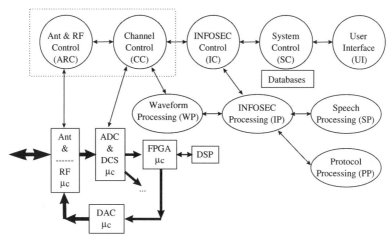

© Mitola's STATIS*faction*, used by permission.

**Figure 11-9**   Software radio architecture objects "bubble chart."

functions in the object class hierarchy. Characterizing the issues that shape SDR class hierarchies is therefore the focus of this chapter.

Second, enterprise architectures should promote migration paths among object representations. Different teams may like to express the same concept in different ways, but that alone may not add value. The maintenance of an enterprisewide SDR architecture provides the standard classes that should be shared, facilitating constructive object reuse. License to abuse the enterprise architecture is as important as having one, so that creative alternatives are not inappropriately suppressed.

## III. SOFTWARE ARCHITECTURE ANALYSIS

Given the above introduction to top-down object-oriented techniques, one may analyze existing software to determine its architecture implications.

### A. SDR Software Architecture

Iterative top-down design and bottom-up implementation processes refine objects. The resulting objects are then structured into aggregates. This process yields the generic software architecture illustrated in Figure 11-9. The top row of high-level objects control the system while the lower rows implement the radio channels and related services. The mix of antenna, waveform, and channel processing front-end hardware and software shown is representative of a contemporary mobile or base station node. (The INFOSEC hardware module and back-end processor/bus hardware is not shown in the figure for simplicity.)

**TABLE 11-1   Characteristics of Radio Software Objects**

| Radio Objects | Object Methods and Slots |
|---|---|
| Antenna and RF Control (ARC) | TX/RX, Power, Polarization |
| Channel Control (CC) | Allocate Resources, Configure, State Machines |
| Waveform Processor (WP) | Generation, Timing, Fault Detection, Mod/Demod |
| INFOSEC Control (IC) | Key, Control Bridge to Black Side, Authenticate |
| INFOSEC Processing (IP) | Encrypt, Decrypt, TRANSEC |
| System Control (SC) | Initialize/Shutdown, Test, System Status |
| User Interface (UI) | Commands and Displays |
| Speech Processing (SP) | Echo Cancellation, Voice Coding |
| Protocol Processing (PP) | Packetization, Routing, VGC Modems |

Each of these high-level object classes has a fine-structure which ultimately consists of primitive single-function radio objects like filters, modulators, interleaving, clock recovery, bit-decision objects, etc. The protocol and speech processing objects implement protocol stacks such as ATM, TCP/IP, Mobile IP, etc. Consequently, internetworking to the wireline infrastructure consists of a few relatively monolithic/predefined (e.g., COTS) software objects.

Continuing with Figure 11-9, it is clear that the top-level objects of the software radio strongly reflect the characteristics of the hardware. Nodes organized around such objects exhibit the behaviors summarized in Table 11-1.

## B. SPEAKeasy I Software Architecture

The SPEAKeasy I system was developed in Ada according to DoD criteria for software quality. Accordingly, the lowest-level Ada packages are generally small—less than 100 lines of code (with a couple of notable exceptions such as built-in-test). The SPEAKeasy I software system as built consists of the modules described in Table 11-2.

Like any other software suite built on a schedule, the as-build code has some strong features—such as the handling of timing and the real-time performance —and some weak ones. Since an Ada implementation was mandated, there is no real-time executive except the Ada run-time kernel and library. In addition, the state machines were apparently hand-coded. Such code is less reusable than a Z.100 SDL equivalent. The degree of reverse engineering required to understand the code varies from package to package as a function of the style of the programmer. As a result, some packages had redundant comments such as |A = B + C; Add B and C together to get A|, when it would have been more helpful to say "The net timing offset, A, is the sum of the base system time, B, and the network offset, C." Nevertheless, studying as-built code reveals design patterns.

**TABLE 11-2   SPEAKeasy I Software Architecture**

| Module | LOC | *Module Descriptions/*Functions |
|---|---|---|
| At | (127 kB) | *C040 interprocessor communications* |
| BIT | (318 kB) | *Built-in-test packages,* including CRC, EEPROM, PID, I/O regiesters, interrupts & DMA |
| Cm | (1.29 MB) | *Configuration management* |
| ALE | (125 kB) | *Automatic Link Establishment Rx & Tx functions* |
| ALE_Rx1 | (378 kB) | *ALE receiver modules* |
| Hvq | (645 kB) | *Have Quick communications ensemble* |
| Hvq_Ct | (109 kB) | *Control modules* (Initialization, Mode Control, Errors) |
| Hvq_Glob | (25 kB) | *Globals* |
| Hvq_Rx | (379 kB) | *Receive* mode (Synchronize, TOD, Rx, Active…) |
| Hvq_Tx | (131 kB) | *Transmit* mode |
| Work | (299 kB) | ALE packages & specs |
| Hfm | (518 kB) | *HF modem communications ensemble* |
| Hfm_ctrl | (58 kB) | *Controls* waveform start/stop messages; protocol events; PM query; TX/RX Done (local); |
| Hfm_dc | (22 kB) | *Data control* packages, source messages, error checking |
| Hfm_rx | (289 kB) | *Receiver* bit & message operations, text I/O, Rx utilities, data correlation tables, filters, queues |
| Hfm_tx | (149 kB) | Squelch, *TX/RX mode, TX* templates, RF Control, Timing |
| Nbg | (334 kB) | *Narrowband frequency hopping* |
| Nbg_ct | (49 kB) | State Machine, Sync Loss, TX/ RX, Waveform, PTT State… |
| Nbg_glob | (105 kB) | *Globals* for NBG package |
| Nbg_hp | (57 kB) | *Hop Packages*—timing, data request/processing, PTT ack, crypto processing… |
| Nbg_rx | (73 kB) | *Receiver packages* MFSK, Preamble, Galois (FEC), Dead Bits, Flags, Bitsync, RX flush, Det/Track |
| Nbg_t | (49 kB) | *TX*: Amplitude, Preamble fill, IQ Samples, AM on Voice, Filter, Inter-Process Communications (IPC) Messages, SSB, DSB, QAM, OQPSK, Event & Constraint Checking |

© Mitola's STATIS*faction*, used by permission.


## C.  Characteristics of Top-Level Objects

For example, typical military SDR objects include agents, databases, and channels. Databases store the load modules and parameter sets that constitute personalities. This includes filter parameters, lookup tables, and other data sets to be loaded into a personality at run-time. If the objects are partitioned into generic objects and a parameter database, the software need not be modified for minor changes. The code management or configuration management system keeps track of revisions and manages multiple personalities. An SDR node that has an associated database of personalities and parameters it can be managed, maintained, and supported in the field.

Channel objects are good abstractions around which to organize radio modes (e.g., HAVE QUICK). A channel object may be organized as a collection of agents, software objects that perform delegated subsystems-level functions of RF control, modem processing, INFOSEC, and internetworking. The channel object obtains system resources for the waveform. These include data flow and signal processing paths for its threads. This object installs its personality on these resources to implement a mode. It then keeps track of the state of its processing threads. Applications-level threads are needed to construct information services from COTS applications, node services (e.g., location finding), and radio applications (e.g., the waveform objects). The installation of the personality consists of assigning system resources to subordinate agents and then keeping track of the top-level state of the radio application. System control ensures that the channel object releases system resources and removes itself from the system when so instructed.

Agents, the functional objects that implement the personalities of the channel objects, may be organized around the top-level system functions of RF control, modem processing, etc. Other agents may serve as hosts for buses, manage IO processes, access timing and positioning data (e.g., from GPS), and control the radio.

The RF control object(s), for example, determine RF direction (i.e., transmit or receive), the RF mode (e.g., linear or nonlinear amplification); pre-emphasis for predistortion, and frequency of transmission. For FH radios, RF control can be fairly sophisticated, involving the use of fast-tuning synthesizer hardware with transmission security features. The modem methods include modulation (AM, FM, QAM, USB, MSK...), demodulation, AGC to avoid saturating or losing a signal, loop bandwidth control, and related data packing and unpacking accomplished on protected bits. The system also keeps track of the status of the mode, number of receivers employed, volume, data rate throughput, network parameters such as network number, and assigned time slot(s). The system may also perform a loopback function for network testing or local diagnosis. It has to accomplish its tasks with associated priorities in force such as network priority, user priority, and priority overrides.

The back-end objects include message processing, internetworking, and managing protocol stacks. System control handles system boot-up, initial TOD, current hop, calibration, status requests, and minimum security level. Data structures used include base types, messages, buffers, addresses, error condition flags, and error messages.

## D. Specialized Tasks

Specialized tasks include network synchronization and waveform-unique protocols. Standard protocols (e.g., TCP/IP) are embedded in the protocol processing object. Timing methods manipulate the system clock. Time Of Day (TOD) is the type of day-time format in use. Timing resolution is sometimes measured in integer nanoseconds, but accurate only to a few hundred nanoseconds.

Modes may have special requirements. HAVE QUICK, for example, has a Word of the Day (WOD) and a training list. SINCGARS employs "cue_frequency" messages and manages complete sets of designated frequencies called *hopsets*. Modem methods also monitor channel states including detect, fade, receiving (data, voice, carrier), transmitting, and lost carrier. INFOSEC methods manage keys, generate cipher, select mode, status, or algorithm, and generate TRANSEC patterns for the transmitter.

### E. SPEAKeasy II Code

SPEAKeasy II code accomplished the three distinct classes of work illustrated in Figure 11-10. Communications services were supplied by linking waveforms. Java could construct services like radio relay across two modes (mode bridging). A Java applet could invoke two installed radio modes to construct such a bridge. Alternatively, a script could express the linking of the two modes. In SPEAKeasy II, bridging code called the waveform agents directly. The radio applications, the waveforms, consisted of collections of agents that performed radio communications tasks. Each distinct mode or waveform was a distinct radio application. In SPEAKeasy I, HF ALE, NBG, etc. were the radio applications.

About 30 to 40% of the as-built SPEAKeasy II code does nothing but set up paths, check to see that all the processors are powered up, move data around, and establish what time it is in each subscriber channel. This collection of functions may be called *radio infrastructure*. Time offsets define the difference between the system master clock and the time understood by the channel. Without such time normalization, one could not resolve TOD differences between independently drifting SINCGARS and Have Quick networks. Time and frequency distribution, system initialization and error recovery, data movement, and related utility functions may all be aggregated into infrastructure.

### IV. INFRASTRUCTURE SOFTWARE

Figure 11-11 lists the function-calls required to set up and control the physical and logical data flows inside an SDR that are the underpinnings of the higher-level objects and services. Infrastructure code manages control flow paths; signal flow paths; and timing, frequency, and positioning information. In addition to the software that accomplishes these functions, a resource manager is needed to set up the paths and see that software objects know what path to use.

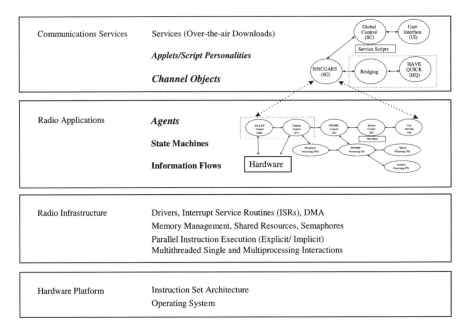

**Figure 11-10**   Services built on channel objects and agents in SPEAKeasy II.

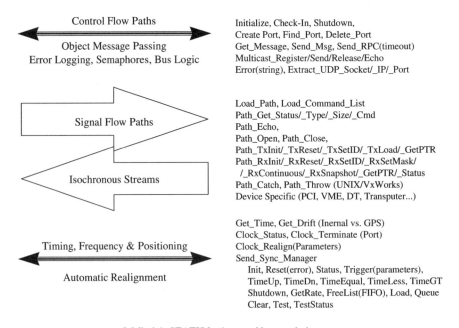

© Mitola's STATIS*faction*, used by permission.

**Figure 11-11**   Infrastructure software function calls.

## A.  Control Flows

The control flow paths pass messages among objects in the system. Error logging, semaphores for shared resources, and bus access protocol messages are examples of such control data. The infrastructure functions initialize the system, create and manipulate ports, move messages, invoke remote procedure calls (RPC), generate multicast messages, and log error messages. Standard protocol interactions such as Internet Protocol (IP) and User Datagram Protocol (UDP) used internally may also be included in infrastructure.

Port manipulation includes finding ports because in a distributed processing environment, each processor creates ports at initialization which other processors and processes use or refer to later. The FindPort method returns a binding of the functional port (e.g., control channel 1) to the logical port (e.g., Com1) on a given processor. Multicast is necessary to simplify the programming of multichannel operations such as initializing 100 subscriber channels distributed among 25 DSP chips. A single multicast message will accomplish this once the multicast has been set up to the 25 DSPs. The infrastructure software handles the logical association of replies from the multicast recipients for the resource manager object that issued the multicast message. The control flow method listed in Figure 11-11 constitutes a reasonably complete set of control flow functions necessary for this aspect of software radio infrastructure.

## B.  Signal Flows

The signal-flow methods listed in Figure 11-11, similarly, set up and manage signal flow paths among processes on the same or on different processors. These are the isochronous streams that must be complete within a short timing window. Due to the overhead associated with path setup and teardown, these paths must be opened and closed multiple times without being set up and torn down again. For example, one may open Path 34 when the user of AMPS channel 34 is speaking. One may close it when the speech epoch has ended or when the call is terminated. The path remains set up, however, so it can be opened on every new call or speech epoch (a very efficient process), but it need not be set up and torn down for each call (a more computationally demanding process).

## C.  Standardizing Flows

Because of the heterogeneous nature of the SDR hardware platforms, researchers and developers have identified the standardization of signal flows as a critical step forward in SDR technology. Consider the code of send_SimpleControlMsg ("sendSimple") that passes a simple control message, for example, as illustrated in Figure 11-12.

This routine declares a message object, fills in its slots, initiates a send, and instructs the operating system to return to the doorbell_interrupt( ) statement

```
static int send_SimpleControlMsg(unsigned subtype) {
    int status;
    /* declare a message object, msg, to be of the appropriate class*/
    sys_tSimpleControlMsg msg;
    /* fill in the message slots */
    msg.Type   = SYS_MSG_SIMPLECONTROL;
    msg.Agent  = remote_rf_agent;
    /* send the message to the rf-agent to let it know we are alive*/
    status = commSend(rf_agent,&msg,comm_sizeof(msg));
    /* set a wakeup call so that this is a wait-free task */
    doorbell_interrupt();
    return(status);

}
```

**Figure 11-12**   C-code that sends a simple control message.

when the commSend( ) function returns a value. The message type is a property of the data exchange among the objects. The definition of message types is supported by a header file that declares SYS_MSG_SIMPLECONTROL to have a specific numeric value for the message header. But other than this limited degree of visibility, the interfaces among the objects are buried in the code. In addition, commSend has to be written for every class of hardware. The description and pseudocode of commSend is provided in Figure 11-13.

Good programming practice requires one to establish the ground rules for allocating the message buffers, for making transfers efficient, and for invoking the driver software, as is accomplished in the programmer's notes. In addition, parameters are declared in the comments in a structured way in part because the operating environment did not provide higher level tools that take care of this. It is good practice to embed such comments in the code for the convenience of (e.g., maintenance) programmers who may not have the development-level software tool suite, yet who must maintain the system in the field.

The associated pseudocode is straightforward (Figure 11-14). This packet interface code handles data setup details, errors, and transmission in a straightforward way. Similarly, the code itself, in this case written in C, raises no surprises (Figure 11-15). In fact, this code is so boring that it is a colossal waste of programming talent to have to write such code. The presence of DEBUG in the source code is a reminder that it also has to be debugged. This boring code has to be written and rewritten again and again for every

```
/************************************************************
commSend   Description:  Sends a message to a remote endpoint.

Parameters: Agent Slot - Identifies the connection;  pMsgBuffer - Pointer to the message buffer to be
sent.;  Size  - Size, in bytes, of the message.

Returns: None.  Visibility: External linkage, part of the comm API.

Side Effects: Sends data to a remote endpoint.

Notes:      The message buffer can be allocated anywhere: on the stack, statically, dynamically
allocated, whatever.  Some systems have special memory that can be used to quickly transfer data to
remote buffers.  If the system has such memory, then it may be useful to have the application allocate
and build message buffers from this special area to avoid unnecessary data movement.  This routine calls
a hardware-specific send() routine.  All that is required to send a message to a remote buffer is the
remote description.  That is passed to Send().
 ************************************************************/
```
SPEAKeasy II ©US DoD, reprinted with permission.

**Figure 11-13**   Description for commSend definition comments.

```
 *****************************************************************/

Begin Pseudocode
*          Obtain a pointer to the agent's connection.
*          IF there is a comm layer error to catch THEN
*              Handle the error wit a call to the error handler
*          ENDIF
*          Using the remote buffer description, call the internal send routine.
*      End Pseudocode
 *****************************************************************/
```
SPEAKeasy II ©US DoD, reprinted with permission.

**Figure 11-14**   Pseudocode for commSend.

new hardware platform and software environment. Furthermore, the data exchange code, sendSimple( ), has to be written for every pair of object classes.

Researchers at the MITRE corporation [31, 216] realized that the U.S. DoD could save a lot of work coding interfaces among applications modules if this whole process were simplified. In its simplest form, the idea was to define a single object to serve as the software equivalent of a hardware backplane. Each object could send messages to and from that object, which would translate the format of its requests into the form other applications need. The standard infrastructure object is the *object request broker* (ORB). Standardized ORB interfaces reduce the number of chunks of code like sendSimple from $O(N^2)$, one for each pair of object classes, to $N$, one between each class and the ORB. Interfaces to the Common ORB Architecture (CORBA) are declared in the Interface Definition Language (IDL) designed for that purpose. One industry-standard implementation of ORBs is CORBA. CORBA was intro-

```
int commSend(int Agent, const void * pMsgBuffer, ubin32 Size)
{ tConn * pConnection = &Conn[Agent];
  if (!Conn[Agent].bInUse)
   { if (pErrorNotify != NULL)
      (*pErrorNotify)(NOT_IN_USE);
     return 0;   }
  if (Size == 0) {  if (pErrorNotify != NULL)
      (*pErrorNotify)(ZERO_LENGTH);
     return 0; }
#ifdef DEBUG
  /* *shared_index = store_index; */
  store[store_index++]=0x55555555;
  store[store_index++]=*(int*)pMsgBuffer;
  if (store_index>=5000) store_index=0;
#endif
  return Send(Agent, &pConnection -> Remote, pMsgBuffer, Size);
```

**Figure 11-15**   C code for commSend.

duced briefly early in this text. Many groups are now developing real-time CORBA to support streaming audio and video on the Internet. CORBA makes specific contributions to infrastructure software.

## D. CORBA

The Common Object Request Broker Architecture (CORBA) has been defined by the Object Management Group (OMG), an industry association of 700 companies dedicated to open-architecture software. CORBA includes distributed real-time audio and video with the opportunity to leverage COTS products into SDR environments. Could such audio streams provide reliable isochronous voice channels for software radios?

*1. CORBA IDL*   As illustrated in Figure 11-16, CORBA IDL allows one to define interfaces in a group called a module. The IDL compiler processes declarations. They may be mapped to C, C++, etc. OMG CORBA includes extensive error checking, and a rich set of exception handlers. The IDL allows one to declare local types for a given interface, which is equivalent to a class in its support of inheritance. Exception handlers are structured as member functions of the Interface class. Instead of dealing with the low-level details of getting pointers, installing slots on message objects, sending packets with

```
module <identifier>              -Module provides scoping for related interfaces
// Comment 1.              -Any characters on a line after the // symbol are comments
// Comment 2.     -Comments are treated as white space, to be ignored by a compiler
{
    <types>;                     -Various declarations with module-wide scope
    <constants>;
    <exceptions>;

    interface <identifier> [:inheritance]              -Equivalent to a class
        {
            <types>;            -Declarations with scope limited to this interface
            <constants>;
            <attributes>;
            <exceptions>;

                        -Operation, equivalent to a class member function or method
            <return_type> <identifier> ([<direction parameter>])
                [raises (exception_name);
        }
}
```

© Copyright SDR Forum, Technical Report 2.0 v1, March 1999 (reprinted with permission).

**Figure 11-16**  CORBA IDL.

byte-counts, etc., as required for sendSimple and commSend, a programmer using IDL specifies the behavior and constraints of the interface. The IDL compiler handles most of the related details.

The SDR Forum is using CORBA IDL in its architecture, as in Figure 11-17. In this example, the Link_Command interface has the attribute "frequency." In addition, two values of ModulationType are declared: AM and FM. The Xmit interface inherits the frequency attribute and the related exception handling from the Link_Command interface. This interface supports the setting of transmit channels and the command to transmit. These interface functions are like ORB methods. Since they are public, any other object can invoke them to accomplish a task. Visibility is obtained into the functional details of the interface by the use of IDL, but nonessential details (like setting block pointers) are hidden. IDL can declare streams, packets, etc., so it is functionally compatible with SDR needs. Computational efficiency has not been a feature of CORBA implementations until recently.

**2. Real-Time CORBA**  Multiple web sites describe the performance of CORBA implementations [355–357]. The layering of CORBA between the operating system and the target object (Figure 11-18) challenges computational efficiency. Performance benchmarks (Figure 11-19) show that less than 1 ms is required by some ORBs to move 1 kByte arrays in Windows NT. Since performance is steadily increasing, one cannot take these as definitive limitations of specific products. The benchmarks are encouraging, however, since blocks of 1 kByte of speech data (500 samples) represent a 62.5 ms epoch. Movement of that amount of speech data in 1 ms uses only 1.6% of

```
module AM_FM_Virtual_Link {          // Namespace for FM/AM VHF/UHF

  interface Link_Command {
    attribute float frequency;
  };

  enum ModulationType { AM, FM };

  exception OutOfRange {          // System can't do it
    string errormsg;
  };

  exception IllegalFrequency {    // Frequency not authorized
    string errormsg;
  };

  interface Xmit : Link_Command { // Transmit inherits from Link_Command
    void set_xmit_channel (in float frequency, in ModulationType Modulation)
                          raises (OutOfRange, IllegalFrequency);

    void transmit ();             // PTT is asserted
  };

  interface Rcv : Link_Command {  // Receive inherits from Link_Command
    void set_rcv_channel (in float frequency, in ModulationType Modulation)
                          raises (OutOfRange);
    void receive ();              // Initial condition, PTT has been released
  };
};
```
© Copyright SDR Forum, Technical Report 2.0 v1, March 1999 (reprinted with permission).

**Figure 11-17**   SDR Forum example of IDL.

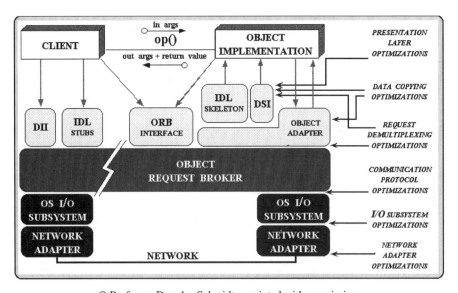

© Professor Douglas Schmidt, reprinted with permission.

**Figure 11-18**   ORB interface requirements.

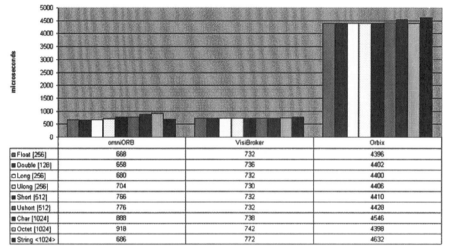

| | omniORB | VisiBroker | Orbix |
|---|---|---|---|
| ▣ Float [256] | 668 | 732 | 4396 |
| ▥ Double [128] | 658 | 736 | 4402 |
| ▢ Long [256] | 680 | 732 | 4400 |
| ▢ Ulong [256] | 704 | 730 | 4406 |
| ▥ Short [512] | 766 | 732 | 4410 |
| ▣ Ushort [512] | 776 | 732 | 4428 |
| ▥ Char [1024] | 888 | 738 | 4546 |
| ▢ Octet [1024] | 918 | 742 | 4398 |
| ▥ String <1024> | 686 | 772 | 4632 |

**Figure 11-19**　Encouraging performance benchmarks.

the isochronous window. While not insignificant overhead, the performance brings real-time speech processing mediated by CORBA into the realm of possibility.

***3. Alternatives to CORBA***　　As illustrated in Figure 11-20, there are numerous alternatives to CORBA [358]. One may employ one of CORBA's direct competitors like DCOM or DCE. Or, one may use alternative approaches like general-purpose communications protocols.

Since the overhead of communications protocols is high, it is usually extremely inefficient to employ a communications protocol like TCP/IP to link objects that are running on the same computational platform. An alternative not illustrated in Figure 11-20 is the Message Passing Interface (MPI) developed for supercomputing applications [359]. CORBA's computational efficiency and acceptance in (DoD-oriented segments of) industry makes it a strong candidate for SDR implementations. In addition, the NOSES commercial telecommunications environment used CORBA [360] for switching software. Technologies like Java and XML may provide future alternative middleware, however. Enterprise-level SDR architecture should adopt some middleware standard. Lighter coupling to specific middleware would ease the transition to future alternative middleware. The US DoD has adopted CORBA as its middleware standard in its Software Communications Architecture (SCA) 1.0 [437].

## E.  Timing, Frequency, and Positioning

In addition to efficient data movement, infrastructure software provides timing, frequency, and positioning support. Timing, frequency, and positioning can be

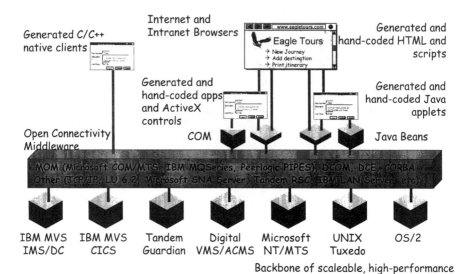

**Figure 11-20**   Alternatives to CORBA.

complex processes, a complete discussion of which is beyond the scope of this brief treatment. The purpose is to highlight the nature of the software facilities needed in SDR infrastructure software.

Time references obtained during network synchronization must be maintained on a per-network basis. Since SDRs generally participate in multiple networks simultaneously, they must maintain absolute time per network. This is accomplished not by changing the SDR clock, but by defining time offsets for each network. Timing facilities are needed to establish exact time of transmission of frequency-hopped packets or CDMA bursts with nominal accuracy of tens to hundreds of nanoseconds. Synchronization of two channels with each other (e.g., to synchronize a traffic channel to a network control channel) requires the establishment of a trigger event. When detected on one channel (e.g., in an FPGA) the event should immediately (or with precisely known time delay) cause a transmission event in an associated channel. One must also have the ability to perform logic and arithmetic on time references and offsets. In particular, one may compute the drift rate between the system clock and each network's TOD. Offset and drift correction accomplished in the infrastructure software then need not be re-implemented for each new waveform.

In addition, mobile nodes may have one or more sensors that assist the node in determining its location. GPS and Glonass positioning satellites, for example, provide precise positioning information if the unit is outdoors and propagation conditions are acceptable. In high-rise urban settings, satellite-based location estimates can be in error by kilometers. In addition, the node

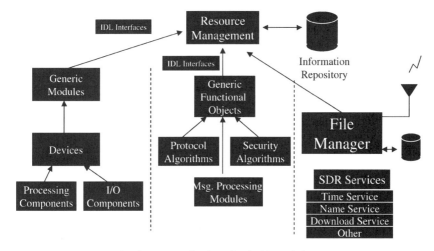

© Exigent International, reprinted with permission.

**Figure 11-21**   Resource management maps applications to components.

can determine the likely validity of a position estimate by analyzing the re-lationship between a sequence of position estimates and physically feasible motion of the node. This kind of autonomous analysis of position is fa-cilitated by intelligent agent technology. Embedding increasing degrees of computational intelligence in radio is part of the research area called cogni-tive radio [361, 438]. PDAs with location sensors are termed *location-aware* [78]. With or without computational intelligence to validate position esti-mates, the close linkage among time standards and position estimation meth-ods groups positioning with time and frequency standards in infrastructure software.

### F.  Resource Management

Finally, the infrastructure software should include a capability to access wave-form images from bulk storage, to install them on appropriate processors, to install the packet and flow primitives, and to arbitrate conflicts that arise, both at load time and during operations. This element of infrastructure software is called the *resource manager*. The resource manager, then, instantiates software objects on physical platforms, as illustrated in Figure 11-21.

The granularity suggested in Figure 11-21 is not necessarily fixed. The resource manager may load a complete binary image over which it has no control. Alternatively, it may load objects incrementally, analyzing conflicts via a constraint language. Early deployments of over-the-air downloads will no doubt be structured monolithically. This approach simplifies type certifica-tion. Over time, however, the congestion of download resources may lead to increased use of incremental downloads, ultimately leading to the download

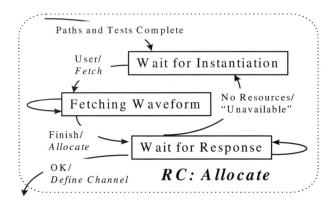

**Figure 11-22**   Waveform-instantiation state machine.

of generic functional objects as suggested in the figure. A resource manager called the domain manager is part of the US DoD's standard radio middleware [437].

## V. SDR STATE MACHINES

Another pattern that is common in radio software is the state machine as illustrated in Figure 11-22. This state machine consists of three states: *waiting for instantiation, fetching the waveform*, and *waiting for a response*. Conditions for transitioning from one state to another are part of the state machine. This state machine could be part of the infrastructure function of managing system resources. The arcs are labeled with conditions that cause a transition from one state to the next and with actions performed upon such a transition. The result is a simple control structure that organizes the object into a set of control slots (the states) and attached procedures (methods). Methods (a) test for state transitions and (b) perform associated actions.

State machines occur in many software radio resources. For example, one state machine controls access to transmit and receive channels; another controls fault recovery actions such as whether to "flywheel" through a fade or attempt to reestablish crypto-sync. These state machines may be aggregated into parameterized forms (templates).

To better appreciate the significance and computational limitations of this type of control structure, the next section reviews the fundamentals of automata theory upon which state machines are based.

### A.  Finite State Automata

A state machine **M** is a finite set of distinct states **S**, a set of inputs, **X**; a set of outputs, **Y**, a next-state map, $\mathbf{A} : \mathbf{X} \times \mathbf{S} \to \mathbf{S}$; and an output map $\mathbf{C} : \mathbf{X} \times \mathbf{S} \to \mathbf{Y}$. There is a distinguished state *So*, the initial state to which the current state of the machine is initialized. One can think of the machine as an automaton that remembers exactly one thing: its current state. For each input, delivered at discrete time point $i$, it looks up the input in the next-state map $A$ to determine what state to assume at the next time instant $i + 1$. In the process of going to that state, it produces output $y$ (a member of $Y$) according to its output map $C$. Both the next state and output maps are conveniently implemented as table lookups.

A machine $M$ is a finite state machine (FSM), or finite state automaton (FSA). It can compute only those functions that are primitive recursive. It can recognize only those languages that are primitive recursive. From a practical perspective, this means that FSA $M$ cannot parse expressions with an arbitrary number of nested parentheses. Nor can it search for complex conditions that may require an unbounded number of iterations or recursions, such as in a While or Until loop. Even with such limitations, FSAs can wait indefinitely for a specific condition to occur (e.g., start of message), and they can recognize any fixed data structure. So FSAs are ideal design patterns to control radio channels. FSAs look for conditions within the channel (carrier, start of message, data, end of message, etc.) and reflect the sequence of conditions in a single (highly meaningful) channel state.

### B. Push-Down Automata

One may augment an FSA with a push-down stack and with Push( ) and Pop( ) operations to create the push-down automaton (PDA), a state machine that now may have an unlimited number of states (e.g., until the stack "overflows"). A PDA will recognize any context-free language. PDAs are good tools for writing compilers, but they can create problems in software radios. In particular, one cannot predict in advance how much stack space a PDA will consume. One cannot predict how long (in terms of numbers of distinct system states traversed) it will take to reach a specified state (one can provide such guarantees with FSAs). So generally, the computationally more powerful PDAs are avoided in favor of FSAs in channel monitors, system status and control machines, etc. The CCITT (Now ITU-T) recommended a Specification and Description Language (SDL) in its Z.100 series recommendations, which provides a precise telecommunications perspective on the definition and use of state machines. There are SDL compilers, and ETSI, for example, will not accept a proposed channel protocol as valid unless there is a working SDL model of the protocol to accompany the textual description [439].

### C. Channel-Control State Machines

FSMs that control channel processes in a channel object operate on several levels. At the top level, each channel object, mode, or personality (e.g., HAVE QUICK) has a top-level channel-setup state machine. This FSA keeps track of whether the logical path over which a channel is defined has the resources

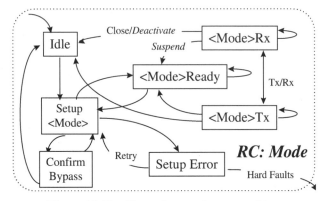

**Figure 11-23**   Channel-control state machine.

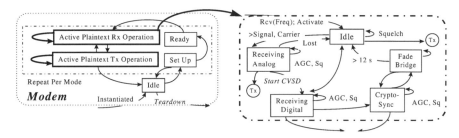

**Figure 11-24**   Modem state machine.

it needs in order to operate. This is generally an infrastructure state machine. Success from this state machine results in the operation of the FSM that establishes the channel as illustrated in Figure 11-23. The channel is initially idle. When it has been set up (e.g., in terms of the use of INFOSEC or INFOSEC bypass), it enters the Ready state. If the user (or network) initiates a "talk" sequence, the transmit (Tx) state is entered (transition not shown in the figure). The radio alternates between transmit and receive states until conditions are met which either suspend or deactivate the channel.

This overall state of the channel reflects the status of all of the agents that support the services of the channel object. Each subordinate agent, e.g., in a different processor, has its own state machine. They reflect the overall structure of the channel, but also carry additional details unique to the agent.

## D.  Agent State Machines

For example, Figure 11-24 shows the specific modem states for active plaintext receive and transmit. The details of the subordinate state machine are resolved to the individual states on the right side of the figure. The active plaintext machine on the left side of the figure is one of two such machines. The active cipher-text machine is shown on the right side of the figure. The plaintext machine has no crypto-sync and fade bridging states but is otherwise identical to the cipher text machine shown. The SPEAKeasy II agent state machines shown above look for RF carrier and appropriate message initiation data (preamble, signaling bits, etc.). They also include states that attempt to maintain communications in spite of channel faults. The one shown maintains crypto-sync through fades by "flywheeling" the current clock through the fade. They have specific timing and error recovery procedures built into the state machines. For example, if the stream cipher part of the crypto unit/algorithm does not reestablish sync within a specified time window, then the channel falls back to an attempt to reinitialize crypto-sync. AGC and squelch control (Sq) are adjusted on all transitions in the lower-level state machines. Thus, the state machines serve as a convenient way to schedule both routine processes such as squelch and AGC and to schedule processes driven by channel conditions such as bridging across fades. Numerous states reflect failure modes like sync failure, loss of carrier, and loss of system resources needed for the call.

The modem state machine manages a real-time stream, while the waveform-instantiation thread manages the infrastructure function of loading executable images into the right DSPs. State machines therefore apply both to radio applications and infrastructure. Such state machines are probably *the* central mechanism for controlling system and radio resources in the SDR.

Software tools for the creation of state machines at high levels of abstraction include Object Geode from Verilog—now TeleLogic (SDL tool) [362] and Rational Rose (UML support) [363].

## VI. ARCHITECTURE IMPLICATIONS

The preceding sections have described software design patterns employed in the construction of SDRs. The distributed layered virtual machine organizes these patterns into a hierarchy of independent virtual machines. Virtual machine interfaces are specified, but the implementation details are hidden within the layers. The SDR architecture that results from the layering process is illustrated in Figure 11-25. The layers are as follows.

### A. Communications Services Layer

Communications services support generic applications like Internet access and radio-specific applications like the downloading of a new vocoder. Table 11-3 identifies communications services of a high-quality architecture. In the software factory, mutually consistent protocol profiles are defined so that combinations of services and applications can work together.

Interfaces from the services layer to the radio applications layer are sequences of calls that set up radio applications and link them to information services. Bridging is one example. Military applications bridge across dissim-

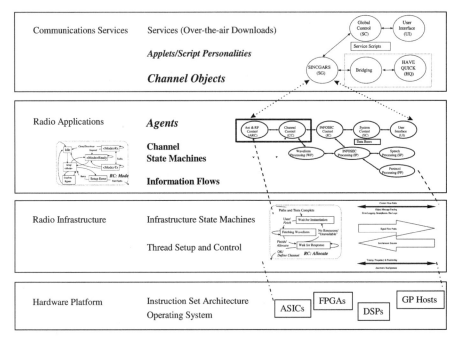

**Figure 11-25**   Software radio (SWR) distributed layered virtual machine.

ilar modes to create gateways among organizations. Law enforcement applications bridge across similar bands and modes to connect agencies with dissimilar radio equipment. Linking a GPRS mode to a web agent and a database on a PDA is another example of a services layer script.

The software tradeoffs at this level of the architecture concern the user. If the SDR is a PDA, then the user should experience seamless access to facilities in any layer. If the SDR is a base station, then the user is the network operator. The tradeoffs concern rapid time to market, for example, by providing interfaces that facilitate the use of COTS software products and protocols (e.g. WAP). If the SDR is a mobile military node, then the tradeoffs concern enhancing capability or interoperability within defined constraints.

Dynamic networking across multiple waveforms occurs at the services layer. Thus, there must be a standard internal representation of speech and data. DSO could suffice for speech. ASCII text with escape sequences could suffice for data. Protocol interactions, congestion management, message prioritization, and related tasks are also addressed in this layer.

## B. Radio Applications Layer

This layer knits distributed objects together into waveforms, the virtual channels that provide wireless (and optional wireline) access. Due to the multiband-

**TABLE 11-3  Services-Layer Standards and Default Facilities**

| Standard or Facility | Overview |
| --- | --- |
| WAP | Translates data from laptop-level displays to handheld; accommodates cell phone data rates. |
| CORBA (Including IDL) | Default object request broker, needed at this level for the interoperation of applications. The software factory might map CORBA services and interfaces to host-environment equivalents such as MexE, Windows, or UNIX calls. |
| Edit | Default notepad composer with choices of ASCII, XML/HTML, and Unicode output. |
| View | Default object viewer (e.g., for images, formatted files). Edit, View, and the other default applications should support mutual cut/copy/paste. |
| Simple SQL | SQL database with a minimal interface to the default GUI. |
| SNMP | Network management default protocol. |
| Local Control | A minimum subset of control capabilities would consist of reading local state information (e.g., file access; hardware, software resources), to start/stop/suspend any application or task (e.g., an extended Windows NT Task Manager), and to initialize any executable file. |
| Access Control | This would provide a minimum password-level of protection in the spirit of at least Windows NT 4 for radio applications. |
| GUI | There should be low-, medium-, and high-end default GUIs, plus a developer's GUI. Commonality with other standards is essential (e.g., the low-end GUI could be the WAP default browser). Local control would use the default GUI. |
| PTT | A default push-to-talk voice communications service. This service should be implemented on all the legacy standards, including the data services, so that there is a minimum order-wire capability among any architecture-compliant nodes. |
| Data Exchange | This primitive e-mail capability should consist of addressless exchange of unformatted text packets. This service ensures that there is a minimum capability to exchange data among architecture-compatible nodes. This service can be used for nonsecure capability exchange. |
| Binary Exchange | This primitive file transfer capability would allow nodes to exchange files with error protection for low bit error rate. With the data exchange and binary exchange capabilities, users could bootstrap essentially any capability known/trusted by either user. They would exchange capability via Data Exchange, then exchange binaries via Binary Exchange. Finally, they would initiate the common application or mode by invoking the $*$.exe via the Local Control capability. |
| TCP/IP | Default transport facilities. A path to Ipv6 should be defined in the evolution plan, along with alternatives to full v6. |

**TABLE 11-3**   (*Continued*).

| | |
|---|---|
| Location | Default facility by which the system knows its own location on the surface of the earth in (latitude, longitude, altitude) coordinates. If there is no local system (e.g., GPS), this facility will provide space for locally generated location. This facility makes location available to software that needs to know it (e.g., in order to manage the network, filter messages, etc.). |
| *OPTIONAL* | This layer of the architecture should flag services that are recommended but optional. |

**TABLE 11-4   Radio Applications Layer Facilities**

| Facility | Overview |
|---|---|
| Waveform Definition | Specifies those channels, agents, and memory allocations that constitute a waveform. Includes parameter sets to tailor the waveform to bands, modes, and applications. |
| Channels | Physical and virtual radio channels constructed from paths provided by the infrastructure layer. |

multimode nature of SDR, the ideal architecture will accommodate both legacy waveforms and new technology in an extensible framework. See Table 11-4.

The interfaces from the radio applications layer to the infrastructure layer could be constructed as a set of object interfaces using CORBA IDL. Alternatively, they could be represented as an infrastructure API, a set of routines with specified calling parameters and behavior that sets up and controls signal flow and packet flow paths, adjusts timing, and supplies frequency and location information.

In SDR base stations, the tradeoffs at this layer concern reducing cost of ownership. The partitioning appropriate for this goal may be the simplest top-level approach that splits the node into a front-end reconfigurable/smart signal processing system and a back-end bitstream processing subsystem. Back-end switching, protocol stacks, and OA&M are software-intensive functions in which the hardware platform supports near-real-time operations. The signal processing front end is hardware-intensive with hard-real-time constraints. This layer may be used to formulate a migration strategy that insulates the applications and services from the layers below.

## C.  Infrastructure Layer

This layer manages the resources of the distributed multiprocessor environment. It includes the capabilities present in a software factory, as well as the ability to null out and streamline interfaces and entities for efficient deployment. The minimum functions assigned to this layer are listed in Table 11-5.

**TABLE 11-5  Infrastructure Layer Facilities**

| Facility | Overview |
|---|---|
| Synchronization | Real-time clock, time standard data per physical channel, timing algebra that includes offset, drift, and comparisons for equality and inequality. |
| Signal Path Package | Set up, initialize, suspend, tear down high-speed (e.g., DMA) signal paths. |
| Control Packet Package | Set up, initialize, suspend, tear down delay-sensitive control packets for message passing among objects, and for SS-7. |
| Resource Manager | Map software objects to hardware/OS platforms; detect and resolve conflicts. |
| Frequency | Frequency standards, phase locking, etc. |
| Position Generation | Estimation of own position from satellite or local information. |

The interface between the infrastructure layer and the hardware/OS includes drivers, interrupt service routines, and calls to operating systems facilities. COTS CORBA packages may provide the connectivity services for non-time-critical threads. This includes most back-end or *red* side functions. COTS CORBA packages bear watching as efficiencies improve to meet real-time applications like streaming video. Over time real-time CORBA will become increasingly suitable for high speed front-end or *black* side functions.

The tradeoffs at this layer concern the efficiency of the signal and packet flows and the complexity of the resource manager. CORBA may be adopted on wireless base stations because access to the power grid relaxes constraints on the computing platform. CORBA may not be suitable for handset applications for a few generations of computing technology, due to computational demands of the IDL/layering process. A resource manager may be simple, consisting mostly of state machines that load object code onto processors. Alternatively, it may be more complex, including the use of a constraint language. Simple approaches that work for single-personality radios may not work for multi-band, multimode radios because of the combinations of object parameter sets. Ultimately, then, one may expect resource management to migrate to the more complex approach.

## D. Hardware Platform Layer

The hardware platform supplies the physical resources and associated software facilities. It includes antenna, RF, interconnect, processing, and storage facilities. It also includes the operating system and drivers required to link hardware components. (See Table 11-6.)

**TABLE 11-6   Hardware Platform Layer Facilities**

| Facility | Remarks |
|---|---|
| RF Facilities | Antenna(s), RF conversion. |
| Digital Conversion | ADCs and DACs. |
| High-Speed Interconnect | Paths among ADCs and processors (dedicated and shared). |
| ASICs | This layer must encapsulate ASICs such that their services may be accessed transparently. |
| FPGAs | This layer not only encapsulates the FPGA, but also parametrically describes it sufficiently to support the download of new personalities in the field. |
| DSPs | Profile memory, MIPS/MFLOPS, and direct I/O channels. |
| CPUs | Profile memory, MIPS/MFLOPS, direct I/O, and special I/O such as speakers, displays, consoles, keyboards, etc. |
| Medium Interconnect | Buses. |
| Connectivity Profile | This facility specifies how interconnect links hardware resources. |

The tradeoffs at this layer concern the affordability of the hardware platform. This is a major procurement cost driver for SDRs. One therefore is wise to overkill the hardware resources in building prototypes, but to tailor the production hardware platform closely to the needs of the product family to keep recurring costs low. One challenging aspect of this tradeoff is the estimation of growth capacity. To incorporate too much room for growth drives the price up. If the anticipated growth of applications complexity fails to materialize, the investment has been wasted.

The layered virtual machine and the functional model introduced in the first chapter are different perspectives on a single architecture framework. Within this framework, one may define a high-level subset that can be embraced by industry as an open architecture for plug-and-play of radio capabilities in wireless devices. This is the path being taken by the SDR Forum. Within this framework, one may also define enterprise-level architectures. These architectures are more detailed and demanding. Their objective is to focus intellectual capital on value-added results by standardizing well-established capabilities and giving management insight into the impact of alternate paths of technology evolution.

## VII. EXERCISES

1. Define software architecture. Differentiate it from radio architecture.

2. What contribution does the functional model of software radio make to the analysis of software architecture tradeoffs?

3. Define encapsulation. Give examples of encapsulated objects that define the personality of a radio. Give an example of a lowest-level object and of a highest-level object.

4. How does message passing facilitate distributed processing? How does it facilitate partitioning software into objects? What types of messages are used in radio software?

5. Define polymorphism. Give an example of polymorphic behavior of an antenna; of an RF conversion subsystem; of an IF processor. What capability does a polymorphic object need that two dedicated-function objects, called differently and in different contexts to accomplish similar tasks, do not need?

6. How does CORBA reduce programming effort? How does it increase visibility into object-to-object interfaces?

7. Define the top-level context diagram of a location-aware wireless PDA. Describe how its environment layer might differ from that of a conventional cell-phone handset. What about its interface layer? What about the objects with which it would interact?

8. Identify the classes of software object in a wireless PDA. Which objects are likely to be implemented in ASICs? in FPGAs? in DSPs? in general-purpose processors?

9. What top-level software objects often occur in radio software? What functions are assigned to each? What design rules might be associated with these objects?

10. Why do radio functions benefit from a layered approach to software components? How are logical layers related to physical code?

11. What four layers comprise the layered virtual machine architecture developed through the analysis of the past few chapters? What sublayers are appropriate to the hardware platform layer? Why are these layers differentiated? What functions do not fit neatly into this layer, but are associated with it? Present arguments for and against a five- or six-layer model; a three-layer model.

12. Consider the disaster-relief application. What software architecture might one envision for such an application? Which objects would be concerned with linking a police department using an analog VHF FM radio with the FBI using a frequency hopped digitally vocoded radio? Which layer would most readily implement the FBI radio? the bridging?

# 12 Software Component Characteristics

This chapter addresses the structure and function of low-level software components. These include algorithms, modules (e.g., Ada packages, C++ objects), and APIs. The perspective is bottom-up, with the emphasis on computational complexity. Low-level algorithms may be simple at first, but complexity can increase over time. The increases in complexity can occur with research advances. Measures taken to compensate for a performance problem in one area (e.g., noisy voice channel) can increase complexity of an algorithm (e.g., dithering the digital LO to spread homodyne artifacts over the voice band, improving voice SNR). Sometimes algorithms have to be restructured to integrate new advances. This chapter introduces low-level algorithms and complexity, core aspects of software component tradeoffs. It also describes APIs useful in implementing the layers defined above.

## I. HARDWARE-SOFTWARE INTERFACES

The SDR engineer must ensure that services are robust. That is, services should be available in spite of the challenges of maintaining isochronism in a distributed multiprocessing environment. External effects of radio propagation, noise and interference, impede the delivery of such services. The SDR accesses multiple bands and modes simultaneously. The advanced implementations manage spectrum use on behalf of the user—band and mode selection, power levels, error-control coding, and waveform choice. In some cases, the services include bridging across modes so that dissimilar legacy systems can intercommunicate. In other cases, users may need special applications encapsulated as scripts or Java-applet like structures, which may be defined via (secure) over-the-air downloads.

As listed in Figure 12-1, these services demand that radio applications include shared resources and interleaved "multithreaded" information flows. In addition, the radio applications must keep track of the state of each such information flow. Infrastructure software includes specialized interrupt-service-routines (ISRs). This software needs efficient use of memory including programming direct memory access (DMA) hardware. Managed access to shared resources includes the use of semaphores. In addition, parallel execution of instructions occurs on multiple levels. One may assign independent information

| Communications Services | Robustness, Isochronism , Multiple Services (Bands, Modes), Bridging Applets/Scripts, Low-Cost Upgrades (Over-the-Air Downloads) |
|---|---|

| Radio Applications | Shared Resources, State Machines, Information Flows |
|---|---|

| Radio Infrastructure | **Interrupt Service Routines** <br> **Memory Size, Block Structures, Double Buffering** <br> **DMA Initialization and Completion** <br> **Semaphores** <br> **Parallel Instruction Execution (Explicit/ Implicit)** <br> **Multithreaded Single and Multiprocessing Interactions** |
|---|---|

| Hardware Platform | **Execution Timing** <br> Initialize, Load <br> Application Process, Waiting Time <br> Termination, Handoff, Dispatching <br> **Operating System Overhead** |
|---|---|

© **Mitola's STATIS*faction*,** used by permission.

**Figure 12-1**   Hardware-software interaction viewed by level of abstraction.

streams to distinct boards, chips, or pipelines. The result is a multithreaded software system with multiprocessing.

Handsets usually have the simplest software environments, limited to only two or three bands and independent information streams at a time. But even such simple SDRs require algorithms to generate air interface waveforms of specified spectral purity. They include digital carrier tracking, demodulation, and protocol stacks. And they must deliver the required QoS in spite of radio channel impairments on a given band and mode. Finally, they must do this within the constraints of the RF and digital processing platform. This chapter therefore begins by considering hardware-software interactions in SDR algorithms. It goes on to characterize SDR algorithms and APIs.

## A. DSP Extensions

Consider first the software interactions with the hardware platform(s) (e.g., Figure 12-2). One DSP may be allocated to a modem algorithm per RF carrier. General-purpose (GP) black processors may be dedicated to link-level processing software, while GP red processors support the higher levels of the protocol stack and the user interface. The DSP platforms have extended hardware instruction sets, real-time operating system kernels, run-time libraries,

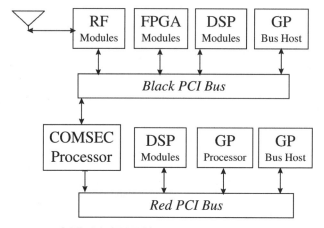

© Mitola's STATIS*faction*, used by permission.

**Figure 12-2** Illustrative SDR hardware platform.

and other software tools that reduce software development time. Instruction timing will have a first-order impact on one's ability to deliver robust performance in the isochronous streams. Timing described in DSP manuals may underemphasize the overhead associated with setting up pipelines (e.g., for digital filtering). Performance may degrade due to cache misses and other factors related to context switching such as termination, handoff from one task to another, and resources used by applications-level dispatching code.

Digital signal processors therefore define much of their value-added in terms of significantly faster execution of computationally intensive algorithms such as filtering, demodulation, and sin( )/cos( ) arithmetic processing. These are facilitated by extensions to instruction sets, which include the following:

- **Instruction set extensions**
  - Register, direct, indirect, immediate addressing
  - Bit-reversed addressing
  - Circular (modulo-$N$) addressing
  - Hardware push/pop, semaphore
  - Repeat-$N$ (no loop overhead)
  - Multiply-accumulate (load, multiply, add, increment, iterate)
  - Parallel multiply-add
- **Data format extensions**
  - Fix, float, double- and triple-precision integer

Address modes such as register, indirect, and immediate allow one to accomplish software tasks entirely within the register set. This avoids the increased latency of memory access ("register" types). DSPs perform very efficient table-

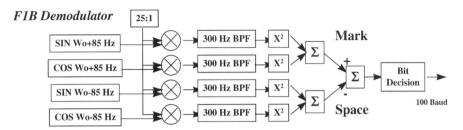

**Figure 12-3**   Illustrative multiply-accumulate algorithm structure.

lookup operations ("indirect" types); and include operands in the same fetch as the instruction, again avoiding memory accesses ("immediate data" types). Bit-reversed addressing allows one to extract the results of the ubiquitous fast Fourier transform (FFT) from an in-place array without suffering the multiple-instruction overhead of calculating the address of the next sample. Instead, one simply reads the in-place FFT in bit-reversed address order to shuffle the results to normal time or frequency domain order. Isochronous tasks include many double-buffer operations in which words or blocks are written into a shared buffer by one task while they are read from the same buffer by another task. If there are $N$ words in the shared buffer, hardware modulo $N$ addressing resets the buffer pointer to zero in hardware whenever it reaches $N$. This avoids the overhead of checking this condition in software and thus speeds up short loops by factors of 2 or more. In addition, the DO loop has a hardware equivalent, repeat $N$, in which loop indexing and testing occurs in hardware in parallel to the execution of the substantive instructions in the loop, again significantly speeding up loops.

Multiply-accumulate instructions are critical to digital filters that typically have an algorithmic structure similar to that illustrated in Figure 12-3. DSP hardware speeds up the bandpass filters (BPF in Figure 12-3), which individually include multiply-accumulate steps. It also speeds up the overall demodulator algorithm by efficient execution of weighted multiply-accumulate steps implicit in the summing junctions of the figure. In addition, DSP chips generally have sin/cos lookup tables or sin/cos approximation algorithms. Hardware lookup tables (CORDIC) speed up the generation of reference waveforms such as the sin and cos ($Wo + 85$) of the algorithm in the figure.

Finally, DSPs generally offer data format extensions such as 32-, 48-, and 64-bit integer, fixed and floating point arithmetic formats. Certain algorithm structures that arise naturally in SDRs require such formats. DSPs with fewer bits of precision are smaller and require less power. They therefore demand less average battery drain in critical handset applications than their larger-precision cousins. A clock that is counted down from a fast crystal may require double- or triple-precision integer arithmetic. Periods of long data transmission may require this arithmetic, such as on a microwave radio link which is expected to operate for months before being reset. Multiple-data fetch instruc-

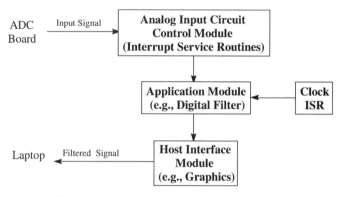

**Figure 12-4**   Illustrative real-time DSP task.

tions fill filtering registers or floating point pipelines quickly. These instruction extensions avoid much software overhead at the expense of increased complexity of the processor core.

Intel's Multi-Media Extensions (MMX) for the Pentium processors extend the standard Intel architecture by including multiple-data fetch and other instructions to enhance multimedia operations needed in today's desktop systems. These extensions have begun to blur the line between general-purpose Complex Instruction Set Computers (CISC) and DSP chips. For the moment, DSP chips provide greater parallelism and ISA extensions to reduce the total number of processors by a factor of two to ten compared to MMX Intel chips for most SDR applications. This is not to say that general-purpose processors cannot be used for software radio research. For example, MIT has used the DEC Alpha chip for their virtual radio [178].

DSP teams apply skill in the use of DSP chips to enhance QoS, or to reduce the hardware footprint. Details are available in texts on programming DSPs [364, 365]. Those details are not necessary for the architecture-level analysis.

### B. Execution Timing

Execution timing techniques ensure that the timing constraints imposed by isochronism are met. Figure 12-4 illustrates the low-level software structures associated with a typical real-time DSP task. When the informal term *real-time* is used in an SDR context, one generally means that the software must be executed within some timing window. This window is defined by the average data rate of a continuous information stream and the maximum size of the buffer that introduces tolerable delay through the processor. For example, a 64 kbps voice channel delivers 8000 8-bit samples to a DSP per second (125 $\mu$sec between samples). Listeners can tolerate up to about 100 ms of end-to-end delay before beginning to perceive a time delay; end-to-end delays of 250–500 ms become uncomfortable. Since there may be many processing steps in

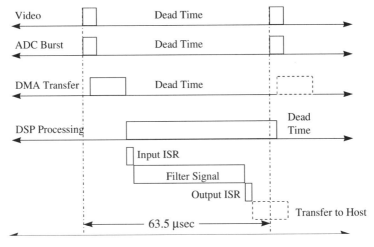

© Mitola's STATIS*faction*, used by permission.

**Figure 12-5**   Illustrative timing diagram.

an end-to-end path, a given DSP task may be allocated 10 ms of time delay. This means that the processor may accumulate $8000*(10 \text{ ms}/1000 \text{ ms}) = 80$ samples in the input buffer. One may allocate two buffers with DMA programming that immediately switches buffers when one is full ("Ping-Pong" or "double buffering"). The DSP supports a continuous input stream while accumulating $80*125$ $\mu$sec of time (i.e., 10 ms) for software processing. As the buffer size increases, the software overhead associated with initializing the processing, setting up and controlling the processing loops, etc. is distributed over more samples, increasing efficiency and hence throughput.

Figure 12-5 illustrates this process for a 150 kHz ADC with overhead that reduces the time available between ten sample blocks to 63.5 usec (the "block" window). The ADC analog input is called the video signal. Although samples are taken continuously, they are transferred to the DSP only when ten have been accumulated in a (double) buffer in the ADC board. This results in the ADC burst that becomes available periodically as shown in the timing diagram of the figure. Since the DMA transfer may not be as fast as the ADC burst, but may begin before all ADC samples are ready, there is overlap of the DMA transfer with the ADC burst. The software—all of it including interrupt service routines—then has processing time which is the difference between the DMA window and the block window. The DMA ties up memory so that processing effectively cannot be accomplished during the DMA burst. Some DSPs segment memory so that there is hardware parallelism that reduces this encroachment of the DMA onto the software tasks.

The input interrupt service routine (ISR) recognizes the DMA complete, switches the pointer between buffers, sets a flag to wake up the associated processing software, and terminates. The ISR should run to complete with a minimum of instructions, limited to pointer manipulation and error checking, so that the hardware interrupt stacks will not exceed their capacity. A few interrupts may be stacked in hardware, but since many ISRs turn off the interrupts so that they will not be interrupted, there may be only a few hardware levels of interrupt available. In very busy systems, lost interrupts can cause system crashes that are not easy to diagnose. So generally, one tries to drive the probability of lost interrupts to as close to zero as possible by strictly limiting the ISRs. They may be coded for recursive calls and double buffering. Circular buffering requires semaphores and tolerates less timing error than double buffers.

The ISR-complete condition signals the applications to process the ten-sample buffers, which in this example, filters the data and sends it to a host processor (e.g., a laptop computer) for real-time display. While users may be relatively forgiving of display update delays, the loss of buffers of data might be more evident in speech applications. This can happen due to slightly exceeding the allocated execution window so that a buffer-full interrupt cannot be serviced. The timing diagram shows the timing budgets. One may test the filter software by running it on a dedicated processor in a loop, which calls it, for example, 10 million times. One measures the elapsed time and divides by 10 million for an estimated execution time. If the time estimated from this kind of measurement is not greater than 50% of the available processing window, then there is little doubt that the DSP will process the samples on time and robustly.

As the estimated execution time approaches 80 to 90% of the allocated window, there is greater and greater chance that unanticipated events will cause the process to fail to complete on time. Operating system servicing of keyboard interrupts is one example. In order to obtain robust performance, the design must take into account the limited resources and unknown arrival times of external events. One cannot predict when one of 100 users will make a telephone call or use the radio.

## C. Aggregate Software Performance

The SDR engineer estimates the computational complexity of software objects in order to ensure that the software personalities and hardware platform(s) are compatible. Software demand should be allocated to hardware in such a way as to keep the estimated demand for processing resources to less than 50% of processing capacity as a general rule of thumb. This concept is introduced in this chapter and addressed in detail in the sequel. Since SDRs are by definition capable of multiband, multimode behavior, multiple software personalities correspond to multiple waveforms and associated protocols of the air interface(s). Each personality is partitioned into software objects. A simple, illustrative set of objects comprising one software personality is illustrated in Figure 12-6.

Each object has an associated processing demand. Simple rules-of-thumb provide top-down estimates of processing demand as shown in the figure.

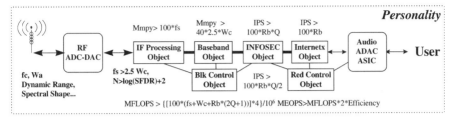

**Figure 12-6** Aggregate software includes all processing regardless of hosting.

Generally, IF processing, the digital filtering required to select a subscriber channel from a wideband IF ADC stream, needs resources that are directly proportional to the IF sample rate, $f_s$. A proportionality constant of 100 multiplies per sample represents the stages of filtering needed to filter a 12.5 MHz IF (30 M samples/sec) to typical cellular subscriber bandwidths of 25 (analog cellular) to 200 kHz (GSM-like). 1.2 MHz CDMA channels require less IF processing but more processing to despread the selected subscriber channel. Once the subscriber channel has been isolated, 40 multiplies per sample times the baseband bandwidth times 2.5 for somewhat oversampled Nyquist criteria yields $100 * Wc$ multiplies needed by the baseband object for demodulation, link processing, and other modem functions. Information security (INFOSEC) processing requires typically fixed point or bit manipulation instructions (MIPS) which are proportional to the baseband data rate times an INFOSEC complexity factor. This factor reflects additional processing for recoding, stream ciphering, etc., which must be accomplished within the INFOSEC module. Black control (on the encrypted side of the radio) and red control (on the unencrypted side) each require additional processing that is directly proportional to the baseband data rate times a fraction of the INFOSEC complexity. Control is generally a passthrough function that consumes fewer resources than subscriber streams. Finally, internetworking consumes integer-processing resources which are directly proportional to the user data rate. The factors of 100 can be combined to yield the simple equation shown in Figure 12-6. This is a first-look, rough order of magnitude estimate of the processing demands of a single subscriber.

The software objects should be supportable in the target distributed processing environment, a simple example of which is illustrated in Figure 12-7. In this case, the IF processing, baseband, and black control processes are all hosted in a front-end processor (FEP). The FEP includes ASICs (e.g., digital filter chip(s)), FPGAs (e.g., for timing and high-speed data control), and DSP chip(s) for baseband processing and control.

The critical measure of performance on the FEP is the number of multiplies supported per second. The INFOSEC processor provides MIPS, but as shown in Figure 12-7, its main contribution may be bus operations (e.g., for delivery of TRANSEC commands to the front end). In addition, INFOSEC

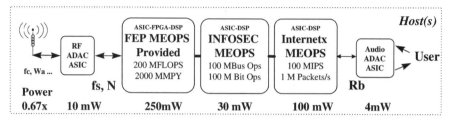

**Figure 12-7**   Target distributed environments provide processing capacity.

typically manipulates bits one at a time. The appropriate characterization of the INFOSEC processor may be bit operations per second. Bus and bit operations will not necessarily fall out of the initial rough order of magnitude estimates, so they will have to be refined using techniques discussed in the chapter on performance management. In addition, some processes such as internetworking may have other measures of processing capacity and demand such as millions of packets processed per second.

Timing and analysis of resource demands and capacity are needed up front. One does not have a good software design until processing demands have been estimated and resources allocated. One does not have a viable unit-level test program unless the estimates have been replaced with measurements. Finally, these estimates and measurements must be maintained throughout the integration process to support optimization, and resource reallocation. In addition, the adjustment of operating system priorities and memory allocation—system tuning—is part of performance management. This is an integral element of software design for SDR. Without such start-to-finish discipline, one runs the risk of building a fragile system which cracks under the slightest load variations and which is incredibly hard to debug as marginal timing conditions impact each other to create intermittent bugs. Performance management is developed in this text in stages. The balance of this chapter describes the software components. A subsequent chapter explains how to estimate resource requirements imposed by software and how to project capacity supplied by hardware so that one may accurately estimate costs, risks, development time, and performance specifications for SDRs.

## II. FRONT-END PROCESSING SOFTWARE

Front-end processing software includes antenna control, diversity selection, and related functions. The SPEAKeasy II applications programmer interface (API) lists the messages in Table 12-1 within RF control. These functions are employed in the phases designated in the table. On power-up, the software requests built-in test (BIT). When the BIT state machines run to complete, the hardware platform has been successfully initialized. The response from the BIT request is the resulting hardware configuration. The front-end control

**TABLE 12-1  Front-End Processing Functions**

| No. | Name | Phase |
|-----|------|-------|
| 1 | ACK | All |
| 2 | Buffer_Complete | All |
| 3 | Buffer_Notify | All |
| 4 | Forward_Message | All |
| 5 | NACK | All |
| 6 | BIT_Request | Power_up |
| 7 | Define_Remote_Child | Power_up |
| 8 | Define_Remote_Parent | Power_up |
| 9 | Allocate_Resources | Instantiation |
| 10 | Connection_Test | Instantiation |
| 11 | Define_Remote_Child | Instantiation |
| 12 | End_Download | Instantiation |
| 13 | File_Download_Complete | Instantiation |
| 14 | File_Download_Start | Instantiation |
| 15 | Initiate_Download | Instantiation |
| 16 | New_Agent | Instantiation |
| 17 | Standard_Data_Msg | Instantiation |
| 18 | Antenna_Select | Params_&_Mode |
| 19 | DeAllocate_Resources | Params_&_Mode |
| 20 | Define_Child | Params_&_Mode |
| 21 | RF_Direction | Params_&_Mode |
| 22 | RX_Calibration | Params_&_Mode |
| 23 | Software_Version_Request | Params_&_Mode |
| 24 | Hop_Strobe | Operation |
| 25 | Initiate_TX_Calibration | Operation |
| 26 | Receive_Mode | Operation |
| 27 | RF_Frequency | Operation |
| 28 | Set_Gain | Operation |
| 29 | Standard_Data_Msg | Operation |
| 30 | T/R_Transmit | Operation |
| 31 | Transmit_Mode | Operation |
| 32 | Destroy_Agent | Teardown |
| 33 | PA_Power | Teardown |
| 34 | Reset_to_Boot | Teardown |

software then creates parents and children which are placeholders for instances of the software entities that control the antenna, transmitter, receiver, and other front-end functions. SDR hop set generation, for example, may be distributed to a microcontroller that controls a fast tuning synthesizer. The hop set parameters would be created by the INFOSEC function, but the details of creating the hops from these parameters might be delegated to the front end (as is the case in Table 12-1).

The instantiation phase then creates the required front-end software entities. First, Allocate Resources requests memory and other processing resources
necessary to instantiate waveform services from a remote parent. Once all resources have been allocated, the download sequence may begin. This API includes separate functions to Initiate the download, start a specific file transfer,
signal the completion of a file, and signal the end of the download. A new
Agent may be declared to manage a specific service. Standard control data messages handle the routine bookkeeping associated with each front-end service.

When instantiation is complete, the parameters and modes are set in the
phase designated for that activity. Antenna, RF calibration, RF direction (TX
or RX), and RX calibration commands control the major front-end resources.
Version request supports software configuration management, ensuring that
versions that are installed are compatible. In addition, resources that are no
longer needed (such as memory in which to stage file transfers) may be deal-
located at this stage. Sometimes it may be necessary to spawn child processes
within a target processor for the parallelism necessary to accelerate this phase.

During the operations phase, the control software can set Mode, Frequency,
Gain, Hop Set, Transmit, or Receive state, and other system parameters. When
the service is to be discontinued, TX and RX modes may be set to suspend
operations without tearing down the service. Since setup is a time-consuming
process, one should defer tearing down a service until the resources are needed
for some other service. When necessary to tear down the system, amplifier
output power (PA) may be turned off, agents may be destroyed, and the host
processor may be rebooted using the functions shown in Table 12-1. In addition to these phase-specific messages, buffer control, acknowledge (ACK)
and NACK, and message forwarding functions are used in any and all phases.
In some APIs, modem software is part of front-end processing (see section
III below).

Enhanced spectral efficiency and improved spatial access are key potential
benefits of SDR. Its inherent flexibility facilitates implementation of advanced
features for dynamic data rates. The variety of ways to approach these aspects
of SDR are surveyed in this section.

## A. Spectrum Management

Techniques for dynamic use of the RF spectrum are listed in Table 12-2. The
simplest way to dynamically manage the RF spectrum is by manual channel
selection. Many large groups of radio users including general aviation, citizens' band (CB), and amateur radio operators employ this approach as the
primary mechanism for spectrum management. The user interface for spectrum management typically consists of the voice channel itself. In the United
States, for example, CB users aggregate on Channel 19 for initial contact with
other mobile users. Since this channel is often congested, they move to other
channels by mutual agreement. Their only mechanism for selecting an alternate channel is the apparent absence of other talkers currently on the channel.

**TABLE 12-2  Techniques for Spectrum Management**

| Need | Approach | Design Issue |
|---|---|---|
| Spectral efficiency | Manual channel selection | User interface |
| | A-priori channel plan | Handoff |
| | Multilayer cells | Handoff vs. demand |
| | Doppler acquisition | Spectrum monitor |
| | Dynamic mode assignment | Spectrum monitor |
| | Data rate management | BER, $\Delta T$ |

Amateur radio operators faced with a similarly anarchistic spectrum management schema, but operating from a fixed site, may use a PC to display the RF spectrum, facilitating choice of channel with a display of energy in the candidate channels. Such an enhanced user interface allows two subscribers to pick a channel that appears clear from both receiving sites.

Military users with AM/FM single-channel radios often have an a-priori channel allocation plan in which each user is given a fixed channel or small set of channels in advance by some central authority. Mobility brings users into conflict in spectrum use leading to a dynamic choice of operating frequency. Some radios facilitate this choice with built-in spectrum displays, again putting the user in the decision loop. Others, like TETRA, pick a clear channel for the user.

Cellular radio systems also manage physical (FDMA) and virtual channels (e.g., TDMA or CDMA) as radio resources. Generally, they have an a-priori set of frequencies per cell site among which a cell handoff algorithm must choose when a mobile subscriber enters the cell. PCS and satellite mobile systems also have to decide when to hand the user "over" to an alternate mode (PCS $\rightarrow$ satellite, for example [366]) or to hand the user "off" to a new cell of the mode currently in use. The handoff algorithms all keep track of which channels are currently in use by the home cell site. Some monitor assigned channels for energy in unused channels to characterize the degree of cochannel interference. Cells with a high rate of transitory traffic, such as near an interstate highway or autobahn in a large city, may employ a hierarchical cell site arrangement with an umbrella cell to handle the fast-moving traffic while conventional cells handle slower-moving vehicular and pedestrian traffic. The handover algorithms may use Doppler to differentiate among fast movers and slow movers [440].

Table 12-2 also shows dynamic mode assignment and data rate management as approaches to QoS management. In the military example, a dynamic mode assignment algorithm could monitor energy in a large number of allocated channels, moving the users from mode to mode as the propagation and interference characteristics indicate. HF Automatic Link Establishment (ALE) employs a channel sounding signal, typically a chirp waveform, to identify the propagation characteristics between a pair of users on a given ionospheric path. The ALE algorithm then chooses the best channel given round-trip characteristics measured by the sounding signal. Although similar approaches are possible in other military bands such as LVHF and VHF/UHF, they have not been widely deployed. There is, of course, a penalty to be paid for the use of such techniques both in terms of the complexity of the transceiver units and in terms of the overhead signals such as the sounders that will appear on the channels, potentially interfering with established users.

Modern receivers almost universally employ embedded microcontrollers which could employ spectrum monitoring and sounding, but the SDR has the DSP power to employ such techniques with little or no incremental impact on cost or complexity. By combining passive monitoring of the spectrum to identify unused channels on alternate modes with a digital sounding and channel coordination waveform, pairs of such military users could enjoy the benefits of dynamic mode assignment without the burden of man-in-the-loop choices. For example, a dynamic channel handoff scheme implemented in the radio could automatically transmit, say, 30 ms bursts of coded data on candidate channels to determine the received SNR on both sides of the link. The radios could autonomously move a pair of users from one channel to the next without user intervention. Such schemes are almost trivial with SDR provided the RF synthesizer tunes fast enough.

Finally, there is a widespread demand for enhanced data rate in military and 3G civilian applications. In order to achieve higher data rate at a given bit error rate (BER), there must be an excess SNR in the channel or there must be multiple channels which may share the aggregate data rate at a lower data rate per channel. Spectrum monitoring can establish the availability of excess BER, which may then be combined with adaptive channel coding (e.g., changing from MSK to QAM) to deliver a higher data rate over a shorter time interval. Spectrum resource management, then, includes spectrum monitoring as a pivotal aspect of autonomous channel, mode, and data rate control. The next section presents two alternative algorithms for monitoring the spectrum in support of such advanced techniques.

## B. Spectrum Monitoring

SDRs with wide IF bandwidths must accommodate different noise levels across the band by using noise-riding squelch algorithms. For example, aeronautical mobile radios operating in VHF and UHF will experience a noise background defined by thermal noise in remote areas such as the arctic and central regions of the Atlantic, Pacific, and Indian oceans. But as the aircraft approaches land masses or heavily populated islands (e.g., Hawaii), the noise backgrounds become dominated by urban noise. This noise is the aggregate of the corona, gap, and ignition noise sources. Some of these sources create synchronous shot-noise (e.g., automobile ignitions). Others are more like intermittent broadband noise with harmonic structure (e.g., electric motors and elevators). The resulting noise has been modeled as Gaussian noise. This sim-

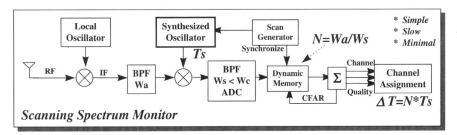

**Figure 12-8**  Scanning spectrum monitor technique.

ple model does not capture the fine structure of this noise. Researchers have characterized the rich time-varying structure of this noise [367]. To achieve best available performance, SDRs need RF squelch algorithms that accommodate the time-varying and nonuniform spectral structure of this noise. Spectral oversampling, narrowband-filtering, and noise-riding threshold squelch algorithms complement more traditional constant false alarm rate (CFAR) squelch algorithms to provide consistent access to the weakest subscribers. Using such techniques, SDRs have the potential to deliver better end-to-end quality with longer reach and greater reliability than analog radios. A first implementation of an SDR may not perform as well as the equivalent analog radio because inadequate attention is paid to the way in which RF/IF monitor algorithms define the effective system sensitivity. Algorithm refinement may include sequential or parallel spectrum monitoring.

***1. Sequential Spectrum Monitor***   Some spectrum management techniques require an estimate of energy in each channel in the access band. The dynamics of this information depend on the rate of change of energy density in the channel which is a function of channel use and multipath. The rate of change of channel use in a spectrum use area (cell) is related to power management, multipath, the speed of the moving users, and the size of the cell sites. For a military scenario there might be 100 users on the average in a use area such as a valley that limits radio propagation to about 20 miles.

A modest rate of change of 6 dB per second per channel can be easily tracked using the sequential scanning spectrum monitor algorithm shown in Figure 12-8. The prototypical SDR has a fixed LO which converts the access band to IF, filtering the access bandwidth $Wa$ for analog-to-digital conversion. Not shown in Figure 12-8, the wideband ADC with sampling rate $> 2.5*Wa$ delivers a wideband stream which is then converted and filtered to select subscriber channels. The scanning spectrum monitor also processes this raw wideband stream, synthesizing a local oscillator digitally, for example, using a tunable bandpass filter (BPF). The subcarrier frequency, $f_c$, is sequentially stepped through the channels so that the output of the BPF represents the energy in the channel. The algorithm synchronizes the stepping of $f_c$ with a memory which retains an estimate of the energy in the channel. This estimate is not just the instantaneous energy in the channel. Such an estimator would be noisy and would not differentiate between a variable-noise background and the presence of a user in the channel. Instead, a *constant false alarm rate* (CFAR) algorithm estimates the background noise while strong differences in CFAR output indicate the onset or departure of a subscriber signal in the channel. The typical CFAR algorithm has the form:

$$X_{i+1} = \alpha Y + \beta X_i$$

where $\alpha < 1$ is the fraction of the current output of the BPF to be included in the power estimate, $\beta < 1$ is the decay rate of the estimator, and $X_i$ is the value of CFAR channel $X$ at time $i$. By adjusting $\beta$, one sets the rate at which the channel will decay, effectively setting the CFAR impulse response. By adjusting $\alpha$, one sets the sensitivity to large fluctuations in the output of the BPF, reducing sensitivity to shot noise. In addition to this energy estimate, one must establish a threshold for noise versus signal. Nonparametric statistical approaches set this threshold at some fraction of total energy distribution across all channels. The idea is that the lowest-power channels contain only noise while the others have interfering signals present.

More complex algorithms keep two estimates of channel energy with different impulse responses. One impulse response is set to decay in a few tens of milliseconds to track the onset of speech energy while the other is set to decay in a few hundred ms to a second or more, tracking the average background noise. When the energy levels in these two estimators differ by some threshold amount, strong interference is present in the channel. When the energy differences between short-term and long-term estimators are reversed, strong interference has left the channel.

The scanner moves from one channel to the next in time $\Delta T$, yielding a complete update in $N * \Delta T$ seconds. If the channel bandwidth is 30 kHz, one must dwell on the channel for at least $1/(30 \text{ kHz}) = 30$ usec; in addition, it will take time to shift subcarrier frequencies. Revisiting $N$ channels sequentially means that each channel is updated only every few milliseconds. Specifically, 100 channels $* 30$–40 usec per channel = 3–4 ms between channel updates. The net effect of the sequential scanner is a reasonably consistent set of estimates across all potentially available channels on which a mode-assignment algorithm operates. Although such scan rates are fairly fast, they cannot track fine-grain channel fading fluctuations which have time constants of tens to hundreds of microseconds.

*2. Parallel Spectrum Monitor*   The parallel spectrum monitor, on the other hand, can track such fine-grain channel characteristics. The structure of this technique is illustrated in Figure 12-9. The parallel spectrum monitor estimates the power spectral density of all channels in bandwidth $Wa$ at once, typically employing an efficient algorithm such as the fast Fourier transform (FFT). The FFT estimates the spectrum of $N$ sample points in $N * (\log_2 N)$ computations,

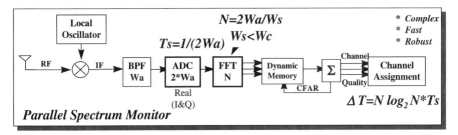

**Figure 12-9**  Parallel spectrum monitor technique.

**TABLE 12-3  Parallel Channel Monitor Parameters**

| Parameter | Value |
|---|---|
| $W_c$ | 25 kHz |
| $N$ | 100 |
| $W_a = N * W_c$ | 2.5 MHz |
| $2 Wa = f_s$ | 5 MHz |
| $T_s = 1/f_s$ | 200 ns |
| $2N$ points | 200 |
| $T_b = 2N * T_s$ | 40 $\mu$sec |

producing $N/2$ nonredundant complex samples. Since the FFT is a block process, yielding results in parallel, its output can feed a parallel CFAR algorithm which computes all CFAR energy estimates "in parallel" between FFT blocks. If the acquisition bandwidth $Wa$ is sampled at exactly the Nyquist rate, $2 Wa$, then $2N$ sample points yields $N$ channel energy estimates if the sample rate is an integer multiple of the channel spacing, $Wc$. The parallel channel monitor parameters for a notional 100-channel FDMA system are given in Table 12-3.

Since the spectrum is updated every 40 $\mu$sec, there are plenty of samples per channel available to track fine fading structure and hence to characterize a channel's stability over time as well as its general energy occupancy. In the limit, each channel may be sampled at a small multiple of the channel's Nyquist rate yielding a sample stream per channel that may be demodulated, having used the FFT as a parallel filter bank. Such an arrangement is sometimes called a transcoder or transmultiplexer.

We may therefore view spectrum monitoring as a family of algorithms for estimating the energy density and related temporal characteristics of channels in an access band. On the low end, the channel-scanning techniques revisit channels sufficiently fast to track user occupancy. As parallelism increases, the rate at which each channel's samples are updated increases. FFT techniques can, in the limit, sample each spectral component fast enough to reconstruct the channel impulse response and subscriber waveforms in the channels in parallel. For the SDR, the wideband ADC architecture supports any of the techniques in this continuum, subject to the availability of processing resources. In fact, such channel scanning can be done in the background in the SDR, employing reserve processing resources in a way that shifts resources to subscriber services as they are needed. Potential dynamic reassignment of processing resources is a key theme of software radio design strategy. Massively parallel hardware platforms may allocate resources in a fixed scheme, wasting large fractions of available processing power. It is possible to reduce hardware costs at the expense of a deliberate increase in software complexity. Antenna diversity and dynamic data rate are two additional areas in which dynamic allocation of processing resources may be appropriate.

## III. MODEM SOFTWARE

The baseband segment imparts the first level of channel modulation onto the signal and conversely demodulates the signal in the receiver. These functions are implemented in the modem software.

### A. Modem Complexity

Predistortion for nonlinear channels and trellis coding are included in baseband modem processing. Soft-decision parameter estimation may also occur in the baseband processing segment. The complexity of this segment therefore depends on the bandwidth at baseband, $W_b$, the complexity of the channel waveform, and related processing (e.g., soft decision support). For digitally encoded baseband waveforms such as binary phase shift keying (BPSK), quadrature phase shift keying (QPSK), Gaussian minimal shift keying (GMSK), and 8-PSK with channel symbol (baud) rates of $R_b$:

$$R_b/3 < W_b < 2*R_b$$

In the transmission side of the baseband segment, such waveforms are generated one sample at a time (a "point operation"). Typically two to five samples are generated for the highest-frequency component so that digital signal processing demand falls between $2*W_b$ and $5*W_b$. Greater oversampling decreases the transmitted power of spectral artifacts, but also linearly increases processing demand. Analog basebands such as FM voice (e.g., in AMPS) may also be modulated and demodulated in the baseband segment, with a processing demand of less than 1 MIPS per subscriber.

### B. SPEAKeasy II API

The functions listed in Figure 12-5 are included in the SPEAKeasy II modem control software API. In addition to the message buffering and control messages, the modem control functions include functions for instantiation, parameter and mode control, and operation (Table 12-4). Instantiation requires a

**TABLE 12-4  Modem Control Functions**

| No. | Name | Phase |
|---|---|---|
| 1 | ACK | All |
| 2 | Buffer_Complete | All |
| 3 | Buffer_Notify | All |
| 4 | Forward_Message | All |
| 5 | NACK | All |
| 6 | Connection_Test | Instantiation |
| 7 | Standard_Data_Msg | Instantiation |
| 8 | Activate_Channel | Params_&_Mode |
| 9 | Adjust_RX_Calibration_Response | Params_&_Mode |
| 10 | TX_Calibration_Complete | Params_&_Mode |
| 11 | Crypto_Status | Operation |
| 12 | Pacing_Indication | Operation |
| 13 | Receive_Mode | Operation |
| 14 | Standard_Data_Msg | Operation |
| 15 | Transmit_Mode | Operation |

©1999 IEEE, reprinted from [30] with permission.

connection test in addition to the standard data messages of front-end control. Channel activation, adjustment of receiver calibration responses, and other transmit calibration are required in the parameter and mode setup phase. The crypto status function allows the modem to report whether the crypto is in sync or not. If not, then the crypto control can flywheel through the loss of sync and resynchronize if necessary. The modem may also report transmit and receive status, as well as accepting the standard data messages.

## C. Modulation/Demodulation Techniques

Modulation in the channel has a significant effect on the quality of the information transfer measured in BER and on the complexity of the receiver. Receiver complexity generally dominates the complexity of the SDR. A receiver is typically four times more complex than a transmitter in terms of MIPS required to implement the baseband and IF processing functions in software. The modem accounts for the majority of processing demand in the isochronous stream after IF processing. Modem algorithm topics include AGC, channel waveforms, coding, and spread spectrum.

*1. AGC*  The AGC algorithm can consume substantial computational resources because it processes every sample on the isochronous streams. AGC may be applied to wideband streams (e.g., implemented in an ASIC). It may be applied to channel-bandwidth streams by a DSP. Or it may be applied to the voice channel. An illustrative AGC algorithm is shown in Figure 12-10.

```
procedure Perform_Automatic_Gain_Control is

    T: constant := 0.005; Attack_Time  : constant := 0.03;
    Decay_Time   : constant := 0.03;

    Alpha_A: constant := 2.3 / Attack_Time;
    Alpha_D    : constant := 2.3 / Decay_Time;

    K1 : constant := (T * Alpha_A) / (2.0 + T * Alpha_A);

    K2 : constant := (T * Alpha_A - 2.0) / (2.0 + T * Alpha_A);

    K3 : constant := (T * Alpha_D) / (2.0 + T * Alpha_D);

    K4 : constant := (T * Alpha_D - 2.0) / (2.0 + T * Alpha_D);

    X_n : Float;   Y_n : Float;
begin
    X_n := Float(Recv_Spl_NB_PTR.RcvPwr_T) / GDP.Scale_Factor;
    if X_n > Previous_Y_n then

      Y_n := K1 * (X_n + Previous_X_n) - K2 * Previous_Y_n;

    else

      Y_n := K3 * (X_n + Previous_X_n) - K4 * Previous_Y_n;

    end if;

    Previous_X_n := X_n;

    Previous_Y_n := Y_n;

    Output_AGC_MSG (Attn_Ctl => Integer16(Y_n));
exception
    when Numeric_Error =>

      put_line("RX_Utility.Perform_AGC - NUMERIC ERROR");

    when Constraint_Error =>

      put_line("RX_Utility.Perform_AGC - CONSTRAINT ERROR");

    when others =>

      put_line("RX_Utility.Perform_AGC - ? EXCEPTION");

    end Perform_Automatic_Gain_Control;
```

**Figure 12-10**   Illustrative AGC algorithm.

*2. Channel Waveform Coherence*   As shown in Figure 12-11, the probability of bit error is a function of channel modulation. Amplitude shift keying (ASK) provides the lowest received signal quality for a given received SNR. Since the receiver does not attempt to lock to the carrier frequency in any way, ASK essentially delivers the performance of a narrowband filter in Gaussian noise. On the other hand, the receiver is exceedingly simple, consisting of a narrowband filter and a threshold circuit.

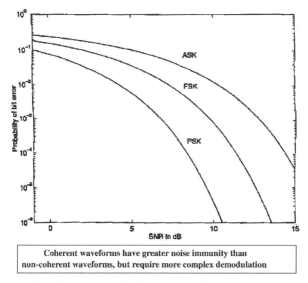

Figure 12-11   Bit error rate (BER) versus signal-to-noise ratio (SNR).

The frequency shift keying (FSK) channel modulation estimates the carrier and forms two filters, generally called the *mark* and *space filters* for binary FSK. In addition, most FSK receivers compute the ratio of the energy of the mark and space filters, deciding on a 1 or 0 as a function of that ratio. Since this ratio is computed continuously as a function of the filter energy in the two filters, there is a transition region between mark and space signals. The algorithm also needs to establish timing. FSK receivers may therefore include initial timing recovery logic that predicts the time of bit transitions and that performs the mark/space decisions near the middle of a channel symbol. The associated data protocols generally include a sequence of repeated reversals between the 1 and 0 states to establish bit timing. There may also be timed energy accumulators that integrate filter energy during each bit period and then reset to zero after a bit decision is made. These are called integrate-and-dump filters. The receiver is more complex than the ASK receiver, but the received BER is the equivalent of about 3 dB better with FSK than with ASK [20].

FSK requires an initial estimate of frequency to determine the parameters of the mark/space filters but the FSK receiver algorithms need not maintain carrier lock at every sample. It is sufficient for an FSK receiver to track Doppler shifts which may be on the order of a few Hz to a few hundred Hz, depending on frequency, speed of the communications nodes, and speed of reflectors such as the ionosphere in HF modes.

Phase shift keying (PSK), on the other hand, detects information as a synchronous change of the instantaneous phase of the carrier [20]. Frequency is the time-domain integral of phase, so the FSK receiver operates on an integral function, while the PSK receiver algorithms operate on its derivative. This introduces the necessity of continuously tracking carrier frequency and phase in order to recognize the phase discontinuities that encode the information. Timing circuits must recognize when phase transitions should occur in order to recognize successive transmission of the same channel symbol. There are differential modes of encoding the information bits and imparting them on to PSK-modulated carriers that enhance the receiver performance. In addition, most PSK signals are randomized in order to ensure sufficient density of bit transitions to keep the clock recovery logic synchronized. The *Costas loop* is a continuous differential feedback method of carrier recovery that provides carrier recovery and tracking [245]. The continuous nature of PSK tracking of phase plus the complexity of the carrier recovery and timing logic make the PSK the most complex receiver of the three in Figure 12-11. Randomization of the transmitted bits and differential encoding also contribute to transmitter complexity. The benefit, however, is another 3–5 dB of enhancement of BER, depending on the quality of the implementation and the degree to which timing recovery algorithms make use of framing and other inherent redundancies present in the aggregate structure of the PSK waveform.

*3. BER, SNR, and Algorithm Complexity*   Seven modulation techniques that are common in current telecommunications practice are illustrated in Figure 12-12. These are amplitude modulation (AM), frequency modulation (FM), FSK, pulsed modulation, phase shift modulation including PSK, minimum shift keying (MSK), and quadrature amplitude modulation (QAM).

AM imparts information to a signal by adjusting the RF amplitude in proportion to the information to be transmitted. Continuously modulated AM voice is audible at negative SNR because of the tonal structure of voice. Voice has a nominal information bandwidth of about 4 kHz. That is, although the speech spectrum extends to about 20 kHz, the words may be understood with a high probability (> .99) if the speech is filtered to 4 kHz. Furthermore, speech includes several temporal-spectral structures, some of which are noiselike and others of which are tonal. Energy in the tonal segments is isolated into a few (1 or 2, generally) strong tones called formants. These formants occupy only a few tens of Hz to perhaps 100 Hz, concentrating the available speech energy into narrow spectral bands. Speech researchers have shown that the first two formants carry most of the information. As a result, an AM SNR of zero dB actually has a positive SNR of the ratio of the concentration of speech energy into formants, which can be 20 Hz/4000 Hz, which is 200 : 1 or about 23 dB. Since the AM receiver's narrowband filter simply tracks the carrier envelope, the sinusoids associated with the first one or two formants are tracked by even the simplest AM algorithm, yielding good speech intelligibility at even negative SNR in the speech channel.

Frequency modulation imparts information by continuously varying the transmitted frequency. An FM receiver's key component is the FM discriminator, essentially a derivative function [20]. The output of the discriminator is

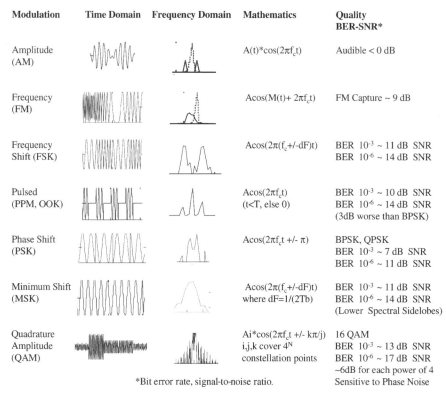

| Modulation | Time Domain | Frequency Domain | Mathematics | Quality BER-SNR* |
|---|---|---|---|---|
| Amplitude (AM) | | | $A(t)*\cos(2\pi f_c t)$ | Audible < 0 dB |
| Frequency (FM) | | | $A\cos(M(t)+ 2\pi f_c t)$ | FM Capture ~ 9 dB |
| Frequency Shift (FSK) | | | $A\cos(2\pi(f_c +/- dF)t)$ | BER $10^{-3}$ ~ 11 dB SNR<br>BER $10^{-6}$ ~ 14 dB SNR |
| Pulsed (PPM, OOK) | | | $A\cos(2\pi f_c t)$<br>(t<T, else 0) | BER $10^{-3}$ ~ 10 dB SNR<br>BER $10^{-6}$ ~ 14 dB SNR<br>(3dB worse than BPSK) |
| Phase Shift (PSK) | | | $A\cos(2\pi f_c t +/- \pi)$ | BPSK, QPSK<br>BER $10^{-3}$ ~ 7 dB SNR<br>BER $10^{-6}$ ~ 11 dB SNR |
| Minimum Shift (MSK) | | | $A\cos(2\pi(f_c +/- dF)t)$<br>where dF=1/(2Tb) | BER $10^{-3}$ ~ 11 dB SNR<br>BER $10^{-6}$ ~ 14 dB SNR<br>(Lower Spectral Sidelobes) |
| Quadrature Amplitude (QAM) | | | $A_i*\cos(2\pi f_c t +/- k\pi/j)$<br>i,j,k cover $4^N$<br>constellation points | 16 QAM<br>BER $10^{-3}$ ~ 13 dB SNR<br>BER $10^{-6}$ ~ 17 dB SNR<br>~6dB for each power of 4<br>Sensitive to Phase Noise |

*Bit error rate, signal-to-noise ratio.

**Figure 12-12**   Quality parameters of channel waveforms.

noiselike until the input SNR exceeds a critical value above which the output becomes a continuous sinusoid. That threshold is about 9 dB, depending on the quality of the filtering before the FM discriminator. As a result, there is essentially no received signal for SNR below 7 dB. The signal quality above 7 dB is generally noise-free due to the noise-suppressing quality of the discriminator. Examining the statistical structure of the output of an FM discriminator is revealing. When no carrier is present, the variance of the energy is large. When a carrier is present, however, the variance collapses significantly, an effect called FM quieting. That variance reaches a critical point between 7 and 9 dB at the onset of FM capture. An algorithm that recognizes FM quieting can be used to identify the presence of man-made interference in a channel with low SNR. This can be useful in autonomous channel negotiations such as in HF Automatic Link Establishment (ALE).

***4. An FSK Demodulation Algorithm***   FSK, PSK and pulsed are discrete or digital modulations that have the characteristics discussed above. In addition, Figure 12-12 shows how BER varies as a function of SNR for each of these channel modulations. The spectra shown in the figure also illustrate the way

```
function Perform_FSK_Demodulation (I : Integer16; Q : Integer16) return Float is
       Y: Float;  Phi : Float;
    begin
       Phi := ML.Atan2(Float(Q), Float(I));    Y := Phi - Previous_Phi;
       Previous_Phi := Phi;
       if Y > FSK_Demod_Threshold then
         Y := Y - 2.0 * NC.Pi;
       elsif Y < -FSK_Demod_Threshold then
         Y := Y + 2.0 * NC.Pi;
       end if;
       return (Y * float(GDP.Sample_Rate));
    exception
       when Numeric_Error =>              handleNUMERIC ERROR;
       when Constraint_Error =>           handleCONSTRAINT ERROR;
       when others =>          handleOtherEXCEPTION");
    end Perform_FSK_Demodulation;
```

**Figure 12-13**  Illustrative FSK demodulation algorithm.

in which each of these channel modulations distributes signal energy in the frequency domain. An illustrative FSK algorithm is shown in Figure 12-13. This algorithm is based on estimating whether phase is greater than or less than a decision threshold. Not included is bit timing that determines when to make a bit decision.

These types of modulation have relatively high-frequency domain sidelobes which result in energy in adjacent channels, creating levels of adjacent channel interference that limit the packing density of users in the spectrum. MSK overcomes this limitation of the discrete modulations somewhat by shifting the phase by the minimum amount necessary to support the data rate. This matching of the keying rate (of the information bandwidth) with the spectral distribution of essentially an FSK waveform (i.e., the separation of the FSK mark/space filters) results in a compact spectral shape in which much more of the spectral energy is concentrated in the central lobe, reducing adjacent channel interference. The computational complexity of the MSK receiver is slightly greater than that of the typical PSK receiver because of additional filtering requirements.

**5. *QAM***  In analyzing BER versus SNR, one differentiates the symbol error rate (SER) from the BER. In antipodal modulation, there are two channel states ("mark" and "space"; or 1 and 0). Each channel symbol is also called a "baud" from voice frequency telegraphy. As the number of channel symbols increases, the number of bits represented per channel symbol increases. Thus, a quaternary phase shift keyed (QPSK) channel symbol with four states conveys two bits of information. The SNR represented as $Eb/No$, then, is reduced by a factor of two (6 dB) for fixed total carrier power, or equivalently for a given $Eb/No$, the total carrier power has increased by 6 dB in QPSK versus BPSK. This relationship may be expressed as a relationship between $Es/No$ and $Eb/No$:

$$Eb/No = Es/No - 10 * \log_2(B),$$

where $B$ is the number of bits per channel symbol.

Examples:    BPSK   with   $Es/No = 10$ dB,

$$Eb/No = 10 \text{ dB} - 10 * \log_2(1) = 10 + 0 = 10 \text{ dB};$$

But for QPSK   with   $Es/No = 10$ dB,

$$Eb/No = 10 \text{ dB} - 10 * \log_2(2) = 10 - 6 \text{ dB} = 4 \text{ dB}$$

These channel symbols having one or two bits per baud or less, work well in moderate and low SNR, but do not take advantage of excess high SNR. QAM symbols, on the other hand, are specifically designed to use excess SNR to improve data rate. That is, there is a tradeoff of BER versus data rate. At a positive SNR of 45 dB, achievable in some circumstances, the BER is essentially zero, but if the modulation type is MSK, there is no flexibility in the signal structure to transmit more information using more bits per symbol. QAM modulates phase and amplitude simultaneously, creating phase-amplitude combinations called channel states. One might, for example, amplitude modulate and stagger the reference phase of four QPSK constellations to create 16 distinct channel states. Modulation is much more complex and the RF power amplifiers have to be linear in order to accurately reflect the amplitude modulation. In addition, the receiver must equalize the amplitude and phase transfer characteristics of the propagation channel in order to recover the transmitted constellation. This equalization is accomplished in a continuous process in which a number of bit intervals (from 2 as a minimum to 20 or more) are stored in a tapped delay line, multiplied by a set of tap weights, and integrated to yield an equalized estimate of the input to the channel.

The computational complexity of the QAM equalizer is at least one to two orders of magnitude more complex than a simple PSK receiver. Generally a period of blind equalization sets the equalizer taps to reasonable values based on the known envelope structure of the waveform. Decision-directed feedback may then fine-tune the weights to drive the amplitude-phase values toward the constellation points that are known a-priori. The benefit of such modulation is to increase the data rate proportional to the log (base 2) of the number of constellation points. QPSK transmits 2 bits per symbol and hence 2b bits per second. At the same baud rate (and hence the same RF bandwidth), 16 QAM transmits $\log_2(16) = \log_2(2^4)$ or 4 bits per symbol.

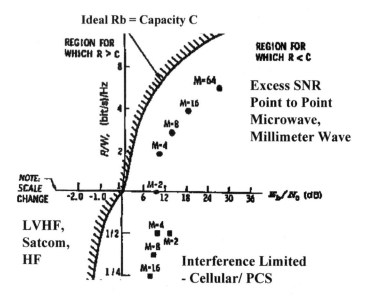

**Figure 12-14**   Modulation efficiency in bits per Hz versus SNR.

*6. Adaptive Channel Coding*   As the number of bits per symbol increases, so does the bandwidth efficiency, $Rb/W$, the ratio of data rate $Rb$ versus occupied bandwidth $W$. The penalty that must be paid in order to achieve this increased packing density is an increase in the SNR. Figure 12-14 shows how $Rb/W$ and $Eb/No$ are related. As more than one channel symbol encodes a single bit (for fractional $Rb/W$), one can transfer information reliably (e.g., BER = $10^{-5}$) with negative SNR. The region of negative per-baud SNR is known as the *power-limited region*. There is a computational penalty to operating in this regime measured in the increased complexity of the FEC encoding and decoding.

Thus, there is a range of digital channel modulation techniques from OOK and BPSK through QAM, FEC encoded or not. Operation on a nominal channel with 10 to 15 dB of SNR is readily accomplished using simple channel modulation techniques, simple transmitters, and unsophisticated receivers. These are typical modes of SDRs. In addition, however, digital (ADC) access in the power-limited regime typically calls for increased transmitter complexity in the form of the FEC encoder or a higher-level retransmission protocol such as automatic repeat request (ARQ), possibly hybridized with FEC. In the SDR, an order of magnitude or so additional processing resources are required for the FEC coding and decoding compared with uncoded modulation. As more SNR becomes available in the channel, QAM modulation can increase Rb for a given channel bandwidth. An additional two or three orders of

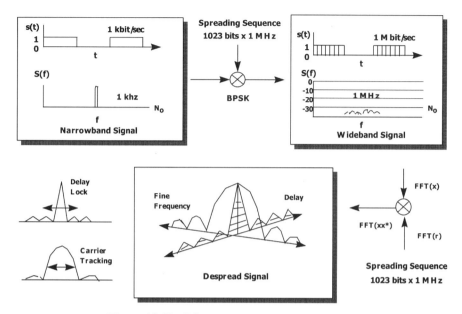

**Figure 12-15**  Direct sequence spread spectrum.

magnitude of equivalent processing resources is required for software implementations of the required equalizers. In addition, the transmitter modulation system must be very linear in order to preserve the amplitude component of the QAM channel modulation. The transmitter and receiver must also have low total phase noise to keep from inducing bit errors in the phase dimension of the QAM constellations.

*7. Spread Spectrum*  Low-SNR conditions can also be addressed by spreading an information bit over multiple redundant channel symbols. For a given information rate, $Rb$ (e.g., 1 kHz), the total channel bandwidth $W$ increases by the number of redundant bits or "chips" per information bit (e.g., 1000). $Rc$ is the chip rate, the product of $Rb$ times the number of chips per information bit. This technique is called direct-sequence, spread-spectrum (DSSS) because one can think of the resulting chip sequence as the direct product of the information bitstream times a repeating pseudonoise sequence of randomized chips. Figure 12-15 illustrates the overall process.

The resulting DSSS channel waveform may be studied in terms of its ambiguity surface, the generalized cross-correlation of the transmitted and received waveforms. This correlation function is also called *Woodward's Ambiguity Function* [441] or the *cross-ambiguity function* (CAF). The two CAF axes are time delay and fine frequency offset, which may be induced by target Doppler shift (as in radar applications), or propagation effects (as in Doppler shift of multipath replicas of the transmitted communications signal).

The delay axis of the CAF surface contains the cross-correlation of the transmitted and received waveforms. With specially designed symbol streams such as Barker codes, the cross-correlation has a large central spike and low time-domain sidelobes as shown in Figure 12-15. In addition, radar designers use CAF to design waveforms that have low sidelobes in time and frequency offset to ensure good target detection properties. Communications designers employ such signal structures for improved despreading. In the time domain, a single strong spike with low time domain sidelobes reduces the probability of a false lock on a sidelobe when attempting to establish timing for despreading. In the frequency axis, low sidelobes avoid false carrier lock. Spread spectrum systems must establish initial time and frequency estimates with small or negative channel carrier-to-noise ratio. As a result, many spread-spectrum systems employ precision timing reference systems to reduce the uncertainty in the time domain and hence to reduce the time-domain search for the signal. DSSS was first developed for military applications for resistance to jamming and to hide the fact of a communications signal from unintended receivers [4].

DSSS signals may use two or more different spreading codes on the same channel at the same time. Codes may have cross-correlation properties that yield small correlation lobes everywhere on the CAF surface. Generally, the energy in a sidelobe on the average cannot be controlled to much less than the square root of the number of chips per bit, or

$$E(t,f) < E(0,0)/(\sqrt{N})$$

where $N$ is the number of chips per bit.

For large $N$ (on the order of 100 to 10,000), more than one user can occupy exactly the same channel at exactly the same time. Receivers set for two orthogonal transmitting codes can receive two independent information streams without undue interference with each other. This effect is the basis for CDMA, in which each subscriber is assigned a different orthogonal sequence. The IS-95 CDMA standard employs a single spreading code which is very long ($10^{15}$ bits before repeating). Independent subscribers are assigned different quasi-orthogonal subsets of the code based on code phase (position from the start of the code at a given initial sequence). A pilot sequence and other techniques simplify the receiver design by providing ancillary information that reduces code search in uplink and downlink directions.

Commercial CDMA receivers employ time-domain correlation to despread the information stream from the channel modulation. Receivers may also use frequency domain techniques. The major components employed in the design of a DSSS receiver that uses frequency domain techniques consist of an ADC, FFTs, and related time and frequency tracking logic. A wideband ADC samples the channel with sufficient oversampling and resolution to preserve the signal properties needed to recover the information stream. A sampling rate of 50 MHz yields a 100 M byte per second input data stream. This stream may then be converted from the time domain to the frequency domain using a fast Fourier transform (FFT). It is then multiplied by a stored frequency-domain replica of the transmitted signal $x(t)$, yielding the Fourier transform of the cross-correlation. An inverse FFT recovers the despread time-domain signal. Given two stored spreading sequences (representing binary 1 and 0 in the information stream), each synchronized FFT block results in a despreading window in which a 1/0 decision is made.

Processing also includes a delay lock loop (implemented in DSP Dly in Figure 12-15); a frequency-lock loop for carrier recovery, Doppler tracking and related signal processing; and other control and interface processing including FEC decoding. Each of the processing modules requires about 200 MFLOPS of processing capacity, yielding about 1 GFLOPS total processing capacity required. Using quad-TMS320C40 class processors in an open-architecture VME enclosure and no particular attention to packaging, such a module would weigh about 40 lbs using technology available in 1995. By 1999, the receiver could be packaged in a Xilinx FPGA with one TI C67xx DSP chip.

The spread-spectrum system provides processing gain. IS-95 employs a 1.2288 MHz chip rate with a 8 kbps information stream for processing gain of $1288/8 = 153.6 = 21.8$ dB. This is the processing gain with respect to the 8 kbps narrowband information signal. Gain with respect to another spread-spectrum signal is the ratio of the processing gain given above (total processing gain) to the inner product residue energy of the two *quasi-orthogonal* spreading codes. The term quasi-orthogonal refers to the fact that no two codes of reasonable length are completely orthogonal for all relative delays. The minimum inner product may be zero, but the residue energy may be as large as the square root of the number of chips per bit, in this case sqrt $(153.6) = 12.4$ $E$, where $E$ is the energy of a chip period. The processing gain, then, would be reduced to $153.6/12.4 = 12.4$, or the gain in dB/2 = 10.9. So, depending on the code design and bit phase, the inner product of two users of a long code yields a cross-correlation processing gain with respect to the other code of on the order of 10 to 20 dB.

At the physical layer, DSSS processing gain can be used in several ways. The military applications tend to drive power down so that the signal energy is less than kTB noise, hiding the fact of a communications event. High-power narrowband signals in the frequency domain become dispersed in the time domain despread signal. As a result, the spread-spectrum waveform rejects narrowband interference whether due to narrowband jamming or to uncooperative use of the spectrum. Finally, spread-spectrum processing gain may be used to differentiate two CDMA subscribers provided the inner product of the RF signals driven by the near–far signal strength does not exceed the relative processing gain of the two quasi-orthogonal codes.

From an SDR perspective, one may also observe that a GFLOPS-class processing suite with reasonable IF bandwidth and dynamic range can operate in DSSS mode, providing the needed processing capacity (e.g., of FFT, IFFT, Dly, and Freq processing described above) can be achieved in processors that have sufficient connectivity, double buffering, etc. required to both sustain

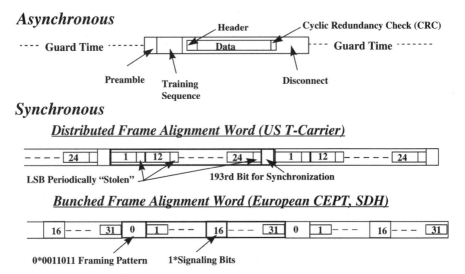

**Figure 12-16**   Data stream synchronization.

throughput. Thus, physical layer access drives the processing requirements of the DSP suite.

## D. Synchronization

Radio links employ both asynchronous and synchronous modes of frame synchronization as illustrated in Figure 12-16. Asynchronous framing is employed in TDMA systems as illustrated in the figure. Each subscriber is assigned a time slot in which the data burst must be presented. Due to the differences in distance from the base transceiver station (BTS), an MS cannot guarantee its time of arrival precisely.

In principle, an MS could know the location of itself (e.g., from GPS) and the location of the BTS from its identity broadcast on a pilot channel. If these locations are known to within, say, 100 meters, the uncertainty of time of arrival at the BTS would be 300–600 ns. Normally, the guard time between TDMA slots reflects the ratio of maximum to minimum range of MS, which might be a few miles, corresponding to 5 to 15 microseconds of propagation delay uncertainty. The MS could, again in principle, adjust its time of transmission so as to arrive within a .6 $\mu$sec window versus a 15 $\mu$sec window, reducing the guard time overhead. In fact, SDRs equipped with GPS receivers and precision timing references could operate in such an enhanced mode.

But in practice, one must keep the cost of the MS low. Thus, the near-term MS generally will not have an on-board GPS receiver and will lack the timing circuitry and calibration necessary to control a transmission to within a few hundred ns. In addition, one would not want to expend MS processing capacity (and hence battery drain) to continuously estimate its precise distance from the BTS. As a result, fairly large guard times are required between TDMA bursts.

Synchronous transmission is also illustrated in Figure 12-16. Frame alignment words are transmitted periodically to ensure proper framing of the received bitstream. Frame alignment words may either be bunched or distributed. The distributed technique is employed in the U.S. T-carrier system. In this case, a least-significant bit (LSB) of subscriber data is "stolen" periodically. The more advanced European E-carrier system employs bunched frame alignment. In this case, a framing octet is transmitted in every frame. The unique signature of the frame alignment word provides for relatively unambiguous frame alignment every $32 * 8 = 256$ bits.

Synchronous framing has been employed widely in terrestrial point-to-point microwave systems. These high-capacity trunking (backbone) systems use dedicated RF spectrum on fixed terrestrial paths to provide high-capacity interconnect. Prior to the advent of gigabit fiber, these radios comprised the bulk of the trunking capacity of developed nations, including the United States. Once the investment has been made in fiber optics, the return on investment is significantly better than microwave radio, however. As a result, microwave radio is now found primarily in developing economies. This technology employed constant modulus (blind) equalization to reduce intersymbol interference (ISI) that is induced by multipath fading in typical microwave radio deployments. This equalizer technology is computationally intensive, but its enhancement to BER makes it advantageous. As a result, many other radio modes (e.g., GSM) employ constant modulus equalizers to enhance BER.

The synchronous multiplex hierarchy is illustrated in Figure 12-17. The United States, Japan, and Europe employ different synchronous hierarchies. International telephone gateways normalize these data rates through the use of "bit-stuffing." This is the periodic insertion of redundant bits as required to synchronize the different clocks on each side of the gateway.

### E. Equalizer Complexity

From the SDR perspective, the choice of synchronization has a major impact on receiver complexity. Synchronous equalizers require equalizer capacity that is a function of the data rate, spectral efficiency of the modulation (bits per Hz), and rate of change of multipath structure. A U.S. T3 carrier, for example, operates at 45 Mbps. With QPSK modulation, the data rate is supported by 22.5 M channel symbols per second. In order to initialize a tapped-delay-line equalizer, one must invert a matrix capable of accommodating the delay spread, typically at least 30 taps. Inverting a $30 \times 30$ matrix requires on the order of $N^3$ multiply operations. The resulting delay line tap weights must be applied on every channel symbol (baud). This requires $30 \times 22.5$ M or over 600 M multiplies per second. To invert a $30 \times 30$ matrix on every baud would be computationally prohibitive. As a result, the taps are estimated infrequently

| U.S., Japanese, and European Multiplex Hierarchies | LEVEL | | | |
|---|---|---|---|---|
| | 1 | 2 | 3 | 4 |
| *United States* (Bell/T-/DS- Level) | | | | |
| Number of Tributaries | – | 4 | 7 | 6 |
| Number of Voice Channels | 24 | 96 | 762 | 4032 |
| Line Rate (Mbit/sec) | 1.544 | 6.312 | 44.736 | 274.176 |
| Designation | T1/DS-1 | DS-2 | DS-3 | DS-4 |
| *Japan* | | | | |
| Number of Tributaries | – | 4 | 5 | 3 |
| Number of Voice Channels | 24 | 96 | 480 | 1440 |
| Line Rate (Mbit/sec) | 1.544 | 6.312 | 32.064 | 97.728 |
| *European* (CEPT) | | | | |
| Number of Tributaries | – | 4 | 4 | 4 |
| Number of Voice Channels | 30 | 120 | 480 | 1920 |
| Line Rate (Mbit/sec) | 2.048 | 8.448 | 34.368 | 139.264 |
| *Synchronous Digital Hierarchy (SDH)** | | | | |
| Number of Tributaries | OC-1 | OC-3 | OC-12 | OC-48 |
| Line Rate (Mbit/sec) | 51.84 | 155.52 | 622.080 | 2448.32 |

*ITU-T Recommendations for the Integrated Services Digital Network (ISDN); OC= optical communications.

**Figure 12-17** Synchronous multiplex hierarchy.

(e.g., once every 2 seconds). But the inverted matrix is updated using linear approximation techniques frequently enough to track the changing channel impulse response. In this case, one might update the weights every 256 baud for a 100 kHz update rate. Even the linearized inversion requires additional processing, driving the aggregate beyond 2 GFLOPS. As a result, many time-domain equalizers employ hybrid analog-digital multipliers and weight-update circuits in lieu of more computationally intensive DSP approaches [368].

The asynchronous TDMA bursts of GSM require equalization on each burst in order to accommodate channel impulse response changes from burst to burst (see Figure 12-18). The first-generation AMPS standards transmitted 20 kbps Manchester-coded bitstreams. As a result, the 1–10 $\mu$sec of delay spread created little intersymbol interference. GSM's 270 kbps data rate is so high that the delay spread causes significant ISI. Equalization per burst requires tens of MFLOPS of sustained capacity, whereas demodulation of the AMPS Manchester-coded low-data-rate stream required only a few MFLOPS per channel. Equalization creates a converged constellation in QAM signals that allows reliable bit decisions.

### F. Demodulation Decisions

Other types of modulation, such as low-speed FSK, may be enhanced using soft-decision decoding. With hard-decision decoding, threshold logic determines whether a received channel symbol is demodulated to the 1 or 0 channel state. An FSK demodulator, for example, has a "mark" filter and a "space" filter, corresponding to each of the transmitted frequencies. The difference in

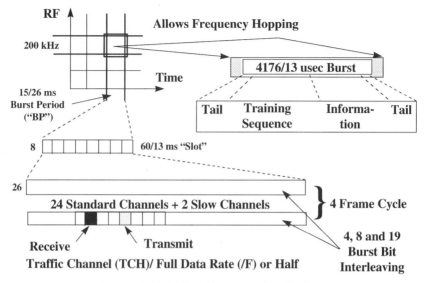

**Figure 12-18** GSM burst synchronization.

energy of the mark filter to the space filter is an example of a hard-decision parameter. (This is not a very good parameter, but its simple mathematical treatment motivates its use here.) If the ratio exceeds 0.0, the channel symbol would be declared a mark (1). Otherwise, it would be declared to be 0. The estimation of the time of baud transitions is obviously key to success with such an approach. Assume, for the moment, that a timing recovery circuit has identified an appropriate place near the middle of the baud at which to make a bit decision. The hard decoder works well when the SNR is high, as illustrated in Figure 12-19.

The $x$-axis of this figure corresponds to the difference of the decision parameter from the decision value. Clearly, the probability of significant positive or negative energy from either the mark or space filters yields two clusters of output values shown by the solid probability densities of the figure. Hard decisions are computationally simple and effective at high SNR. As the channel enters a fade, however, the probability densities become more diffuse as the variance of the signal from the mark and space filters increases. Note that fading at the mark and space frequencies need not be uniform. But an extremely negative value of the decision parameter indicates a mark with little probability that the decision is in error. Near the center of the decision space (near 0), the situation is not so clear. The mark and space filter differences could easily "flip" so that the output is negative for spaces and conversely. The soft-decision template beneath the $x$-axis divides the decision space into eight equal regions. The leading bit of the soft decision represents the most likely value of the bit, while the next two bits represent the confidence in such a decision.

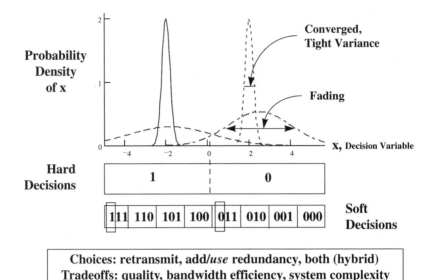

**Figure 12-19**   Hard vs. soft decision error control.

How could such soft-decision data be used? A bunched frame alignment word (synchronous case) and a training sequence (asynchronous case) are known bit patterns. If either pattern is received during a shallow fade, its approximate time of arrival will be known. If 8 out of 10 bits correspond to the expected pattern and the two bits that are in error have low confidence, it maximizes the a-posteriori probability of the sequence to change the low confidence bits to the appropriate state. The value here is that the overall bitstream remains synchronized in spite of a poor SNR. As a result, the subscriber may experience low-quality data, but the channel will not suddenly "go away" as would be the case for lost synchronization. Resynchronizing a synchronous stream typically requires the equivalent of thousands to millions of bits of lost data. Thus, soft-decision decoding can be a powerful technique for maintaining synchronization of the air interface. It increases the computational burden on the SDR, however. In the case of the simple hard-decision threshold logic bit decision, only a few instructions (less than 10) need be executed for each bit decision. In the case of the soft-decision algorithm, however, the computational burden may increase 10 to 100-fold, depending on the richness of the framing information in the data stream and on the implementation approach (e.g., high-level language versus optimized microcode).

## G. Forward Error Control (FEC)

Other techniques for error control are illustrated in Figure 12-20. Automatic repeat request (ARQ) is among the simplest. This algorithm keeps track of packets, acknowledging those that are received. The acknowledgments may

| Technique | | Complexity Drivers |
|---|---|---|

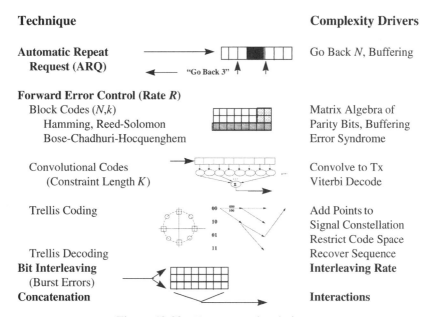

**Figure 12-20** Error control techniques.

be explicit packet-by-packet "ACK" messages. Or they may be grouped into a single ACK of multiple received packets. Or the receiver may instruct the transmitter to "Go Back $N$" packets [6]. The complexity of a software implementation of this technique includes providing storage for $k*N$ packets (where $k$ is a factor that reflects end-to-end turnaround of NACKed packets). It also includes keeping track of packets received and waiting for retransmission.

Block codes embed redundancy into the transmitted packets for systematic FEC. Hamming codes structure the redundancy so that the bit position of single errors is computed in an "error syndrome." The knowledge of which bit is in error allows that single bit to be corrected [276]. More powerful codes include Reed–Solomon and BCH codes [21]. These codes require polynomial multiplication and division for both generation and decoding of the protected packets. As a result, these codes may be implemented in a dedicated chip set or coprocessor on a core-based ASIC. Convolutional codes insert parity bits that are produced by dividing the data stream by a fixed polynomial. The maximum a-posteriori decoder can be implemented using threshold logic as first proved by A. Viterbi, after whom the decoding technique is named [277].

There is usually a mismatch between the structure of the bitstreams of convolutional codes and the structure of DSP chips. Convolutional codes work best when the polynomial has a prime order. This results in feedback shift registers with, for example, 11, 17, or 23 bits. These are not the highly composite powers of 2 of DSP instruction lengths. As a result, the packing and unpacking of bitstreams consumes large computational resources. Again, the least

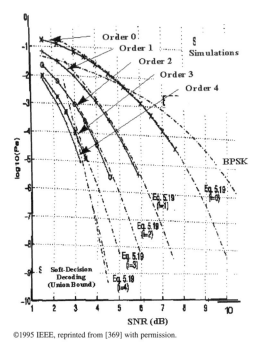

**Figure 12-21**   Error protection vs. algorithm complexity.

size, weight, and power realization of such codes will typically use FPGAs or digital ASICs.

Trellis codes [9] combine the constellation, coding, and decoding over time for better performance than block and convolutional codes at the cost of increased complexity.

## H.  Error Protection Complexity Tradeoffs

In addition to these well-established FEC techniques, the research literature offers algorithms for achieving further improvements, again at the cost of increased complexity. Fossorier and Lin, for example, describe a method of soft-decision decoding based on order statistics [369]. As shown in Figure 12-21, there is significant improvement over conventional techniques using their approach. A BPSK curve of BER versus SNR is shown for reference. The details of the algorithm are not important from an SDR perspective. This decoding technique can be considered a black box that yields improvements in BER for a given SNR as a function of the *order* of the decoding technique. The higher the order, the lower the BER. As illustrated in Figure 12-22, however, the computational demands of the increasing code order can be significant. In this case, there is more than a four order of magnitude difference between the highest-order decoder in the worst SNR conditions and the lowest-order

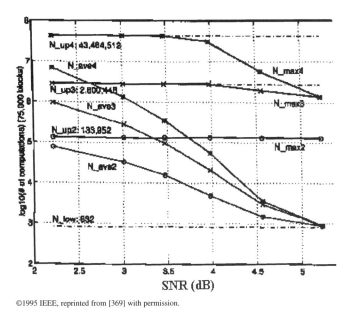

**Figure 12-22**  Computational demand vs. SNR by algorithm family.

decoder in the best SNR conditions. In addition to the change in complexity from one order of decoder to the other, there is a variation of computational complexity with SNR.

But not all users need maximal protection afforded by FEC, soft-decision decoding, and the most complex decoder. In a typical cell site application, for example, 20% of users are so close to the site or have such favorable multipath that they have excessive SNR. They may be subject to power reduction through power management techniques. Another 40 percent of users may experience nominal SNR conditions. Only the remaining 20% may need BER enhancement. Consequently, an SDR implementation that provides processing capacity in a flexible DSP pool may employ that capacity for those users who need it. In SDR design, one must identify in advance the range of computational demand that error control algorithms will present to the DSP pool. If the interconnect topology is flexible enough, the capacity may be redirected among users as needs increase and decrease over time. The net effect is that those users that were previously disadvantaged experience better information services due to the enhancement of the BER-SNR profile.

## I. Multiple Data Rates

The designs of commercial and military networks may be further understood from the constraints that they place on SDR nodes in the seven-layer ISO/OSI protocol stack. In 1997, some GSM networks allocated two TDMA time slots

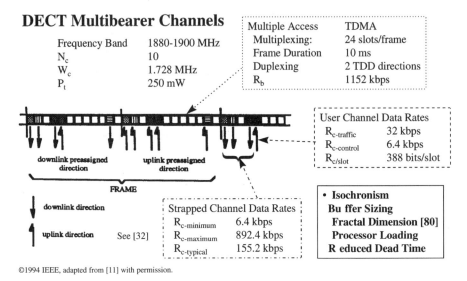

**Figure 12-23**   Channel strapping.

for fax and file transfer, for a net throughput of 14.4 kbps [371]. In 1998, Ericsson demonstrated 115 kbps data rates using the General Packet Radio Service (GPRS) [372], using channel strapping. Each GSM channel offers 13.3 kbps of voice, or 14.4 kbps of data traffic when suitably compressed. Eight such TDMA channels comprise an RF carrier, which can support $14.4 \times 8$, or 115.2 kbps [373].

The DECT multibearer channel scheme illustrated in Figure 12-23 delivers variable data rates based on channel strapping [11]. Channels are used together to deliver bits in parallel on several channels at once, increasing the effective bandwidth from a basic rate per user of 32 kbps to an adjustable range from 6.4 kbps to 892 kbps, sufficient for Internet, multimedia, and video-teleconferencing applications.

The impact on the SDR is also suggested in the figure. Isochronism must be preserved in the data paths. A DSP architecture designed to support the 32 kbps maximum may not be able to support over 892 kbps per subscriber without I/O code optimization or new interconnect hardware. In addition, buffer size must be increased to accommodate greater amounts of data per fixed processing interval. Processor loading will also increase to reflect the demands of the channel-strapping algorithm. Generally, the processor loading for a multichannel system will not increase much due to channel strapping if aggregate data rate of all strapped channels does not exceed the aggregate data rate of all unstrapped channels. But DECT was originally designed as a single user per handset system for cordless telephony. Therefore, the processing architectures optimized for standard DECT rates cannot accomodate channel strapping. The SDR architecture for a DECT base station could be

| Function | Network Application | | | |
| --- | --- | --- | --- | --- |
| | Simple Peer | Control Station | Single Hierarchy | Multiple Hierarchy |
| Addressing | Null | Local | Network | Global |
| Handoff | Null | Null | Simple | Complex |
| Framing | Null | Simple | Control | Frame |
| | Async | Sync | Channels | Remapping |
| Error | ARQ | ARQ | Optimized | ARQ/FEC |
| Control | FEC | FEC | ARQ/FEC | Interactions |
| Multiplexing | Push-to | TDM | TDM | Mux |
| & Access | Talk | CDMA | C/T/FDMA | Remapping |
| Modulation | AM,FM | Digital | Digital | Digital |
| | Digital | FH/DS | Analog | Multiple |
| Timing | Per Msg | Per Session | Per Cell | End-to-End |

**Figure 12-24**   Radio link function complexity.

sized for an aggregate data rate (e.g., for $N$ users) that also supports channel strapping for some users. The design chapters to follow explain how to design software radio architectures with the attributes of reduced hardware cost in some applications and enhanced flexibility in others.

Recent research has shown that LAN traffic has fractal properties [80]. Fractal properties are also evident in other types of communications systems including SDRs. This can mean that the worse-case demand predicted to be extremely rare for DECT multibearer channels can in fact occur much more frequently and with longer persistence when supporting strapped Internet traffic. Consequently, processing margins may be less than the standard exponential distributions predict [380]. This text therefore emphasizes techniques that manage processing resources in the presence of such fluctuations in processing demand.

## J. Link-Level Complexity Drivers

The radio link functions that drive software complexity include synchronization, timing, modulation, demodulation, and error control. Other aspects of the radio link layer of the protocol stack are illustrated in Figure 12-24.

In simple peer networks, addressing, framing, and handover will be null. Packet networks may employ simple ARQ or FEC-ARQ hybrids for low-SNR conditions. The modulation types for these simple peer networks generally are limited to simple AM or FM techniques or their digital equivalents (e.g., machine Morse, PSK, FSK). Timing in such networks is driven by push-to-talk bursts.

If a network control station is present, as is the case with JTIDS and commercial cellular standards, then timing is established by the control station. Each new subscriber must conform to that timing standard while participating in the network. In SPEAKeasy I, substantial effort was expended defining timing references for each network. Since SPEAKeasy I supported up to four networks simultaneously, it could be facing up to four different network time standards. The solution was to define a timing offset and a drift rate that could be set for each network. Each network could then be timed with respect to the internal SPEAKeasy clock, modified by the timing offset and drift rate.

Single hierarchy networks are complicated by handoff algorithms. Handoff allows a mobile user to transition from one fixed base station to another without losing connectivity. First-generation handoff algorithms commanded the mobile to switch frequencies at an appropriate instant. Although there is some uncertainty about exactly when to transition the mobile from one cell to the next, the decision is not complex. Second-generation (e.g., PHS) handoff is complicated by power management and by the time-multiplexed nature of the control channels [32]. GSM is forgiving of power differences due to its 90 dB near–far requirement. The near–far ratio is the ratio of the strongest signal to the weakest signal that are processed together. With a ratio of 90 dB, it is possible for a distant subscriber to have a very weak signal and yet not be lost from the cell site. The CDMA standard IS-95, on the other hand, requires the power received at a base station to be within 30 dB for the weakest and strongest signals received. To implement this, IS-95 employs aggressive power management. Thus, not only is the CDMA despreader the most complicated receiver, the power control requirement increases the sensitivity of the radio to time delays in the power control loop and to impairments in the cell handoff [141, 142, 159].

## IV. BITSTREAM PROCESSING SOFTWARE

The bitstream segment digitally multiplexes and frames source-coded bitstreams from multiple users (and, conversely, framesynchs and demultiplexes them). The bitstream segment also imparts FEC onto the bitstream, including bit interleaving and block and/or convolutional coding. Frame alignment, bit-stuffing, and radio link encryption would also occur in the bitstream segment. FEC decoding, automatic repeat request (ARQ) protocols, and final trellis-coded modulation (TCM) decisions occur in the bitstream segment. Final TCM converts soft/delayed decision parameters from the baseband segment to final bit decisions. The complexity of this segment depends on multiplexing, framing, FEC, privacy, and related bit-manipulation operations.

Signaling, control, and data bases (e.g., for operations, administration, and maintenance functions) are also provided in the bitstream segment. The processing demand associated with these functions depends on the signaling, control, and operations systems. It increases linearly with the number of active subscribers. These functions are event-driven and typically impart an order of magnitude less computational demand than baseband processing. These functions, may, however, require access to distributed databases, not all of which will be local to the base station. Thus, although the processing demand is relatively small, the timing requirements may be severe. In addition, the related protocols may account for 80% of the LOC of an SDR waveform.

As mentioned earlier, it is useful to partition the bitstream segment of a radio system with encryption facilities into black (modem), INFOSEC, and red (user interface and internetworking) components. Although the data rates across these components are very similar, the physical separation of these subsegments is necessary to avoid contaminating the encrypted transmitted signal with artifacts of the unencrypted data that one is trying to protect. Power lines, signal lines, and any source of parasitic modulation must be controlled from an EMI perspective to ensure red/black isolation and hence the integrity of the INFOSEC.

## V. INFOSEC SOFTWARE

INFOSEC software from the SPEAKeasy II API is shown in Table 12-5. The general-purpose messages of this API include an error status message. This facilitates determining the detailed condition of the INFOSEC function threads so that appropriate actions may be taken, such as discontinuing transmission if necessary. In addition to connection tests, the INFOSEC threads must be able to define red channels through the INFOSEC function. During parameter and mode setup, INFOSEC manages the cryptographic keys using Load_Key, Key_Tag and Mode-Req(uest) functions. Plaintext confirmation permits exchange of unencrypted (red) data. The INFOSEC function also determines when to start traffic, so as to avoid inadvertent transmission of unencrypted traffic.

## VI. INTERNETWORKING SOFTWARE

Internetworking software from the SPEAKeasy II API is as shown in Table 12-6. Internetworking also uses the generic ACK, NACK, buffer management, and message forwarding functions. Instantiation and power-up messages are also common to other segments. The bridging function connects two front-end channels. Typically, these channels would be operating in different modes and/or bands. One might, for example, bridge a standard mobile radio with a SINCGARS frequency-hopped radio using the bridging function. Internetworking messages also carry the setup information for digitally vocoded voice, including the type of vocoder and the associated data rate. External access tones needed to access remote networks are also generated in this API. Loopback tests may be performed using the corresponding message. Push-to-talk (PTT) priority determines which channel will receive service if two PTT users key their channels simultaneously. Side tones may be generated in the internetworking API in order to make two different networks compatible. This

**TABLE 12-5   INFOSEC Messages**

| No. | Name | Phase |
|-----|------|-------|
| 1 | ACK | All |
| 2 | Buffer_Complete | All |
| 3 | Buffer_Notify | All |
| 4 | Error_Status | All |
| 5 | Forward_Message | All |
| 6 | NACK | All |
| 7 | BIT_Request | Power_up |
| 8 | Define_Remote_Child | Power_up |
| 9 | Define_Remote_Parent | Power_up |
| 10 | Agent_Description | Instantiation |
| 11 | Connection_Test | Instantiation |
| 12 | Connect_Children | Instantiation |
| 13 | Define_Red_Channel | Instantiation |
| 14 | Define_Remote_Child | Instantiation |
| 15 | End_Download | Instantiation |
| 16 | File_Download_Complete | Instantiation |
| 17 | File_Download_Start | Instantiation |
| 18 | Initiate_Download | Instantiation |
| 20 | Standard_Data_Msg | Instantiation |
| 21 | Activate_Channel | Params_&_Mode |
| 22 | Crypto_Algorithm | Params_&_Mode |
| 23 | Define_Child | Params_&_Mode |
| 24 | Key_Tag | Params_&_Mode |
| 25 | Load_Key | Params_&_Mode |
| 26 | Mode_Req | Params_&_Mode |
| 27 | Plaintext_Confirm_Response | Params_&_Mode |
| 28 | Software_Version_Request | Params_&_Mode |
| 29 | Suspend_Channel | Params_&_Mode |
| 30 | Channel_Status | Operation |
| 31 | Deactivate_Channel | Operation |
| 32 | Pacing_Indication | Operation |
| 33 | Receive_Mode | Operation |
| 34 | Standard_Data_Msg | Operation |
| 35 | Start_Traffic | Operation |
| 36 | Transmit_Mode | Operation |
| 37 | Disconnect_Children | Teardown |

API can also suspend and resume a channel without tearing it down. During operation, the internetworking API controls the audio volume.

## A.   Open Systems Interconnect Protocol Stack

The ISO/OSI protocol stack is illustrated in Figure 12-25. SDR functions appear at all levels of this stack as highlighted in the figure. In military and

**TABLE 12-6   Internetworking Messages**

| No. | Name | Phase |
|-----|------|-------|
| 1 | ACK | All |
| 2 | Buffer_Complete | All |
| 3 | Buffer_Notify | All |
| 4 | Forward_Message | All |
| 5 | NACK | All |
| 6 | BIT_Request | Power_up |
| 7 | Define_Remote_Child | Power_up |
| 8 | Define_Remote_Parent | Power_up |
| 9 | Connection_Test | Instantiation |
| 10 | Define_Remote_Child | Instantiation |
| 11 | End_Download | Instantiation |
| 12 | File_Download_Complete | Instantiation |
| 13 | File_Download_Start | Instantiation |
| 14 | Initiate_Download | Instantiation |
| 15 | New_Agent | Instantiation |
| 16 | Standard_Data_Msg | Instantiation |
| 17 | Activate_Channel | Params_&_Mode |
| 18 | Alert_Tones | Params_&_Mode |
| 19 | Bridging | Params_&_Mode |
| 21 | Default_Voice_Rate | Params_&_Mode |
| 22 | Default_Voice_Type | Params_&_Mode |
| 23 | Define_Child | Params_&_Mode |
| 24 | External_Access_Tone | Params_&_Mode |
| 25 | Loopback_Test_Mode | Params_&_Mode |
| 26 | Mode_Req | Params_&_Mode |
| 27 | Mode_Resp | Params_&_Mode |
| 28 | PTT_Priority | Params_&_Mode |
| 29 | Sidetones | Params_&_Mode |
| 30 | Software_Version_Request | Params_&_Mode |
| 31 | Suspend_Channel | Params_&_Mode |
| 32 | Test_Mode | Params_&_Mode |
| 33 | User_Tone | Params_&_Mode |
| 34 | Channel_Status | Operation |
| 35 | Deactivate_Channel | Operation |
| 36 | Standard_Data_Msg | Operation |
| 37 | Start_Traffic | Operation |
| 38 | Volume_Level | Operation |
| 39 | Destroy_Agent | Teardown |

peer networks, the radio node supports all layers of the stack, while in a hierarchical network, the BTS, BSC, and MSC nodes each support different layers. Layer 7, the applications layer, supports applications, which include voice, fax, data, and multimedia services. For mobile SDR nodes, applications

| Layer | # | Definition | SDR Functions |
|---|---|---|---|
| Application | 7 | User Services | Voice, Fax, Data, Multimedia, Location Finding Services* |
| Presentation | 6 | Translation and Remapping Syntax, Control Code Translations | Encryption, Virtual Terminals Object Request Broker* |
| Session | 5 | Connection Between Applications Overload Control, Checkpointing | Connection Allocation Radio Network Reconfiguration* |
| Transport | 4 | Group & Broadcast, Peer-Peer (5 ISO Classes of Transport) | Reliable Message Transfer Bandwidth Management* |
| Network | 3 | Segmenting, Routing & Integration | Transparent Data Transfer Background Packet Modes* |
| Datalink | 2 | Framing, Addressing & Error Control | Link (Point-to-Point) Interface Adaptive Quality Management* |
| Physical | 1 | Mechanical, Electrical Functional, Procedural and Timing | Transmission Segments ALE, Peer Contact* |

\* Unique to software radio and software-defined network architectures.

Stallings, *Handbook of Computer-Communications Standards, Volume 1, The Open Systems Interconnection (OSI) Model* [7].

**Figure 12-25**  ISO protocol stack.

generally include location-finding services. For many military applications include video-teleconferencing or battlefield video.

Layer 6, the presentation layer, provides translation and mapping with control code translations. Encryption is also provided at this layer. Object request broker (ORB) technology for interconnecting applications provided by diverse vendors is an appropriate layer 6/7 interface service. Layer 5 provides reliable connections between applications, including overload control and checkpointing. SDRs may conduct different sessions on different bands and modes, transparently delivering connectivity to the upper layers in spite of the unavailability of a particular lower-layer communications path. Layer 4 provides reliable message transfer including group addressing, broadcast, and peer-to-peer connectivity. ISO defines five classes of transport [7], a mix of which may be supported by an SDR. They may deliver bandwidth agility in layer 4, trading off data rate for BER or SNR as appropriate to QoS and propagation.

The lower layers of the protocol stack have significant implications for SDR architecture. Layer 3, the network layer, segments the messages, routes them across the network, and reintegrates the packets into messages in the receiver. This corresponds to the Internetworking function and API. SDRs not only support the standard services, they can provide background routing services, which the military calls *transparent bridging*. Layer 2, the data link layer, frames the messages into data bursts appropriate for the air interface. It also adds forward error control and retransmits packets as necessary. Typically, this is accomplished in black-side data processing, the link-layer aspect of the abstract modem function. Layer 1, the physical layer, provides the RF me-

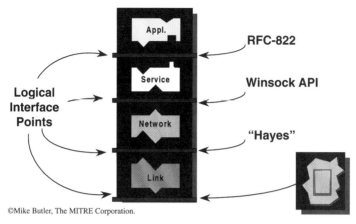

Figure 12-26 Layering an information system.

chanical and electrical interface with appropriate data sequencing and timing. The SDR accommodates the anomalies of the physical layer through propagation channel estimation and equalization. The physical layer maps to the modem function.

Within an SDR with heterogeneous processing, the infrastructure of hardware and software may be interconnected by X.900 Open Distributed Processing, CORBA, or equivalent.

## B. Layering Network Access

Butler [374] advocates the architecture for radios shown in Figure 12-26. He envisions applications at the top layer. His concept of services is akin to radio applications, services in the sense of software facilities. At this point, he adopts two ISO layers, the network and link layers. His identification of intermediate interfaces is informative. This approach maps readily to the functional model. Services are radio applications. The network layer maps to the Internetworking function/API. The link layer and below map to the modem. The Butler model sheds little light on how to insulate these software-defined functions from the rapidly evolving heterogeneous multiprocessing hardware of the SDR, however.

## C. Mode Handover

Cellular service quality may be enhanced by employing monitor and control processes such as mode handover algorithms. Reference [375], for example, describes three different algorithms for microcell/macrocell handover.

The priority and overlay algorithms are relatively simple while the priority + overlay algorithm employs the priority algorithm when offered load is

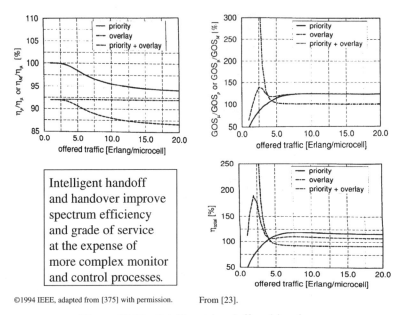

Intelligent handoff
and handover improve
spectrum efficiency
and grade of service
at the expense of
more complex monitor
and control processes.

**Figure 12-27**    Intelligent handoff and handover.

low, switching to the overlay algorithm gracefully as offered load increases.
The net effect of the more complex algorithm is the enhanced performance
shown in Figure 12-27. This tradeoff among relatively simple algorithms with
less robust performance versus more complex algorithms with higher performance and higher processing demands represents one of the key design issues
in the SDR. In this particular case, neither the priority handover algorithm nor
the overlay algorithm need to monitor offered demand. The priority + overlay
algorithm adapts to offered demand, however. Consequently, the more robust algorithm must measure and track aspects of the air interface that the
simpler algorithms ignored. The emphasis here is on managing the processing demand that such increased algorithm complexity is readily accommodated.

## VII. SOURCE SEGMENT SOFTWARE

The *source segment* differs between the mobile unit and the base station. In the
mobile, the source segment consists of the user and the source encoders and
decoders (voice, video, data, facsimile, and multimedia). In this segment, the
relatively narrowband voice and video ADCs and DACs are typically located
very close to the subscriber (e.g., in the handset, palmtop, or workstation). In
the base station, on the other hand, the source segment consists of the interface
to the PSTN for access to remote source coding. Conversion of protocols

required for interoperability with the PSTN creates processing demand in the base station's source segment. Conversion of DS0 64 kbps PCM to RPE-LTP (GSM), for example, creates 1 to 10 MIPS of demand per subscriber.

## A. Voice Processing Software

Voice processing software modules are available in commercial and government third-party libraries for purchase and reuse. Some vocoders are defined as algorithms historically hosted on ASICs. *Companding*, for example, is the process of nonlinear compression and expansion of the speech waveform to preserve information content in a DS0 channel. Companding may be expressed as an algorithm, but is typically implemented in the microphone hardware as an analog device or digital hard-wired lookup table driven by a 14-bit ADC [376]. Other algorithms have historically been embedded in ASICs, but are now being deployed in software to reduce parts count. The series of low-data-rate (2400 bit-per-second) codecs for military digitally enciphered speech, for example, began with the LPC-10 linear predictive coding algorithm [150] and evolved to codebook-excited linear prediction (CELP) and vector-CELP or V-CELP or sometimes VELP. Mixed Excitation Linear Prediction (MELP) is poised to supersede them all as the DoD's 2400 bps standard [442].

## B. Message Processing Software

Message processing supports system control and user-interface functions. Control functions from the SPEAKeasy II API are shown in Table 12-7. In addition to the generic functions used in all phases, this API may initiate and terminate external channel control (Ext_Channel_Control). In addition to the standard instantiation messages, this API includes the Instantiate_Waveform function. This function orchestrates the creation of internetworking, INFOSEC, modem, and front-end functions. The security level for a channel is also set in this API. Message processing sets up the direction of the channel, its associated crypto algorithm, transmit and receive parameters, and AGC loop bandwidth. Signal strength, automatic gain control, squelch level, volume level, and PA power, set in the message processing API are inherited by the state machines of the modem object(s).

## C. User-Interface Software

API functions for the user interface include those listed in Table 12-8. Using this API, channels may have status determined and controlled externally. This allows the distribution of channel control between a core system processor that implements the user-interface functions and an external control "head" such as a laptop. Distributed in this way, the laptop may request status of a channel and may supply control parameters. The user interface instantiation functions include connection testing, setting the security level and exchanging

**TABLE 12-7   Message Processing Functions**

| No. | Name | Phase |
|-----|------|-------|
| 1 | ACK | All |
| 2 | Buffer_Complete | All |
| 3 | Buffer_Notify | All |
| 4 | End_Ext_Channel_Control | All |
| 5 | Error_Status | All |
| 6 | Ext_Channel_Control | All |
| 7 | Forward_Message | All |
| 8 | NACK | All |
| 9 | Add_Agent | Power_up |
| 10 | BIT_Request | Power_up |
| 11 | BIT_Response | Power_up |
| 12 | Define_Remote_Parent | Power_up |
| 13 | Connection_Test | Instantiation |
| 14 | Download_Complete | Instantiation |
| 15 | Download_Files | Instantiation |
| 16 | Instantiate_Waveform | Instantiation |
| 17 | Resources_Allocated | Instantiation |
| 18 | Security_Level | Instantiation |
| 19 | Standard_Data_Msg | Instantiation |
| 20 | Activate_Channel | Params_&_Mode |
| 21 | Alert_Tones | Params_&_Mode |
| 22 | Antenna_Select | Params_&_Mode |
| 23 | Bridging | Params_&_Mode |
| 24 | Channel_Direction | Params_&_Mode |
| 25 | Crypto_Algorithm | Params_&_Mode |
| 27 | External_Access_Tone | Params_&_Mode |
| 28 | Key_Tag | Params_&_Mode |
| 29 | Loopback_Test_Mode | Params_&_Mode |
| 30 | Modem_AGC_Loop_Bandwidth | Params_&_Mode |
| 31 | Modem_TX_Power | Params_&_Mode |
| 32 | Mode_Req | Params_&_Mode |
| 33 | Mode_Resp | Params_&_Mode |
| 34 | Plaintext_Confirm_Request | Params_&_Mode |
| 35 | Plaintext_Confirm_Response | Params_&_Mode |
| 36 | PTT_Priority | Params_&_Mode |
| 37 | Sidetones | Params_&_Mode |
| 38 | Software_Version_Request | Params_&_Mode |
| 39 | Software_Version_Response | Params_&_Mode |
| 40 | Test_Mode | Params_&_Mode |
| 41 | User_Tone | Params_&_Mode |
| 42 | Channel_Status | Operation |
| 43 | Crypto_Status | Operation |
| 44 | Deactivate_Channel | Operation |
| 45 | Receive_Mode | Operation |
| 46 | Signal_Strength | Operation |
| 47 | Squelch_Level | Operation |
| 48 | Standard_Data_Msg | Operation |
| 49 | Transmit_Mode | Operation |
| 50 | Volume_Level | Operation |
| 51 | PA_Power | Teardown |

**TABLE 12-8   User Interface Functions**

| No. | Name | Phase |
|-----|------|-------|
| 1 | ACK | All |
| 2 | Buffer_Complete | All |
| 3 | Buffer_Notify | All |
| 4 | End_Ext_Channel_Status | All |
| 5 | Error_Status | All |
| 6 | Ext_Channel_Status | All |
| 7 | Forward_Message | All |
| 8 | NACK | All |
| 9 | BIT_Response | Power_up |
| 10 | Connection_Test | Instantiation |
| 11 | Security_Level | Instantiation |
| 12 | Standard_Data_Msg | Instantiation |
| 13 | Alert_Tones | Params_&_Mode |
| 14 | Antenna_Select | Params_&_Mode |
| 15 | Channel_Direction | Params_&_Mode |
| 16 | Crypto_Algorithm | Params_&_Mode |
| 17 | External_Access_Tone | Params_&_Mode |
| 19 | Loopback_Test_Mode | Params_&_Mode |
| 20 | Modem_AGC_Loop_Bandwidth | Params_&_Mode |
| 21 | Modem_TX_Power | Params_&_Mode |
| 22 | Plaintext_Confirm_Request | Params_&_Mode |
| 23 | PTT_Priority | Params_&_Mode |
| 24 | Sidetones | Params_&_Mode |
| 25 | Software_Version_Response | Params_&_Mode |
| 26 | Test_Mode | Params_&_Mode |
| 27 | User_Tone | Params_&_Mode |
| 28 | Channel_Status | Operation |
| 29 | Crypto_Status | Operation |
| 30 | Receive_Mode | Operation |
| 31 | Signal_Strength | Operation |
| 32 | Squelch_Level | Operation |
| 33 | Standard_Data_Msg | Operation |
| 34 | Transmit_Mode | Operation |
| 35 | Volume_Level | Operation |

standard data messages. Most of the other user interface functions are similar to the functions described above.

The class of user interfaces provided that support the API is illustrated in Figure 12-28. Each of six channels may be displayed using the function buttons. Location information from GPS is read out continuously at the bottom of the display. Specific waveform parameters are shown in windows designed for each waveform. There is consistency in the presentation of transmit and receive frequencies.

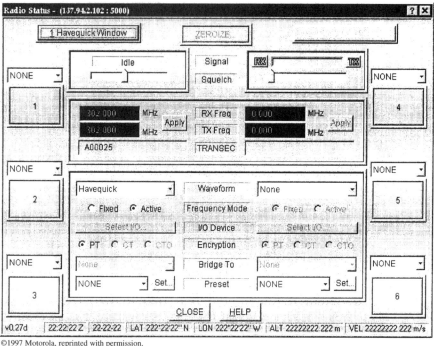

©1997 Motorola, reprinted with permission.

**Figure 12-28**   Illustrative user interface.

## VIII.  OTHER SOFTWARE ISSUES

A range of protocol techniques from elementary to advanced air-interfaces are
summarized in Figure 12-29. Advanced air-interface protocols take advantage
of redundancy in existing packet structures to enhance link effectiveness. ATM
over the air, for example, can be enhanced by using the redundancy of the ATM
packet header. ATM packet headers have 24 bits for virtual circuit and path
identification on wireline networks. In many air-interface conditions, however,
there are only a few dozen addressees at most. The LANET protocol [84,
85] recodes the VPI/VCI bits from the ATM header with redundancy such
that the ATM frame structure and addresses are recoverable via forward error
correction in BER conditions up to $10^{-1}$. As a result, the ATM stream need
not be resynchronized after a deep fade because the framing and addressing
sustains synchronization through the fade. The price that must be paid for
such enhanced performance is increased complexity of data generation and a
more complex receiver.

Other advanced techniques such as those listed in Figure 12-30 further
enhance performance at the cost of additional complexity. This complexity
is manifest in the design issues shown. Intelligent or "smart" antennas form
beams and directional nulls which enhance C/I of cell site subscribers through

Automatic Repeat Request (ARQ)
    Go-Back-N     Packet-Size x N
    Interleaving     Interleaving Array Size

Trellis-Coded ARQ Hybrids
    Selectively Transmitted Redundancy
    Depth and Order of the Trellis
    Type and Sampling Rate of Channel State Information
    Complexity of the ARQ Algorithm

ATM Over The Air (LANET)
    Frame Synchronization and Recovery
    Reliable Header (VPI, VCI) Extraction
    Commercial Products [84]
    Military Applications [85]

**Figure 12-29**   Other protocol areas.

| Technique | Design Issue(s) |
| --- | --- |
| Intelligent Antennas | |
|    Direction Finding | Array Parallelism (Phase, Amplitude) |
|    Beam Forming | Matrix Inversion |
| Equalization | |
|    Constant Modulus | Number of Taps, Feedback, Bandwidth |
|    Decision Directed | Known Bits, Nonlinear Feedback |
| Interference Rejection | |
|    Spectrum Spreading | Chip-Rate Processes, Timing Recovery |
|    Filtering | Analog Linearity, Digital SFDR, #Taps |

**Figure 12-30**   Other design issues.

reducing the energy received from distant cells. Additional complexity arises in the parallelism of the multiple antenna elements across which phase and amplitude must be matched in order for beams to be pointed in the right direction. In addition, the beam-former weights are determined through computationally intensive matrix inversion.

Equalization is designed into second-generation mobile radio systems such as GSM, but equalization techniques can be employed on most waveforms to enhance SNR at the expense of increased computational complexity. In addition, waveforms designed to reject interference, such as direct-sequence, spread-spectrum waveforms, require more complex generation and receiver techniques such as precision timing of the generated waveform with the de-spreading receiver.

## IX. ARCHITECTURE IMPLICATIONS

This chapter has presented further insights into the software components of the SDR. The software must be designed with insight into the implications of instruction set architecture on timing and flow of the software through its implementing threads. In addition, an SDR may be designed using a combination of top-down and bottom-up techniques. Object-oriented technology provides the methods for the top-down aspect. Bottom-up design centers on the integration of existing software components into the top-down structure. In order to do this in a commercially viable way, we have to have an open architecture that defines interfaces that facilitate the creation of components that may be reused commercially. Open architectures such as that emerging through the work of the SDR Forum, promise to move industry in that direction.

The fundamentals of software design for SDR flow from good software design principles. Object-oriented design, followed as a set of organizing principles, is the best contemporary starting point. In addition, SDRs are real-time signal processing systems. The design of radio software, therefore, emphasizes the estimation and management of processing demand in the isochronous streams. The additional critical aspect is to ensure that the software team has an in-depth understanding of digital signal processing. As surprising as this seems, teams sometimes have only one or two "experts" in signal processing who are assigned to design specific algorithms. The experts may also serve as signal processing consultants to less knowledgeable team members. This approach sometimes dilutes the attention of the signal processing designers from the task of algorithm design. Programmers who write operating systems modules such as drivers are often an asset to an SDR team. These programmers understand real-time code and efficiency issues. In addition, they can help the signal processing experts understand how to take advantage of the native instruction set and operating system.

The API presented above provides an initial point of departure for structuring reusable objects, whether for commercial resale or to manage costs of inserting SDR technology within a product line. APIs mask the internal structure of a software library. This is an advantage with associated challenges. The advantage, of course, is that a COTS library does not have to be written and debugged. The challenge is that it has to be understood and fully characterized. Thus, one should test reuse libraries and API facilities to determine computational resource use with representative calling parameters before one commits to a specific reuse library.

Often, one does not have the luxury of testing in advance of design. The techniques of the next chapter address this common situation.

## X. EXERCISES

**1.** What are the points of departure for SDR design? How do these differ?

2. Differentiate the power-up phase from the instantiation phase of setting up a software radio. What parameters are defined for front-end processing functions in SDRs? Which of these are infrastructure related?

3. Why is built-in test necessary in initializing an SDR? How can software partitioning assist one in accommodating hardware faults? How can spare capacity in one processor translate into fault recovery (fall-back or fail-soft modes) for another processor?

4. What unique capabilities of SDR are supported by monitoring the radio spectrum? How is this similar to HF ALE? Describe two methods of scanning the radio spectrum. Which method can estimate channel impulse response? Which method can transform one multiplexed FDM channel into a large number of parallel narrowband channels? What specific parameter combinations are required to accomplish this? How can this technique be used to enhance quality, quantity, or timeliness of information?

5. What modem control functions defined in the SPEAKeasy II API are likely to be useful for 3G as well? Which are not?

6. If the SNR is 10 dB, what is the ideal BER of an FSK (2-state) waveform? Is the BER of ASK better or worse? by how much? Answer the same question for PSK.

7. Besides ASK, OOK, FSK, and PSK, what channel modulations are in wide use in radio? Which of these is most complex? Which is simplest? Which modulation formats do not require vocoding? Which of these can be understood at negative SNR?

8. Differentiate SER from BER. When is 64 QAM an appropriate channel waveform? When is it necessary to use noncoherent orthogonal and coded waveforms?

9. Define DSSS processing gain. What is the processing gain associated with spreading a binary waveform with a 1023-bit sequence? Can this be used to extract the signal from below the noise floor of white noise? At what SNR will it no longer be recoverable? What is the computational burden of such an approach? What is the impact on spectrum allocation if a DSSS signal is employed in a 12.5 MHz AMPS spectrum, say for one 4 kHz voice channel centered in the band? (Will the signal fit into the band, or will it occupy additional spectrum?) How much processing capacity is needed to implement such a mode? How many high-performance chips are necessary to implement a despreader (to a rough order of magnitude)?

10. What types of synchronization may be employed in an air interface? Which types are used in TDMA systems? What SDR function associated with burst TDMA (e.g., GSM) demands the most computational resources?

11. Differentiate soft-decision decoding from hard-decision decoding. Why does soft-decision decoding yield lower BER? How much additional computational demand is likely to accrue from the use of this technique?

12. What are the popular methods of error control? Which method is most suitable to FPGA implementation? Why is this method less suited to DSP implementations? If such an FEC algorithm is implemented in an FPGA with a 1 MHz clock, and consumes 10,000 gates, how many equivalent MIPS are needed to implement the algorithm on a DSP without suffering performance degradation?

13. Discuss the tradeoff between computational complexity and BER as a function of SNR for soft-decision decoding based on order statistics.

14. What aspects of an SDR personality come under stress when implementing multiple data rates using strapped channels? Give an example.

15. Is addressing an issue in a peer network? In a point-to-point network? How does the issue become more complex in single and multiple hierarchies?

16. Describe the difference in timing a waveform on a point-to-point transmission versus in a single hierarchy cellular network. Suppose the node has to operate on multiple hierarchies. What infrastructure facilities are needed to keep the two separately drifting networks synchronized?

17. What parameters control internetworking software? How is the control of such parameters related to the protocol stack of a COTS TCP/IP package?

18. What correspondence can be drawn between the OSI protocol stack and the unique features of SDR?

19. What kinds of control functions are needed to support user interfaces?

20. What other design issues may drive complexity of SDR software? Why is it important to understand the relationship between QoS and computational complexity?

# 13 Performance Management

The material covered in this chapter can reduce DSP hardware costs by a factor of 2 : 1 or more. Thus it is pivotal and in some sense the culmination of the SDR design aspects of this text.

## I. OVERVIEW OF PERFORMANCE MANAGEMENT

Resources critical to software radio architecture include I/O bandwidth, memory, and processing capacity. Good estimates of the demand for such resources result in a well-informed mapping of software objects to heterogeneous multiprocessing hardware. Depending on the details of the hardware, the critical resource may be the capacity of the embedded processor(s), memory, bus, mass storage, or some other input/output (I/O) subsystem.

### A. Conformable Measures of Demand and Capacity

MIPS, MOPS, and MFLOPS are not interchangeable. Many contemporary processors, for example, include pipelined floating point arithmetic or single instruction FFT butterfly operations. These operations require processor clock cycles. One may, however, express demand in the common measure of millions of operations per second (MOPS), where an operation is the average work accomplished in a single clock cycle of an SDR word width and operation mix. Although software radios may be implemented with 16-bit words, this requires systematic control of dynamic range in each processing stage (e.g., through automatic gain control and other normalization functions). Thus 32-bit equivalent words provide a more useful reference point, in spite of the fact that FPGA implementations use limited precision arithmetic for efficiency. The mix of computation (e.g., filtering) versus I/O (e.g., for a T1 multiplexer) depends strongly on the radio application, so this chapter provides tools for quantitatively determining the instruction mix for a given SDR application. One useful generalization for the mix is that the RF conversion and modem segments are computationally intensive, dominated by FIR filtering and frequency translation. Another is that the INFOSEC and network segments are dominated by I/O or bitstream functions. Those protocols with elaborate networking and error control may be dominated by bitstream functions. Layers of packetization may be dominated by packing and unpacking bitstreams using protocol state machines.

MIPS and MFLOPS may both be converted to MOPS. In addition, 16-bit, 32-bit, 64-bit, and extended precision arithmetic mixes may also be expressed in Byte-MOPS, MOPS times bytes transformed by the operation. Processor I/O, DMA, auxiliary I/O throughput, memory, and bus bandwidths may all be expressed in MOPS. In this case the operand is the number of bytes in a data word and the operation is store or fetch.

A critical resource is any computational entity (CPU, DSP unit, floating-point processor, I/O bus, etc.) in the system. MOPS must be accumulated for each critical resource independently. Finally, software demand must be translated rigorously to equivalent MOPS. Benchmarking is the key to this last step. Hand-coded assembly language algorithms may outperform high-order language (HOL) code (e.g., Ada or C) by an order of magnitude. In addition, hand-coded HOL generally outperforms code-generating software tools, in some cases by an order of magnitude. Rigorous analysis of demand and capacity in terms of standards MOPS per critical resource yield useful predictions of performance. Initial estimates generated during the project-planning phase are generally not more accurate than a factor of two. Thus, one must sustain the performance management discipline described in this chapter throughout the project in order to ensure that performance budgets converge so that the product may be delivered on time and within specifications.

## B. Initial Demand Estimates

Table 13-1 illustrates how design parameters drive the resource demand of the associated segment. The associated demand may exceed the capacity of today's general-purpose processors. But, the capacity estimates help identify the class of hardware that best supports a given segment. One may determine the number of operations required per point for a point operation such as a digital filter. One hundred operations per point is representative for a high-quality frequency translation and FIR filter, for example. One then multiplies by the critical parameter shown in the table to obtain a first cut at processing demand. Multiplying the sampling rate of the stream being filtered times 100 quickly yields a rough order of magnitude demand estimate.

Processing demand depends on a first-order approximation on the signal bandwidths and on the complexity of key operations within IF, baseband, bitstream, and source segments as follows:

$$D = D_{if} + N^*(D_{bb} + D_{bs} + D_s) + D_{oh}$$

Where:

$$D_{if} = W_a^*(G1 + G2)*2.5$$
$$D_{bb} = W_c^*(G_m + G_d)$$
$$D_{bs} = R_b^*G3^*(1/r)$$

**TABLE 13-1    Illustrative Functions, Segments, and Resource Demand Drivers**

| Application | Radio Function | Segment | First-Order Demand Drivers |
|---|---|---|---|
| Analog | Companding | Source | Speech bandwidth (Wv) and Sampling rate |
| Speech | Gap suppression | Bitstream | Gap identification algorithm complexity |
| | FM modulation | Baseband | Interpolation required (Wfm/Wv) |
| | Up conversion | IF | IF carrier and FM bandwidth: fi, Wi = Wfm |
| Receiver | Band selection | IF | Access bandwidth (Wa) |
| | Channel selection | IF | Channel bandwidth (Wc) |
| | FM demodulation | Baseband | fi, Wi |
| | DS0 reconstruction | Bitstream | Speech bandwidth; vocoder |
| TDMA | Voice codec | Source | Voice codec complexity |
| TDM | FEC coding | Bitstream | Code rate; block vs. convolutional |
| | Framing | Bitstream | Frame rate (Rf); bunched vs. distributed |
| | MSK modulation | Baseband | Baud rate (Rb) |
| | Up conversion | IF | fi, Wi + Rb/2 |
| | Band selection | IF | Access bandwidth (Wa) |
| | Channel selection | IF | Channel bandwidth (Wi = Wc) |
| | Demodulation | Baseband | Baud rate (Rb) or channel bandwidth (Wc) |
| | Demultiplexing | Bitstream | Frame rate (Rf) |
| | FEC decoding | Bitstream | Code rate |
| CDMA | Voice codec | Source | Choice of voice codec |
| | FEC coding | Bitstream | Code rate |
| | Spreading | Baseband | Chip Rate (Rc) |
| | Up conversion | IF | Wc, fi , Rc |
| | Band selection | IF | Wc, fi, Rc |
| | Despreading | Baseband | Chip rate (Rc) |
| | FEC decoding | Bitstream | Code rate |

$D$ is aggregate demand (in standardized MOPS). $D_{if}$, $D_{bb}$, $D_{bs}$, and $D_s$ are the IF, baseband, bitstream, and source processing demands, respectively. $D_{oh}$ is the management overhead processing demand. $W_a$ is the bandwidth of the accessed service band. $G1$ is the per-point complexity of the service-band isolation filter. $G2$ is the complexity of subscriber channel-isolation filters. $N$ is the number of subscribers. $W_c$ is the bandwidth of a single channel. $G_m$ is the complexity of modulation processing and filtering. $G_d$ is the complexity of demodulation processing (carrier recovery, Doppler tracking, soft decoding, postprocessing for TCM, etc.). $R_b$ is the data rate of the (nonredundant) bitstream. The code rate is $r$. $G3$ is the per-point complexity of bitstream processing per channel (e.g., FEC). Table 13-2 shows how parameters of processing demand are related in an illustrative application.

This real-time demand must be met by processors with sufficient capacity to support real-time performance. At present, most IF processing is off-loaded to special-purpose digital receiver chips because general-purpose processors with sufficient MOPS are not yet cost effective. This tradeoff changes approximately every 18 months in favor of the general-purpose processor. Aggregate baseband and bitstream-processing demand of 4 to 10 MOPS per user is within the capabilities of most DSP chips. Therefore, several tens of subscribers may

**TABLE 13-2   Illustrative Processing Demand**

| Segment | Parameter | Illustrative Value | Demand |
|---|---|---|---|
| IF | $Ws$ | 25 MHz | |
| | $G1$ | 100 OPS/Sample | $Ws*G1*2.5 = 6.25$ GOPS[a] |
| | $G2$ | 100 OPS/Sample | $Dif = Ws*(G2+G2)*2.5 = 12.5$ GOPS[a] |
| | $N$ | 30/ cell site | |
| | $Wc$ | 30 kHz | |
| | $Gm$ | 20 OPS/Sample | $Wc*Gm = 0.6$ MOPS |
| Baseband | $Gd$ | 50 OPS/Sample | $D_{bb} = Wc*(Gm+Gd) = 2.1$ MOPS |
| | $R$ | 1 b/b | |
| | $R_b$ | 64 kbps | |
| Bitstream | $G3$ | 1/8 FLOPS/bps | $Dbs = G3*Rb/r = 0.32$ MOPS |
| Source | $D_s$ | 1.6 MIPS/user | $N*G4 = 4.02$ MIPS per user |
| | | | $N*(Wc*(Gm+Gd)+Rb*G3/r+G4) = 120.6$ MOPS per cell site |
| | $D_o$ | 2 MOPS | |
| Aggregate | $D$ | | 122.6 MOPS per cell site (excluding IF) |

[a]Typically performed in digital ASICs in contemporary implementations.

be accommodated by the highest performance DSP chips. Aggregate demand of all users of 122.6 MOPS, including overhead is nominally within the capacity of a quad TMS 320 C50 board. When multiplexing more than one user's stream into a single processor, memory buffer sizes, bus bandwidth, and fan-in/fan-out may cost additional overhead MOPS.

## C. Facility Utilization Accurately Predicts Performance

The critical design parameter in relating processing demand to processor capacity is resource utilization. Resource utilization is the ratio of average offered demand to average effective capacity. When expressed as a ratio of MOPS, utilization applies to buses, mass storage, and I/O as well as to CPUs and DSP chips. The bottleneck is the critical resource that limits system throughput. Identifying the bottleneck requires the analysis and benchmarking described in this chapter. The simplified analysis given above applies if the processor is the critical resource. The SDR systems engineer must understand these bottlenecks in detail for a given design. The SDR architect must project changes in bottlenecks over time. The designer should work to make it so. Sometimes, however, I/O, the backplane bus, or memory will be the critical resource. The SDR systems engineer must understand these bottlenecks in detail for a given design. The SDR architect must project changes in bottlenecks over time. The following applies to all such critical resources.

   Utilization, $\rho$, is the ratio of offered demand to critical resource capacity, $\rho = D/C$, where $D$ is average resource demand and $C$ is average realizable capacity, both in MOPS. Figure 13-1 shows how queuing delay at the resource varies as a function of processor utilization. In a multithreaded DSP, there

$R(\rho, \sigma) = t/s$

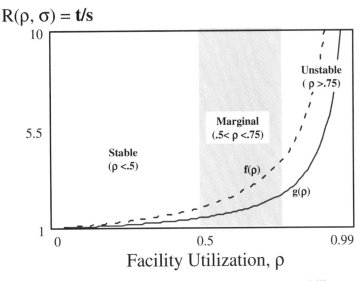

**Figure 13-1** Facility utilization characterizes system stability.

may be no explicit queues but if more than one thread, task, user, etc. is ready to run, its time spent waiting for the resource constitutes queuing delay. The curve $f(\rho)$ represents exponentially distributed service times, while $g(\rho)$ represents constant service times. Simple functions like digital filters have constant service times. That is, it takes the same 350 operations every time a 35-point FIR filter is invoked. More complex functions with logic or convergence properties such as demodulators are more accurately modeled with exponentially distributed service times.

Robust performance occurs when $\rho$ is less than 0.5, which is 50% spare capacity. The undesired events that result in service degradation will occur with noticeable regularity for $0.5 < \rho < 0.75$. For $\rho > 0.75$, the system is generally unstable, with queue overflows regularly destroying essential information. Systems operating in the marginal region will miss isochronous constraints, causing increased user annoyance as $\rho$ increases.

An analysis of variance is required to establish the risk that the required time delays will be exceeded, causing an unacceptable fault in the real-time stream. The incomplete Gamma distribution relates the risk of exceeding a specified delay to the ratio of the specification to the average delay. Assumptions about the relationship of the mean to the variance determine the choice of Gamma parameters. Software radios work well if there is a 95 to 99% probability of staying within required performance. A useful rule-of-thumb sets peak predicted demand at one-third of benchmarked processor capacity: $D < C/4$. If $D$ is accurate and task scheduling is random, with uniform arrival rates and exponential service times, then on average, less than 1% of the tasks will fail to meet specified performance.

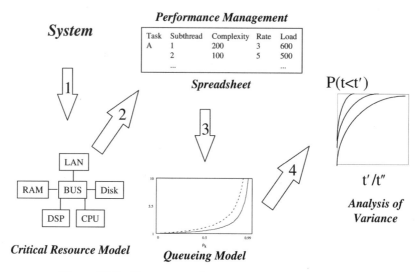

**Figure 13-2**   Four-step performance management process.

Simulation and rapid prototyping refine the estimates obtained from this simple model. But there is no free lunch. SDRs require three to four times the raw hardware processing capacity of ASICs and special-purpose chips. SDRs therefore lag special-purpose hardware implementations by about one hardware generation, or three to five years. Thus, canonical software-radio architectures have appeared first in base stations employing contemporary hardware implementations. The process for managing performance of such multichannel multithreaded multiprocessing systems is now defined.

## II. PERFORMANCE MANAGEMENT PROCESS FLOW

The performance management process consists of the four steps illustrated in Figure 13-2. The first step is the identification of the system's critical resources. The critical resource model characterizes each significant processing facility, data flow path, and control flow path in the system. Sometimes, the system bottleneck can be counterintuitive. For example, in one distributed processing command-and-control (C2) system, there were two control processors, a Hardware Executive (HE) and a System Controller (SC). The system had unstable behavior in the integration laboratory and thus was six months late. Since I was the newly assigned system engineer for a derivative (and more complex) C2 system, I investigated the performance stability problems of the baseline system. Since the derivative system was to be delivered on a firm-fixed price commercial contract, the analysis was profit-motivated. The timing and loading analyses that had been created during the proposal phase of the project over two years earlier were hopelessly irrelevant. Consequently, we had to create a system performance model using the developmental equipment. The HE produced system timing messages every 100 milliseconds to synchronize the operation of a dozen minicomputers and over 100 computer-controlled processors, most of which contained an embedded microcontroller. The SC stored all C2 messages to disk. The disks were benchmarked at 17 accesses per second, net, including seek and rotational latency and operating-system-induced rotational miss rate. Ten accesses per second were needed for applications at peak demand. System instability occurred as the demand on that disk approached 20 transactions per second. The solution was to pack 100 timing messages into an SC block for storage. The demand on the disk due to timing messages dropped from 10 to 0.1, and system demand dropped from 20 to 10.1

Since the SC development team had no critical resource model, they were unaware of the source of the buffer overflows, interrupt conflicts, and other real-time characteristics of the "flaky" system. In the process of developing the resource model, the team gained insight into overall system performance, ultimately solving the performance problems. Before we analyzed the data flows against critical resources, the management was prepared to buy more memory for the SC, which would have cost a lot and accomplished nothing. The quick fix of packing system-time messages more efficiently into disk blocks solved the major performance problem almost for free.

Instead of creating a resource model to help cure a disaster in progress, one can create the model in advance and maintain it throughout the system life cycle. This approach avoids problems through analytical insights described below. The second step, then, of performance management characterizes the threads of the system using a performance management spreadsheet. This spreadsheet systematizes the description of processing demand. It also computes facility utilization automatically, given estimated processor capacity. The first estimate of processing demand should be done early in the development cycle. The challenge is that much of the software has not been written at this point. Techniques described below enable one to make good estimates that are easily refined throughout the development process.

Given a spreadsheet model of each critical resource, the SDR systems engineer analyzes the queuing implications on system performance. In this third step of the performance management process, one can accommodate a mix of operating system and applications priorities. Finally, analysis of variance yields the statistical confidence in achieving critical performance specifications such as response time and throughput. This fourth step allows one to write a specification that is attainable in a predictable way. For example, one may specify "Response time to operator commands shall be two seconds or less." Yet, given an exponential distribution of response times and an estimated average response time of one second, there is approximately a 5% probability that the two-second specification will be violated on a given test. As the system becomes more heavily loaded, the probability may also increase, causing a perfectly acceptable system to fail its acceptance test. On the other hand, the specification could state "Response time to operator commands shall be two seconds or less 95% of the time given a throughput of $N$." In this case, the system is both functionally acceptable and passes its acceptance test because the throughput condition and statistical structure of the response time are accurately reflected in the specification. The author has been awarded more than one cash bonus in his career for using these techniques to deliver a system that the customer found to be stable and robust and that the test engineers found easy to sell off. These proven techniques are now described.

## III. ESTIMATING PROCESSING DEMAND

To have predictable performance in the development of an SDR, one must first know how to estimate processing demand. This includes both the mechanics of benchmarking and the intuition of how to properly interpret benchmarks. The approach is introduced with an example.

### A. Pseudocode Example—T1 Multiplexer

In the late 1980s, there was a competitive procurement for a high-end military C2 system. The evolution of the proposal included an important case study in the estimation of processing demand. The DoD wanted 64 kbps "clear" channels over T1 lines. There was a way to do this with a customized T1 multiplexer board, a complex, expensive hardware item that was available from only one source. The general manager (GM) wanted a lower-cost approach. I suggested that we consider doing the T1 multiplexer (mux) in software. Literally on a napkin at lunch, I wrote the pseudocode and created the rough order of magnitude (ROM) loading analysis shown in Figure 13-3.

The T1 multiplexer is a synchronous device that aggregates 24 parallel channels of DS0 voice into a single 1.544 Mbps serial stream. The companion demultiplexer extracts 24 parallel DS0 channels from a single input stream. DS0 is sampled at 8000 samples per second and coded in companded 8-bit bytes. This generates 8000 times 24 bytes or 192,000 bytes per second of input to a software mux. The pseudocode consists of the inner loop of the software mux or demux. Mux and demux are the same except for the addressing of the "get byte from slot" and "put byte" instructions. Adding up the processing in the inner loop, there are 15 data movement instructions to be executed per input byte. Multiplying this complexity per byte times the 192,000-byte data rate yields 2.88 MIPS. In addition to the mux functions, the multiplexer board maintained synchronization using a bunched frame alignment word in the spirit of the European E1 or CEPT mux hierarchy. The algorithm to test and maintain synchronization consumed an order of magnitude fewer resources than the mux, so it was not included in this initial analysis.

Again, MIPS, MOPS, and MFLOPS are not interchangeable. But given that this is being done on a napkin over lunch, it is acceptable to use MIPS as a ROM estimate of processing demand. The capacity of three then-popular

```
Object:  T1 Demultiplexer
    8000 Frames per second
    192000 Bytes per second
```

### T1 Demux Pseudocode

|                          | Instructions |
|--------------------------|--------------|
| For Each Frame           |              |
|   For Each Slot = 1,24 | 5       |
|     Get byte from slot | 3 Inner |
|     Put byte into output slot | 5 Loop |
|   End For       | 2            |
| End For                  |              |

```
ROM Loading Analysis:
    15 Instructions/byte x 192000 Bytes/sec = 2.88 MIPS
    VAX (8 MIPS): 2.88/8 = .35 CPU Seconds per second
    Sun-3 (4 MIPS): 2.88/4 = .7 CPU Seconds per second
    Gould (2 MIPS): 2.88/2 > 1 => Real Time Not Possible
```

**Figure 13-3**   Specialized T1-mux ROM feasibility analysis.

single-board computers is also shown in Figure 13-3, also in MIPS as published by the manufacturer. Dividing the demand by the capacity yields the facility utilization. This is also the amount of time that the central processing unit (CPU) accomplishes useful work in each second, CPU seconds of work per second. The VAX was projected to use 350 milliseconds per second, processing one real-time T1 stream more or less comfortably. The VAX was also the most expensive processor. The Sun also was within real-time constraints, but only marginally. The Gould machine was not in the running. However, that answer was unacceptable because the Gould processor was the least expensive of the single-board computers. The GM therefore liked the Gould the best. But the process of estimating software demand is like measuring a marshmallow with a micrometer. The result is a function of how hard you twist the knob, so we decided we were close enough to begin twisting the knobs.

To refine the lunchtime estimates, we implemented benchmarks. The prototype mux pseudocode was implemented and tested on each machine. The first implementation was constrained by the rules of the procurement to "Ada reuse." This meant that we had to search a standard library of existing Ada code to find a package to use to implement the function. The software team identified a queuing package that could be called in such a way as to implement the mux. The Ada-reuse library is very portable, so it ran on all three machines. We put the benchmark code in a loop that would execute about ten million bytes of data with the operating system parameters set to run this function in an essentially dedicated machine. The time to process 193,000 bytes (one second of data) is then computed. This time is shown in Figure 13-4 for each of the series of benchmarks that resulted. If it takes 10 seconds to process one second of data, then it would take at least 10 machines working in parallel to process the stream in real-time. The ordinate (vertical axis), which shows the facility utilization of the benchmark, can therefore be viewed as the num-

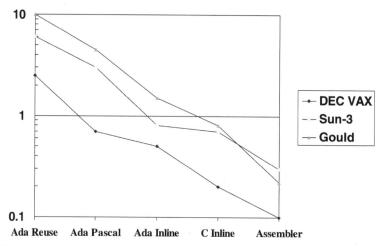

**Figure 13-4**   Five benchmarks yield successively better machine utilization.

ber of machines needed for real-time performance. The Ada-reuse approach
thus required 2.5 VAX, 6 Sun, and 10 Gould machines for notional real-time
performance. Of course, one could not actually implement this application
using that number of machines in parallel because each machine would be
100% utilized, which, as indicated above, cannot deliver robust performance.
With proper facility utilization of 50% , it would actually take 5 VAX, 12 Sun,
and 20 Gould processors. The purpose of the benchmarks is to gain insights
into the feasibility of the approach. Thus one may think of the inverse of
the facility utilization as the number of full-time machines needed to do the
software work, provided there is no doubt that twice that number that would
be required for a robust implementation.

The reuse library included run-time error checking with nested subroutine
calls. A Pascal style of programming replaces subroutine calls with For-loops
and replaces dynamically checked subroutine calling parameters with fixed
loop parameters (e.g., Slot = 1, 24). An exemption to the Ada-reuse mandate
of the procurement would be required for this approach. The second column
in Figure 13-4 shows 0.7 VAX, 3.2 Sun, and 4.5 Gould machines needed for
the Ada Pascal style. A VAX, then, can do the job this way in real-time, if
somewhat marginally. Pascal style is not as efficient as is possible in Ada,
however.

The In-line style replaces For-loops with explicit code. Thus, for example,
the For-loop code segment:

For Slot = 1, 24

X = Get (T1-buffer, Slot);

Put (X, Channel[Slot]);

End For

would be translated to In-line more or less as follows, trading space for speed:

$$X = Get\ (T1\text{-buffer},\ 1);$$

$$Put\ (X,\ Channel\ [1];$$

$$X = Get\ (T1\text{-buffer},\ 2);$$

$$Put\ (X,\ Channel\ [2];$$

...

$$X = Get\ (T1\text{-buffer},\ 24);$$

$$Put\ (X,\ Channel\ [24];$$

Consequently, the loop parameters are set at compile time in the In-line style whereas they were created at execution time in the For-loop style. This improves the performance as shown in the third column of Figure 13-4. Real-time performance now requires 0.5 VAX, 0.8 Sun, or 1.5 Gould machines. The VAX implementation would be robust using the In-line programming style, provided that the additional processing, such as synchronization control, did not consume more than 0.1 of the machine. The Sun implementation would be unreliable and the Gould cannot be used. Note that the difference between Pascal and In-line is more significant for the Sun and Gould Ada compilers than for the VAX. The VAX Ada optimization facilities precomputed loop variables, to the degree that it could given the limited number of registers in the machine. The other Ada compilers were not nearly as efficient.

At the time, other aspects of the efficiency of the Ada compilers were also suspect. In particular, the C compilers included register optimization not available in Ada. It was easy to recompile the In-line Ada code into C on each machine. The result, shown in column four of Figure 13-4, is that real-time performance could be achieved in 0.2 VAX, 0.7 Sun, or 0.8 Gould machines with the In-line C style of programming. The difference between Ada In-line and C In-line is more pronounced on the VAX than on the other machines. This improvement reflects the greater optimization of the C compiler with respect to the available instruction set. The VAX had immediate operand modes for small integer arguments that the C compiler used to reduce execution time, for example. The Sun and Gould lacked extended instruction types. In addition, the C compilers were not as highly optimized as the VAX C compiler.

By now, of course, the GM was convinced that we could use the Gould machines with just a little more work. So our best superhacker assembly language programmer was retained to make it happen. He did. The right-most column of Figure 13-4 shows utilization of 0.1 for the VAX, 0.22 for the Sun, and 0.3 for the Gould. Each processor can deliver robust performance for a single T1, in addition to doing may other tasks. In particular, the least expensive machine can be used as a robust T1 platform. At this point, the GM was happy. After we explained our idea to the T1 mux board vendor, we got such a great price on that board that it was better to buy the board than to use the software-based approach.

The experience gained through the benchmarks was enlightening and remains relevant today. Processor performance ranged over two orders of magnitude as a function of programming style, compiler, and machine instruction set architecture. The quickest to implement and most robust implementation, Ada reuse, also proved to be the most computationally demanding. Multiplatform COTS packages like enterprise Java beans may exhibit similar inefficiencies today. Extensive run-time error checking is safer than the more efficient implementations. Over time, compiler technology has evolved to the point where even early releases of C compilers for new machines are relatively efficient. Design choices that trade throughput for robustness are outside the scope of the compiler, however. In addition, the skills of programmers range over orders of magnitude as well. In any large project, there will be a mix of computationally efficient and inefficient designs that even the best compiler cannot overcome. It is therefore essential to systematically test the processing efficiency of each software object as early as practical in a project. A good rule of thumb is to include time-resources along with code and data space resources in the initial characterization of each module during unit test. These values may then be accumulated systematically using techniques explained below in order to determine where to optimize the implementation later in the integration phase.

## B. Quantified Objects

To generalize from this example, one must quantify the way in which processing requirements and resources interact in a software radio. Quantified objects are software configuration items that are encapsulated as objects, are accessed via message passing, consume specified data and program memory, and consume specified processing resources. Figure 13-5 lists resource demands for illustrative telecommunications applications [304, 307]. These resource requirements apply to the ADSP 21xx series of processors. That is, the MIPS listed are not generic MIPS, but MIPS on the 21xx instruction set architecture (ISA). A 21xx processor with a faster clock and with no memory access bottleneck will supply proportionally more MIPS. But rehosting these objects to another processor changes the MIPS required as a function of the new ISA.

Since new DSP chips are announced every year, one must plan on rehosting radio software every few years. Companies that support a variety of wireless and related DSP-intensive products may rehost core functions a few times in a year. To determine how many of the new processors are required for a large-scale application, one must be able to estimate how the MIPS change in rehosting scenarios. There are three steps in this process. One must first determine the instruction mix of the library object. One must then translate this mix to the new processor. Finally, one must aggregate demand on a per-

| Application | Function | MIPS | Data | Program |
|---|---|---|---|---|
| Telephone | DTMF Generation | 0.7 | 160 | 140 |
| | Caller Identification | 1 | 200 | 500 |
| | Acoustical Echo Canceler | 7.2 | 600 | 750 |
| | Line Echo Canceler | 2.7 | 200 | 400 |
| Answering Machine | DTMF Detection | 0.7 | 350 | 500 |
| | Playback Speed Modification | 2.5 | 300 | 1000 |
| | AGC | 0.1 | 80 | 440 |
| | Speech Codec | 3.5 | 1000 | 2000 |
| Modem | Call Progress Mon | 0.7 | 60 | 200 |
| | V.21 | 3 | 500 | 500 |
| | V.22 | 5 | 800 | 1500 |
| | V.22bis | 7 | 1000 | 1500 |
| | V.32 | 10 | 1300 | 1700 |
| | V.32bis | 12 | 2500 | 3000 |
| Fax | V.27ter | 5 | 500 | 1000 |
| | V.29 | 7 | 600 | 1100 |
| | V.17 | 10 | 2300 | 2500 |

**Figure 13-5**   Analog devices ADSP 21xx quantified objects.

processor basis to manage resource utilization. These steps are now explained in greater detail.

Referring again to Figure 13-5, each function listed is assumed to be implemented in a library object class. This object may be run on a dedicated machine using a test stub with the operating system set to preclude virtual memory. Most operating systems now include process-monitoring software that statistically monitors the type of instruction used by the processor. One must partition the ISA into convenient subsets based on the use of the ISA by the application. A few simple partitions are nearly as effective as a more complex treatment. An ISA may, for example, be partitioned into data movement and floating point arithmetic instructions. Although this is an oversimplification for modern processors, the types of instruction included in each partition differ more across SDR segments than within a segment. The goal is to pick a partition for the SDR segment that reflects the difference in ISA from the host processor of the library object to the new host.

The equation of Figure 13-6 then applies as follows. The proportion of load reflected in the object is determined from the benchmarking on the library host processor. The time to execute an average instruction from the partition is estimated for the target processor. One then adds up the capacity of the processor across the partitions ("types of instructions" in the figure). In addition, one may need to normalize across partitions. The number of bytes of data changed by the instruction may be used to normalize across partitions. If data movement instructions on the average access 2 bytes, but floating point arithmetic

$$C = \sum_{\substack{\text{Types} \\ \text{of} \\ \text{Instructions}}} \left( \begin{array}{c} \text{Proportion} \\ \text{of} \\ \text{Load} \end{array} \right) \frac{1}{\substack{\text{Time to} \\ \text{Execute}}} \left( \begin{array}{c} \text{Equivalent} \\ \text{Instruction} \end{array} \right)$$

**Figure 13-6**   Quantifying performance of a new processor.

accesses 4 bytes, multiplying as shown in the figure normalizes both types to byte-operations per second so they may be added without inconsistency. The result is a representative capacity estimate for the target processor. Consider the following example. Suppose caller identification uses 60% data movement and 40% floating point instructions on the library host processor. Suppose further that the new processor requires an estimated 10 ns for data movement and 40 ns for floating point arithmetic. Capacity, C, of the processor is:

C :     Data Movement :   $60\% \times 1/10$ ns $\times 2$ Bytes/instruction

Floating Point :     $40\% \times 1/40$ ns $\times 4$ Bytes/instruction

$C = .6 \times 100$ MHz $\times 2 + .4 \times 25$ MHz $\times 4 = 160$ MByte Ops/sec

This estimate of capacity is compatible with the way in which the object uses the ISA. The estimates may be validated by computing the MByte Ops/sec for the original processor and for the original object. Inefficiencies introduced by cache misses, for example, can be calibrated out this way. Suppose a data movement instruction should take 30 ns on the library host processor. Computing the MIPS of the library object yields a total wall-clock time in the validation step equivalent to 40 ns used per data movement instruction. This means that the clock time of the target processor should also be multiplied by 4/3. This additional factor times the nominal execution time yields an effective execution time on the new processor that is closer to what one will experience with the rehosted software. The net effect of this analysis is to establish a processing capacity estimate for a new host processor that is compatible with the way in which the library object uses machine resources.

### C. Thread Analysis and Object Load Factors

Often, however, one will not have a comparable library object on which to base the preceding analysis. In such cases, code inspection yields the necessary insights. Recall the T1 benchmarking example above. At the conclusion of the benchmarking work, each implementation style used a specified fraction of a given processor. This fraction is readily converted into equivalent instructions per invocation by the equation:

Instructions = (Processor_MIPS $*$ Processor_fraction)/Invocation_rate

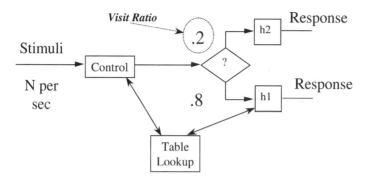

**Figure 13-7**   Visit ratios represent the impact of decisions on resources.

The T1 benchmark that took 10% of an 8 MIPS VAX with one invocation per second takes 800,000 instructions each time it is invoked. Alternatively, if the module is structured to be invoked 8000 times per second (e.g., once per T1 frame), it would use 100 instructions per invocation. The number of instructions per invocation is called the execution complexity, thread complexity, subthread complexity, or just the complexity. In SDRs, processors rarely perform just one function. That would require a large number of processors, rendering the configuration too costly. Instead, each processor generally supports a mix of tasks, possibly depending on the state of some control parameter. One might, for example, have to run a synchronization algorithm every hundredth frame. Or statistically, an equalizer might converge quickly 80% of the time and fail to converge (using more instructions) 20% of the time.

Figure 13-7 illustrates such a situation. Two lower-level objects h1 and h2 perform the bulk of the processing. These objects are akin to the core T1 processing object and the frame synchronization objects required for the T1 application. Or they could represent a GSM equalizer converging quickly versus the equalizer failing to converge. In the generic example of the figure, flow of control is represented by single-headed arrows. Data reference is represented by double-headed arrows. Each of $N$ stimuli per second invokes the control object. This object performs a table lookup in order to compute some notional system state. The state is tested in the decision box. Eighty percent of the time, the state indicates that object h1 should be invoked. The other 20% of the time, h2 is invoked. These two objects, of course, have different complexity. Thus, the complexity of the top-level object illustrated in the figure is a kind of weighted average of the complexity of the subordinate objects h1 and h2. The object load factor reflects the number of times, on the average, that an object is invoked for each incoming stimulus. The table-lookup object is invoked each time control is accessed. In addition, it is invoked by object h1, but not by object h2. Thus the table-lookup load factor is 1.8 as illustrated. Each external stimulus causes table lookup to be invoked an average of 1.8 times. Thus, the load factor is not constrained to be less than one.

| Object | Rate | Load Factor | Demand |
|--------|------|-------------|--------|
| Table Lookup | N | 1+.8 | 1.8 N |
| h1 | N | .8 | .8 N |
| h2 | N | .2 | .2 N |

**Figure 13-8**   Subthread load factors may be greater than unity.

$$D = \sum_{Processor} \left(\begin{array}{c} \text{Thread} \\ \text{Activation} \\ \text{Rate} \end{array}\right) \text{Complexity} \left(\begin{array}{c} \text{Subthread} \\ \text{Load Factor} \end{array}\right)$$

**Figure 13-9**   Demand is the product of object complexity, activation rate, and load factor.

**TABLE 13-3   Useful Complexity Estimates**

| Function | Complexity |
|----------|------------|
| Primitive (check, flag, DMA setup) | 30 |
| Simple subroutine | 100 |
| Simple matrix operation | 300 |
| Operating system call | 500 |
| Sort $N$ elements | $6.17\,N^2$ |
| Interrupt service | 1800 |
| Database access (in RAM) | 3000 |
| When MIPS are known | MIPS/Clock |

The total software demand is the product of object complexity, thread activation rate, and subthread (or object) load factors. A thread is an end-to-end path from stimulus to response. A subthread is the intersection of a thread on a processor. A subthread may consist of a single object invocation, or multiple objects may be invoked as illustrated in Figure 13-8. The fundamental equation for average processing demand is shown in Figure 13-9. Demand is defined with respect to a processor. Thus, if a processor supports multiple threads, demand per thread is aggregated for that processor.

The complexity of an object is often not known in advance. But as was seen for the T1 multiplexer, simple estimates are often accurate to within a factor of two. Table 13-3 provides a list of illustrative complexities. The point is not to use these, although these have been useful in the past. The point is to develop your own ROM complexity estimates that reflect software in your radio applications. The first seven estimates in the table suggest the types of primitive function categories for which one typically needs ROM complexity estimates. The last item is a reminder of the equation for estimating instructions of equivalent complexity on a per-second basis from the processor's clock rate

| Resource Management Spreadsheet | | | | | | | |
|---|---|---|---|---|---|---|---|
| Copyright Mitola's Statisfaction, All Rights Reserved @1996 | | | | | | | |
| | | | | | | | |
| Thread | Object (S/R) | Processor | Event | Event Rate | Complexity | Visit Ratio | Demand |
| | | | | | | | MIPS |
| T1 | Control | A | Frame | 8000 | 100 | 1 | 0.8 |
| | Table Lookup | A | Frame | 8000 | 250 | 1.8 | 3.6 |
| | h1 | A | Frame | 8000 | 1000 | 0.8 | 6.4 |
| | h2 | B | Frame | 8000 | 2500 | 0.2 | 4 |
| | | | | | | | |
| Processor | Demand | Capacity | Utilization | | | | |
| A | 10.8 | 25 | 0.432 | | | | |
| B | 4 | 25 | 0.16 | | | | |

**Figure 13-10**   The resource management spreadsheet.

if the demand in MIPS has been measured. This measurement step is critical. No matter how trivial the object or module, each significant parameter set should be timed on a dedicated machine with no paging. This may be done when creating unit development folders (UDFs). UDFs that include timing data may be used to keep the resource management spreadsheet up to date as the project proceeds. The next section describes this spreadsheet.

## D.  Using the Resource Management Spreadsheet

The resource management spreadsheet accumulates the average software demand and the average processing capacity available per processor. The organization of the spreadsheet is illustrated in Figure 13-10. The upper section aggregates resource requirements (processing demand), while the bottom three lines summarize processing capacity and resource utilization. Consider the upper section further. For each object, a line in the spreadsheet identifies the thread to which it is contributing, the processor on which it runs, the event that triggers that thread, and the stimulus rate of those events. Complexity and visit ratios are as presented above. The spreadsheet computes the appropriate products and aggregates the total demand by processor. The aggregate demand per processor is carried to the lower section of the spreadsheet where the ratio of demand to capacity is computed and presented as utilization.

From the resource utilization, the spreadsheet computes the queuing factor $R(\rho)$. This factor is multiplied by the service time to yield the wall-clock time that will transpire, on the average, in the execution of that object, given queuing delays. Response time is the sum of expected service times across a thread as illustrated in Figure 13-11. The resource management spreadsheet available from the author's includes illustrative values of each of the above parameters for SPEAKeasy-class radio applications (see [379]).

$$T_r = \sum_{\text{Thread}} T_s \qquad \begin{array}{l} T_r \text{ is Response Time} \\ T_s \text{ is Service Time} \end{array}$$

$$Ts = \left( \frac{\displaystyle\sum_{\text{Processor}} \text{Object Loads}}{\text{Processor Capacity}} \right) R(\rho_s)$$

**Figure 13-11**  Response time is the sum of expected service times.

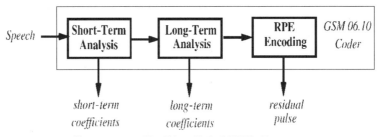

Figure courtesy of Dr. Thierry Turletti, INRIA, France.

**Figure 13-12**  GSM vocoder functions.

## IV. BENCHMARKING APPLICATIONS

This section provides benchmarks from the literature. The principal source of this material is the *IEEE Journal on Selected Areas in Communications (JSAC) on the Software Radio,* for which the author served as editor in chief. Early drafts of most of the papers did little to quantify computational demand. An algorithm that promises some SDR function without quantifying the resources has limited insertion potential. Thus, the JSAC papers were revised for greater emphasis on concrete computational complexity. Dr. Thierry Turletti, now of INRIA, France, however, was focused from the outset on benchmarking the GSM basestation.

### A. The GSM Basestation

Turletti was among the first to quantify the resource requirements of the software radio [108]. He characterized the processing requirements of a GSM software radio basestation in terms of industry standard benchmarks, the SPEC-marks. The GSM vocoder illustrated in Figure 13-12 is one of the fundamental building blocks of the GSM basestation. This vocoder includes short-term analysis which yields short-term coefficients. Long-term analysis yields long-term average coefficients on the residue from short-term analysis. Finally, the regular pulse excitation module extracts a model of the excitation pulse.

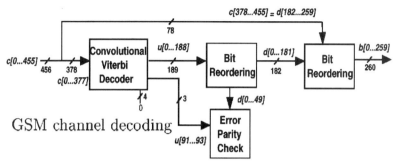

Figure courtesy of Dr. Thierry Turletti, INRIA, France.

**Figure 13-13**   GSM channel decoding.

The channel decoder is illustrated in Figure 13-13. The modem yields a bitstream which is the input to the decoder. The Viterbi decoder computes a maximum likelihood decoding of the channel bitstream. Since bits had been interleaved for transmission, bit reordering is required to de-interleave the bits and to reconstruct the parity bits for the parity error check.

Figure 13-14 provides highlights of Turletti's paper. The SPECmarks include integer ("int") and floating point ("fp") benchmarks. The definition of the benchmark was changed in 1995 from the SPECmark established in 1992, so both are provided to allow comparisons. The CPU used in these benchmarks was a DEC Alpha processor. In some cases, the benchmarks were run on Sun and Pentium processors as well as the Alpha. Turletti characterizes the minimum and maximum requirements per user channel. The polyphase transformation is the combination of frequency translation and filtering that converts a digitized IF signal to baseband. The Alpha can convert one channel without exceeding the 50% utilization target, but capacity is limited to a few channels per processor. A base station, then, requires multiple heterogeneous processors.

Figure 13-14 also shows the processing capacity of the four processor types that were benchmarked. Although the data lacks the Intel XXpress Pentium SPECfp92, the other benchmarks quantify the degree to which GSM modules can be implemented on these processors. Such quantification is essential for the transition of software radio technology from research to engineering development and product deployment. One of the drivers for the increased emphasis on associating benchmarks with code modules is the work of the SDR Forum. The Forum's download API includes a capability exchange. This exchange specifies the resources required to accept the download. Some radio functions are so demanding of resources that they are unlikely to be downloaded. Smart antenna algorithms, for example, may require a separate front-end processor and antenna array infrastructure. Benchmarks of such capabilities allow systems designers estimate processing capacity for a smart antenna applique. But as smart antenna infrastructure proliferates, product developers and service

| GSM Modules (for one GSM logical channel) | Performance (metric/user-channel) | | | | | | | |
|---|---|---|---|---|---|---|---|---|
| | SPECint92 | | SPECfp92 | | SPECint95 | | SPECfp95 | |
| | min | max | min | max | min | max | min | max |
| Speech coder | 11 | 20 | 17 | 28 | 0.15 | 0.18 | 0.31 | 0.73 |
| Channel coder | 0.50 | 0.76 | 0.68 | 0.85 | 0.011 | 0.019 | 0.013 | 0.018 |
| Interleaver | 0.11 | 0.30 | 0.13 | 0.23 | 0.002 | 0.007 | 0.003 | 0.005 |
| Cipher | 0.16 | 0.22 | 0.14 | 0.30 | 0.004 | 0.004 | 0.003 | 0.007 |
| Modulator | 5.3 | 7.3 | 4.8 | 8.4 | 0.12 | 0.18 | 0.11 | 0.19 |
| Decipher | 0.16 | 0.22 | 0.14 | 0.30 | 0.004 | 0.004 | 0.003 | 0.007 |
| Deinterleaver | 0.11 | 0.32 | 0.12 | 0.32 | 0.003 | 0.008 | 0.003 | 0.005 |
| Channel decoder | 3.1 | 4.8 | 4.2 | 5.9 | 0.09 | 0.12 | 0.07 | 0.15 |
| Speech decoder | 6.7 | 11 | 5.9 | 15 | 0.17 | 0.21 | 0.13 | 0.40 |
| Total | 30 | 45 | 33 | 59 | 0.85 | 1.1 | 0.68 | 1.51 |

| Polyphase Transform 200 kHz channels | Perf. (%cpu) # frequency channels | | | |
|---|---|---|---|---|
| | 1 | 2 | 4 | 8 |
| SUN Ultra M170 | 20 | 60 | 180 | 620 |
| DEC 3000/800 | 25 | 80 | 220 | 670 |
| Pentium 166 | 40 | 110 | 340 | 1200 |
| Pentium Pro 200 | 25 | 70 | 210 | 720 |

| SPEC'92 Benchmarks Platform | SPECint 92 | SPECfp 92 |
|---|---|---|
| SUN Ultra M170 | 252 | 351 |
| DEC 3000/800 | 138.4 | 187.6 |
| Intel XXpress Pentium | 197.5 | – |
| Intel Alder PentiumPro | 318.1 | 283.2 |

Figure courtesy of Dr. Thierry Turletti, INRIA, France.

**Figure 13-14**   Significant GSM benchmarks.

providers will be looking for smart antenna capabilities that differentiate the product or service. In this situation, the infrastructure may have a 1.1 GFLOPS smart antenna front-end processor, so the issue of whether the algorithm requires .9 or 1.3 GFLOPS can be a significant one. The algorithm that needs only .9 GFLOPS may, in fact, be a download candidate. In the near term, speech coders and INFOSEC algorithms are the more likely candidates for downloading.

## B. Benchmarking Partial Interference Cancellation Receivers

Neiyer Correal et al. [377] describe a partial interference cancellation CDMA receiver, with benchmarking results. Figure 13-15 is the block diagram of this receiver. The received signal is processed to extract the CDMA codes of K subscribers. The air interfaces are then regenerated at IF and subtracted from the aggregate received signal, leaving a residue signal. Matched filters are then applied in Stage 2 to yield an error signal stream which is applied to the stream recovered in Stage 1 to yield a corrected stream. The relationship among BER and delay-estimate error is also defined for this receiver. For

r(t)

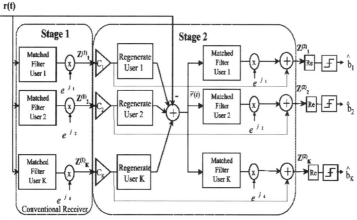

Figure courtesy Neiyer Correal, University of Florida, ©1999 IEEE from [377].

**Figure 13-15**　The partial interference cancellation receiver.

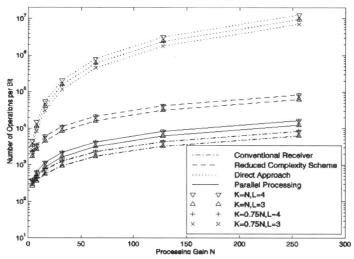

Figure courtesy Neiyer Correal, University of Florida, ©1999 IEEE from [377].

**Figure 13-16**　Operations per bit versus processing gain.

near-zero RMS delay estimate, 20 users, $Eb/No = 8$ dB, spreading gain of 31, the Stage 1 receiver has a BER of $2 \times 10^{-2}$, while the Stage 2 processing reduces this to $5 \times 10^{-3}$. This improved performance is not without its costs, however. Figure 13-16 shows that about 800 operations per bit are required with a conventional algorithm. The direct approach to implementing partial interference cancellation requires around a million operations per bit. Their reduced complexity scheme reduces this demand to about 10,000. With such

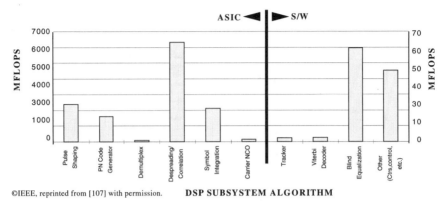

**DSP SUBSYSTEM ALGORITHM**

**Figure 13-17**   Processing requirements of a CDMA software radio.

quantification, this 100 : 1 increase in computational complexity can be traded against the performance improvements.

### C. Benchmarking Handsets

Gunn et al. [378] developed low-power digital signal processor (DSP) sub-system architectures for advanced SDR platforms. They considered the re-quirements of a portable radio for flexible military networks. Their functional taxonomy identified the following CDMA radio functions:

1. Pulse Shaping: Filtering the waveform to be transmitted for spectrum occupancy.
2. PN Code Generation: Generating the direct-sequence spreading code.
3. Demultiplex: De-interleaving constituent bitstreams.
4. Despread: Correlating the received waveform to a reference.
5. Symbol Integration: Combining chips into an information symbol.
6. Carrier NCO: Numerically controlled oscillator creates the (IF) carrier.
7. Tracker: Carrier tracking and delay-lock loops.
8. Viterbi Decoder: Maximum likelihood majority decoder.
9. Blind Equalization: Weighted tapped-delay line filter.
10. Other: Counters, timing, control, etc.

These functions are listed in Figure 13-17 from left to right.

Gunn partitions these functions into ASIC and software implementations as shown in the figure. (Note the change of scale between the left half of the figure in GFLOPS and the right half in MFLOPS.) The strategy for this parti-tioning is to put the most computationally intensive functions on the ASIC to reduce total system power. In addition, items that require flexibility such as control are allocated to the DSP. Functions requiring chip-speed processing are strong candidates for the ASIC, while functions on the de-spread bitstream more naturally occur on the DSP. The analysis leading to this partitioning also includes the minimization of internal interconnect on the ASIC. Due to the nature of the waveform, despreading requires 6.3 GFLOPS, the most computationally intense function in the radio. Pulse shaping and symbol integration are next with 2.3 and 2.1 GFLOPS. The envisioned ASIC delivers about 12 GFLOPS, so that the DSP core need deliver only 110 MFLOPS (usable, or 220 peak from the resource-utilization model).

## V. SPECIFYING PERFORMANCE PARAMETERS

Throughput and response-time are the key performance parameters of transaction-based systems. Queuing models have been used to estimate these parameters of transaction-based systems such as airline reservation systems since the 1970s. In that time-frame, I began to experiment with modeling distributed radio control software as if each discrete task were a transaction. The goal of such modeling was to write performance specifications for high-end command-and-control systems. The first such system so modeled was an airborne command-and-control system that included 12 minicomputers, 127 computer-controlled "smart" boxes (such as transmitter and receiver controllers), and 1.2 million lines of FORTRAN and assembly language. The system was delivered on a firm fixed-price contract on time and on budget. The throughput and response time specifications were rigorously tested and every one was sold off. As a result of that success, I taught in-house courses on real-time executive control software and performance modeling. Subsequently these techniques were refined on ground-based telecommunications, global (air and ground) communications, message processing, and automatic voice mail systems. As the high-end SDR technology began to emerge in DoD programs in the 1980s, the methods for specifying throughput and response time were refined. Recently, I have evolved this baseline through analysis of JTRS technology pathfinders and hands-on rapid prototyping. The following sections provide lessons learned.

### A. Facility Utilization

Facility utilization is the key to accurate throughput and response time specification. There will always be some bottleneck to each service thread in an SDR. Not all threads have the same bottleneck. A simple FSK packet data modem object, for example, that is part of a message store-and-forward thread, may not be the system bottleneck. If, for example, the message formats are complex and the messages and metadata are stored on secondary storage, the storage medium or I/O channel may be the bottleneck. At the same time, in the same system, a 64QAM modem object may be the bottleneck of the tactical data link, even though the tactical data link has a lower data rate than the

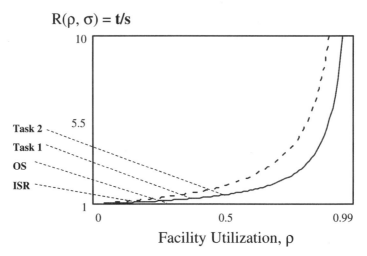

$R(\rho, \sigma) = \mathbf{t/s}$

Facility Utilization, $\rho$

**Figure 13-18**   Modem facility utilization example.

packet data network. The level of resources that an application experiences given its priority with respect to any processes that may execute at the same time is its operating point. By examining the operating point of the service on the serving facility, one may identify bottlenecks throughout a heterogeneous SDR system.

Consider, for example, the modem facility utilization curve of Figure 13-18. Suppose the serving facility is a front-end DSP chip. The interrupt service routines (ISR) and operating system consume some average resources as indicated in the figure. In this case, the total for both of these infrastructure tasks is approximately 27% of the maximum deliverable to a single task under ideal conditions. This maximum may be established by benchmarking a small (20–100-line) sequence of code that can loop forever, creating known internal processing demand. It is important that this module not demand any operating system services and that such services be shut off to the degree possible. This provides a benchmark of peak deliverable MIPS. One then measures the time of a benchmark with representative demand on the real and virtual memory system, I/O, etc., representative of peak system loading. Since such small modules may be locked into memory and clocked against the wall-clock (as opposed to the system clock), one may infer the computational resources the system is capable of in a representative mix. One counts instructions in the inner loop, runs these benchmarks a billion times, and then divides the total instructions by elapsed wall-clock time to determine effective MIPS. Facility utilization is then computed using wall-clock time of each module of the system as it is tested in isolation during software development. Using wall-clock time overcomes the inaccuracies of system clocks on virtual-memory machines.

The operating point is the level of facility utilization achieved when a given task and all higher priority tasks are loading the system at the specified throughput. In Figure 13-18, let Task 1 be the packet modem. Suppose the system test conditions specify (directly or indirectly) that two of these channels *may* operate at once. Although there is a duty cycle associated with a packet radio, it is likely that both of these high-priority tasks will be operating at once. Thus, although the packet radio consumes only 12% of the system, the second packet radio (Task 2) drives the system to 50% loaded instantaneously. Suppose the tactical data link takes 30% of capacity when operated in isolation. If this modem executes 20 times per second, it has 50 ms to run to complete. In addition, if it executes for 10 ms when running in isolation (35% of CPU peak demand), it has an average demand of only 1/5 (10/50) of 35% or 7%.

One must know for sure whether the tactical data link and these two packet radio channels might operate at the same time, and how statistically independent they are. If simultaneous operation is absolutely impossible because there are hard limits in the control software (such as not loading them into the modem chip at the same time), then one may consider the operating point of the tactical data link to be 25% (OS/ISR) + 7%, or 31%. The performance will be within a factor of 1.5, on the average, of performance tested in software integration. The resource management spreadsheet may have amalgamated this average with the 24% demand of the two packet radio tasks to yield about 56% total demand—again, well within solid average performance numbers.

But, if both tactical data link and packet radios may operate at once, then although one would like to think that the tactical data link and the packet radios are statistically independent, this may be very risky. Suppose the packets are being derived from the tactical data link. In this case, whenever a tactical data link burst is to be processed, two outbound packets may also be created. This violates the assumption of statistical independence. In this case, one must specify performance that is the *worse-case average*. That is, one must add the 50% operating point of the two packet radio modes plus the 35% load of the tactical data link to get a worse-case for the average loads of 85%. Note that the performance factor is now read from the $y$-axis as about 6 : 1. That means that if the modem task took, say, 10 ms per block of data in a dedicated machine, it will take an average of 60 ms when it is contending for resources with the packet radio threads. In this case, the radio network may crash. As a result, the modem chip that is just flying along on any one of these tasks will in fact be creeping when these tasks randomly align. And if they statistically align too often, the radio crashes. These kinds of faults are often extremely difficult to diagnose because they are intermittent and thus nearly unrepeatable.

Although the theory of nonlinear processes has not been fully applied to software radio, it is my experience that once something like this happens, it keeps happening or happens at a rate that is not predicted by conventional statistics. You can convince yourself of the power of nonlinear interactions by experimenting with a double pendulum. When the pendulum is touched lightly at the right time, it may enter a chaotic state which causes small internal forces to reinforce each other, yielding dramatic behavior such as wild oscillations. Somehow, through the operating system, ISR delays, timing delays, etc., if these tasks can happen at once, they often mutually reinforce. In addition, that simultaneous behavior may become the dominant mode of the system, rather than a mode that averages out. The situation where the information content of the tactical data link is in fact driving the packet radios is analogous to pinging on the pendulum just in sync so as to cause potentially catastrophic behavior.

Thus, one has to carefully examine the performance management spread-sheet to determine whether there are possibly nonlinear modes present. In such cases the average load can be adjusted to be the worse-case average to yield the appropriate operating point.

### B. Response Time Estimation

The response time is the sum of the time delays experienced in each proces-sor that supports the thread. So in the packet radio example above, the FSK modems operate at a low multiplier, but the message processing database may operate at a higher multiplier. This multiplier yields an average time delay for the packet. The sum of these two time delays would be the total if one is either transmitting or receiving messages. The end-to-end turnaround for input via one packet stream through the message processor and out via the other packet stream will be the sum of the two delays plus any processing required for bridging.

The resource management spreadsheet computes these sums for all threads defined in the system. One has to compare the traces of the threads through the network to ensure that the calculations are accurate. In the case of a one-way thread that intersects each processor only once, the performance management spreadsheet generally computes the correct result provided the data in the spreadsheet represents the loads in isolation. But if the thread winds its way onto and off of a processor and then back onto the processor, the spreadsheet will be slightly optimistic (and this is bad if you want to meet spec). It will compute the delay through the processor that it visits more than once as if the operating point were the sum of the two visits. In fact, the thread may experience a lower (and thus better) operating point on its first visit, for a given time delay; it will then experience the higher delay on its second visit. But the spreadsheet employs a single-visit model. So it will compute the total delay through the processor as the single-visit delay at maximum operating point. This ignores the earlier, lower operating point visit. The estimate is therefore optimistic by this amount. This particular element will thus be in error by roughly a factor of two. If the total time spent is only 1% of the end-to-end time, this error may be safely ignored. But if the visit to this processor represents more than 10 to 20% of the end-to-end budget, this error must be corrected.

## C. Throughput Estimation: How Much Hardware?

Throughput is the value of input rate $N$/sec for which the probability of exceeding the delay (queue, etc.) limits is acceptable. That is, there is some rate at which events occur. In the resource management spreadsheet, the event rates are to be entered for each event. For those events that are variable, it is best to use the spreadsheet in a way that the value of $N$ is entered only once and is carried around the spreadsheet by reference. In Excel, one may name the spreadsheet cell $N$ and then enter the equation "$= N$" in characterizing objects that are invoked as a function of $N$. One may also enter fractions of $N$, or functions of $N$ that represent the way in which the load is presented. For example, one might have a number of voice channels of $N$. The event rate for processing that is synchronous with the 8000 samples per second of a voice input channel could be entered as "$= 8000 * N$". One then must assure that the end-to-end delay is within limits.

For transaction-based throughput such as supporting a packet radio, isochronism means that the end-to-end delay of processing each packet does not exceed the timing window allowed by the protocol. Thus, the number of packets offered to the system per second is the independent variable and the end-to-end delay is the dependent variable. After completing the resource management spreadsheet, one varies $N$ over an appropriate range representing the possible demand on the system. If during this process the facilities begin to exceed safe operating points, the spreadsheet (as designed) will flag the excessive utilization. If it does not, then the user may have inadvertently written over the expressions that perform these tests. In any event, the SDR engineer should understand the spreadsheet in terms of events that cause work to be done and the related processing demand on each system resource. The result will be a reasonable estimate of average throughput.

But a specification that merely specifies the average value of such a critical parameter is not a complete spec. Throughput will not be met some fraction of the time. End-to-end delays will be exceeded some fraction of the time. What amount of such failure to meet spec is acceptable? This varies by customer and application. If the spec can never be exceeded, then one has to provide sufficient hardware to preclude any worst-case situations. This is generally not affordable, because one has to provide 2 to 5 times the hardware that would be required with some degree of timesharing and statistical multiplexing of tasks. To see this one may compute the processing capacity required to guarantee worst-case performance. This may be done in the resource management spreadsheet by driving the product of event-rate times execution-time of all tasks to the worse-case maximum. That is, if an event normally happens once per second but the worse case is 5 times per second, set the event rate to 5. If the task takes 30 ms on the average but

100 ms worse case, then set the task duration to 100 ms (e.g., by increasing the MIPS accordingly). The product of event rate of 5 times worse-case execution time of 100 ms yields a worse-case timing budget of 500 ms per second. If there are no other tasks on the machine, then this task may approach 90% before the system will have problems. That is, there is no statistical structure to the offered demand, so only the incidental statistical structure of execution times (e.g., I/O latency) will tend to drive the performance beyond this worse-case average. The task, in short, operates on a dedicated machine.

The slightly more aggressive approach described earlier computes average behavior and requires total demand to be kept below 50% of realizable peak for a high quality, robust system. Nevertheless, the system will occasionally exceed its throughput or response time, so one must specify the fraction of the time that this is allowed.

### D. Probability of Exceeding Specifications

For such transaction-based tasks, then, the overall process of establishing a specification consists of: (1) determining the average performance, (2) selecting a variance ratio $V$, and (3) using the Incomplete Gamma Function. The average performance specification is derived as described above. For the sake of exposition, consider packet delay in a software radio. Suppose that on the average, one may achieve a total processing delay of 100 ms. If the specification is written as such, then it will be exceeded about half the time because, as we said, this is the average delay. An average is made up of values, half of which are above the average and half of which are below the average, so half the time you fail. Failing to meet spec half the time is not good at system sell-off. Customers and users expect the system to behave appropriately "most" of the time.

Two things determine how badly the spec will be exceeded. The first is the variance or amount of scatter of the specified parameter. If the average delay is 100 ms, how often will it exceed 200? This depends on the statistical structure of the service-providing software. If the software is simple and has few logical tests, then there will be little variation in the time required to execute the module. A FIR filter exemplifies this case. If the object is doing something complex and open ended like equalizing a signal that may not converge, then there may be a specified number of iterations that represents worse case. And the variation from average may be a function of, for example, SNR. One may then specify an SNR under which to test the system so that the average is representative. The variance ratio, $V$, is the ratio between the variance and the mean service time. An FIR filter has a low variance, so $V$ may be near zero. A more complex function like a channel modem may have a variance ratio more like 1, where the variance is about the size of the average. This is characteristic of the Poisson distribution. As the variance ratio gets bigger, so does the variance. If there

Incomplete Gamma: $1-\Gamma(a,c)$

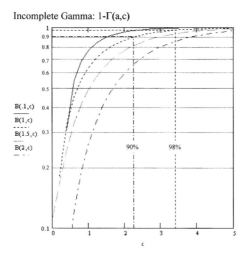

**Figure 13-19**   Analysis of variance using the incomplete gamma function [380].

is any suspicion that a task has combinatorially explosive logic, then one must characterize its variance ratio by measuring the task over a range of input parameters that excites the task into consuming large resources. But most of the SDR tasks presented in this book are not combinatorially explosive (unless explicitly presented as such, as in some of the coding techniques). Thus, a useful starting point is to assume that the system is modeled by $V = 1$.

The second factor that determines the probability of exceeding the spec is the degree to which the spec exceeds the average performance. Again, if the average packet delay is 100 ms and the specification is also 100, tests will fail 50% of the time (more or less). But what if the spec is increased to 200 ms? The Gamma Function of Figure 13-19 provides an estimate of the probability of meeting the relaxed specification. The ratio of the specification, $T$, to the average time delay, $t'$, estimated using the resource management spreadsheet is the abscissa. The probability that the time experienced in the system, $t$, exceeds this spec, $T$, is the ordinate. Thus, for example, if one wants to have a 98% probability of meeting spec, one must set $T/t'$ equal to 3.5. Solving for $T$, the spec should be 350 ms. In order to truly test such a specification, one must not only test the system some statistically significant number of times, one must also estimate the fit of the distribution to the data. Statistically significant numbers will be on the order of 100 to 1000 events in a range of SNR and protocol conditions. Given that packet delays are being tested, one may readily obtain statistically significant numbers of events automatically. If the system operates at high packet rates for a long period of time, the number of test faults ($t > T$) will approach 2% asymptotically. Other parameters such as user response time may take more test time to yield a statistically significant number of test events.

## VI. ARCHITECTURE IMPLICATIONS

The process of estimating the critical performance parameters, verifying them throughout the development process, and using them to decide where to fine-tune the system yields a reliable system that performs according to specifications on affordable hardware. Performance management is the discipline of allocating processing capacity to demand and monitoring that allocation throughout the development cycle. Quantified functions best serve COTS product selection and performance management. Inspection of critical code yields insights. The resource management spreadsheet simplifies resource utilization estimates. The performance management steps outlined in this chapter therefore constitute a key aspect of enterprise-level SDR architecture.

The resource management function of infrastructure middleware can be implemented in a way that addresses performance management as a constraint. For example, the performance manager could include a performance management spreadsheet. The spreadsheet should capture the fundamental relationships among processing demand and computational capacity as the one defined above. With such a strategy, each plug-and-play component must express to the resource manager either its computational capacity or its processing demand. The resource manager would aggregate demand as it assigns software components to processing platforms. It could thus establish the quality of performance to be expected. Predictions of unacceptable performance could then be dealt with by the network and/or by the user.

## VII. EXERCISES

1. What metric best characterizes the capacity of a PCI bus? of a TIC62 DSP? Suppose a bit-serial protocol is to be distributed on two C62s using the PCI bus. Are these still appropriate metrics? Suppose a domain manager permits user-defined constraints. Should the metrics be defined per processor class or per application class?

2. What SDR parameters drive the computational complexity of IF processing? Of the receiver? What additional parameters may be important for TDMA? CDMA?

3. Define facility utilization. What is the utilization of an Analog Devices 21xx running one DTMF application? One GSM channel? Answer these questions for the latest SHARC processor. How many GSM channels can one processor support with 95% confidence?

# 14 Smart Antennas

Smart antennas are an important application of SDR technology [381]. An in-depth treatment is beyond the scope of this chapter. The objective is to introduce the topic to identify the implications of smart antennas for software-radio architecture.

The smart antenna is a logical extension of antenna diversity described above [382]. Smart antenna arrays integrate the contributions of spatially distributed antenna elements to provide wireless communication systems with larger capacity and higher link quality through frequency reuse and cochannel interference suppression [383, 384]. Since smart antennas require an order of magnitude more IF and baseband digital processing capacity than a conventional receiver, the smart antenna base station is "90% antenna." Contrast this to a conventional base station, which is only "10% antenna," including diversity processing. The rate of proliferation of the smart antenna technology in the commercial sector has been slow because of the cost of this increase in capability.

## I. SMART ANTENNA DOMAINS

Four applications domains attract investments in smart antenna technology as illustrated in Figure 14-1. Historically, military radar and communications jamming laid the foundations of smart antenna technology. Investment leadership has shifted to commercial terrestrial networks, however. For example, a smart antenna with per-subscriber AOA estimation, interference differentiation, and coherent multipath combining was demonstrated for AMPS in 1994 [385, 386]. In addition, GSM infrastructure is amenable to smart antenna applications [387]. In the future, 3G base stations with W-CDMA 1 : 1 frequency reuse also should benefit from this technology [388].

Military applications remain substantial. Technology for beamforming on transmit for communications, for example, was sponsored by DARPA's GloMo program [389]. Academic interest is growing in the area of joint transmission and reception diversity via smart antennas [390, 391]. Since the smart antenna places a null on interference [385], the military could use this COTS technology to reduce jamming effects. In addition, both military [392] and commercial satellite communications terminals benefit from rapid electronic beam steering [393] and other features of smart antennas, such as overcoming light and heavy shadowing [394]. This chapter provides an over-

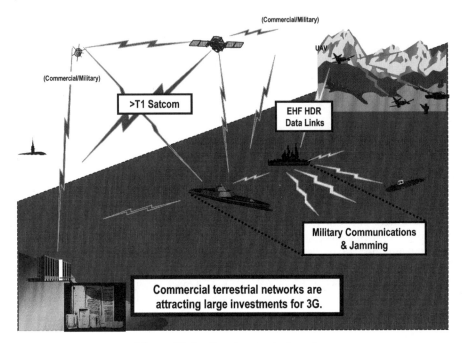

**Figure 14-1** Smart antenna domains.

view of the relationship between smart antenna technology and software-radio architecture.

Levels of smart antenna technology, in order of increasing cost and complexity, include:

1. Multibeam antennas to enhance SNR [395]
2. Null-forming to reduce interference in high traffic density [385]
3. Space-time adaptive processing to jointly equalize the spatially enhanced signals [396]
4. SDMA via joint beamforming, null pointing, and equalization [397]

## II. MULTIBEAM ARRAYS

The concept of operations of a multibeam antenna is illustrated in Figure 14-2. Conventional sectorized antennas cover the bulk of this notional suburban area that includes an interstate highway system. Each conventional antenna has three 120-degree sectors with frequencies assigned according to the air interface standard's frequency reuse plan (e.g., 1/7 for AMPS, 1/3 for GSM, 1/1 for CDMA, etc.). An area between the highways includes a high-density commercial zone that generates high-intensity traffic.

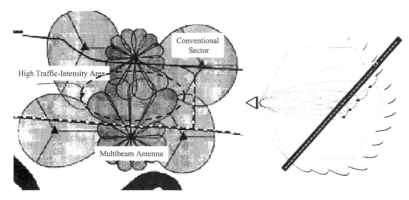

**Figure 14-2** Multibeam array concepts.

The service-provider has only a few alternatives. If the intensity level is several times the design capacity of the conventional sector, additional capacity must be provided. Several additional smaller cells could be provided in the high-intensity area. This requires the acquisition of the sites and establishing connectivity between the new sites and the provider's existing infrastructure. In some areas, the opportunities to establish sites are limited and/or the cost of backhaul from the sites is high. The multibeam antenna alternative creates additional smaller sectors, each of which has a conventional fixed-frequency assignment. The physical layout of the multibeam alternative is as illustrated in Figure 14-2. In the notional highway scenario, the subscriber signal is switched to the beam with the best CIR via high-speed analog or digital beam switching [398]. Such a fixed multibeam antenna may use sector beamforming technology, such as a Butler matrix [399, 400]. Figure 14-3 illustrates the contemporary Butler matrix technology.

In spite of the level of maturity of multibeam array technology, research challenges remain. For example, the complexity of the multibeam array technology is high, keeping costs high. The ADAMO (ADaptive Antennas for MObiles) project, for example, addressed this challenge with a circular array of patch antennas [401] and low-complexity analog processing. The benchmark set for this project is to suffer only small performance degradations compared to (macroscale) digital techniques.

In addition, Thomson-CSF has developed prototype antennas for the evaluation and qualification of the SDMA concept in the field of UMTS radio communications under contract to CNET/France TELECOM [402]. Figure 14-4 shows prototype SDMA hardware. In general, SDMA may employ multibeam arrays, digital beamforming, joint beamforming-equalization, and other smart antenna techniques. As a practical matter, however, the costs of SDMA products must be kept low in order to be affordable to infrastructure operators.

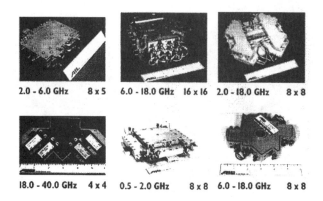

| 2.0 - 6.0 GHz | 8 x 5 | 6.0 - 18.0 GHz | 16 x 16 | 2.0 - 18.0 GHz | 8 x 8 |

| 18.0 - 40.0 GHz | 4 x 4 | 0.5 - 2.0 GHz | 8 x 8 | 6.0 - 18.0 GHz | 8 x 8 |

### Butler Matrix Implementations
Photographs © AIL Corporation, reprinted with permission

**Figure 14-3**   Illustrative multibeam technology.

# SDMA Antennas

Photographs © Thomson CSF, reprinted with permission.

**Figure 14-4**   SDMA antenna prototypes.

## III. ADAPTIVE SPATIAL NULLING

If the multibeam array has a dozen beams, it may not be feasible to assign a complete frequency-reuse plan to each beam. This is because of interference with adjoining sectorized antennas. In such situations, it may be useful to

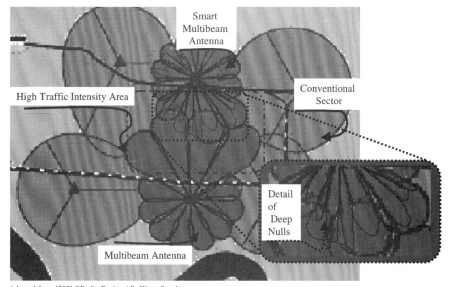

Adapted from [395] ©Radio Design AB, Kista, Sweden.

**Figure 14-5** Smart antennas complement conventional sectors.

cancel interference by creating spatial nulls in the direction of nonsubscriber signal components.

Figure 14-5 illustrates the deployment concept for a smart antenna with spatial nulling. As subscribers that are on the same frequency (cf. cochannel subscribers) move through the high-traffic-intensity area, nulls track their movement and cancel their path components. The architecture of such a spatial nulling subsystem (e.g., [385]) is illustrated in Figure 14-6. This smart antenna replaces a conventional sectorized array, interfacing to the cell site via the existing RF distribution system. The three 3-element sectors of a conventional sectorized base station have been replaced with eight circularly disposed antenna elements. The signal is preamplified and converted to digital form by a bank of eight wideband ADCs. The angle of arrival of all incoming signal components is estimated by a super-resolution DF algorithm [403, 404]. Since the DF algorithm requires a few milliseconds to compute its estimates, the eight raw ADC streams are delayed so that the digital beamformer weights correspond exactly to the received signal. Subscriber channels are then isolated (e.g., using a bank of digital filter ASICs). The measurement of the *supervisory audio tones* (SAT) is one of the AMPS-specific baseband algorithms implemented in a pool of DSPs. The out-of-band SAT generated by the basestation is transponded by the mobile. The basestation can therefore differentiate its subscribers from cochannel interference based on SAT. The cross-correlation process determines the delay-azimuth parameters needed for the final beamforming-equalization stage. The resulting 100 signals from

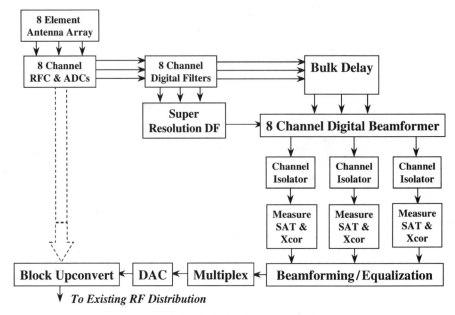

**Figure 14-6**   Spatial nulling architecture.

the base station's subscribers exhibit enhanced CIR. These are digitally multiplexed by adding the signals in a high-dynamic-range numerical process. Finally, they are converted to analog and sent to the base station.

## A. Algorithm Operation

This section illustrates the operation of such spatial-nulling antenna systems. The exposition is similar to that of Kennedy and Sullivan [385]. The spatial distribution of a wavefront arriving at a smart antenna is illustrated in Figure 14-7. The power-delay profile (a) shows the autocorrelation of a single, direct-path wavefront arriving from a single direction (b). Multipath reflections will generally exhibit some time-delay with respect to this principal component. The azimuth display helps visualize the distribution of energy in space.

When multipath components are present, they are delayed with respect to the principal component as illustrated in Figure 14-8a. In addition, the multipath components are not collinear with the direct path and they usually have less signal strength than the direct path as seen in Figure 14-8b.

When interference is present, it is mixed with the multipath components as illustrated in Figure 14-9. In this case, the interference is not on the same azimuth as the direct-path, so it may be suppressed by pointing a null in the appropriate direction.

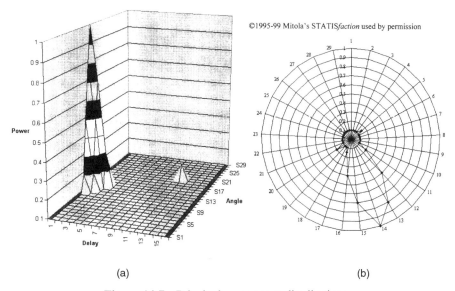

**Figure 14-7**   Principal component distributions.

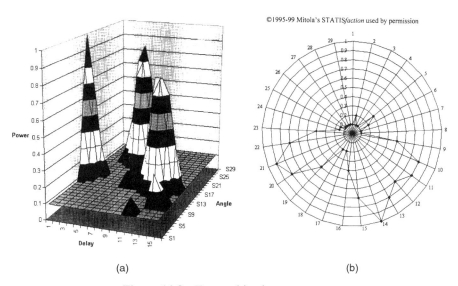

**Figure 14-8**   Two multipath components.

Adjustments to the weights of the beamforming matrix yield the kind of response illustrated in Figure 14-10. Although the depth of the null exceeds 30 dB, a residual remains. The CIR, however, has been improved by 3 to 6 dB.

The simple beamformer does not equalize the received multipath components. Such a process would delay the signal components with respect to each

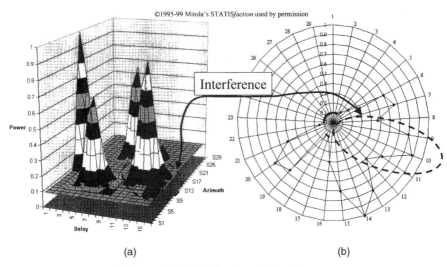

©1995-99 Mitola's STATIS*faction* used by permission

(a)                    (b)

**Figure 14-9**   Multipath and interference.

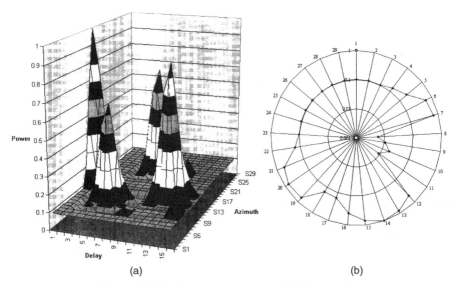

(a)                    (b)

**Figure 14-10**   Illustrative array manifold response.

other so that they may be combined, further enhancing the SNR. The smart antenna described by Kennedy includes baseband equalization in each subscriber channel. This is an example of a spatial beamformer followed by a temporal equalizer in which each stage operates independently. In space-time adaptive processing (STAP) the beamforming and equalization parameters are calculated jointly.

**TABLE 14-1   Beamforming Algorithm Complexity**

| Algorithm | Multiplications | Divisions | Additions |
|-----------|-----------------|-----------|-----------|
| LMS | $2Q + 1$ | 0 | $2Q$ |
| RLS | $2Q^2 + 7Q + 5$ | $Q^2 + 4Q + 3$ | $2Q^2 + 6Q + 4$ |
| FTF | $7Q + 12$ | 4 | $6Q + 3$ |
| LSL | $10Q + 3$ | $6Q + 2$ | $8Q + 2$ |

Adapted from [405] ©IEEE 1999, with permission.

## B.  Beamforming Algorithm Complexity

Cellular systems structure signals such that base stations can differentiate subscribers from cochannel interference. In the case of AMPS, the interference would have a different SAT frequency. In the case of CDMA, the interference has a different placement on the long-code. In the case of GSM, the burst has different header bits. In both of these latter cases, the individual path components could be demodulated in order for the system to differentiate signal from interference. This would be computationally expensive, but might be unavoidable. Researchers have therefore sought less computationally intensive algorithms.

In particular, Razavilar et al. [405] analyzed the computational aspects of beamforming algorithms that use training sequences. *Direct matrix inversion* (DMI) is the simplest method for calculating beamforming weights based on a known training sequence of length $Q$. Its complexity is on the order of $Q^3$, where $Q$ is the length of the training sequence. Adaptive algorithms iterate the weights as the training sequence is received, yielding an estimate at the conclusion of the training sequence. Razavilar characterized the complexity of the following algorithms: *least mean square* (LMS), *recursive least square* (RLS), *fast transversal filter* (FTF), and *least squares lattice* (LSL). Complexity in terms of $Q$ is given in Table 14-1.

## IV.  SPACE-TIME ADAPTIVE PROCESSING

At times, a cochannel interferer will also be collinear with the subscriber and the base station. This situation cannot be corrected spatially: deep nulls cancel both the interference and the desired signal. These two signals are not likely to be mutually coherent in the time domain, however. Joint space-time adaptive processing (STAP) uses this lack of coherence to separate the signals in parameter-space. This allows one to cancel such collinear interference.

A STAP array includes a tapped delay line in each antenna element's processing channel [406, 407], as illustrated in Figure 14-11. The matrix $W_{ij}$ transforms the signal from multiple antenna elements into a space-time-equalized signal. Those weights in part reflect array normalization, weights

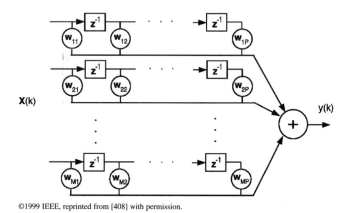

**Figure 14-11**   Conceptual structure of STAP.

that correct differences in the magnitude and phase transfer functions between antenna elements. Such differences arise because the corresponding
analog signal processing paths are not perfectly matched. Those weights
also reflect the placement of spatial nulls. Moreover, they reflect the inversion of matrix equations that compensate for relative time delays (or equivalently for relative phase differences) of the multipath components. STAP
therefore generally requires computationally intensive matrix factorization
[406].

Matrix inversion substantially increases the processing requirements, but
yields improved performance. Consequently, many techniques have been investigated either to reduce the computational burden of optimal STAP algorithms, or to enhance the cancellation capability of simpler algorithms. A
taxonomy of smart antenna techniques is provided in Figure 14-12. In the figure, array algorithms require more than one statistically independent antenna
element. Highlights of the techniques are as follows.

In *sequential interference cancellation* (SIC), the highest-power signal is demodulated to estimate its bitstream. The bitstream is then remodulated and
filtered to form an idealized replica of the analog signal in the channel.
The subtraction of this replica from the composite input stream yields a
residue that includes the remaining users. The cochannel interference from
the strongest interferer has been reduced substantially. This recovery process
continues, recovering multiple users in turn. In CDMA applications, most of
the signals thus recovered are likely to be from viable users, because of soft-
handoff.

*Minimum mean squared error* (MMSE) processes estimate signal parameters using a Gaussian noise error model [407]. *Maximum likelihood* (ML) techniques formulate the likelihood ratio (for a sequence of channel symbols), the
maximum of which determines the signal parameter estimate. One variation
of this is also called *maximum likelihood sequence estimation* (MLSE) [408].

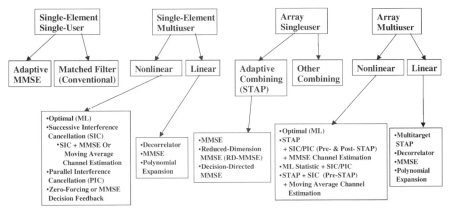

Figure developed for the author by T. Slocumb, P. Zablocky, J. Liberti, D. Kilfoyle, P. Monticiollo, and K. Forsythe.

**Figure 14-12** Scope of smart antenna algorithms.

SIC, ML, and MMSE may be employed on a single channel, on a multichannel array antenna, and in conjunction with other techniques.

The simplest STAP algorithm is the joint equalizer-beamformer. Joint beamforming and equalization maximizes received signal strength by coherently combining the multipath components while placing spatial nulls on sources of interference [409–411]. Advanced STAP techniques include multichannel SIC [412] and other multichannel adaptive techniques [413, 414]. The performance of smart antennas degrades as a function of the structure of the multipath environments [415]. *Parallel interference cancellation* (PIC) techniques structure the computations so that multiple subscriber signals are processed at the same time. The parallelism uses an amplitude estimate for each user as a soft-decision metric. The decision-biases thus introduced can be canceled using *partial interference cancellation* (also PIC) [416]. Parallel algorithms distribute well onto parallel DSP hardware [418].

In general, multichannel SIC and MMSE outperform the other approaches, but they are one to four orders of magnitude more computationally intensive. Additional treatment of the relationship between concrete computational complexity and interference cancellation effectiveness in real environments is needed. The further quantification of resource demands of candidate algorithms is essential to insertion in software-radio architectures.

## V. ARCHITECTURE IMPLICATIONS

The STAP algorithms are the most computationally intensive algorithms investigated to date for canceling cochannel interference. These algorithms require several orders of magnitude more processing capacity than digital beamformers, which require an order of magnitude more processing capacity than single-channel architectures. Thus, digital circuits of envisioned receivers for

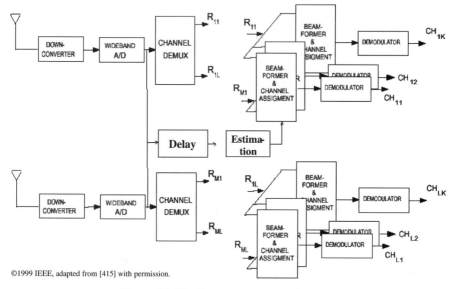

©1999 IEEE, adapted from [415] with permission.

**Figure 14-13**   Smart antenna architecture.

some STAP algorithms may require a minimum of 50 GOPS of processing capacity, with ASIC gate-counts of 100,000 or more [417]. Murotake's analysis determined that 12.3 GFLOPS is required for each channel of a 5 MHz W-CDMA modem [418]. The conclusion was that a 60 GFLOP configuration would support a W-CDMA smart antenna using PIC algorithms. Because of such high computational burdens, both the reduction of computational complexity and the enhancement of processing platforms have received attention. Low-cost DSP architectures are suitable for laboratory investigations of such smart antenna algorithms [419].

## A. Smart Antenna Components

The introduction of smart antennas into SDR implementations is accommodated by software radio architecture. Figure 14-13 illustrates the functional organization of a smart antenna. Delay and estimation processes vary from algorithm to algorithm, so these blocks would be connected into the signal flow paths as a function of algorithm.

The organization of DSP components for smart antennas may be based on the diversity platform developed in prior chapters. The block diagram of the (physical components) reference platform for this class of smart antennas is given in Figure 14-14. Many possible signal flows may be implemented on such a reference platform. In an $N$-element array, the channel isolation filters extract channels for each of $K$ subscribers on each of $N$ elements. These are processed to form beams and to extract first-stage soft-decision

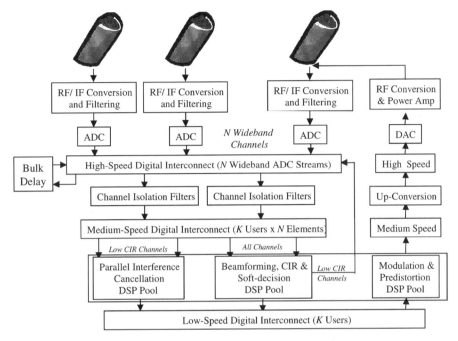

**Figure 14-14**  Smart antenna reference platform.

parameters. The channels with low CIR are thus identified. Their bulk-delayed signals may be isolated for the second and higher stages of interference cancellation, performed in a logically separate segment of the DSP pool. This pool also provides the processors for modulation and predistortion, which can include beamforming on transmission. The node's switching functions may be implemented by addressing on the low-speed bus, sized to interconnect the $K$ users locally and/or to the PSTN via other elements of the bitstream segment.

## B.  Design Rules

*1. Digital Base Station Interface*   The analog RF interface to the cell site is one of the problematic aspects of the introduction of smart antenna technology. The DAC, filtering, block up-conversion, and the RF distribution steps of Figure 14-6 all add noise to the enhanced signals from the smart antenna subsystem, countering some of the gains of the smart antenna. To help overcome this problem, the SDR Forum has defined a digital interface between a smart antenna and the core base station. This interface is defined between the Modem functional block and the INFOSEC functional block of the architecture. More precisely, this Decoded Channel Bits Interface (DCBI) defines a point in an SDR that divides a future base station into a smart antenna subsystem and a core base station. The smart antenna includes the primary RF of the base station, the wideband ADC and DACs, IF processing, smart antenna processing (e.g., beamforming, interference cancellation, and equalization), and demodulation. Any soft-decision decoding, Trellis coding/decoding, etc. required to decode the channel bits is in the domain of the smart antenna. This leaves the majority of the bitstream processing (bit interleaving, FEC, turbocoding, etc.), speech processing (e.g., GSM TRAU processing), and data processing (e.g., billing-related, logging, operations support, maintenance diagnostics, etc.) in the domain of the core base station.

*2. Business-oriented Design Rules*   The identification and adoption of such interface standards promotes open architecture. During the deliberations of the SDR Forum on this interface, base station manufacturers became resistant to the publication of the DCBI standard. This reflects the business reality that the introduction of smart antenna technology could restructure the basestation marketplace. The smart antenna is currently positioned in a conflict between business-oriented design rules. One rule that base station suppliers have to follow is to protect and enhance their business base. A rule that service providers like to follow is to sustain competition. The DCBI sustains competition, but at the possible expense of established base station incumbent suppliers. It seems likely that established suppliers will incorporate smart antenna technology into next-generation base stations, introducing the "smart base station." Such products reduce the need for a digital interface between smart antennas and the core base station.

At some point, these issues will be resolved. Adaptive nulling provides some enhancement to CIR, while STAP techniques improve CIR further. The better interference suppression may be necessary to achieve the information densities planned for 3G. It therefore seems likely that smart CDMA base stations will proliferate with 3G deployments.

## VI. EXERCISES

1. Define smart antenna. Differentiate STAP from other smart-antenna techniques.

2. How can SIC be used in a single-antenna configuration? What additional complexities arise from multiple-antenna applications of SIC?

3. Express the smart antenna reference platform in a table (i.e., do not use a block diagram). Be sure to differentiate classes of DSP pool.

4. How can smart antennas enhance quality, quantity, or timeliness of communications?

5. Can smart base stations introduce asymmetries into the link to the mobile subscriber? What are the alternatives for closing the link in spite of the asymmetries? What are the advantages and disadvantages of each alternative?

6. What are the alternatives to smart antennas for enhancing CIR?

7. Describe a smart antenna architecture that overcomes the business issues of technology insertion.

8. Apply a smart antenna to the Disaster Relief case study. In which scenarios is the added cost worth it?

# 15 Applications

This chapter develops illustrative applications, including the design of a contemporary SDR infrastructure product, the disaster-relief system.

## I. THE DESIGN PROCESS

The implementation of SDR applications can be structured into an SDR design process. This process begins with the definition of a *concept of operations* (CONOPS), in which functions of the product are identified. The next stage, system definition, includes rapid prototyping and benchmarking. The third stage, system development, includes the implementation of hardware-software components. Acquisition and integration of COTS components and/or system-on-a-chip IP characterizes this stage. The expense of coding and documenting software for reuse also may be borne in this stage. The final stage, system deployment, includes platform upgrades and software downloads, with multiple incremental enhancements.

The CONOPS provides the foundation for the development of use-cases of object-oriented design with UML. The top-level design constraints must be expressed as an initial set of design rules. The design rules include the degree of openness of the architecture. If the design has an open architecture that supports industry standards, then there may be third-party suppliers of hardware and/or software for the product. If the design is proprietary, the product should be unique, because it will not have the value-added features of a robust third-party supplier program.

The functions then must be allocated to hardware and software components that can be procured or developed within a market-driven timetable, and within the design rules. The node design process is illustrated in Figure 15-1. Physical design addresses the choice of components from the antenna to the user interface. These components may be hardware intensive in one implementation (e.g., in a handset) and software-intensive in another (e.g., in a base station). The CONOPS establishes a list of RF bands and modes that the product has to support, both initially and over its life cycle. The state of RF technology determines how many parallel antenna-RF-IF-conversion chains of hardware have to be included in order to support these bands. The maximum number of *simultaneous* subscribers in each RF band and GoS define the number of traffic channels supported per band. The number of channels plus the

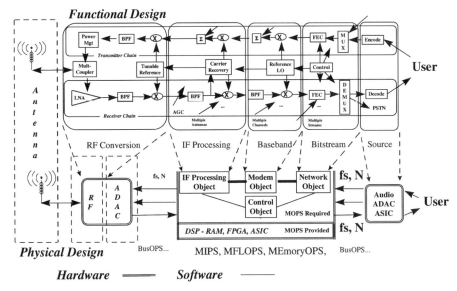

**Figure 15-1**  From functional design to node design.

allocated bandwidth and other parameters of the air interface define the minimum bandwidth of the RF or IF ADC. Over time, ADC technology continues
to advance, so one ADC may cover multiple RF bands. If the design decisions include the use of wideband ADCs (e.g., with hundreds of MHz or
GHz sampling rates), then digital interconnect of the ADC data streams to the
processing channels becomes a high-visibility design issue.

Air interfaces and services define the software that has to be supplied. Once
the software components have been identified, the digital processing architecture of ASICs, FPGAs, DSP chips, and/or general-purpose processors may
be defined. In addition, the designer must balance the need to satisfy computational demands of the software against the competitive pressures of cost-
effective design. Computational demands argue for larger DSP pools while
competitive pressures tend to argue for smaller ones.

The design example illustrates the development of a CONOPS. It then
describes the methods of selecting hardware and software components, and
sizing them for a well-engineered SDR design. Subsequent analysis highlights
those features of the design necessary for it to be part of a robust software-
radio architecture.

## II. THE DISASTER-RELIEF SYSTEM DESIGN

Consider the disaster-relief case study introduced previously. This section elab-
orates that case study into an SDR development project. The first step in the

**Figure 15-2**  Concept of operations.

applications development process is to establish a vision or top-level concept that motivates the creation of SDR applications. The vision may be a high-level statement of a challenge (e.g., "Put a man on the moon by the end of this decade"). Or it may be a statement of an abstract goal ("We want to own the night"). The vision for disaster relief might imply notions like "connecting diverse relief organizations," "reconstituting communications," "assisting the stricken," etc. To provide the foundation for a technical approach, one should support the ideas behind the vision with a CONOPS as illustrated in Figure 15-2. The CONOPS should identify:

· The customer for the product, system, or service
· The scenarios in which the system will be employed
· The benefits of the system in those scenarios
· The people who will benefit

The following is an illustrative CONOPS.

### A.  FEMA Concept of Operations (CONOPS)[26]

In addition to the material provided previously a CONOPS could include the material in Exhibit 15-1.

---

[26] Any relationship between this concept and any actual project is purely coincidental.

The U.S. Federal Emergency Management Agency (FEMA) provides a national-level pool of resources that come to bear on major disasters like the destruction of much of Holmstead, Florida by Hurricane Andrew a few years ago. FEMA would like to acquire a mobile system that is capable of reconstituting local cell phone service while enhancing communications among emergency relief personnel. The assumption is that the wireless service has been wiped out in the disaster area.[27]

In a typical scenario, disaster relief comes from 50 or more teams drawn from dozens of federal, state, and local organizations, including police, fire, and rescue. These groups have a diverse set of communications capabilities. The Federal Bureau of Investigation (FBI) has speech-privacy radio technology as do a few of the largest municipalities in the United States. Special operations groups such as the Florida Department of Law Enforcement (FDLE) also employ special systems which, of course, tend to be incompatible with almost everyone else. If the U.S. National Guard or reserves are called upon, the Army uses the SINGCARS system discussed earlier. The Air Force components are typically equipped with Have Quick (I and/or II) in addition to simultaneous VHF and UHF transmission for air traffic coordination. In addition, airlift of equipment and supplies to the disaster area requires the use of aeronautical mobile bands such as the 100 MHz air traffic control band, and the 225–400 MHz military band.

Figure 15-2 illustrates the organizations that may participate in the disaster-relief operation. Several scenarios are contemplated as follows. The system must flexibly support all the scenarios.

In the Hurricane/Tornado scenario, a large area has been ravaged by a category 4/5 hurricane, or a series of simultaneous tornadoes. There is one large damage area in which a population center (e.g., a small city) is located, and up to three additional smaller areas (e.g., towns or hamlets). Large fractions of the population are victims. The terrain is flat to hilly, presenting few serious natural obstructions to radio propagation.

In the Mud-Slide/Avalanche scenario, the disaster occurs in very hilly or mountainous terrain, breaking the disaster area up into a dozen or more isolated valleys in which relatively small numbers of victims and bystanders are located. The terrain provides natural impediments to radio propagation.

Populations range from a few thousand distributed in a rural setting to 50,000 or more in densely a populated area. Assume that 10% of the population are victims and that 50 organizations send relief workers.

**Exhibit 15-1** National FEMA CONOPS.

---

[27]The author apologizes to any mobile phone companies offended by this notion. To set the record straight, wireless is generally very reliable, even in natural disasters. But in order to make this design study interesting, and relevant to both military and commercial markets, we need a motivation for designing mobile infrastructure. This happened in the earthquake in Turkey because of the lost power grid.

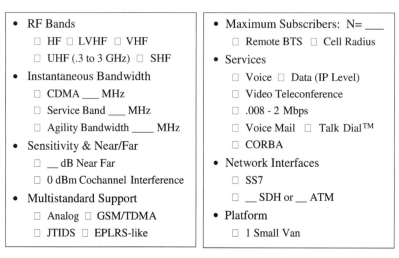

**Figure 15-3**   Illustrative project requirements.


## B. Requirements Analysis

A notional list of requirements for such a system is provided in Figure 15-3. A service provider such as FEMA must decide on the RF bands and modes, maximum number of subscribers, and services to be provided.

In addition, the mobile radio equipment must fit in a mobile vehicular platform. For the sake of this example, assume that each radio node is to be configured in a commercial four-wheeled sport-utility vehicle (SUV). This SUV may be equipped with a kerosene-powered electric generator and an electric or hydraulic mast with a limited maximum height (e.g., 10 meters). The number of SUVs should be decided analytically based on GoS, Erlangs of traffic offered per subscriber, and spatial area covered by the mast-based antenna. On the other hand, one may focus on the reconstitution of cellular service and derive a number top-down through similarities and differences. Typical cell sites support 100 simultaneous subscribers. If an arbitrary mix of VHF/UHF, HF, and cellular subscribers is envisioned, then there might be 100 to 200 potential subscribers in each of the two major bands for a total of 200 to 400 users. The peak capacity of each van could then be set at some number between 100 and 400 parallel voice or data channels. The lower the number, the lower the cost of the system. Consider each of the additional requirements in turn.

RF bands are selected based primarily on the requirement for "seamless interoperability" of the emergency teams. LVHF is required for SINGCARS, while VHF and UHF are needed for federal, state, and local law enforcement push-to-talk radios and for Have Quick. HF was not explicitly called out. However, if the emergency occurs in a mountainous region, one of the most effective ways of connecting teams operating in adjacent mountain valleys is

HF *near vertically incident* (NVI) skywave. HF will therefore be included. In addition, the restoration of the cellular telephone service in the United States requires UHF modes in the 850–900 MHz band, as well as PCS modes from 1700 to 2500 MHz.

An additional driver for RF coverage is the need to link vans to each other efficiently. Suppose two vans are operating 10 miles apart, each supporting 100 local users. Some fraction of these users will need to communicate with users supported by the other van. The choices include HF, fiber, VHF/UHF, satellite communications, and SHF point-to-point radio relay. HF may provide the connectivity for a small number of channels. But if on the average there will be 20 to 30 calls between the two vans, T-1 (24 channel) or E-1 (30 channel) service is warranted. This level of cross-connect capacity is at the limits of viable HF communications. Suppose we have 10 vans, raising the cross-connect traffic to 200 to 300 channels? HF clearly is not viable in this case. Although it is possible to lay field fiber, this mode is subject to breakage, especially in emergencies. The dynamics of a forest fire, for example, render fiber impracticable. If the system simply allocates VHF/UHF channels for cross-connect (which is possible), the number of subscribers supported in these RF bands at each node decreases proportionally, This may not be a problem for a small number of users, say, 100 per node. Satellite communications historically is expensive, costing from $1 to $3 per minute. At a rate of $1,000 per minute for 300 satellite channels, even the smallest disaster could accrue a large satcom phone bill. Terrestrial microwave, on the other hand, is essentially free (especially to the government who owns the allocated spectrum). It also easily provides T-1 to T-3 levels of service with relatively modest bandwidths and subsystem complexity. For the sake of this example, assume that SHF in the 4, 6, or 11 GHz microwave bands is the high-capacity cross-connect mode. Satcom, on the other hand, might be best for a widely distributed disaster where SHF LOS connectivity cannot be maintained.

The subscribers must also be connected to the PSTN. In some areas, the PSTN may employ SHF microwave to protect primary fiber infrastructure. But most service providers in the United States now protect (back up) fiber with other fiber paths. So the vans should have a fiber interconnect port compatible with the SDH and SS7 for interoperability with the PSTN. Since the design of such interconnect is not central to software radios, the sequel will reflect the assumption that the physical interconnect and the necessary driver software are available as commercial off the shelf (COTS) products. The SDR node will have to deliver isochronous streams to the interface and route streams from this interface to radio users. But the design of the interface itself is not central to the SDR.

Instantaneous bandwidth, sensitivity, and dynamic range (near–far ratio) are driven by the commercial standards. Most state and local police, fire, and rescue units employ push-to-talk VHF/UHF AM/FM radios. The instantaneous bandwidths range over the set {4, 8 1/3, 12.5, 25, 50, 100} kHz. Commercial cellular standards, on the other hand, now include the IS-95 CDMA system with its 1.2288 Mchip/second spreading rate with 1.25 MHz bandwidth. GSM only requires 200 kHz of instantaneous bandwidth per burst, but the FH modes can hop over 25 MHz. A software radio implementation of the FH mode requires 25 MHz bandwidth on transmit and on receive. This drives the DAC and ADC requirements. In addition, the dynamic range is set by the 90 dB near–far ratio.

To complete the design, each of the areas listed in Figure 15-3 must be analyzed in detail. Chapters 6–15 refer to this example to motivate the discussion.

## C. System Description

The system description expresses design decisions. An exemplar for the disaster-relief system is as follows.

*1. Communications Services* The UMC-2000 mobile infrastructure system will integrate the communications capabilities of diverse police, fire, and rescue organizations. In addition, it will bridge communications of national and international relief agencies with disparate communications equipment into the local disaster-recovery operations. Finally, it will integrate military and national guard communications.

Services consist of voice, data, and video-telemedicine. Voice services include voice mail with Enhanced TalkDial$^{TM}$ [443] (ETD) capabilities. Rescue personnel therefore need know only the name of the person or the general category of function in order to get the right person on the line. The UMC-2000 system manager assigns a virtual telephone number to each participant, and then tracks that participant's location and communications mode for seamless connectivity. Data services include wireless e-mail. In addition, UMC-2000 establishes a gateway to the PSTN via microwave LOS T- or E-carrier SDH interfaces. It also can link to a local office via fiber.

The most important contribution of UMC-2000 is the reduction of confusion. Since each UMC-2000 van is equipped with a 30-foot mast, it establishes a cell within which commercial cellular handsets (e.g., of the victims) can operate even when local cellular service has been interrupted by the disaster conditions. Organic radios of disaster support teams talk to the local UMC-2000 van for bridging, and they obtain frequency assignments for local communications from the spectrum management authority using UMC's SmartSpectrum spectrum management tool suite.

*2. RF Bands and Modes* UMC-2000 has capabilities in HF, LVHF, VHF, UHF, and SHF. HF AM and ALE provide voice and data circuits using NVI modes in mountainous regions. LVHF coverage integrates contributions of military and national guard units. VHF coverage of the 100 MHz air traffic control band permits coordination with aircraft and the reconstitution of communications at an airfield. VHF/UHF push-to-talk AM, FM, and TETRA dig-

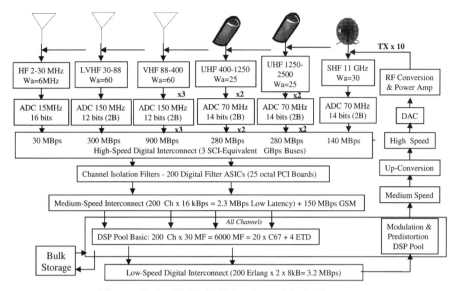

**Figure 15-4** UMC-2000 hardware block diagram.

ital radio modes are supported in all bands, subject to frequency coordination with the UMC-2000 spectrum managers. Additionally UHF cellular coverage includes 1G, 2G, and 3G air interface modes. The RF LANs operate on the 2.5 MHz ISM band so that wireless laptops can be used in the vicinity of the UMC-2000 vans for status monitoring and coordination displays. Telemetry modes permit UMC-2000 to uplink patient status data via wireless and PSTN links to remote medical personnel. Streaming video supports telemedicine.

Switching of voice channels is accomplished in software under the control of ETD. The interface to the PSTN employs SS-7 and SDH Levels 1, 2, or 3 trunking through microwave or fiber optic media.

***3. Capacities*** UMC-2000 supports 2000 emergency personnel per node, with up to five vans. The internal capacity of each van is 200 Erlangs of traffic. Band coverage consists of ten subbands from six antenna channels.

## D. Illustrative Design

The design of UMC 2000 includes hardware and software components.

***1. Hardware Components*** An illustrative hardware design is provided in Figure 15-4. HF supports a 6 MHz subset of the HF band, with that 6 MHz tunable between the LUF and MUF. LVHF is accessed in parallel using a 150 MHz ADC. This limits near–far, but this can be operationally controlled by placement of the 5 vans within the disaster area. There are three tunable subbands in the VHF range from 88 to 400 MHz. The low subband would access commercial broadcast and air traffic. The medium and high subbands are placed for maximum support of emergency personnel, given the capabilities of their equipment. Operationally, to reduce cochannel interference, emergency units are assigned separate uplink and downlink bands to the UMC-2000 nodes, but can communicate among each other using conventional push-to-talk TDD. The two lower UHF subbands are tunable to 1G and 2G allocations; 3G bandwidths of 20 MHz are supported, but only for one CDMA overlay, traded off against the 2G capacity. The upper UHF band supports one PCS band and one RF LAN band. The 11 GHz SHF band was chosen because it minimizes antenna size on the mast for van–van trunking at E1 or E2 rates.

The high-speed digital interconnect, in a current implementation, would require three separate SCI-equivalent gigabyte per second buses to interconnect ADC streams to the 200 channel-isolation filters. These are organized into three shelves with the wideband ADCs. Medium-speed Raceway-class interconnect switches these signals to appropriate C67 DSPs. GSM-class voice and data traffic requires 30 MFLOPS (MF) per Erlang or 6000 MFLOPS. Assuming 60% efficiency of deliverable MFLOPS, 10 C67 DSPs could provide this capacity. The operating point of 50% capacity used then requires 20 C67 chips. Since there are ten bulk streams, two chips (one dual C67 board with local and global memory) are nominally associated with each bulk stream, for 10 Erlangs of traffic per chip. These 10 boards are organized into one DSP pool shelf. For simplicity, the bus hosts are not shown. The DSP pool requires one shelf, and the bulk storage, LANs, hosts, etc. require an additional shelf.

In addition, the transmission facilities (DSP pool, up-conversion, etc.) are sized as requiring 25% of the capacity of the receivers, or 4 C67 chips. The digital up-conversion could be based on Intersil/Harris HSPs or Graychip GC4114 quad digital up-converter chips. A shelf of 8 octal boards provides 64 transmission channels switched to 10 RF amplifier boards.

The system is configured into the van with a control rack (CTL) in the front, the receiving rack (RX) on the left side, and the transmission rack (TX) on the right side of the van. One operator position is provided on the CTL rack for local technical control and mission planning. Five additional wireless laptops are packed for use near the van via RF LAN. Shelf allocations are as follows.

| Shelf | Number | Rack |
|---|---|---|
| Rx ADC, filter | 3 | RX |
| DSP pool | 1 | RX |
| Bulk storage | 1 | CTL |
| Tx DAC, filter | 1 | TX |
| Tx amplifiers | 2 | TX |

The back of the van has a swingaway auxiliary power unit kerosene-powered UPS to supply the substantial power requirements of the DSPs and

RF transmission system. Antenna-mast design minimizes EMI/RFI with physical separation, insulators, etc. This is one of the highest-risk areas of the system.

*2. Software Components* Software consists of AM, FM, and vocoder algorithms for voice, AMPS for 1G cellular, Digital AMPS and IS-136 licenses, and a GSM suite for the cellular bands. TETRA, DECT, and PHS/PDC modules have been identified that are compatible with the DSP platforms, and are licensed as needed. HF ALE, GPRS, and V.xx modem software provide data connectivity. MS Office with Access and Outlook provides word processing, database, and e-mail. Netscape and Internet Explorer are included for Internet services. WAP and the default SDR services recommended in the layered virtual machine architecture are provided as COTS packages as well. RF CAD [444] is used for propagation prediction to site the vans and to assist in managing spectrum allocations.

## III. ARCHITECTURE IMPLICATIONS

The above reference design is just one of a family of designs ranging from much more conservative to very aggressive. It is representative of the level of technology available in the 1999–2000 time frame. Given the extensive tradeoffs associated with each aspect, this brief treatment cannot do justice to the design of such a system. It is provided as an integrated example of one point in the large, complex evolutionary path of software radio technology.

In order to support an enterprise architecture, the hardware components of Figure 15-4 need to be identified with a migration plan. The 200 discrete digital isolation filters, first of all, could be replaced with multi-channel ASICs within 2–3 years. The digital interconnect and ADCs/DACs could be upgraded in 3–5 years.

To support industry-standard open architecture, the software components could be based on CORBA, following the SDR Forum. Since the C6 does not yet support CORBA, one might work with a third-party software supplier and/or TI toward this goal.

## IV. EXERCISES

1. Work through a use-case scenario with the design of Figure 15-4. What questions arise that would revise the design substantially?

2. Address question 1 using a UML tool such as Rational Rose. What were the benefits of UML? Of the tool?

3. Develop the rack elevations for the design of Figure 15-4. What engineering issues arise? What bus did you choose (VME, PCI, other)?

**4.** Represent the waveforms of II.D.2 as objects. Can the performance of II.C be achieved? If not, how can this be fixed?

**5.** Re-do the ADC tradeoffs for 90 dB near-far in all bands without loss of RF coverage. How many parallel channels are needed? What is the impact on rack elevations? On software?

# 16 Reference Architecture

This chapter provides a consolidated view of the software-radio architecture models developed throughout the text for convenient reference. This consists of a radio platform view (Figure 16-1) and a software components view (Figure 16-2).

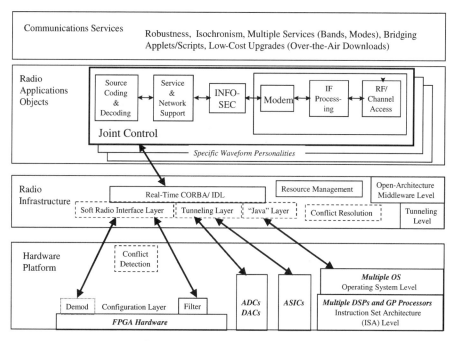

**Figure 16-1**    Radio platform view.

The radio platform view shows how radio infrastructure can be based on CORBA/IDL. Using wrappers for FPGA personalities and ASICs, it can also deliver cutting-edge performance efficiently.

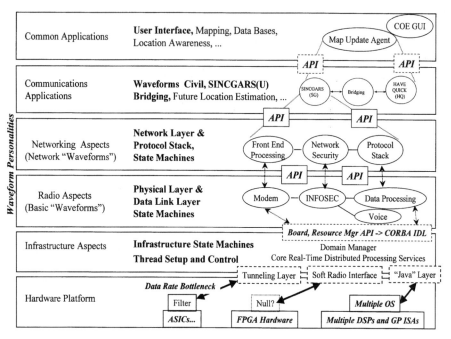

**Figure 16-2**   Software components view.


The software components view shows how objects may be associated with hardware and protocol stack layers. This facilitates the insertion of hardware and/or software components that enhance radio functionality without causing ripple-effects into other objects in the system.

# REFERENCES

1. McGarth, et al., RFIC technology for wireless consumer products—Trends in GaAs, M/A-COM LOUD & Clear (Lowell, MA: M/A-COM, 1995).

2. Kennedy and Sullivan, Direction finding and smart antennas using software radio architectures, *IEEE Communications Magazine* (New York: IEEE Press, May 1995).

3. Upmal and Lackey, SPEAKeasy, the military software radio, *IEEE Communications Magazine* (New York: IEEE Press, 1995).

4. Nicholson, D., *Spread Spectrum Signal Design LPE and AJ Systems* (Rockville, MD: Computer Science Press, 1988).

5. *Reference Data for Engineers* (Carmel, IN: Sams Publishing, 1993).

6. Stallings, *Handbook of Computer-Communications Standards, Volume 1: The Open Systems Interconnection (OSI) Model and OSI-Related Standards* (New York: Macmillan, 1987).

7. Pickholtz and Hill, *Adaptive Beamforming for Interference Reduction*, GW University PW3312A, Dec. 31, 1990.

8. Zoltowski, et al., Blind 2-D rake receivers based on space-time adaptive MVDR processing for IS-95 CDMA system, *MILCOM 96* (New York: IEEE, Oct. 1996).

9. Belzer et al., *Joint Source Channel Coding of Images with Trellis Coded Quantization and Convolutional Codes* (Los Angeles: UCLA, 1998).

10. Ferguson and Huston, *Quality of Service* (New York: Wiley, 1998).

11. Paradells, et al., DECT multibearer channels, Proc. 44th IEEE Vehicular Technology Conference (New York: IEEE Press, 1994).

12. Strom and Shaula, Optimistic recovery in distributed systems, *ACM Transactions on Computer Systems* (New York: Association for Computing Machinery, 1985).

13. Pesonen, Object-based design of embedded software using real-time operating systems, IEEE 1068-3070/94 (New York: IEEE Press, 1994).

14. Cummings, M., and Heath, S., Mode switching and software download for software defined radio: The SDR forum approach, *IEEE Communications Magazine* (New York: IEEE Press, Aug. 1999).

15. Recommendation H.320, Narrow-band visual telephone systems and terminal equipment (www.itu.int/publications/itu-t/ituth13.htm: International Telecommunications Union, 1998).

16. Coding of analogue signals by pulse code modulation (G.711–G.712) and by methods other than PCM (G.720–G.729) (Geneva, Switzerland: International Telecommunications Union, 1998).

17. ietf references to internetworking.

18. Mouly and Pautet, Evolution of the GSM system, *IEEE PCS Magazine* (New York: IEEE, Oct. 1995).

19. Storer, J., *Data Compression* (Rockville, MD: Computer Science Press, 1988).

20. Ziemer and Peterson, *Digital Communications and Spread Spectrum Systems* (New York: Macmillan, 1985).

21. Peterson, W., and Weldon, E., *Error-Correcting Codes* (Cambridge, MA: MIT Press, 1972).

22. Edited by Simmons, G., *Contemporary Cryptography* (New York: IEEE Press, 1992).

23. Razavilar, J., et al., Software radio architecture with smart antennas: A tutorial on algorithms and complexity, *IEEE JSAC* (New York: IEEE Press, April 1999).

24. Recommendation H.320 (Geneva: ITU-T, 1998).

25. *Random House Unabridged Webster's Dictionary* (New York: Random House, 1999).

26. DII Strategic Enterprise Architecture (Washington, DC: DISA, 1994).

27. Technical Architecture for Information Management (TAFIM) (Washington, DC: US DoD, 1996).

28. Gamma, E., et al., *Design Patterns: Elements of Reusable Object-Oriented Software* (Reading, MA: Adisson-Wesley, 1994).

29. Gardner, K., et al., *Cognitive Patterns: Problem Solving Frameworks for Object Technology* (Cambridge: Cambridge University Press, 1998).

30. Cook, P., An architectural overview of the speakeasy system, *IEEE JSAC* (New York: IEEE Press, April 1999).

31. Mowbray, T., and Zahavi, R., *The Essential CORBA* (New York: Wiley, 1995).

32. Bensley, et al., Introduction to parallel supercomputing, (Bedford, MA: MITRE Corporation, 1988).

33. Mouly and Pautet, *The GSM System for Mobile Communications* (Plaiseau, France: Published by the authors, 1992).

34. Pentek Corporation Product Technical Description (Laurel, MD: SPSI, 1994).

35. The Common Object Resource Broker: Architecture and specification, OMG 91.12.1 (Maynard, MA: Digital Equipment Corp. et al., 1992).

36. Kondo, Matsuo, and Suzuki, A software-defined architecture concept for telecommunication information systems, ICC 94 (New York: IEEE Press, 1994).

37. Voruganti, A global network management framework for the 90's, ICC 94 (New York: IEEE Press, 1994), p. 1094.

38. Walden, R., Analog to digital converter survey and analysis, *IEEE JSAC* (New York: IEEE Press, April 1999).

39. *The Software Defined Radio*, BellSouth, Dec. 1995.

40. *Harris Digital Channelizer Application Note: Channelized Receiver* (Melbourne, FL: Harris Corporation, 1990).

41. Zangi, K., and Koilpillai, R., Software radio issues in cellular base stations, *IEEE JSAC* (New York: IEEE Press, April 1999).

42. Dick, C., Configurable logic for digital communications: Some signal processing perspectives, *IEEE Communications Magazine* (New York: IEEE Press, August 1999).

43. www.motorola.com.

44. Bose, V., et al., Virtual radios, *IEEE JSAC* (New York: IEEE Press, April 1999).

45. Withington, P., Impulse radio overview (www.timedomain.com: Time Domain, 1999).

46. Turletti, T., Towards the software realization of a GSM base station, *IEEE JSAC* (New York: IEEE Press, April 1999).

47. Government Open System Interconnect Protocol (GOSIP) (Bedford, MA: MITRE Corporation, 1985).

48. Mitola, J., *Cognitive Radio: Model Based Competence for Software Radios*, Licentiate thesis (Stockholm, Sweden: KTH (The Royal Institute of Technology), August 1999).

49. Van Tuyl, et al., FCC transmitter certification requirements: Issues related to software defined radio, *Proc. SDR Forum* (Rome, NY: SDR Forum, June 1999).

50. Mitola, J., Software radio architecture: A mathematical perspective, *IEEE Journal on Selected Areas in Communications* (New York: IEEE Press, April 1999).

51. Tebbs and Garfield, *Real Time Systems* (Berkshire, UK: McGraw-Hill, 1977).

52. Ellison, K., Developing real-time embedded software in a market-driven company (New York: Wiley, 1994).

53. Mitola, J., The software radio architecture, *IEEE Communications Magazine* (New York: IEEE, May 1995).

54. Mitola, J., Cognitive radio for mobile multimedia communications, *Proceedings of the IEEE Mobile Multimedia Communications (MoMuC) Workshop* (San Diego, CA: IEEE Press, Nov. 1999).

55. *Random House Unabridged Webster's Dictionary* (New York: Random House, 1999).

56. TEAL WING Report (Arlington, VA: DARPA, 1970).

57. European Council Decision 94/572/EC of 27 July 1994, ACTS 1997 CD-ROM (Brussels: EC, 1997).

58. Qualcomm, The technical case for convergence of third generation wireless systems based on CDMA, (www.qualcomm.com, March 1999).

59. Nicholson, D., *Spread Spectrum Signal Design: LPE and AJ Systems* (Rockville, MD: Computer Science Press, 1988).

60. Reference Data for Radio Engineers (New York: Sams & Co., 1993).

61. Wireless broadband communication systems, Feature Topic, *IEEE Communications Magazine* (New York: IEEE Press, Jan. 1997).

62. Petersen and Zeimer, op cit., Appendix on JTIDS.

63. Lee, William C. Y., *Mobile Communications Design Fundamentals* (Indianapolis, IN: HW Sams, 1986).

64. Mouly and Pautet, Evolution of the GSM System, *IEEE PCS Magazine* (New York: IEEE, Oct. 1995).

65. Evaluation report for the cdma2000 RTT by U.S. TG8/1 Ad Hoc Evaluation Group, Annex 3 (New York: TIA, 1999).

66. UMTS reference *IEEE Communications Magazine* citation.

67. Mitola, J., Cognitive radio for flexible mobile multimedia communications, *Proceedings of the Mobile Multimedia Communications Conference (MoMuC)* (San Diego, CA: IEEE Press, Nov. 1999).

68. Astago, J., and Russel, T., *CDPD* (New York: MCGraw-Hill, 1997).

69. Rappaport, T., *Wireless Communications* (Upper Saddle River, NJ: Prentice Hall PTR, 1996).

70. *The Software Defined Radio Request for Information* (Atlanta, GA: BellSouth, Dec. 1995).

71. Colombo, G., *Proceedings of the International Conference on Universal Communications* (Italy: CSELT, 1998).

72. *1st International Workshop on Software Radio, Proceedings of the ACTS Mobile Summit* (Rhodes, Greece: European Commission, June 1998).

73. RSS-130 Annex 1, CT2Plus Class 2: Specification for the Canadian Common Air Interface for Digital Cordless Telephony, Including Public Access Services, *Spectrum Management Radio Standards Specification* (Ottawa: Industry Canada, January 23, 1993).

74. MacNamee, G., et al., *Personal Wireless Communication With Dect and Pwt* (www.artech-house.com: Artech House, 1999).

75. Padgett, et al., Overview of wireless personal communications, *IEEE Communications Magazine* (New York: IEEE, Jan. 1995).

76. Nowicki, D., Fulfilling the promise of WLL with PHS, ArrayComm, Inc. (www.watmag.com: *WAT* Magazine On Line, July 1998).

77. DaSilva, J., Remarks before the First European Workshop on the Software Radio (Brussels, European Community, April 1997).

78. *ACTS Mobile Communications Summit '98* (Rhodes Greece: European Commission, June 1998).

79. *4th ACTS Mobile Communications Summit '99 CD-ROM* (Sorrento, Italy: EC, June 1999).

80. Kohno, Software radio and software antenna: Spatial and temporal communication theory using software antenna (Yokohama, Japan: Yokohama National University, 1998).

81. Shinagawa, Y., Software radio technologies—From the viewpoint of terminal manufacturer, *Proceedings of the 1st International Workshop on Software Radio*, (Rhodes, Greece: European Commission, June 1998).

82. Kinoshita, K., and Nakagawa, M., Japanese cellular standard, in *Mobile Communications Handbook*, J. Gibson, Editor (Salem, MA: The CRC Press, 1996).

83. Akaike, M., and Muraguchi, M., C6: Software radio for future communications—Including developments in high frequency integrated-circuit technologies for portable/personal communications, *Program of the URSI* (Toronto: URSI, 1999).

84. Muraguchi, M., RF IC technologies for wearable terminals with software radios, *Program of the URSI* (Toronto: URSI, 1999).

85. Araki, K., Research activities of software defined radio in Japan, *Program of the URSI* (Toronto: URSI, 1999).

86. Mitola, J., Research foundations for the software radio I: Global perspective on software challenges, *Program of the URSI* (Toronto: URSI, 1999).

87. Nickols, P., *C4I Integrated Idea Team In Progress Review* (Ft. Monmouth, NJ: U.S. Army CECOM, Feb. 1999).

88. As illustrated in the topics of papers in the *Proceedings of MILCOM 97–99* and of *ICC 97–99*.

89. BATTLEFIELD AUTOMATION performance uncertainties are likely when army fields its first digitized division, Report to the Chairman, Subcommittee on Defense, Committee on Appropriations House of Representatives (Washington, DC: United States General Accounting Office, July 1999).

90. MMITS forum technical report (www.mmitsforum.com), 1997.

91. Airnet presentation, minutes of the MMITS Forum, June 1996.

92. Eriksson and Penker, *UML Toolkit* (New York: Wiley, 1998).

93. *Reference Data for Engineers*, E. Jordan, Editor (Indianapolis, IN: Howard W. Sams, 1986).

94. *Recommendations of the S and T Series, Volume VII—Facsicle VII.2* (Geneva: CCITT, 1988).

95. Asatani, K., et al., Standardization for GII and multimedia communications, *IEEE Communications Magazine* (New York: IEEE Press, September 1998).

96. Sasaki, A., and Yabusaki, M., The current situation of IMT-2000 standardization activities in Japan, *IEEE Communications Magazine* (New York: IEEE Press, September 1998).

97. Tsui, J., *Digital Techniques for Wideband Receivers* (Boston: Artech House, 1995).

98. Tuttlebee, W., Software radio: Developments in Europe, *Proceedings of the First International Workshop on Software Radio, Rhodes, Greece* (Athens: Trochos Technical Editions LTD, 1998).

99. Mitola, J., Software radios: Survey, critical evaluation and future directions, *Proc. National Telesystems Conference* (New York: IEEE Press, May 1992).

100. Bose, et al., Virtual radios, *IEEE JSAC* (New York: IEEE Press, April 1999).

101. Srikanteswara, S., et al., Soft radio architecture for reconfigurable platforms, *IEEE Communications Magazine* (New York: IEEE Press, February 2000).

102. Weissler, A., and Jondral, F., Software radio structure for second generation mobile communications systems, *IEEE Transactions on Vehicular Technology* (New York: IEEE Press, 1998).

103. Kohno, Software radio and software antenna: Spatial and temporal communication theory using software antenna (Yokohama, Japan: Yokohama National University, 1998).

104. Walden, R., Analog-to-digital converter survey and analysis, *IEEE JSAC* (New York: IEEE Press, April 1999).

105. Zangi, Software radio issues in cellular base stations, *IEEE JSAC* (New York: IEEE Press, April 1999).

106. Oh, et al., On the use of interpolated second order polynomials for efficient filter design in programmable downconversion, *JSAC* (New York: IEEE, April 1999).

107. Gunn, A low power DSP-core based software radio architecture, *IEEE JSAC* (New York: IEEE Press, April 1999).

108. Turletti, et al., op. cit. [46].

109. Correal, N., et al., A DSP-based multiuser receiver employing partial parallel interference cancellation, *IEEE JSAC* (New York: IEEE Press, April 1999).

110. Patti, Husnay, and Pintar, A smart software radio: Concept development and demonstration testbed, *IEEE JSAC* (New York: IEEE Press, April 1999).

111. Cook, P., An architectural overview of the SPEAKeasy system, *IEEE JSAC* (New York: IEEE Press, April 1999).

112. Razavilar, J., et al., Software radio architecture with smart antennas: A tutorial on algorithms and complexity, *IEEE JSAC* (New York: IEEE Press, April 1999).

113. Lee, J-Y., and Samueli, H., Adaptive antenna arrays and equalization techniques for high bit-rate QAM receivers, *IEEE JSAC* (New York: IEEE Press, April 1999).

114. Evans, J., et al., The rapidly deployable radio network, *IEEE JSAC* (New York: IEEE Press, April 1999).

115. Lach, S., et al., Broadband interference excision for software-radio spread spectrum communications using time-frequency distribution synthesis, *IEEE JSAC* (New York: IEEE Press, April 1999).

116. Li, X., and Ritchey, J., Trellis-coded modulation with bit interleaving and iterative decoding, *IEEE JSAC* (New York: IEEE Press, April 1999).

117. Smith, J. R., et al., Code-division multiplexing of a sensor channel: A software implementation, *IEEE JSAC* (New York: IEEE Press, April 1999).

118. Cummings, J., Software radios for airborne applications, *IEEE JSAC* (New York: IEEE Press, April 1999).

119. Wagner, L., Goldstein, J., Meyers, W., and Bello, P., The HF skywave channel: Measured scattering functions for midlatitude and auroral channels and estimates for short-term wideband HF rake modem performance, *MILCOM 89* (New York: IEEE Press, Oct. 15–18, 1989).

120. Feature topic on military communications, *IEEE Communications Magazine* (New York: IEEE Press, Oct. 1995).

121. *Jane's Military Communications 1992–93* (Surrey, UK: Jane's Information Group, 1992).

122. MIL-STD-188 C-220 (Washington, DC: U.S. Department of Defense, 1988).

123. Digital cellular technologies (as of April 1, 1999) (http://home.intekom.com, Sept. 1999).

124. Crochier, *Bell Systems Technical Journal* (Whippany, NJ: AT&T, 1982).

125. Kaiser, *A Friendly Guide to Wavelets* (Berlin, Germany: Birkhauser, 1994).

126. Tanaka, et al., Urban multipath propagation delay characteristics in mobile communications, *Trans IEICE (B-II)*, Nov. 1990, pp. 772–778.

127. Affes and Mermelstein, Spatio-temporal array-receiver for multipath tracking in cellular CDMA, *ICC 97* (New York: IEEE Press, 1997).

128. AN/GRC 103 (Toronto: Marconi Canada, 1995).

129. FHM9104 (Paris: SAT Communications, 1996).

130. TFH950S Product Description (Paris: Alcatel, 1997).

131. GRC-461 (Tel Aviv: Tadiran, 1996).

132. Comparetto, G., Satellite communications—Current features and future trends, *WESCON 96*, p. 233.

133. Leopold, R., Low-earth orbit global cellular communications network, *ICC 91* (New York: IEEE Press, 1991).

134. Dong-Hee Lee, et al., A network architecture for the integration of IRIDIUM and CDMA Systems, *ICC 97*.

135. Matricciani, Transformation of rain attenuation statistics from fixed to mobile satellite communication systems, *IEEE Transactions on Vehicular Technology* (New York: IEEE, Aug. 1995).

136. Wexler, R., Tactical satellite communications (TACSATCOM) terminal single versus multicarrier operation investigation, *MILCOM 96* (New York: IEEE, Oct. 1996).

137. Hayward, W., Switching networks and traffic concepts, *Reference Data for Engineers*, E. Jordan, Editor (Indianaplois, IN: Howard W. Sams, 1986).

138. Fractal dimensions of LAN traffic, *IEEE Network* (New York: IEEE Press, May 1996).

139. Crombie, D., Electromagnetic-wave propagation, *Reference Data for Engineers*, E. Jordan, Editor (Indianaplois, IN: Howard W. Sams, 1986).

140. Meche, P., *UWC-136 Self Evaluation* (Irving, TX: Nokia Mobile Phones, Aug. 1998).

141. Dennett, S., *The cdma2000 Candidate Submission (0.18)* (Washington, DC: TIA, July 1998).

142. Ramasastry, J., et al., *IMT-2000 CDMA Chip Rate Selection* (Washington, DC: Quacomm, Aug. 1998).

143. Inoue, et al., Performance analysis of microcellular mobile communications systems, IEEE 0-7803-1927-3 (New York: IEEE Press, 1994).

144. Haas, S., et al., *Edge Node to Remote Node Topology Optimization in the Rapidly Deployable Radio Network (RDRN)* (Lawrence, KS: University of Kansas, May 1997).

145. Haas, S., *A Consistent Labeling Algorithm for the Frequency-Code Assignments in a Rapidly Deployable Radio Network (RDRN)* (Lawrence, KS: University of Kansas, Jan. 1994).

146. Bertonie, et al., UHF propagation prediction for wireless personal communications, *Proceedings of the IEEE* (New York: IEEE Press, Sept. 1994).

147. Erceg, V., An empirically based path loss model for wireless channels in suburban environments, *IEEE JSAC* (New York: IEEE Press, July 1999).

148. Erceg, V., Comparisons of a computer-based propagation prediction tool with experimental data collected in urban microcellular environments, *IEEE JSAC* (New York: IEEE Press, May 1997).

149. Pagani, E., and Rossi, G., Reliable broadcast in mobile multihop packet networks, *Proc. Mobicom 97* (Budapest, Hungary: ACM, 1997).

150. *Reference Data for Engineers* (Carmel, IN: Sams Publishing, 1993).

151. Lin, P., et al., Improving GSM call completion by call reestablishment, *IEEE JSAC* (New York: IEEE Press, July 1999).

152. Akyildiz, I., et al., Mobility management in next-generation wireless systems, *Proceedings of the IEEE* (New York: IEEE Press, August 1999).

153. Kessler, G., *ISDN*, 2nd ed. (New York: McGraw-Hill, 1993).

154. Atatani, et al., *Introduction to ATM Networks and B-ISDN* (New York: Wiley, 1997).

155. Chelian, M., The role of ATM in multimedia computation, *WESCON 96*, p. 566.

156. Mikkonen, J., Quality of service in radio access networks (Tampere, Finland: Tampere University of Technology, May 1999).

157. TR45.3/97.12.15.08, UWCC input for the next revision of 136, *Overview of the IS-136 Air-Interface* (New York: TIA, 1998).

158. TR45.3/98.04.06.07R4 (TR45/98.03.03.19R6) UWC-136 RTT Update (New

York: TIA, Feb. 1999).

159. Gilhousen, et al., On the capacity of cellular CDMA system, *IEEE Transactions on Vehicular Technology* (New York: IEEE Press, May 1991).

160. Interim Standard-95, Mobile station-base station compatibility standard for dual-mode wideband spread spectrum cellular system (New York: TIA, July 1993).

161. *Ericsson in Wideband CDMA*, EN LZT 123 4023 (Stockholm: Ericsson Radio Systems AB, 1997).

162. Lyberopoulos, G., et al., On the dimensioning and efficiency of the UTRA air-interface modes, *Proceedings of the 4th ACTS Mobile Communications Summit '99* (Sorrento, Italy: EC, 1999) (W-CDMA and TD-CDMA).

163. GSM GSM 01.02 Technical Specification (Sophia Antipolis Cedex, France: ETSI, March 1996).

164. Mouly and Pautet, Evolution of the GSM System, *IEEE PCS Magazine* (New York: IEEE, Oct. 1995).

165. Carney, R., AirNet architecture, remarks before the *2nd Meeting of the MMITS Forum* (http://troi.web.rl.af.mil/mmits/) June 2, 1996.

166. Barsoum, Y., *Tactical Internet Course* (Eatontown, NJ: The MITRE Corporation, 1999).

167. O'Keefe, S., MPLS: High octane IP (www.hottech.com, 1998).

168. Voice, data combine over phone lines, *PC Week*, June 12, 1995.

169. Bezar, D., *LAN Times Guide to Telephony* (Berkley: Osborne McGraw-Hill, 1995).

170. *Enabling the Wireless Internet* (Redwood City, CA: Unwired Planet, Feb. 1999).

171. Wireless Applications Protocol (www.wapforum.org), 1999.

172. COMNET III (www.caciasl.com), 1999.

173. *Signal Processing WorkSystem* (Foster City, CA: Alta Group of Cadence Design Systems, 1994).

174. Perkins, C. E., Mobile-IP, ad-hoc networking, and nomadicity, 0730-3157/96 (New York: IEEE Press, 1996).

175. Ogasawara, G., et al., Experiments with tactical network simulation, routing and management, *MILCOM 96* (New York: IEEE, Oct. 96).

176. Sanchez, R., et al., Networking on the battlefield: Challenges in highly dynamic multi-hop wireless networks, *MILCOM 99* (New York: IEEE Press, Nov. 1999).

177. IETF MANET Working Group Charter (www.ietf.org/html.charters/manet-charter.html, 1999).

178. Racherla, G., MAdSim—A simulation environment for ad-hoc mobile networks, *ICPWC97* (New York: IEEE Press, 1997).

179. UCLA Parallel Comp. Lab. and Wireless Adaptive Mobility Lab., Glo-MoSim: A scalable simulation environment for wireless and wired network systems (pcl.cs.uclo.edu/projects/domains/glomosim.html), 1999.

180. Lee, S., et al., A simulation study of table-driven and on-demand routing protocols for mobile ad hoc networks, *IEEE Network* (New York: IEEE Press, July–August 1999).

181. Omar, H., Support for fault tolerance in local registration mobile-IP systems, *MILCOM 99* (New York: IEEE Press, Nov. 1999).

182. Li, Adopting ATM in low speed environments, *MILCOM 1996*.

183. Yourdon, E., *Structured Design* (Englewood Cliffs, NJ: Yourdon Press, 1978).

184. Mills, H. D., Stepwise refinement and verification in box-structured systems, *IEEE Computer Magazine* (New York: IEEE Press, June 1988).

185. Selic, et al., *Real-Time Object-Oriented Modeling* (New York: Wiley, 1994).

186. Lemay, L., and Cadenhead, R., *Sam's Teach Yourself Java 2* (Indianapolis: Sams Publishing, 1999).

187. Ellis, J., *Objectifying Real Time Systems* (New York: SIGS Books, 1994).

188. Recommendation H.320, Narrow-band visual telephone systems and terminal equipment (www.itu.int/publications/itu-t/ituth13.htm: International Telecommunications Union), 1998.

189. Telephone Applications Programmers Interface (TAPI); (http://www.microsoft.com/ntserver/communications/tapi_wp.htm), 1998.

190. Douglass, B., *Real-Time UML* (Reading, MA: Addison-Wesley, 1998).

191. Jacobson, et al., *Object-Oriented Software Engineering: A Use Case Driven Approach* (Workingham, England: Adisson-Wesley, 1992).

192. Booch, G., *Object-Oriented Design with Applications* (Menlo Park, CA: Benjamin/Cummings, 1991).

193. Coad, P., and Yourdon, E., *Object-Oriented Design* (Englewood Cliffs, NJ: Prentice-Hall/Yourdon Press, 1991).

194. Ellis, J., *Objectifying Real-time Systems* (New York: SIGS Books, 1994).

195. Kaufman, S., *The Emergent Properties of Information Overload* (Santa Fe, NM: Bios Group LP, 1997).

196. Mitola, J., Software radio architecture: A mathematical perspective, *IEEE Journal on Selected Areas in Communications*, April 1999.

197. Hoest and Shavit, Towards a topological characterization of asynchronous complexity, *PODC 97* (Santa Barbara, CA: ACM, 1997).

198. Milewski, *The Topology Problem Solver* (Piscataway, NJ: Research and Education Asociation, 1994).

199. Ono, *Introduction to Point Set Topology* (Baltimore, MD: Johns Hopkins University, 1974).

200. Mitola, J., Software radios: Technology and prognosis, *Proc. National Telesystems Conference* (New York: IEEE, May 1992).

201. Mitola, J., Software radio architecture: A mathematical perspective, *IEEE JSAC*, April 1999.

202. Adams, M., *SDR Mobile Working Group Framework Subgroup Baseline SDR Forum Submission* (Rome, NY: SDR Forum, December 1999).

203. Cummings, M., and Heath, S., Mode switching and software download for software defined radio—The SDR Forum approach, *IEEE Communications Magazine* (New York: IEEE Press, Aug. 1999).

204. Chairperson, Working Group 5, Radio Communication Study Groups: Preliminary Draft New Recommendation ITU-R [IMT.RKEY]: Key Characteristics for the IMT-2000 Radio Interfaces (Version 1), Document 8-1/230-E (Geneva: ITU-R Feb. 12, 1999).

205. Prill, R., and Antonesco, M., Programmable channelized digital radio, *IEEE National Telesystems Conference* (New York: IEEE Press, May 1992).

206. Rinard, W., and Vujcic, D., Implementation of a programmable digital receiver multi-chip module, *IEEE National Telesystems Conference* (New York: IEEE Press, May 1992).

207. Solowey, G., *Programmable Digital Radio* (New Jersey: GEC Marconi, Inc., 1994).

208. ITT Corporation, *Digital Radio Architecture* (Rome, NY: MMITS Forum, 1996).

209. Carney, R., *The AirNet Architecture* (Rome, NY: SDR Forum, 1996).

210. Girard, L., et al., Presentation of the measurement and demonstration campaigns with the FRAMES TDD TD-CDMA demonstrator, *Proceedings of the 4th ACTS Mobile Communications Summit* (Sorrento, Italy: European Commission DG XIII-F, June 1999).

211. Speakeasy II Home Page (http://www.rl.af.mil/Technology/Demos/SPEAKEASY), 1997.

212. Fredrikson, F., et al., A software testbed for performance evaluation of adaptive antennas in FH-GSM and Wideband-CDMA, *Proceedings of the ACTS Mobile Communications Summit 98* (Rhodes, Greece: European Commission, June 1998).

213. Taylor, C., *A DSP System for a Software Radio Testbed* (Rhodes, Greece: European Commission, June 1998).

214. Margulies, A., and Mitola, J., III, Software defined radios as a migration strategy, *ISSTA 98* (New York: IEEE Press, 1998).

215. Joint Combat Information Terminal (ttdl.nrl.navy.mil/products/jcit), Dec. 1999.

216. Mowbray, T., and Malveau, R., *CORBA Design Patterns* (New York: Wiley, 1997).

217. Siwiak, *Radiowave Propagation and Antennas for Personal Communications* (Norwood, MA: Artech House, 1995).

218. Time difference of arrival technology for locating narrowband cellular signals (www.trueposition.com: Pittsburgh, PA: True Position, Inc., 1999).

219. Fattouche, M., and Zaghloul, H., Estimation of phase differential of signals transmitted over fading channels (www.cell-loc.com: Cell-loc, Inc., 1997).

220. Zaghloul, H., Morrison, G., and Fattouche, M., Frequency response and path loss measurements of the indoor channel, *Electronics Letters*, 27 (12), 1991, 1021–1022.

221. Fishler, E., and Bobrovsky, B., Anti multipath cellular radio location for DS/CDMA systems using a novel EKF subchip RAKE tracking loop, *MILCOM 99* (New York: IEEE Press, Nov. 1999).

222. Pozar, D., Microstrip antennas, *Proceedings of the IEEE* (New York: IEEE Press, Jan. 1992).

223. Siwiak, *Radiowave Propagation and Antennas for Personal Communications* (Norwood, MA: Artech House, 1995).

224. Burkhardt, M., and Kuster, N., Review of exposure assessment for handheld mobile communications devices and antenna studies for optimized performance, *Radio Science 1996–1999* (Oxford: Oxford University Press, 1999).

225. Chuang and Chen, Computer simulation of the human-body effects on a circular-loop-wire antenna for radio-pager communications at 152, 280 and 400 MHz, *IEEE Trans Vehicular Tech* (New York: IEEE Press, Aug. 1997).

226. Commission K: Electromagnetics in biology and medicine, edited by S. Ueno, *Radio Science 1996–1999* (Oxford: Oxford University Press, 1999).

227. Adey, W., Cell and molecular biology associated with radiation fields of mobile telephones, *Radio Science 1996–1999* (Oxford: Oxford University Press, 1999).

228. Smith, J., et al., Code-division multiplexing of a sensor channel: A software implementation, *IEEE JSAC* (New York: IEEE Press, April 1999).

229. Slocumb, T., Novel antennas kickoff meeting (Washington, DC: DARPA, Jan. 11, 1999).

230. Loundu, et al., Estimating the capacity of a frequency-selective fading mobile radio channel with antenna diversity, 0-7803-1927-3 (New York: IEEE Press, 1994).

231. deSeze, Influence of slow frequency hopping and antenna diversity techniques in European GSM system, Proc. ICC97 (New York: IEEE Press, 1997).

232. Morgensen and Petersen, Practical considerations of using antenna diversity in DECT, 0-7803-1927-3/94 (New York: IEEE Press, 1994).

233. Effect of system imperfections on BER performance of a CDMA receivrer with multipath diversity combining, *IEEE Trans on Vehicular Technology* (New York: IEEE Press, Nov. 1996).

234. *Model 204 IF Diversity Combiner* (Sunnyvale, CA: Applied Signal Technology, 1990).

235. Schell and Shrimpton, Super-exponentially convergent blind fractionally spaced equalization and cochannel interference rejection, *MILCOM 96* (New York: IEEE, Oct. 1996).

236. Drago, D., Optically reconfigurable antennas, *SPIE Advanced Sensor Technology Workshop* (Fairfax, VA: SPIE, Dec. 1995; no proceedings).

237. Brown, E., RF-MEMS switches for reconfigurable integrated circuits, 0018-9480/ 98 (New York: IEEE Press, 1998).

238. Feher, K., *Wireless Digital Communications* (Upper Saddle River, NJ: Prentice Hall, 1995).

239. Digital VXI VHF/UHF Receiver WJ-8629 (Gaithersburg, MD: Watkins-Johnson Company, 1994).

240. Orsak, Optimum receivers, in *The Mobile Communications Handbook* (Boca Raton, FL: CRC Press, 1996).

241. Abidi, Low power radio frequency IC's for portable communications, *Proceedings of the IEEE* (New York: IEEE Press, April 1995).

242. Tsui, J., *Digital Techniques for Wideband Receivers* (Boston: Artech House, 1995).

243. Rolfes, M., Field trial results of superconducting products in wireless communicaions, *WESCON 96*, p. 109.

244. Tsurumi, H., and Suzuki, Y., Broadband RF stage architecture for software defined radio for handheld terminal application, *IEEE Communications Magazine* (New York: IEEE Press, February 1999).

245. Rhode, U., et al., *Communications Receivers* (New York: McGraw-Hill, 1997).

246. *RF IC Design for Wireless Communication Systems* (Corvallis, OR: Mead Microelecronics, 1996).

247. Clarke, Siemens boosts RF integration: Mobile-phone transceiver, *EE Times* (Buffalo, NY: CMP Publications, Aug. 29, 1994).

248. Robinson, Process builds silicon RF chips, *EE Times* (Buffalo NY: CMP Publications, Nov. 25, 1996).

249. Brown and Sward, Digital downconversion test results with a broadband L-Band GPS receiver, 0-7803-2425-0/94 IEEE (New York: IEEE Press, 1994).

250. Lammers, TI tunes into communications at 0.18 micron, *EE Times* (Buffalo NY: CMP, June 16, 1997).

251. Abidi, A., CMOS wireless transceivers: The new wave, *IEEE Communications Magazine* (New York: IEEE Press, Aug. 1999).

252. www.phillipsvision.com.

253. Piezo resonators broaden RF design palette, *EE Times*, Jan. 27, 1997.

254. MEMS process bets on nickel, *EE Times* (Buffalo, NY: CMP Publications, June 16, 1997).

255. MCM technology puts passives in wireless apps, *EE Times* (Buffalo, NY: CMP Publications, Dec. 29, 1997).

256. Brown, Novel substrates scale RF integration, *EE Times* (Buffalo, NY: CMP Publications, Oct. 27, 1997).

257. Ohr, DARPA sows the seeds of a telcom revolution, *EE Times* (Buffalo NY: CMP Publications, Aug. 4, 1997).

258. Ribas, R., et al., Micromachined planar spiral inductor in standard GaAs HEMT MMIC technology, *IEEE Electron Device Letters* (New York: IEEE Press, August 1998).

259. Fan, L., Universal MEMS platforms for passive RF components: Suspended inductors and variable capacitors, Proc. of the 11th Annual International Workshop on Micro-Electro-Mechanical Systems (New York: IEEE Press, 1998).

260. Harsh, K., Flip-chip assembly for Si-based RF MEMS, Provided through US Air Force Contract F33615-98-C-5429, 1998.

261. Duewer, B., et al., MEMS-based capacitor arrays for programmable interconnect and RF applications, IEEE Proc. 20th Anniversary Conference on Advanced Research in VLSI (New York: IEEE Press, 1999).

262. Pisano, A., *MEMS Principal Investigator's Meeting* (Washington, DC: System Planning Corporation, Jan. 1999).

263. Grant, P., et al., Switches for microwave routing: A comparison of MEMS switches with PIN, FET, and optical switches, *Proceedings of the Canadian Workshop on MEMS* (Ottawa: National Research Council of Canada, August 13, 1999).

264. Yao, J. J., et al., RF MEMS switch, *Transducers 95—Eurosensors IX* (New York: IEEE Press, 1995).

265. Franzon, P., Low-power, high-performance MEMS-based switch fabric (www.ncsu.edu: North Carolina State University, 1999).

266. Pottenger, et al., MEMS: The maturing of a new technology, *Solid State Technology*, Sept. 1997.

267. EDA Software—Suite supports signals in the electrical, mechanical and thermal domains, *Electronic Engineering Times*, May 25, 1998, p. 50.

268. Hogan, Superconductor products poised for market, *Industrial Physicist* (Woodbury NY: C. Harris, March 1997).

269. Pengelly, R., Low cost GaAs MMIC chip set for dual mode AMPS/CDMA phones, *WESCON 96*, p. 96.

270. Wirbel, Anadigics develops multiple 1.9 GHz PCS devices, *EE Times* (Buffalo NY: CMP, July 14, 1997).

271. Wilson, Philips respins RF transistor, *EE Times* (Buffalo NY: CMP, Feb. 10, 1997).

272. IMP50E10 EPAC (San Jose, CA: IMP, Inc; 1995).

273. Programmable IC delivers analog designs with digital ease, *Electronic Products* (Buffalo, NY: CMP Publications; http://electronicproducts.com, July 1997).

274. Oppenheim and Schaffer, *Digital Signal Processing* (Cambridge, MA: MIT Press, 1972).

275. Ziemer and Peterson, *Digital Communications and Spread Spectrum Systems* (New York: Macmillan, 1985).

276. Peterson and Weldon, *Error Correcting Codes* (Cambridge, MA: MIT Press, 1972).

277. Blahut, R., *Theory and Practice of Error Control Codes* (Reading, MA: Addison-Wesley, 1983).

278. Steer, Antialiasing filters reduce errors in A/D converters, *Electronics Design News Design Feature* (New York: Cahners Publishing, March 30, 1989).

279. Wepman, J., Analog to digital converters and their applications in radio receivers, *IEEE Communications Magazine* (New York: IEEE Press, May 1995).

280. Walden, ADC integrated circuits, *Proceedings of the IEEE GaAs IC Symposium* (New York: IEEE Press, Dec. 1995).

281. Walden, R., Analog-to-digital converter survey and analysis, *JSAC* (New York: IEEE Press, Feb. 1999).

282. Ridgley, R., *Digital Receiver Revolution* (Arlington, VA: DARPA, August 26, 1999).

283. Zakhor and Oppenheim, Reconstruction of two-dimensional signals from level crossings, *Proceedings of the IEEE* (New York: IEEE Press, Jan. 1990).

284. Thao and Vetterli, Optimal MSE signal recovery in oversampled A/D conversion using convexity, *Proceedings of ICASSP-92* (New York: IEEE Press, 1992).

285. Candy and Temes, *Oversampling Delta-Sigma Data Converters* (New York: IEEE Press, 1992).

286. Rolfes, M., Field trial results of superconducting products in wireless communicaions, *WESCON* (New York: IEEE Press, 1996).

287. Chen and Francis, Design of multirate filter banks by H (sub-infinity) optimization, *IEEE Transactions on Signal Processing* (New York: IEEE Press, Dec. 1995).

288. Jayaraman, A., et al., Bandpass delta-sigma modulator with 800 MHz center frequency, *IEEE GaAs IC Symposium Technical Digest*, vol. 19, pp. 95–98 (New York: IEEE Press, Oct. 1997).

289. D/A converter fit for cell base stations, *EE Times PRODUCT File* (Buffalo NY: CMP, Feb. 1997).

290. Miller, M., An analysis of the dynamic range of HF radio receivers, *MILCOM 96* (New York: IEEE, Oct. 1996), p. 679.

291. Mitola, J., op. cit. [99].

292. Mitola, The software defined radio, *WESCON 96* (New York: IEEE Press, Oct. 1996).

293. Turletti, et al., op. cit. [43].

294. Chassaing, *Digital Signal Processing with C and the TMS320C30* (New York: Wiley, 1992).

295. www.motorola.com.

296. www.lucent.com/micro.

297. www.analogdevices.com.

298. Texas Instruments home page (www.ti.com).

299. TMS320C54x family description on Internet site http://www.ti.com.

300. New wave of DSP silicon targets cellular, PCS, *Electronic Enginering Times*, October 2, 1995.

301. pASIC 2 FPGA family, White Paper (Santa Clara, CA: QuickLogic Corporation, 1995).

302. V.34 modem built with integrated DSP, *Electronic Product Design*, Nov. 1994).

303. Motorola readies major wireless-IC thrust, *Electronic Engineering Times*, September 25, 1995.

304. Mercury achieves record performance with Analog Devices' SHARC chip, (Boston, MA: Mercury Computer, Oct. 1996).

305. Mercury Computer Systems Inc. home page (www.mc.com).

306. Lieberman, Multiprocessor systems drive Mercury's raceway, *EE Times* (Buffalo, NY: CMP, Oct. 13, 1997).

307. Analog Devices' home page ([297]), June 2000.

308. Sky Computer home page (www.sky.com), 1999.

309. CSPI home page (www.cspi.com), 1999.

310. Baugh, e-mail subject: Massively parallel LINPACK on the Intel Touchstone Delta machine (www.cs.cmu.edu/ ∼ scandal/info/intel.linpack).

311. Pentek home page (www.pentek.com), 1999.

312. Clark, R., SCI interconnect chipset and adapter: Building large scale enterprise servers with Pentium Pro SHV nodes, *WESCON 96*, p. 389.

313. SkyChannel, ANSI/VITA 10-1995 (New York: American National Standards Institute, 1995).

314. Brown, DARPA looks to optical interconnects, *EE Times* (Buffalo, NY: CMP, May 26, 1997).

315. Brown, High speed optical interconnects arrive, *EE Times* (Buffalo, NY: CMP, June 1997).

316. Robinson, Filter design enables single-chip transceiver, *EE Times* (Buffalo NY: CMP, July 14, 1997).

317. Oh, et al., On the use of interpolated second order polynomials for efficient filter design in programmable downconversion, *JSAC* (New York: IEEE, Feb. 1999).

318. Zangi, Software radio issues in cellular base stations, *IEEE JSAC* (IEEE Press) 1998 (to be published).

319. van Wyk, D. J., and Linde, L. P., Fading correlation and its effect on the capacity of space-time turbo coded DS/CDMA systems, *MILCOM 99* (New York: IEEE Press, Nov. 1999).

320. Miranda, H., et al., BITSPREADER-2000: A software configurable spread spectrum transceiver, *Proceedings of the Fifth Baiona Workshop on Emerging Technologies in Telecommunications* (Baiona, Spain: Universidade de Vigo, Sept. 1999).

321. ITU Recommendation Z.100 (SDL) (Geneva: ITU, 1990.)

322. Fawcet, B., Designing high-performance applications using LogiCore modules, *WESCON 96*, p. 917.

323. pASIC 2 FPGA Family, White Paper (Santa Clara, CA: QuickLogic Corporation, 1995).

324. Xilinx home page (www.xilinx.com), 1999.

325. Oelsner, Implementing a data-convolution algorithm in a QuickLogic FPGA using VHDL, *App Review* (Redmond, WA: Synario, Nov. 1996).

326. Iwami and Tanaka, VLSI systolic array for SRIF digital signal processing algorithm, *IEICE Trans. Fundamentals*, Vol. E77-A, No. 9, Sept. 1994.

327. Fawcett, B., Advancements in SRAM-based FPGA technology, *WESCON 96*, p. 994.

328. Altera readies 2M gate CPLD, *EE Times* (Buffalo, NY: CMP, Oct. 27, 1997).

329. Hickling, R., et al., Meeting the challenge of future high speed systems with multiple technology ASICs, *WESCON 96*, p. 987.

330. Dick, C., Configurable logic for digital communications, *IEEE Communications Magazine* (New York: IEEE Press, Aug. 1999).

331. Srikanteswara, S., et al., A soft radio architecture for reconfigurable platforms, *IEEE Communications Magazine* (New York: IEEE Press, Feb. 2000).

332. Bittner, R., *Wormhole Runtime Reconfiguration: Conceptualization and VLSI Design of a High Performance Computing System* (Blacksburg, VA: Virginia Tech, 1997).

333. Reeves, et al., Reconfigurable hardware accelerator for embedded DSP, *Proceedings of Photonics East '96* (SPIE, citation).

334. Mitola, J., op. cit. [50].

335. Feher, K., op. cit. [238].

336. deGaris, H., CAM-brain: Growing an artificial brain with a million neural net modules inside a trillion cell cellular automata machine, *Journal of the Society of Instrument and Control Engineers* (SICE), Vol. 33, No. 2 (Japan: Society of Instrument and Control Engineers of Japan, 1994).

337. deGaris, H., Review of proceedings of the First NASA/DoD Workshop on Evolvable Hardware, *IEEE Transactions on Evolutionary Computation* (New York: IEEE Press, Nov. 1999).

338. Fogel, D., *Evolutionary Computation: Toward a New Philosophy in Machine Intelligence* (New York: IEEE Press, 1995).

339. Davis, L., *Handbook of Genetic Algorithms* (New York: Van Nostrand Reinhold, 1991).

340. Genetics evolves gigascale systems, *EE Times*, Jan. 27, 1997.

341. Oh, S., DSP technology for a hands-free car cellular phone, *WESCON 96*, p. 821.

342. Brooks, T., Planning for PCS: The TI strategy for upbanded GSM, *WESCON 96*, p. 1042.

343. Motorola and Lucent unveil first offering from Star*Core joint DSP design team (www.starcore-dsp.com), Apr. 1999.

344. Clarke, Virtual components a real issue at Euro forum, *EE Times* (Buffalo, NY: CMP, Nov. 3, 1997).

345. Comm port in, FFT, Comm port out, *SPSinfo* (Laurel, MD: SPSI, Winter 1995).

346. Threads of RISC architecture follow three paths, *EE Times* (Buffalo, NY: CMP, Mar. 17, 1997).

347. RISC twists, IA-64 parallelism vie at $\mu$P Forum, *EE Times* (Buffalo, NY: CMP, Oct. 20, 1997).

348. Clark, STM, Hitachi to build 64-bit SH processors, *EE Times* (Buffalo, NY: CMP, Dec. 15, 1997).

349. Pentium® III Processor for the SC242 at 450 MHz to 800 MHz Datasheet (www.intel.com: Intel Corp, 1999).

350. Gunn, J., op. cit. [107].

351. Ojanpera, T., and Prasad, R. (Eds.), *Wideband CDMA for Third Generation Mobile Communications* (Artech House Universal Personal Communications Series, 1999).

352. Miller, L., and Lee, J. S., *CDMA Systems Engineering Handbook* (Artech House, 1998).

353. Mouly and Pautet, *The GSM System for Mobile Communications* (Plaiseau, France: Authors, 1992).

354. Lee, William C. Y., *Mobile Communications Engineering: Theory and Applications* (New York: McGraw-Hill Series on Telecommunications, 1997).

355. TAO home page (www.cs.wustl.edu/ ~ schmidt/ACE_wrappers/TAO) (St. Louis: Washington University, 1999).

356. VisiBroker home page (www.visigenic.com/prod/orbpd.html), 1999.

357. Visigenic home page (www.iona.com/), 1999.

358. Simpson, The duel over object models, *Client/Server Today*, Aug. 1994.

359. Message Passing Interface (MPI) (Internet: www.mcs.anl.gov/mpi/mpi-report-1.1/mpi-report.html), Feb. 1998.

360. Sunaga and Yamada, NOSES distributed communication switching software structure based on CORBA, *ICC 97* (New York: IEEE Press, 1997).

361. Mitola, J., III, *Cognitive Radio*, Licentiate thesis (Stockholm, Sweden: KTH, Royal Institute of Technology, 1999).

362. *Object GEODE* (Paris, FR: Verilog, 1998).

363. *Rational Rose* (Santa Clara, CA: Rational Software Corp., 1999).

364. Chassaing, *Digital Signal Processing with C and the TMS320C30* (New York: Wiley, 1992).

365. Lapsley, et al., *DSP Processor Fundamentals* (New York: IEEE Press, 1997).

366. Karabinis, P., et al., Dual-mode cellular/satellite hand-held phone technology, *WESCON 96*, p. 208.

367. Miller, M., An analysis of the dynamic range of HF radio receivers, *MILCOM 96* (New York: IEEE, Oct. 1996), p. 679.

368. Sampath, B., Li, Y., and Liu, K. J. R., A subspace based blind identification and equalization algorithm, *Proc. 1996 IEEE International Conference on Communications* (New York: IEEE Press, 1996).

369. Fossorier and Lin, Soft-decision decoding of linear block codes based on ordered statistics, *IEEE Transactions on Information Theory* (New York: IEEE, Sept. 1995).

370. Fossorier and Lin, Soft-decision decoding of linear block codes based on ordered statistics, *IEEE Transactions on Information Theory* (New York: IEEE, Sept. 1995).

371. Emmerson, B., and Gretham, D., GSM's extraordinary growth (www.byte.com), Aug. 25, 1997.

372. Ericsson world first with live demo of GPRS services, *Ethos* (www.tagish.co.uk: Tagish, LTD), Mar. 12, 1998.

373. Rysavy, P., General packet radio service (GPRS), *PCS Data Today* (www.pcsdata.com), 1998.

374. Bultler, M., Providakes, G., and Blythe, G., The layered radio, *MILCOM 98* (New York: IEEE Press, 1998).

375. Inoue, et al., Performance analysis of microcellular mobile communications systems, IEEE 0-7803-1927-3 (New York: IEEE Press, 1994).

376. See [3] for a-law and mu-law companding.

377. Correal, N., et al., A DSP-based multiuser receiver employing partial parallel interference cancellation, *IEEE JSAC*, Feb. 1999.

378. Gunn, J., op. cit. [107].

379. Mitola, J., III, home page (www.ourworld.compuserve.com/homepages/jmitola), 1999.

380. Tebbs and Garfield, *Real Time Systems* (Berkshire, UK: McGraw-Hill, 1977).

381. Liberti, J. C., Jr., and Rappaport, T. S., *Smart Antennas for Wireless Communications: IS-95 & Third Generation CDMA Applications* (Prentice Hall, 1999).

382. Kohno, Software radio and software antenna: Spatial and temporal communication theory using software antenna, (Yokohama, Japan: Yokohama National University, 1998).

383. Mitola, J., et al., Special issue on software radios, guest editorial, *IEEE JSAC* (New York: IEEE Press, April 1999).

384. Biedka, T., et al., Implementation of a prototype smart antenna for low tier PCS, *IEEE Vehicular Technology Conference* (New York: IEEE Press, May 1999).

385. Kennedy and Sullivan, Direction finding and smart antennas using software radio architectures, *IEEE Communications Magazine*, May 1995.

386. ARGOSystems smart antenna system in trials at Bell Atlantic Mobile cell site (www.boeing.com), 970826.

387. Godara, L., Applications of antenna arrays to mobile communications, part I: Performance improvement, feasibility, and system considerations, *Proceedings of the IEEE* (New York: IEEE Press, July 1997).

388. Beach, M., et al., Candidate calibration architectures for use in UTRA adaptive antenna basestations, *Proceedings of the 4th ACTS Mobile Communications Summit '99* (Sorrento, Italy: EC, 1999).

389. Evans, J., et al., The rapidly deployable radio network, *IEEE JSAC* (New York: IEEE Press, April 1999).

390. Lu, I.-T., and Choi, J.-S., Sensitivity study of smart antenna systems with both transmission and reception diversities, *MILCOM 99* (New York: IEEE Press, Nov. 1999).

391. Kwak, J., and Lu, I., Blind adaptive space-time receiving and transmitting diversities for multiuser DS-CDMA systems, *MILCOM 99* (New York: IEEE Press, Nov. 1999).

392. Thomas, T., A new EHF ground terminal, *Proceedings of the Tactical Communications Conference 92* (New York: IEEE Press, 1992).

393. McIlvenna, J. F., and Schindler, J. K., EHF monolithic phased arrays—a stepping-stone to the future, *MILCOM 88* (New York: IEEE Press, 1988).

394. Lin, H.-P., and Jeng, S.-S., Smart antenna system and its application in low-earth-orbit satellite communication systems, *IEEE Proceedings on Microwave Antennas and Propagation*, Vol. 146, No. 2, April 1999.

395. RD 4500 product description (http://www.radiodesign.se/home.htm: Radio Designs AB), 1999.

396. Li, Y., et al., Spatial-temporal equalization for IS-136 TDMA systems with rapid dispersive fading and cochannel interference, *IEEE Transactions on Vehicular Technology* (New York: IEEE Press, July 1999).

397. Ariyavisitakul, S., Optimum space-time processors with dispersive interference: Unified analysis and required filter span, *IEEE Transactions on Communications* (New York: IEEE Press, July 1999).

398. Overview (www.radiodesign.se/home: Radio Designs AB), 1999.

399. *Reference Data for Engineers* (Carmel, IN: Sams Publishing, 1993).

400. AIL has developed state of the art Butler matrices covering UHF through Ka-band (www.ail.com), Nov. 1999.

401. ADAMO project overview (www.tcc.thomson-csf.com/antennas/europe/adamo/), 1999.

402. SDMA prototype (www.tcc.thomson-csf.com/antennas), 1999.

403. Smith, J. O., and Abel, J. S., Closed-form least-squares source location estimation from range-difference measurements, *IEEE Transactions on Acoustics, Speech, and Signal Processing* (New York: IEEE Press, Dec. 1987).

404. Subleti, R., et al., Separation and bearing estimation of co-channel signals, *MILCOM 98* (New York: IEEE Press, Oct. 1989).

405. Razavilar, et al., Software radio architecture with smart antennas: A tutorial on algorithms and complexity, *IEEE JSAC*, Feb. 1999.

406. Yang, Y., A matrix factorization approach to signal copy of constant modulus signals arriving at an antenna array, *Proceedings of the 1994 Conference on Information Science and Systems* (Princeton: Princeton University, 1994).

407. Lo, B., Adaptive equalization and interference cancellation for wireless communication systems, *IEEE Transactions on Communications* (New York: IEEE Press, April 1999).

408. Schlegel, C., Multiuser projection receivers, *IEEE JSAC* (New York: IEEE Press, Oct 1996).

409. Martone, M., Hybrid nonlinear moments subspace processing for wireless communication systems using antenna arrays, *Proc. 1996 IEEE International Conference on Communications* (New York: IEEE Press, 1996).

410. Liu and Xu, Smart antennas in wireless systems: Uplink multiuser blind channel and sequence detection, *IEEE Transactions on Communications* (New York: IEEE, February 1997).

411. Martone, M., Cumulant-based adaptive multichannel filtering for wireless communications systems with multipath RF propagation using antenna arrays, *IEEE Trans. on Vehicular Technology* (New York: IEEE Press, May 1998).

412. Sate, T., et al., Sequential interference cancellation system applying to wideband CDMA systems, *Proceedings of the IEEE Vehicular Technology Conference* (New York: IEEE Press, 1996).

413. Lo, B. C. W., and Letaief, K. B., Equalization and CCI cancellation for wireless communications using blind trellis search techniques, *Proceedings of the IEEE Vehicular Technology Conference* (New York: IEEE Press, 1998).

414. Nelson, L., and Poor, V., Iterative multiuser receivers for CDMA: An EM-based approach, *IEEE Transactions on Communications* (New York: IEEE Press, Dec. 1996).

415. Kim, S., On the performance of adaptive smart antenna systems in multipath environments, *MILCOM 99* (New York: IEEE Press, Nov. 1999).

416. Correal, N., et al., A DSP-based multiuser receiver employing partial parallel interference cancellation, *IEEE JSAC*, Feb. 1999.

417. Lee, J.-Y., and Samueli, H., Adaptive antenna arrays and equalization techniques for high bit-rate QAM receivers, *IEEE JSAC* (New York: IEEE Press, April 1999).

418. Murotake, D., Oates, J., and Fuchs, A., Real-time implementation of a reconfigurable, adaptive Imt-2000 base station channel modem using software defined radio methods, *IEEE Communications Magazine* (New York: IEEE Press, Feb. 2000).

419. Qi, P., et al., *Smart Antenna Based on Parallel DSP Architecture Developed Using Matlab* (Sydney, Australia: University of New South Wales, 1999).

420. www.Qualcomm.com (Search=Omni TRACKS).

421. Caffrey, J. and Stuber, G., Vehicle location and tracking for IVHS in CDMA micro-cells, 5th IEEE Symposium on Personal, Mobile, and Indoor Communications (New York: IEEE Press, 1999).

422. Karlsson, P., Bergljung, C., Thomsen, E., and Borjeson, H., Wideband measurement and analysis of penetration loss in the 5 GHz band VTC 1999–Fall, IEEE VTS 50th Vehicular Technology Conference (New York: IEEE Press, 1999).

423. Acosta, R. J., Bauer, R., Krawczyk, R. J., Reinhart, R. C., Zernic, M. J., and Gargione, F., Advanced Communications Technology Satellite (ACTS): four-year system performance, *IEEE Journal on Selected Areas in Communications* (New York: IEEE Press, Feb. 1999).

424. SciComm Viper (www.scicomm.com), 1998.

425. Mitola III, J. , Cognitive radio for flexible mobile multimedia communications, *Proc. IEEE Workshop on Mobile Multimedia Communications* (New York: IEEE Press, Nov. 1999).

426. Xetron Interference Cancellation Module (www.xetron.com), 1999.

427. Lucent completes calls over wideband radio platform designed to evolve TDMA to third-generation (3G) services (Murray Hill, NJ: Lucent, 1999).

428. Sanchez, R. J., Wahhab, F. F., Evans, J. B., Frost, V. S., and Minden, G. J., Design and evaluation of an adaptive data link control protocol for wireless ATM networks, Global Telecommunications Conference, 1998, GLOBECOM 1998.

429. Lakshman, T. V., Madhow, U., and Suter, B.,Window-based error recovery and flow control with a slow acknowledgment channel: a study of TCP/IP performance, Proc. INFOCOM '97. Sixteenth Annual Joint Conference of the IEEE Computer and Communications Societies. Driving the Information Revolution (New York: IEEE Press, 1997).

430. Jern, M., "Thin" vs. "fat" visualization client, Proceedings, 1998 Computer Graphics International (New York: IEEE Press, 1998).

431. Simpson, The duel over object models, Client/Server Today, Aug. 1994.

432. Luenberger, Optimization by vector space methods (New York: John Wiley & Sons, 1969).

433. Mitola, J., and Zvonar, Z., *Software Radio Technology: Selected Readings* (New York: IEEE Press, Oct. 2000).

434. Moon, P. J., Lee, J. W., Jun, M. S., and Lee, C. H., The new high-speed digital signature, Proceedings, 17th Conference on Local Computer Networks (New York: IEEE Press, 1992).

435. Brock, et.al., A dynamically programmable ADC for multifunction digital receivers, Proc. 2000 Government Microcircuit Applications Conference (GOMAC'00) (Anaheim, CA: IEEE Press, March 2000).

436. Mitola, J., Testimony in the Federal Communications Commission public forum on secondary spectrum markets, May 31, 2000 (www.fcc.gov).

437. www.jtrs.sarda.mil.

438. Mitola, J., Cognitive radio: An integrated agent architecture for software defined radio, Doctoral Dissertation (Stockholm: KTH (The Royal Institute of Technology), May 2000).

439. Methods for Testing and Specifications (MTS), Strategy for the use of formal SDL for descriptive purposes in ETSI products, TR 101 081 V1.1.1 (Sophia Antipolis Cedex, FR: ETSI, 1997).

440. Holtzman, J. M., and Sampath, A., Adaptive averaging methodology for handoffs in cellular systems, *IEEE Transactions on Vehicular Technology* (New York: IEEE Press, Feb. 1995).

441. Rhiaczek, A., *High Resolution Radar* (Prentice Hall, 1970).

442. Bharucha, B., et. al., Editors, Future Voice Technologies, *IEEE JSAC* (New York: IEEE Press, Jan. 1999).

443. TalkDial™ User's Guide (Herndon, VA: BellAtlantic Communications, 1999).

444. RF CAD 2.3 Demo CD (www.comm-data.com), 1999.

# GLOSSARY

**1G**  First-generation mobile cellular radio.

**2G**  Second-generation mobile cellular radio.

**3G**  Third-generation mobile cellular radio.

**4G**  Fourth-generation mobile cellular radio.

**A1A**  A standard for HF radiotelephony.

**ACI**  Adjacent-channel interference.

**ADC**  Analog-to-digital converter.

**ADPCM**  Adaptive differential pulse code modulation, a means of compressing a 64 kbps PCM audio channel to typically 32 or 16 kbps by encoding the differences between successive 8 bit samples of the PCM stream.

**AGC**  Automatic gain control.

**ALE**  Automatic Link Establishment, a method for automatically establishing HF links by probing the ionosphere/propagation paths for clear stable channels. This standard was defined by the Mitre Corporation for the U.S. Air Force MIL-STD-188-141, Appendix A.

**AM/FM**  Amplitude modulation (AM) and frequency modulation (FM), generic voice and data modes in which an analog modulated waveform is simulated digitally for compatibility with existing radio equipment, as demonstrated in SPEAKeasy I.

**AMPS**  Analog mobile phone system (AMPS), the first-generation mobile cellular radio (MCR)-based telephone system in which 30 kHz analog channels provide voice telephony while shared 20 kbps control channels moderate channel usage.

**AN/GRC**  AN stands for Army/Navy, a U.S. designation of electronic equipment. GRC refers to general-purpose radio communications equipment.

**ANSI**  American National Standards Institute.

**AOA**  Angle of arrival, an emitter-location-related estimate of the angle at which the radio wave is arriving with respect to an arbitrary reference on the antenna or antenna platform (e.g., boresight); AOA is usually based on RF phase estimates.

**APCO**  APCO-25 is a telecommunications standard under which among other things air carriers must accommodate channel spacings denser than 25 kHz per voice radio channel (e.g., $8\frac{1}{3}$ kHz).

**API**  Applications programmer interface.

**Architecture**   Functions, components, and design rules that comprise a comprehensive framework for system definition, design, development, deployment, and support in the field.

**ARIB**   Association of Radio Industries and Businesses, the radio standards association of Japan.

**ASIC**   Application-specific integrated circuit.

**ATM**   Asynchronous Transfer Mode, a network protocol in which traffic is transported asynchronously in 53 octet packets via an underlying synchronous network such as an SDH or OC network.

**AWGN**   Additive white Gaussian noise.

**Baud**   Traditional term for the period during which a channel symbol is sent and/or for the rate at which such symbols are transmitted; 75 baud = 75 bits per second. A 1 ms baud in FSK indicates a 1 kbps stream of 1 ms-duration channel symbols.

**BCH**   Bose-Chadhuri-Hocquenghem codes, an efficient systematic block code.

**BER**   Bit error rate: the proportion of bits that are received in error for a given number of bits transmitted. BER approximates the bit error probability to the degree that the probability model represents the channel.

**Biconical**   A type of reflector antenna.

**Block process**   A process that has memory, such as an FFT or a Reed–Solomon FEC block coding algorithm.

**Bluetooth**   2.4 GHz FH desktop area link (30 meter range) to replace IR; IBM, Ericsson, et al.

**bps**   Bits per second.

**BPSK**   Binary phase shift keyed; use of antipodal phases (e.g., 0, 180 degrees) in a channel symbol in order to encode one bit per channel symbol.

**BSC**   Base station controller.

**BT**   British Telecommunications, the PTT of England.

**BTS**   Base transceiver station.

**BW**   Bandwidth, the amount of radio spectrum used by a signal, typically measured between upper and lower points at which power is 3 dB of peak; or bandwidth allocated to a channel.

**CAD**   Computer-aided design.

**CAM**   Computer-aided manufacturing.

**CAS**   Close air support, a military mission.

**CASE**   Computer-aided software engineering or computer-aided systems engineering.

**CCITT**   The Consultative Committee International Telephonie and Telegraphie; French term for the former international standards body now aggregated into the ITU.

**CDMA**  Code division multiple access, a technique in which users employ the same spectrum at the same time (typically a wideband channel) but avoid interference through the (approximate) orthogonality of assigned codes with minimum cross-correlation.

**CDPD**  Cellular Digital Packet Data (CDPD), the modem protocol in which 1G mobile cellular radio channels are allocated to wide area exchange of packet data between computers (typically laptop mobile computers).

**CELP**  Codebook Excited Linear Prediction, a speech coding technique that often refers to the algorithm described in USFS 1016 (U.S. Federal Standard 1016).

**CF**  Core framework, part of the JTRS SCA.

**CFAR**  Constant false alarm rate; an algorithm that estimates the noise floor, detecting energy that exceeds the noise floor by a fixed amount, resulting in what would be a constant false alarm rate in Gaussian spectrally uniform noise.

**Ch**  Channel(s), abbreviation used in this text.

**CIR**  Carrier-to-interference ratio.

**CISC**  Complex Instruction Set Architecture, the category of ISAs characterized by special instructions and additional addressing modes. Compare RISC.

**CMOS**  Complementary metal-oxide semiconductor.

**Codec**  Coder-decoder, an electronics subsystem that translates speech from analog to digital and back, generally employing some kind of compression and expanding (e.g., A-law or mu-law).

**COFF**  Common object file format, a method of storing object files for DSP.

**COMSEC**  Communications security, the encryption of transmissions to preclude unauthorized access.

**cont**  Abbreviation for continuous, used in the propagation quad charts.

**CONUS**  The Continental United States.

**CORBA**  Common Object Request Broker Architecture.

**COSSAP**  A block-diagram-oriented DSP programming environment from Synopsis, Inc.

**cPCI**  Compact PCI, a version of the PCI bus/card format.

**CPU**  Central processor unit.

**CSF**  Thompson CSF, a French radio manufacturer.

**CT**  Cordless telephone.

**CVSD**  See Delta Mod.

**DAC**  Digital-to-analog converter.

**DARPA**  Defense Advanced Research Projects Agency, a research agency of the U.S. Department of Defense.

**dB**  Abbreviation for deci-Bells, 20 log (voltage ratio) or 10 log (power ratio).

**dBc** Abbreviation for deci-Bells, relative to full scale.

**dBi** Abbreviation for deci-Bells, with respect to the gain of an isotropic (spatially uniform) antenna.

**DBS** Direct broadcast satellite, a satellite system with sufficient power and gain in the spacecraft and uplink that the receive antenna can be very small (typically 18 inches).

**DDS** Direct digital synthesis, a method for frequency synthesis.

**DECT** Digital European cordless telephone.

**DEF-133** A UK telecommunications standard.

**Delta Mod** A technique for digitally encoding a waveform in which the changes of the waveform are encoded into the bitstream, possibly with constant slope or possibly with continuously variable slope (CVSD).

**DF** Direction finding.

**DII** Defense information infrastructure; the aggregation of all telecommunications and information processing systems in use by the U.S. DoD at any point in time.

**Directivity** Providing a directional capability; omnidirectional antennas provide equal gain in all directions while directional antennas provide gain (or nulls) in a desired direction.

**DISA** Defense Information Systems Agency of the U.S. DoD.

**Discone** A reflector-type antenna.

**Diversity** The process of combining signals from two or more independent propagation paths in order to enhance received signal quality, typically through antenna diversity and IF analog or digital combining.

**DMA** Direct memory access, an input/ouptut hardware architecture in which dedicated circuits mediate the exchange of blocks of data between memory and the external environment. DMA hardware typically has block/word count registers and interrupts.

**DME** Distance measuring equipment.

**DMI** Direct matrix inversion, a beamforming technique associated with smart antennas.

**DNR** Dynamic range.

**DOA** Direction of arrival, the result of expressing AOA as a bearing relative to true North.

**DOMSAT** A domestic satellite.

**Doppler** The person who discovered that approaching objects shift frequency up while receding objects shift frequency down compared to stationary objects, and the effect so named.

**Downconvert** To translate a (typically analog) signal from a carrier frequency to IF or baseband (at which information and signal bandwidths are nearly equal).

**DPSK** Differential phase shift keying, a BPSK technique in which the bits transmitted indicate whether the data sequence state has changed or not (e.g., 1 indicates a change and 0 indicates no change). DPSK tolerates carrier phase ambiguity.

**DS** Direct sequence, a technique in which each data bit is multiplied by a spreading sequence consisting of $N$ bits, often of a pseudonoise shift register sequence. Despreading is accomplished by an $N$-bit correlator, yielding $10\log N$ db processing gain.

**DS0** Digital Signal Level Zero (DS0); 64-kbit per second channel consisting of 8-bit octets (typically voice samples) at 8 kHz synchronously coded (ISDN B or bearer channel).

**DSB** Defense Science Board, a group of leaders of U.S. industry and academia who annually consider key issues facing the U.S. DoD.

**DSB** Double side band, an AM modulation technique.

**DSCS** Defense Satellite Communications System.

**DSP** Digital signal processing; the use of discrete time sampling and amplitude quantization to represent signals for processing via algorithms whether implemented in hardware, firmware, or software.

**DSSS** Direct sequence spread spectrum (see direct sequence).

**E1** European Standard 2.048 Mbps PCM data rate; 30 channels at 64 kbps clear channels (no LSB stealing), but 2 (of 32 potential 64 kbps) channels are dedicated to frame synchronization and control.

**Eb/No** The ratio of energy per bit, Eb, to the single-sided noise power spectral density No in watts per Hz.

**EC** European Community.

**ECCM** Electronic counter-counter measures, ways of defeating jamming and other Electronic counter-measures (ECM), e.g., by increasing the redundancy of a transmitted waveform.

**EDAC** Generic error detection and correction; voice and data mode demonstrated in SPEAKeasy I.

**EDGE** Enhanced data-rate for GSM evolution, up to 384 kpbs over GSM.

**EEPROM** Erasable Electronically Programmable Read Only Memory.

**EERLs** External events/response lists, characterizations used in the ROOTSA design approach.

**EHF** Extremely high frequency (30 GHz–300 GHz).

**EID** External interface descriptions, used in the ROOTSA method to specify data exchange among objects.

**ELOC** Effective lines of code, the number of executable lines of code that meet specific criteria defined in a software metric standard.

**EMC** Electromagnetic compatibility, the control of EMI.

**EMI** Electromagnetic interference.

**EP** Electronic protection (e.g., jamming).

**EPAC**  Electronically Programmable Electronic Circuits, a type of analog chip that combines standard analog parts such as amplifiers, capacitors, etc. with facilities for clock generation and EEPROM interconnect on chip.

**EPROM**  Electronically Programmable Read Only Memory.

**ERM**  Entity Reference Model, used by the JTRS and SDR Forum.

**ETSI**  European Telecommunications Standards Institute, Sophia Antipolis Cedex, France (www.etsi.org).

**EUROCOM**  A European military communications standards suite.

**F1B**  A telegraphy standard used on HF radio.

**F2D2**  Functional Flow Diagrams and Descriptions, a systems design approach widely used in the 1980s for defense (mission-critical) embedded computer systems.

**Fault**  Refers to any failure. In communications faults may be caused by propagation, handoff failure, unavailability of DSP resources, failure to meet a timing requirement, etc.

**FDD**  Frequency domain duplexed.

**FDM**  Frequency division multiplexing, a multichannel technique in which each subscriber has an analog frequency offset subcarrier multiplexed with respect to a common carrier frequency.

**FDMA**  Frequency division multiple access, a spectrum sharing technique in which each subscriber is allocated a specific narrowband channel for which a unique RF carrier is generated.

**FEC**  Forward error control (FEC), the introduction of redundant information into transmitted data so that errors can be detected and corrected by the receiver.

**FFT**  Fast Fourier transform, a method of computing Fourier coefficients (representing a signal in the frequency domain) that uses a minimum number of multiplies, nominally $N^* \log_2(N)$.

**FH**  Generic frequency-hopped voice and data mode, e.g., demonstrated in SPEAKeasy I.

**FIR**  Finite impulse response filter; a filter that uses only feed-forward summations on a tapped delay line (also called *transversal*).

**FM**  Frequency modulation (FM); a communications method in which an information signal is imparted to a carrier by continuously modifying the frequency (the integral of the phase) of the carrier. Discrete FM is called *frequency shift keying*, FSK.

**FPAA**  Field Programmable Analog Array.

**FPGA**  Field Programmable Gate Array, a semiconductor chip with input, output, programmable interconnect and storage typically used for state machines including very fast DSP functions.

**FPLMTS**  Future Public Land Mobile Telecommunications System, the initial ITU name for next-generation mobile cellular and personal communications systems now called IMT-2000.

**Fresnel** The discoverer of properties of optical gratings and knife edges; the propagation zones named after him.

**FSK** Frequency shift keyed; a modulation technique in which a channel symbol is transmitted by shifting a carrier frequency by a small amount (e.g., positive RF shift = 1 and negative shift = 0).

**FTF** Fast transversal filter, a beamforming technique associated with smart antennas.

**G.711** An ITU-T standard for compressing audio signals from 16-bit samples at 8 kHz to 64 kbps using A-law or mu-law compression and expanding (companding).

**G.721** An ITU-T standard for compressing audio signals using ADPCM.

**GBOPS** Billions of bus operations per second, 1000 MBOPS.

**Gbps** Gigabits per second, $10^9$ bits per second.

**GEO** Geosynchronous orbit, that is, an orbit at about 22,500 miles altitude such that the satellite's rotational period and the rotational period of the earth are nearly exactly the same, yielding apparent stationary location above the suborbital point on earth.

**GFLOPS** Billions of floating point operations per second, 1000 MFLOPS.

**GII** Global Information Infrastructure, the aggregate of all telecommunications and information processing systems in use worldwide at a specified point in time.

**GloMo** Global Mobile, a DARPA communications research project.

**GOPS** Billions of operations per second, 1000 MOPS.

**GPRS** General Packet Radio Service, an enhanced GSM network mode with up to 115.2 kbps per GSM carrier.

**GSM** The Global System for Mobile communications; originally this acronymn referred to the Gruppe Speciale Mobile, the committee that created the GSM standard beginning in 1983.

**Handoff** The process of transfering control from one cell base station to another in mobile cellular radio or, more generally, from one radio access point to another.

**Handover** The process of transferring control from one band or mode to another to achieve seamlessness among, for example, an in-building microcellular system at EHF, a macrocellular land mobile system at 950 MHz, and a satellite mobile system at SHF.

**HAVE QUICK II** A U.S. frequency-hopped voice and data radio access interface standard for UHF.

**HAVEQUICK** Very slow frequency hop mode for TRANSEC equipment compatibility for UHF demonstrated in SPEAKeasy I.

**HDB-3** A European communications standard.

**HDR** High data rate, typically 155 Mbps or higher.

**HEO** Highly elliptical orbit.

**HF** High frequency (nominally 3–30 MHz RF).

**HIPO** Hierarchical input-process-output design and documentation approach pioneered by IBM in the 1970s.

**Homodyne** Another name for a direct conversion receiver with an IF of zero Hz.

**Host** Processor or other physical thing that supports a software-defined behavior.

**Hps** Hops per second, a measure of the speed of frequency hop.

**HW** Hardware.

**Hybird** In radio air interfaces, the use of direct sequence spread spectrum and frequency hop spread spectrum at the same time.

**I&Q** In-phase and quadrature; a technique for signal processing in which two parallel channels use sin and cos references to obtain complex sampled waveforms.

**I/O** Input/output, the ports that connect processors to peripherals.

**IBW** Instantaneous bandwidth, the bandwidth used by the signal at a given instant. For frequency-hopped signals, the IBW may be 30 kHz while the full hopped bandwidth could be 60 MHz.

**IC** Integrated circuit.

**ICAO** International Council on Aircraft Operations.

**ICNIA** Integrated communications, navigation and identification architecture.

**IEA/AIE** Proceedings of the (*n*th) international conference on industrial and engineering applications of artificial intelligence and expert systems.

**IEEE** Institute of Electrical and Electronics Engineers.

**IEEE P1149.1** A standard small serial interface for on-chip debugging (JTAG).

**IETF** Internet Engineering Task Force.

**IFF** Identification Friend or Foe; originally for aircraft, a protocol in which friendly vehicles are identified as such by response to an interrogating radar pulse.

**IIR** Infinite impulse response filter; a filter that uses feedback summations on a tapped delay line.

**IMT-2000** International Mobile telecommunications for the year 2000 and beyond (see FPLMTS).

**INFOSEC** Information security. Liberally, TRANSEC plus COMSEC. More strictly, INFOSEC includes authentication, nonrepudiation, privacy and data integrity.

**INMARSAT** International Maritime Satellite communications consortium.

**Intersatellite** Links between two satellites with no intervening ground station.

**Ionosphere** The upper layers of the earth's atmosphere that consist of molecules in an electrically excited state due to the Sun's energy.

**IP**   Intellectual property.

**IP**   Internet Protocol, associated with TCP.

**IPT**   Integrated Product Team, an abbreviation used by the U.S. DoD.

**IS-136**   North American second generation (TDMA) air interface.

**IS-54**   North American Digital TDMA Standard precursor to IS-136.

**IS-95**   North American CDMA cellular/PCS air interface standard.

**ISA**   Instruction set architecture; the organization of instructions and related memory of a central processor or microprocessor.

**ISI**   Inter-symbol interference, the distortion of temporally adjacent channel symbols by the impulse-response of the channel.

**ISM**   Instrumentation, Scientific and Medical (ISM), unlicensed radio bands.

**ITU-R**   International Telecommunications Union, Radio standards organization.

**ITU-T**   International Telecommunications Union, (fixed plant) Telephony standards organization.

**J2A**   A standard for HF radiotelephony.

**J3E**   A standard for HF radiotelephony.

**JCIT**   Joint Communications Interoperabiity Terminal, a U.S. Naval Research Laboratory SDR technology pathfinder.

**JEDEC**   Joint Electron Device Engineering Council.

**JTAG**   Joint Test Action Group, the committee that created the IEEE P1149.1 standard for on-chip test ports and debug facility.

**JTIDS**   Joint Tactical Information Dissemination System (JTIDS), an air interface, protocol and/or the radio hardware used by the U.S. military; includes DS and FH TDMA data exchange.

**JTR**   Joint Tactical Radio (JTR), Mission Element Needs Statement (MENS) establishes a need for a consolidated radio program for US DoD. JTR System (JTRS), a joint PM office and program to procure such radios, based on the recommendations of the PMCS IPT.

**Ka**   The millimeter wave frequency band between 33 and 36 GHz.

**kbps**   Thousands of bits per second.

**KG-84**   COMSEC equipment compatibility mode for HF-UHF.

**KGV-10**   TRANSEC equipment compatibility mode for HF-UHF.

**Ku**   Satellite band in which the downlink is from 11 to 12 GHz and the uplink is from 13 to 14 GHz.

**KY-57**   COMSEC equipment compatibility mode for HF-UHF.

**KYV-5**   COMSEC equipment compatibility mode for HF-UHF.

**LATA**   Local access telephone area, an area within which local telephone access is allocated to a specific monopoly provider under the consent decree that broke the U.S. Bell System into the Regional Bell Operating Companies (RBOCS).

**LBR** Low bit rate, typically less than 9600 bits per second.

**LEO** Low earth orbit.

**LMDS** Local multipoint distribution service, essentially wireless local loop.

**LMS** Least mean square, a beamforming technique associated with smart antennas.

**LNA** Low noise amplifier.

**LO** Local oscillator.

**LOC** Lines of code (see ELOC).

**LOS** Line of sight; a propagation mode in which radio waves pass directly from the transmitter to the receiver without reflection.

**LPC** Linear predictive coding, a method of approximating the speech waveform using a small number of bits—typically 2400 bits per second.

**LPD** Low probability of detection.

**LPI** Low probability of intercept.

**LSB** Least significant bit; the binary number whose weight is unity.

**LSB** Lower side band: see SSB.

**LSL** Least squares lattice; a beamforming technique associated with smart antennas.

**LUF** Lowest useable frequency; in HF communications, that frequency below which the transmitted signal fails to reflect from the ionosphere.

**LVHF** Lower very high frequency (30–88 MHz).

**MBOPS** Millions of bus operations per second.

**Mbps** Millions of bits per seconds.

**MCM** Multichip module; an electronics substrate accommodating a number of unpackaged dies.

**MCR** Mobile cellular radio; general term used to refer to terrestial mobile telephony systems that reuse RF as propagation conditions permit (e.g., AMPS and GSM).

**MEMS** Micro-electromechanical systems.

**MEOPS** Millions of memory operations per second.

**Mercury Talk** HF radio mfg. in People's Republic of China (PRC).

**Method** A procedure attached to a software object.

**MetSat** Meterological satellite.

**Mfg** Manufacturing.

**MFLOPS** Millions of floating point operations per second.

**MFSK** Multifrequency FSK, a generic frequency-hopped voice and data mode demonstrated in SPEAKeasy I.

**MILCOM** The IEEE Military Communications Conference.

**MIL-STD-810C** A telecommunications standard used at HF.

**MIME** Multipurpose Internet Mail Extension.

**MIPS** Millions of instructions per second (MIPS); the number of instructions a processor can execute per second. This will depend on the relationship between the mix of types the application demands vs. the efficiency of the processor.

**MMDS** Multipoint multichannel distribution service, the use of wireless to distribute multimedia telephony, entertainment, and other services to the home from a neighborhood cable/wireless terminus.

**Molnyia** A Russian HEO communications satellite.

**MOPS** Millions of operations per second.

**MPSK** Generic frequency-hopped voice and data mode demonstrated in SPEAKeasy I.

**MPT** Ministry of Posts and Telecommunications (e.g., of Japan).

**MPTT** See PTT.

**MSC** Message sequence chart.

**msec** Milliseconds; thousandths of a second.

**Msps** Millions of samples per second, an ADC/DAC specification.

**MSS** Mobile satellite services (e.g., Iridium or Globalstar).

**MTSO** Mobile telephone switching office.

**MUF** Maximum useable frequency; in HF communications, that frequency above which the transmitted signal fails to reflect from the ionosphere.

**Multichannel** A waveform in which more than one subscriber channel is accommodated. FDM and PCM are typical multichannel waveforms.

**Multipath** A propagation mode in which in addition to a direct path, a distinct reflected path contributes significant energy to a received signal.

**mW** Abbreviation for milliWatts, one thousanth of a Watt of power.

**$N^2$** $N$-squared, a method for checking the completeness and consistency of interfaces by constructing a square matrix $(i, j)$ of $N$ components and characterizing what output of component $i$ provides input to component $j$ at each intersection $(i, j)$.

**Nav** Navigation signals such as Distance Measuring Equipment (DME), LORAN, Omega, etc.

**NCA** National Command Authority.

**Near–far ratio** The ratio of the power of the largest processable subscriber signal to the smallest processable signal.

**NFR** Near–far ratio.

**NG** Nongovernment; an abbreviation for the spectrum bands allocated to nongovernmental use.

**NII** National Information Infrastructure, the aggregate of all telecommunications and information processing systems in use by the U.S. at a specified point in time.

**NMT**   Nordic Mobile Telephony, first generation air interface standards on 450 and 900 MHz.

**ns**   Nanoseconds.

**NVI**   Near vertically incident (NVI) skywave; in HF communications, the use of frequencies that reflect from the ionosphere immediately above the transmitter for short range ($< 30$ km) transmission in rugged terain (e.g., jungles, mountains).

**Obs**   Observation; an abbreviation used in this text.

**OC**   Optical carrier multiplex hierarchy levels from OC-1 (50 Mbps) to OC-3 (155 Mbps) and upward in multiples of 4 times OC-3, e.g., OC-12 (622 Mbps), and OC-48 (2.4 Gbps).

**OCONUS**   Outside of CONUS.

**Octave**   A bandwidth for which the ratio of upper cutoff frequency to lower cutoff frequency is 2 : 1.

**OMG**   Object Management Group, the developers of CORBA.

**Omni**   Omnidirectional; radiating evenly in all directions (360 degrees).

**OOK**   On-off keyed (OOK); a modulation type in which information is represented by the presence of a signal, such as manual Morse code.

**ORB**   Object request broker.

**PABX**   Private Automatic Branch Exchange, a computer-controlled telephone switching center (e.g., for a small business).

**PBX**   Private Branch Exchange.

**PC**   Personal computer.

**PCI**   PC Interface, a circuit card and interconnect bus format.

**PCM**   Pulse Code Modulation, originally a technique for telephony line encoding one of $N$ binary states as a $\log 2(N)$ bit sequence of pulse states (on/off); an adaptation in which any binary state (FSK, PSK, QAM) substitutes for a "pulse."

**PCS**   Personal Communications Systems, the generation of mobile telephony subsequent to cellular systems—this generation was named with the vision that cell phones would become as pervasive as PCs. Many PCS systems use low-power, short-range radio technology.

**PDC**   Personal Digital Cellular, a Japanese standard.

**Personality**   That which characterizes behavior or makes it possible. In particular, software objects provide the personality of digital signal processes (e.g., hosted on DSP chips). Switch settings, as another example, give a specific personality to a modem board.

**PFLOPS**   PetaFLOPS; 1000 TFLOPS.

**PHS**   Personal Handyphone System, a Japanese standard.

**PIC**   Parallel interference cancellation, a smart antenna technique.

**PIC**   Partial interference cancellation, a class of interference-suppression algorithms.

**PID** Processor Identification or Process Identification.

**PIN** Personal identification number.

**Platform** (1) the hardware that supports a radio personality, or (2) the vehicle on which a radio node is operated.

**PMCS** Programmable Modular Communications System, an Integrated Process Team established by the U.S. DoD in Feb. 1997 to define an approach to consolidating military radio programs using software radio technology (see Joint Tactical Radio, JTR).

**Point process** A process that has no memory, such as an FM demodulator.

**PPM** Pulse position modulation, a channel coding technique in which an analog displacement is encoded as a temporal displacement of a pulse from its nominal position in a regular pulse train.

**PRC** The People's Republic of China.

**PSK** Phase shift keyed, a technique in which discrete phases encode digital data. See BPSK.

**PSTN** Public Switched Telephone Network.

**PTT** Ministry of Posts, Telephone and Telegraph (referred to as Ministry of PTT in this text, or MPTT).

**PTT** Push to talk. (This is the standard usage of PTT in this text.)

**PWM** Pulse width modulation, a channel coding technique in which the width of pulses in a regular pulse train is increased or decreased to represent instantaneous source analog amplitude, typically of voice.

**QAM** Quadrature amplitude modulation, a technique in which channel states are differentiated with respect to phase and amplitude, generally for 16 or greater channel states.

**QoS** Quality of service, a contract negotiated between a node offering service and an ATM network upon admission; also a generic term for high data rate with low delay jitter.

**QPSK** Quaternary phase shift keying, a PSK technique in which four carrier phases encode 2 bits of binary data (typically $+/-45$ degrees and $+/-135$ degrees).

**QQT** Quality, quantity, and timeliness of information.

**Radiolocation** See Nav.

**RAM** Random access memory, the memory directly accessible to a processor.

**RAP** Radio access point.

**Rayleigh** A type of fading in which a nearly infinite number of paths of random amplitude and phase contribute equally to the received signal which then has an amplitude that follows a Rayleigh probability distribution.

**RBOC** A Regional Bell Operating Company (RBOC) created in the consent decree that broke the U.S. Bell System monopoly into multiple independent telephone companies.

**Reconfigurable** Having an ability to have a configuration changed in the field, such as reconfiguration of an air interface by downloading a waveform script over the air.

**RF** Radio frequency (RF), the carrier frequency that is modulated to bear information.

**RFI** Radio frequency interference.

**RFI** Request for Information, a formal document provided to industry by a potential customer, such as a cellular service provider.

**Rician** A type of fading (after Rice) involving a principal component plus a variable number of discrete randomly phased components.

**RISC** Reduced Instruction Set Computer, a computer chip with a minimum number of instruction types and modifiers, differentiated from Complex ISCs with specialized instructions.

**RLS** Recursive least square, a beamforming technique associated with smart antennas.

**ROM** Read-Only Memory; memory the content of which has been permanently encoded using a final metalization layer.

**ROM** Rough order of magnitude (cost or performance estimate).

**RPE-LTP** Regular Pulse Excitation—Long-Term Prediction, the standard GSM speech encoding algorithm.

**RS** Reed–Solomon, a block-oriented FEC code.

**RT** Real time; the situation in which a stimulus and the related response occur within a small time window of each other (e.g., microseconds or milliseconds).

**RTOOSA** Real-time object-oriented structured analysis; a contemporary systems/software design approach useful in embedded software design.

**RTOS** Real-time operating system; an operating system with a kernel that has been optimized for speed including very fast context switching and efficient use of memory.

**SAT** Supervisory audio tone, an out-of-band tone used in first generation mobile cellular radio to manage power and spectrum occupancy.

**SAT Paris** A telecommunications company located in Paris, France.

**Satcom** Satellite communications.

**Sats** Satelites; an abbreviation used in this text.

**SCA** Software Communications Architecture, open middleware based on CORBA sponsored by the JTRS program.

**SCPC** Single carrier per channel, or single channel per carrier, a class of modulation formats.

**Script** A sequence of high-level descriptions that can be interpreted or compiled to implement a desired action. A waveform script, for example, describes a waveform in terms of simple structures (e.g., training sequence, body, trailer, etc.).

**SD**   Structured design, a software design approach attributed to Ed Yourdon in which functional and data attributes moderate the design process, leading to the finding that functional cohesion and data coupling in some sense optimize software design.

**SDH**   Synchronous digital hierarchy, a method of synchronizing bitstreams when multiple DS0 channels are multiplexed.

**SDMA**   Space division multiple access.

**SDR**   Software-defined radio.

**SEFT 001A**   A communications standard used at HF.

**SEM-E**   Standard Electronics Module (type E), a U.S. military hardware form factor.

**SER**   Symbol error rate; the proportion of channel symbols that are received in error for a given number of channel symbols transmitted. Approximated by the symbol error probability to the degree that the probability model represents the channel.

**Serial Modem**   HF modem for multirate data (MIL-STD-188-110A).

**SG**   Study Group, an abbreviation used by the ITU.

**SHF**   Super high frequency (3 GHz–30 GHz).

**SIC**   Sequential interference cancellation.

**Sideband**   The part of a signal that is above or below the carrier frequency.

**Signaling**   In telephony, the information exchange by which the placement of calls is requested, dialed numbers are exchanged, charges are billed, and network resources are cleared upon call completion.

**SINCGARS**   Single-Channel Ground-Air Radio System, a U.S. frequency-hopped voice and data radio access interface standard for VHF-JTC3A 9001C.

**SINR**   Signal to interference plus noise ratio, a measure of SNR.

**Skywave**   The wave in HF propagation that reflects from the ionosphere. There may be a direct wave (e.g., from ground to air) and multiple sky-waves.

**SNR**   Signal-to-noise ratio: The ratio of signal power to noise power, generally expressed in deci-Bells, $10^* \log(Ps/Pn)$. See also Eb/No and SINAD.

**SONET**   Synchronous Optical Network (SONET), the standard multiplex hierarchy for fiber optic (OC) digital telephony.

**SpecFP**   An industry standard measure of floating point computational capacity.

**SpecINT**   An industry standard measure of integer computational capacity.

**SpecMark**   An industry standard measure of computational capacity.

**SPOX**   A real-time operating system for DSPs created by Spectron Microsystems and used for multimedia applications.

**Spur**   Spurious response (e.g., of a modulator, IF conversion stage, amplifier, etc.).

**SS-7**  Signaling System 7, the international standard for signaling in telephony (e.g., call setup, tear-down, forwarding, billing, etc.).

**SSB**  Single side band, an AM mode in which a redundant sideband on one side of the carrier is suppressed by filtering while the upper side band (USB) or lower side band (LSB) that remains is transmitted.

**STAJ**  HF frequency hop standard for data and voice, demonstraed software mode in SPEAKeasy I—MIL-STD-188-148A.

**STAP**  Space-time adaptive processing, a family of smart antenna algorithms.

**Superhet**  Superheterodyne receiver; a receiver with multiple RF/IF conversion stages.

**SW**  Software.

**SWR**  Software radio.

**T1**  North American Standard 1.544 Mbps PCM data rate; 24 channels at 64 kbps per channel reduced to 56 kbps clear channel due to stealing of the least significant bit (LSB) for synchronization and control.

**TACAN**  Tactical Air Navigation, a family of international radio products.

**TACP**  Tactical Air Control Party, a military unit.

**TAFIM**  Technical Architecture for Information Management, a top-down object-oriented approach to requirements and design due to U.S. DoD.

**T-Carrier**  North American Digital Telecommunications Multiplex Hierarchy, ranging from T-1 (1.544 Mbps/24 voice channels) and T-3 (45 Mbps) to T-4 (274 Mbps); supplanted by SDH.

**TCP**  Transmission Control Protocol.

**TD-CDMA**  Time Domain Duplexed CDMA (same uplink and downlink RF), ETSI 3G candidate.

**TDD**  Time division duplexing, e.g., used in CT2 cordless telephones by which one 2 ms time slot is used for base to handset while the next is used for handset to base communications.

**TDMA**  Time division multiple access, a spectrum-sharing technique in which each subscriber is allocated a specific time slot in a larger synchronous signal frame.

**TDOA**  Time difference of arrival, an emitter location technique.

**TEC**  Thermoelectric cooler.

**TETRA**  Trans-European Trunked Radio Access, a private mobile radio system that supports voice and data in private/public motor vehicles and trucking fleets.

**TFLOPS**  Trillions of floating point operations per second, 1000 GFLOPS.

**Thread**  Sequence of operations from stimulus to response, including hardware, software, data movement, and processing. Threads typically originate and terminate with external responses such as the PSTN, air interface, and user interface.

**TI** Texas Instruments (see www.ti.com), one of the world's leaders in DSP technology and products, such as the TMS320 series.

**TIA** Telecommunications Industry Association, a U.S. standards-setting body.

**TOC** Tactical Operations Center.

**TOD** Time of Day, a word used to synchronize digital networks.

**TRANSEC** Transmission security; hiding the fact of a transmission event.

**TRAU** Transcoder and Rate Adaptation Unit; a unit that converts 13 kbps GSM vocoded speech traffic to wireline (e.g., DS0) format.

**Tropo** Troposcatter communications, a mode in which a high-power, highly directional radio signal is scattered from a high, distant part of the troposphere, typically at 4–5 GHz RF, for reception by a receiver over the horizon.

**UAV** Unmanned aerial vehicle.

**UHF** Ultra high frequency (300–3000 MHz).

**UML** The Unified Modeling Language.

**UMTS** Universal Mobile Telephone System (UMTS); the Eurpoean name for next-generation mobile cellular and PCS.

**Upconvert** Translate a (typically analog) signal from a baseband (at which information and signal bandwidths are nearly equal) to a carrier frequency.

**USB** Upper side band: see SSB.

**V.xx** CCITT Recommendations for voice channel modems (V.22, V.27, V.32, V.34, and V.42bis address modems, compression and error correction for modems using voice channels of the PSTN.

**Vestigial** Very small; see VSB.

**VHDL** VHSIC Hardware Description Language, specified by IEEE 1076.

**VHF** Very high frequency range (30–300 MHz).

**VHSIC** Very high speed integrated circuits.

**VLIW** Very long instruction word.

**Vocoder** Voice coder.

**VSAT** Very Small Aperture Terminal, a very small remote satellite antenna, typically a parabolic reflector 18 to 36 inches in diameter.

**VSB** Vestigal side band, a channel modulation in which a small residue (vestige) of upper sideband energy and carrier are transmitted along with the lower sideband.

**VSELP** Vector Sum Excited Linear Prediction, the voice compression standard used in IS-54 digital cellular radio.

**VSWR** Voltage standing wave ratio; expresses the relationship between RF power transmitted efficiently through an analog interface and power that is reflected from that interface.

**VTC** Video-teleconference.

**WAP**  Wireless Applications Protocol.

**WB**  Wideband, typically hundreds of kHz to tens of MHz or more; as distinguished from narrowband, in which the signal bandwidth is similar to the information bandwidth, e.g., 4 kHz for voice.

**W-CDMA**  Wideband CDMA (4096 kchips/sec; 5 MHz) ETSI standard for 3G.

**WLAN**  Wireless LAN.

**WLL**  Wireless local loop.

# INDEX